ECOLOGY OF
FRESH WATERS

Man and Medium, Past to Future

ECOLOGY OF FRESH WATERS

Man and Medium, Past to Future

BRIAN MOSS
School of Biological Sciences
University of Liverpool, UK

THIRD EDITION

b

**Blackwell
Science**

© 1980, 1988, 1998 by
Blackwell Science Ltd
Editorial Offices:
Osney Mead, Oxford OX2 0EL
25 John Street, London WC1N 2BL
23 Ainslie Place, Edinburgh EH3 6AJ
350 Main Street, Malden
  MA 02148 5018, USA
54 University Street, Carlton
  Victoria 3053, Australia
10, rue Casimir Delavigne
  75006 Paris, France

Other Editorial Offices:
Blackwell Wissenschafts-Verlag GmbH
Kurfürstendamm 57
10707 Berlin, Germany

Blackwell Science KK
MG Kodenmacho Building
7–10 Kodenmacho Nihombashi
Chuo-ku, Tokyo 104, Japan

Iowa State University Press
A Blackwell Science Company
2121 S. State Avenue
Ames, Iowa 50014-8300, USA

First published 1980
Second edition 1988
Reprinted 1989, 1991, 1992, 1993 (twice)
Third edition 1998
Reprinted 2000, 2001

Set by Semantic Graphics, Singapore
Printed and bound in the United Kingdom
by the Alden Group, Oxford

The Blackwell Science logo is a
trade mark of Blackwell Science Ltd,
registered at the United Kingdom
Trade Marks Registry

DISTRIBUTORS

Marston Book Services Ltd
PO Box 269
Abingdon, Oxon OX14 4YN
(Orders: Tel: 01235 465500
        Fax: 01235 465555)

USA
Blackwell Science, Inc.
Commerce Place
350 Main Street
Malden, MA 02148 5018
(Orders: Tel: 800 759 6102
        781 388 8250
        Fax: 781 388 8255)

Canada
Login Brothers Book Company
324 Saulteaux Crescent
Winnipeg, Manitoba R3J 3T2
(Orders: Tel: 204 224-4068)

Australia
Blackwell Science Pty Ltd
54 University Street
Carlton, Victoria 3053
(Orders: Tel: 3 9347 0300
        Fax: 3 9347 5001)

A catalogue record for this title
is available from the British Library

ISBN 0-632-03512-9

Library of Congress
Cataloging-in-publication Data

Moss, Brian.
    Ecology of fresh waters:
    man and medium, past to future
    Brian Moss. — 3rd ed.
        p.   cm.
    Includes bibliographical references
    (p.   ) and index.
    ISBN 0-632-03512-9
    1. Freshwater ecology.
    I. Title.
    QH541.5.F7M67   1998
    577.6—dc21
97-35549
CIP

For further information on
Blackwell Science, visit our website:
www.blackwell-science.com

FOR MY WIFE, JOYCE, AND DAUGHTER, ANGHARAD, whose love and loyalty make all other happiness possible and the bad things bearable; and for many others who, in one way or another, have made me what I am and therefore this book what it is.

O river, born of penance
Named by laughter,
Your dishevelled streams
Inlay the stone mountains of the Vindhyas
Like ichor gilds the body of an elephant.

And along your riverbanks
The stamens of the green gold Nipa flowers
Tear through their enclosing petals
Desiring you.

Woodlands heavy with wild jasmine
Embrace you with their fragrance.
Hearing your approach
Young plantain trees
Burst into sudden blossom.

Shankaracharya, ninth-century invocation to the Narmada. Translated from the original Sanskrit in Gita Mehta, A River Sutra *(Heinemann, 1993)*

One of the penalties of an ecological education is that one lives in a world of wounds. Much of the damage inflicted on land is quite invisible to laymen. An ecologist must either harden his shell and make believe that the consequences of science are none of his business, or he must be the doctor who sees the marks of death in a community that believes itself well and does not want to be told otherwise.

Aldo Leopold, Round River

Contents

Colour plates fall between pp. 144 and 145

Preface and Acknowledgements

Just before I completed the previous edition of this book, Marian Shoard published *The Theft of the Countryside* [891] detailing the damage that had been wrought to the landscape of Britain by the increasingly intensified agriculture that has prevailed since the Second World War. In 1995, the Council for the Preservation of Rural England (CPRE) and the Countryside Commission produced a series of maps, showing that since the 1960s an area from within England that was the size of Wales had been changed from a tranquil landscape to one sullied by motorways, road widenings, airports, power lines and other intrusions of sight and sound. Just as I was completing this edition, Graham Harvey, agricultural-story editor of a British institution, Radio 4's *The Archers*, to which I am addicted, wrote *The Killing of the Countryside* [388], containing information that underlined the previous messages of Shoard and the CPRE.

You might ask what all this has to do with a textbook on freshwater ecology and the answer would be 'everything'. What happens in fresh waters is determined by what happens in and on the surrounding catchment. And every square metre of land is part of the catchment of one or other freshwater system.

Writing this third edition has been pleasurable – creation always is – but also depressing, for it has become clear just how much freshwater ecosystems continue to be abused by economic interests. This edition continues a tradition of linking the fundamental science of freshwater ecology with its application to practical problems and for this, and other, reasons, that approach has been strengthened. The other reasons are that an ecology book that is centred on a particular group of habitats should concentrate on what is specific to such places and not attempt to be a textbook of theoretical ecology, which should rightly draw its examples irrespective of habitat, and the recent trends in scientific research. These trends in freshwater ecology have been very much towards application, for the funding of fundamental research has dwindled in the political climate of the last two decades.

This is, of course, a grave mistake, for application feeds on pure science, but the reality is that rather little has changed in the last twenty years in our fundamental understanding of fresh waters compared with the advances made in the 1960s and 1970s. A great many details have been added and the menu of case-studies, from which examples might be drawn, has lengthened considerably, but I do not think that any paradigms have been overturned or revolutions spawned. On the other hand, it is very clear to me that practical ecological problems are insoluble unless the political and social contexts in which they are created are considered. Even if Chief Seathl really didn't say that all things are connected, he ought to have done, for they are! The scope of this edition has thus been widened to include a final chapter on fresh waters, the world and the future.

This edition preserves the structure of the previous one but is largely rewritten and bigger. It could have been much larger. There is a huge amount of information available, far more than I can cope with, to the extent that electronic information revolutions, if so they are, are superfluous. There are plenty of excellent specialist treatments of limnology (e.g. Wetzel [1011]; Goldman & Horne [340]; Le Cren *et al.* [560]), stream ecology (e.g. Hynes [459]; Allan [5]), wetlands (e.g. Dugan [223]; Welcomme [1003]) and water-pollution ecology (Hynes [458]; Mason [637]) for those who wish to delve further, and many very specialist books and papers beyond these. I have quoted something over 1000 references. I estimate that there must be at least 100 000 references, and possibly more than twice as many, relevant to the subject area covered by this book. Use of electronic databases will produce lists of well over 1000 each, published since 1981, on topics such as mosquitoes, eutrophication and hydrology.

What I have quoted thus constitutes less than 1% of what I might have quoted. The choice is therefore bound to have elements of randomness and idiosyncrasy in it. Inevitably, work that I have been involved with or done by people I happen to know personally, or who have sent me reprints, will be overrepresented and much that is equally or more worthy will have been left out. The role of a reference list is to provide leads into the rest of the literature and it matters little which particular path is signposted. The abundance of literature is a problem for everyone, not least because much of it is repetitious, as work is repeated or done in a slightly different form, or by a different method, simply because of lack of awareness that it has been done before. A current tendency to use electronic databases covering the period only from the early 1980s will mean that much excellent work done before then will, in future, be ignored. It was the historian, George Santayana, who said that those who cannot remember the past are condemned to repeat it.

My intellectual debts to an increasing number of people continue to mount. To my early mentors, Frank Round and Charles Sinker and that incredibly interesting group, including John Lund, Hilda Canter, Jack Talling, Geoffrey Fryer, John Mackereth and Winifred Pennington, now retired but who established the reputation of the Freshwater Biological Association, I should add many colleagues, visiting workers, postgraduate students and postdoctorals, who have widened my knowledge in many ways. I must thank also those who have made travels to other countries possible, for, although inevitably this book reflects the freshwater ecology of the UK to a greater extent than anywhere else, it does seek not to be parochial and to provide a basis for students in the rest of the world.

1: Introduction: the Pantanal

G. Evelyn Hutchinson (1903–91) was a considerate gentleman, who always promptly and courteously answered the letters I sent him. He was also a limnologist (one who studies fresh waters) and the architect [233] of important ideas that underpin modern ecology. One of these was that of the ecological theatre and the evolutionary play [452], which he created to understand how, through continuously acting natural selection (the evolutionary play), the available environmental 'space' (the ecological theatre) is divided up as niches among a variety of organisms. It is a metaphor (Fig. 1.1) that can also embrace a great deal more of ecology.

1.1 The environmental theatre

The ultimate theatre is the planet itself, and its stage the biosphere, that part of Earth which is capable of containing liquid water, at least for some periods. Such a stage can sustain a living system, whose chemistry, though carbon-based, must operate in an aqueous medium. This platform is a thin skin of moist habitat extending to perhaps 6000 m on the mountains and down to 12 000 m below see level in the deepest ocean trenches. On a planet 12 760 km in diameter and surfaced by crustal plates that fracture and move against one another, the proportions of the stage are those of a thin film of dew covering a cracked egg. The vulnerability of the biosphere is suggested by this analogy.

It is a complex stage, varied in scenery that changes continuously. There are day/night and seasonal changes, determined by latitude and climate in dependable rhythms, to which organisms can easily adjust. There are longer-scale cycles, such as glaciations, which cool the planet as the Earth's orbit around the sun wobbles in a predictable way. These are affected by the placing of the landmasses and ocean, as the crustal plates, on which the continents are borne, move steadily in response to currents and forces in the underlying plastic mantle. The characters in the evolutionary play can change through natural selection and perhaps other mechanisms in reponse to all of these or they may become extinct if their genetic capacities cannot cope.

There are also unpredicatable changes – global catastrophes, which quickly rearrange the scenery and remove, often randomly, enormous numbers of organisms at a stroke. Collision with large meteors may be one.

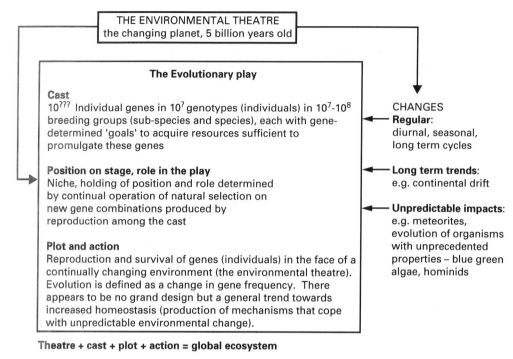

Fig. 1.1 The environmental theatre and the evolutionary play.

Such impacts would create so much dust in the atmosphere as to darken the surface for long periods and severely upset a system dependent on solar energy and photosynthesis.

Another such disturbance has been the evolution of organisms with radically new properties. Those bacteria which, in the late Precambrian Era, began to use water as a hydrogen donor for photosynthesis rather than the hydrogen sulphide or hydrogen which were previously used, caused profound change. They released a highly reactive, and, to the previously entirely anaerobic inhabitants, very poisonous gas, oxygen, into the atmosphere. Anaerobic life became impossible except in the isolated pockets – rich organic sediments, rotting tissues, deep ocean vents or groundwater, where anaerobes now survive as curiosities. The latest such evolved agent of rapid change is ourselves.

1.2 The evolutionary play

The players on the stage live turbulently. The play is about staying on stage. Some authors define ecology as the science of how organisms persist. Organisms stay indefinitely only rarely. Most have walk-on and walk-off parts but, while on stage, they change genetically between generations, as small changes continually occur in their environment. Some of these are physical and chemical; others, probably more important at this scale, involve other organisms seeking space and removing rivals from the stage by competition, predation or parasitism. One view is that the ultimate players are the genes themselves, and that persistent genes are those which most successfully reproduce themselves, through manipulations of the individual organisms which carry them [197]. Resources – energy, water and materials for growth – are variously scarce in all parts of the stage, and such battles are the norm. Nature is not a balanced harmony but the outcome of multiple struggles.

Because resources are scarce, there have developed interacting systems of organisms which, collectively, have shaken down to use these resources efficiently. Organisms have come to occupy niches of greater and greater specialization in parts of the stage that are least changing or change on a regular ('predictable') basis. Specialization has advantages in such situations, because species become adept at gleaning particular foods or occupying particular microhabitats. Combinations of specialists use the available resources very efficiently and give the superficial appearance of an ordered 'ecosystem'. It is ordered, in the sense that all survive so long as the system remains undisturbed, but it is always vulnerable to externally imposed change, which may rapidly destroy it and allow replacement by something different.

1.3 The freshwater part of the stage

Such a system (or rather group of systems) has developed in fresh waters. It is a small part of the global stage in area and volume but a key one. This is because of the role of water in defining the biosphere and of fresh water in supporting that part of it, a third of the planet including the terrestrial systems, which does not live in the ocean.

Freshwater bodies are usually small (at least in relation to the ocean, which covers two-thirds of the planet and which has been a permanent feature, in one shape or another, for 4 billion years) and relatively temporary. They easily dry up, or freeze, as climate changes. Extreme specialization is thus not a notable survival strategy in many freshwater bodies, although some have existed long enough for this to be shown. Moreover, it is through impacts on fresh waters that the global impact of ourselves is shown most seriously. The future of freshwater systems is our own future. If the freshwater players are pushed from the stage, so shall we be.

A piece of the freshwater stage is the Pantanal of Matto Grosso [511, 778], which lies mostly in Brazil, with smaller portions in Bolivia. It is a large basin, perhaps the world's largest wetland area, which collects water from a larger catchment. It is not well known and yet, from that which is known, much else can be deduced. The Pantanal will introduce some general features of freshwater systems.

1.4 The Pantanal

Some 150 million years ago, the crustal plate bearing the continent of South America finally collided, in its drift away from the former supercontinent of Gondwanaland, with the Pacific plate and forced the Andes to even greater than their present impressive height of crumpled rock. At their foot, in the middle of the present continent, a gently sloping shallow basin, at 200 000 km^2 the size of an average European country, remained as part of the ancient supercontinent. It is floored by Precambrian and only slightly younger rocks and bounded on the north and east by lesser highlands, from which numerous rivers drain a catchment two or three times the size of the basin. It has always been swampy, but sometimes much wetter than at present, and occasionally much drier.

In the wet periods, the deltas of the rivers built up and in the drier periods (including the recent glaciation) these deltas became intersected and eroded by smaller rivers. This lumpy bottom topography is further relieved by monadnocks of harder rock, which have not yet been smoothed away. Substantial rivers flow from the highlands and across the basin to join the main southerly-flowing River Paraguay, which flows southwards to join the River Parana and become the River Plate in Argentina.

1.4.1 The physical and chemical environment

The basin is the Pantanal (Fig. 1.2, Plate 1), whose present climate of dry winters and wet summers results in rivers that occupy wide flood-plains in summer, replenishing and deepening thousands of small and some large lakes and huge areas of swampland characteristically covered by floating water plants. Some parts – monadnocks and the higher parts of ancient river deltas – are usually above water level, forming an archipelago of island refuges for terrestrial plants and animals. In winter, as the waters recede, the dry land extends and the rivers and lakes contract to smaller basins. Some disappear altogether.

The lakes are often only 500–1000 m in diameter and often round in shape. How they formed is not certain, but they may be oxbow lakes, isolated from the rivers as their meanders are cut off by erosion and deposition on the bends. Or their basins may have been formed by wind excavation amid sand-dunes on the old river deltas in drier periods. Often they are rimmed by low raised banks, which could be former river levees or old dunes.

An important feature of natural waters is the nature and amount of substances dissolved in them. These may include thousands of different inorganic and organic substances. No natural water has ever been exhaustively analysed. The Pantanal waters are usually neutral or basic and dominated by bicarbonate, calcium and magnesium, reflecting the lime-

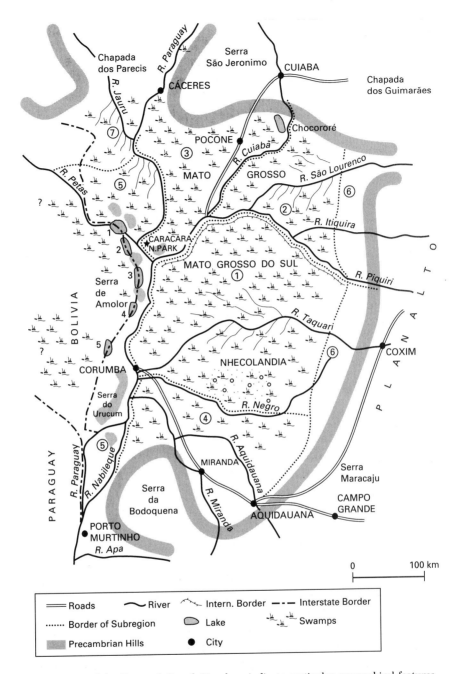

Fig. 1.2 Map of the Pantanal, Brazil. Numbers indicate particular geographical features. 1, 2 and 7 are alluvial fans of ancient rivers. 3 and 4 are extensive flood-plains of sand and silt. 5 is the *varzea* (seasonal swamp) of the River Paraguay, which drains the Pantanal to the south. 6 is the rising upland rim of the Pantanal (Based on Por [778].)

stone rock in much of the catchment. Their concentrations are absolutely low, because the catchment has long been weathered and whatever soil nutrients are now released are mostly retained by the catchment forests.

Total nitrogen concentrations (the sum of all forms of nitrogen compounds dissolved or suspended in the water) are very low, less than $60 \, \mu g \, l^{-1}$, while total phosphorus is about half of this. The concentrations in the Pantanal are apparently very small, especially considering the high production of floating water plants, but there are other sources – the sediments brought in and annually deposited by the rivers and biologically mediated processes, such as nitrogen fixation by microorganisms. Phosphorus and nitrogen are often scarce and many features of freshwater systems are conditioned by their supply and transformations.

Superimposed on this basic chemistry are processes that change it. First, as the floods rise, the decomposition of drowned vegetation leads to deoxygenation. The limited supplies of oxygen dissolved in water, especially at tropical temperatures, are used at higher rates than diffusion from the atmosphere can replace them. Large areas of anoxic water persist, even at high water, in the swamps, as the leaves and roots slough off from the extremely productive aquatic vegetation and accumulate on the bottom. Air-breathing (pulmonate) snails and fish are common in these areas, and the availability of detritus in the flood plains brings many other fish species in upstream migrations, called the *piracema*, to spawn and feed.

Secondly, because the direct rainfall on the Pantanal (about 1100 mm $year^{-1}$) is less than the evaporation that takes place in the high temperatures even of winter, many lake waters and some river waters become more saline with time. Calcium is precipitated and sodium and chloride, which are more soluble, come to dominate the waters Compared with the incoming floodwater or direct rainfall to the lakes most distant from the rivers, the final dry-season salt concentrations may increase 10- or 100-fold and some of the smaller lakes may become salt flats. Even the River Paraguay, leaving the Pantanal basin, is slightly saltier than when it entered.

Areas where annual evaporation is greater than rain- and snowfall are called endorheic; those where it is equal or lower are exorheic. Endorheic areas, which characteristically have saline lakes, occupy almost as great a part of the world's land surfaces as exorheic ones (Fig. 1.3). The Pantanal combines an endorheic basin with a catchment area of well-watered upland which is exorheic, and this gives a range of both exorheic and endorheic areas within the Pantanal itself.

1.4.2 Biogeography of the Pantanal

Geology (the nature of the water derived from the catchment rocks), geomorphology (the topography of the basin floor and its lakes) and climate (variation in the river floods, the effects of evaporation) are crucial in

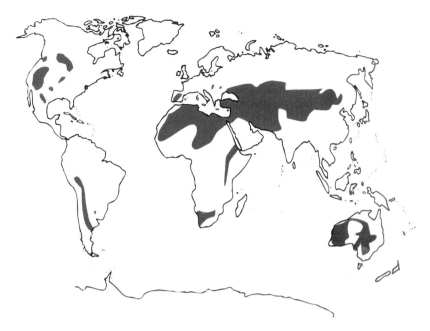

Fig. 1.3 The world distribution of endorheic areas [1016].

understanding the Pantanal and all other freshwater systems. So also are the living organisms. Ecology is not a part of the subject of biology; it is the whole of it. For there is little that living organisms have or do that does not influence their survival and reproduction. All evolved in habitats and none can be understood fully outside those habitats. On the environmental stage, every actor's action has consequences, although many actions may be governed by chance.

The Pantanal lies at a crossroads. To the north, the headwaters of one of its rivers, the Jauru (see Fig. 1.2), share collecting grounds in swamps confluent with the headwaters of the Guapare, a tributary of the Amazon. There has thus been a route for movement of organisms between the basins, although the swamps are shallow and a barrier to large organisms, such as the river dolphins of the Amazon. To the south, the River Paraguay is substantial and connects with the sea, offering a route for the movement of fishes and other organisms ultimately of marine origin. The Pantanal has several species of otherwise characteristically marine stingrays and a flounder. To the west are the Andes, with wet-season torrents discharging water and living jetsam, and on all other sides are rich forested areas – the tropical moist forests and the savannahs of the Brazilian *cerrados* providing a huge reservoir and recipient for mobile species.

Areas with relatively undisturbed geological history, huge size and a high degree of isolation develop a great diversity and many species (endemics) unique to themselves. Lesser size, frequent upheaval and open connections lead to lower diversity and little endemism. As with its hydrology, the mixture of endo- and exorheicity, the Pantanal is a biogeographical *mélange*. It is large enough and old enough to have a number of endemics, although these are many fewer, as is its overall lower diversity, than in the systems of the adjacent Amazon; but it shares many species with the forests, *cerrados* and swamps of a large area of South America.

Among those animal groups for which there are reasonable data (Lepidoptera, birds, amphibians, reptiles and mammals), only 2% are endemic, whilst for 42%, mostly open-savannah grassland or forest species, the Pantanal lies well within their much larger ranges. For a further 40% of species, the Panatanal acts as a barrier and they penetrate into one or other extremity of it but no further. The remainder are aquatic birds of very wide distribution, some shared with the southern part of North America.

Most of the species discussed above, however, are terrestrial or amphibious and associated with the archipelago of islands of grassland or savannah vegetation set among the sea of swampland. There are more endemics among the less mobile, truly aquatic fauna, with perhaps 10% among the fish and aquatic invertebrates. Nonetheless, there are also species of very wide distribution, including an amphipod crustacean, *Hyallela azteca*, which is pan-American, and a conchostracan, *Cyclestheria hislopi*, shared with Africa. The diversity is not meagre, with 657 species of birds and 405 of fish, but it is low compared, for example, with the 2000 fish species found in the huge, varied and ancient complex of basins of the Amazon and its tributaries.

These biogeographical facts again hide principles important in understanding the ecology of all fresh waters. The general appearance of the aquatic biota, in their life-forms and the predominance of particular groups, is similar to that of other freshwater areas. Among the invertebrates, for example, the list of oligochaete worms, beetles, ostracods, cladocerans, copepods, dragonflies, caddis-flies, chironomids, ceratopogonids, bugs rotifers, turbellarians, leeches, snails, bivalves, mayflies and crabs is very familiar. Water is a medium with extreme properties, which strongly shape the nature of the organisms that can survive in it.

Fresh waters are also geologically temporary habitats, prone to frequent upheaval. This limits the time available for evolution of high specialization before the process is disrupted. Such impermanence demands superior abilities for dispersal, and mobility, active or passive, by water or air, is characteristic. Most freshwater organisms have recently evolved from land or marine ancestors (see Chapter 2) and it appears that repeated reinvasions of fresh waters through these routes have occurred in response to the frequent disruption of freshwater systems. Lack of extreme specialization may thus be adaptive in such circumstances, but, nonetheless, freshwater systems are long-lived enough for considerable adaptation to the environment to have evolved.

1.4.3 The present biota

The Pantanal reflects this. With its seasonal rise and fall of water levels, its prominent feature is the extent and biomass of floating vegetation (Plate 1). Indeed, with forty or so species, it has the highest diversity of such vegetation in the world, with genera like *Eichhornia*, the water hyacinths, *Pistia*, the water cabbage, and *Salvinia auriculata*, a floating fern. Some of these are weakly rooted in the bottom sediment, but most are completely free-floating, with roots that trail into the water. The mats rise and fall with the water levels and may become densely tangled and capable of supporting quite heavy mammals and birds. There are invertebrates, such as grass-hoppers, that specialize in feeding on the green parts of these plants and some birds, such as the jacana, with its widely spread toes, that live most of their lives and eventually nest on the mats.

But most of the plant production probably enters the food webs through sloughing off and sinking to the bottom and via the extensive root systems. Among these are the richest invertebrate communities, and on the bottom of soft, deoxygenated, richly organic mud is the greatest diversity of pulmonate snails in South America. Pulmonate snails breathe atmospheric air rather than absorb dissolved oxygen from the water through gills. There are many air breathers among the fish fauna also and this is no coincidence. Swamps are so productive and water dissolves so little oxygen that low concentrations of oxygen or total anoxia are normal in swamps and animals must move periodically to the surface to breathe. For this they require specific adaptations – swim-bladders modified to 'lungs' or gills supported on rigid arches so that they do not collapse in air.

Among the 405 species of Pantanal fish, there is a huge variety of habits and diets. Many come from three groups predominant in South America – the 'tooth carps' or cyprinodonts, many of which are fitted to survive in stressful habitats prone to high temperature and the risk of drying out; the spiny cichlids; and the silurids or catfish – although many other families are represented. Some species are catholic in diet, others specialize – for example the freshwater sponge eater, *Leporinus friderici*, while the Pantanal probably has larger populations of the serrasalmids, or piranha, than elsewhere in South America. There are many piranha species, some innocuous and feeding on other small fish or the scales of larger ones, and some shoal feeders, with a reputation for dismembering any large vertebrate prey to which their attention is directed by fluids oozing into the water from wounds.

Apart from the low-oxygen-tolerant species of the swampy floating mats and the tooth carps of the temporary lakes and pools, there are also riverine fish that move extensively up, down and across the rivers and floodplain in their feeding and breeding migrations. These require conditions of relatively high oxygen concentrations and are disfavoured by stagnant deoxygenated water.

1.4.4 Community structure

The community structure of aquatic systems, particularly standing waters, is determined not only by 'bottom up' processes like the chemistry of the water and the hydrological regime but also by processes acting 'top downwards' through the food webs by predation. Thus, in many temperate lakes, the larger members of the open water zooplankton community are selectively removed by fish with consequences for the amount of photosynthetic phytoplankton, which is then less efficiently grazed, and thence for the clarity of the water and the potential growth of bottom-rooted plants.

The Pantanal is so rich in fish species, and also reptilian and bird predators on the fish, that such top-down effects must be important, but no investigations have yet been made of them. Indeed, little is known of such processes in any tropical waters. However, only in some of the Pantanal lakes that lack fish (due perhaps to physical isolation and past fish kills following deoxygenation or drying out) does a large and brightly blue-coloured calanoid copoepod, *Argyrodiaptomus*, occur, and this may be because its size and visibility make it an easy prey when fish are present.

1.4.5 Vertebrates other than fish

It is said that the Pantanal offers the greatest wildlife spectacle in South America because of its abundance of reptiles, birds and mammals (Plate 1). Truly they are very obvious parts of the system, although their functional roles and interactions with the less obvious, but equally important, smaller animals and vegetation are largely unknown and not revealed by mere lists of species. This is a common gap in the understanding of most freshwater systems, even those well studied in temperate regions, where most limnologists work.

Prominent among the larger animals of the Pantanal is the jacaré, a small crocodilian, *Caiman jacare*, which basks in groups (Plate 1) by the edges of lakes and swamps and grows to about 1 m in length. It feeds on snails, fish, insects and crabs but will take other reptiles, birds and small capybara (a rodent) if it can. Jacaré breed in the wet season between August and April, making nests on the floating mats of plants and suffering predation of their eggs by mammals, such as the coati-mundi and forest dogs. The total population is about 10 000 000, but two other Pantanal caiman species are close to extinction through poaching for their skins. Amphibians and other reptiles are also present but apparently not abundantly in total numbers (except for the constricting snake, the anaconda), perhaps because of intense predation from jacaré, birds and fish.

The jacaré is prominent enough, but the birds are the visually dominant wildlife. The aquatic birds in a list of 650–700 species include 17 species of herons, six of ibis, five of kingfishers and nine of ducks. There are three species of stork, including the spectacular red-and-black-headed jabiru (Plate 1), and a coterie of grebes, cormorants, spoonbills, rails, jacana, snail kites, osprey and bitterns.

The water birds often nest colonially in tall trees, forming rookeries called *ninhais*, which are prominent for their concentrations of black and white birds and guano-strewn branches. The rain of organic matter – excreta, unfortunate chicks and food debris – that falls from them to the water attracts fish, other birds and snakes to feed beneath them, while snakes and climbing mammals, such as howler monkeys, rob nests in the canopy. The dynamics of the constituent populations in these communities must be fascinating but are largely unknown.

The Pantanal is mostly grazed by cattle now (see below) but there remain several species of deer, including well-adapted swamp deer (*Blastocerus dichotomus*), with large spreading feet, wild pigs, tapir and the capybara (Plate 1). The latter is a large (50–70 kg) rodent, grazing in groups of 10 or 20 animals among the plant mats and the gallery forest bordering the rivers. In turn, it is preyed upon by the jaguar, while two species of otter, four of wild dog, some small cats, the puma and at least one fish-eating bat show that predators are also prominent.

1.4.6 Humans

The Pantanal is not unusual in this richness of large vertebrate species. At the hearts of many of the national parks and nature reserves of the Earth are large rivers that expand in the wet season to flood and fertilize grassy plains, whose complex topography embraces a huge range of potential habitats. The key to retention of the huge diversity of players that the evolutionary play has provided is to allow the stage to keep changing in the natural ways that have been its normal features for more than millennia. But there has stridden on to this stage a new character. Human beings now dominate the play and many other players are having to leave the stage. The Pantanal is no exception but it is as yet less damaged than many of the world's freshwater systems, especially in temperate regions.

An ability to alter the stage to suit human convenience, as opposed to accepting the scenery as it comes, is at the heart of what it is to be human. But there are many indigenous human societies that have exploited their environment without destroying its diversity or sustainability [643]. It is largely the technologies developed in the last 100 years or so, coupled with peculiarly western social arrangements, which encourage the individual to exploit the society [991] (Chapter 11), that are responsible for our environmental problems.

We know little about the former indigenous peoples of the Pantanal. There were apparently seventy nations of a language group called Tupi-Guarami, who cultivated a wild water rice (*Oryza perennis*), hunted deer and constructed artificial islands among the swamps upon which to live. The anthropologist, Claude Levi-Strauss [569] painted a romantic picture of the lives of remaining groups who occupy lands reserved for them, but anthropological approaches that look at interactions with the environment, as well as social structures within the group, have not yet been widely used.

In the early 1800s, gold was discovered near Cuiaba (see Fig. 1.2) to the north of the Pantanal, and this led to division and acquisition of the land by the Portuguese colonial government and the consequent provision for purchase of blocks of land by individuals. Today the whole area is divided into a series of vast cattle ranches, often with absentee owners and managed by local *pantaneiros*. The cattle stock (Plate 1) is a mixture of about 4 000 000 zebu with local breeds. It rests reasonably well with the natural ecosystem and the feral herds of cattle, pigs, horses and water-buffalo, which have escaped from farms and ranches in the past, and perhaps represents a reasonable compromise among a variety of potential human interests in use of the area.

There are future threats to the integrity of the Pantanal. Natural high floods in the 1970s led to the drowning of many cattle, which could not find sufficient high ground for refuge. This stimulated the building of many ditches and dams to abate and drain future floods, but these have resulted only in interference with the natural water movements essential for fish migrations and in creation of stagnant, deoxygenated areas, in which such fish die.

Agriculture is not yet prominent in the Pantanal but there are suggestions that the huge biomass of aquatic plants could be used as fuel for the generation of methane (biogas) and that rice culture should be expanded. Rice and soy-bean growing are supported by the use of pesticides and herbicides, which may accumulate in a system that is not vigorously flushed out. The rivers are becoming clogged by soil eroded from the higher islands, which have been deforested by burning to create more grazing for the cattle. Gold is mined in parts and the process of purifying it is primitive. It is amalgamated with mercury, which is then evaporated and there is some evidence of mercury pollution interfering with the reproduction and behaviour of birds.

The greatest foreseeable threat, however, is serious disruption of the hydrology of the system by the proposed deepening and channelling of the River Paraguay, the Hidrovia project, to create a river navigable to seagoing ships from Argentina to Caceres in the north of the Pantanal. Not only would this prevent much of the present seasonal flooding on which the Pantanal system depends, it would also open up to further exploitation an area that currently survives because of its remoteness and inaccessibility.

On the other hand, ecotourism, a much less destructive exploitation of the land and the wildlife, and the interest engendered by a television soap opera *Pantanal*, screened in the late 1980s, may promote preservation rather than destruction. Control of land and resource exploitation, in the interests of all, in a huge country is not easy, however, and it has been past global experience that, where a resource could be ruthlessly exploited, it was.

1.5 End of the overture

The Pantanal, in its range of flowing- and standing-water habitats, its biological communities, its present use by people and the threats to its future, epitomizes the freshwater systems of the Earth. In this book, their components will be taken, one by one, and examined in detail. But these parts form a whole, which has additional properties. The chemistry of sediment and water, the genetics of a bacterium, the diet of a fish have no meaning away from the grand environmental stage. There is a greater human meaning also. The way we treat our natural resources is an index of how we are likely to treat our fellow human beings. The prognosis is at present very bad. But stand quietly in a great swamp system like the Pantanal and watch and listen and perhaps you may know that the description of structure and the revealing of mechanism are only small aspects of a greater reality.

2: On Living in Water

Earth has much water: a continuous ocean covers over seven-tenths of its surface. The fresh waters, in contrast, are divided and small (Table 2.1). There is, nonetheless, a large annual turnover – inflow and outflow – from this 'pool'. On average the surface fresh waters are naturally replaced in days or months, compared with residence times of thousands of years for the oceans. Although a few, very large, deep lakes, such as Baikal, Malawi and Tanganyika, are exceptions, this is the first main difference between the ocean and the fresh waters.

There is another, more obvious, difference. The ocean is salty (about $35 \, g \, l^{-1}$ of salts) and the fresh waters (on average less than $1 \, g \, l^{-1}$) are not. Water flows into the ocean but leaves only by evaporation; while some salts are precipitated, the more soluble ions dissolved in the inflowing water have been concentrated over time. In fresh waters, the outflow is mostly as liquid rather than as vapour and, although there is concentration of salts by evaporation in endorheic areas (see Fig. 1.3), there is little opportunity for salt accumulation in exorheic regions.

These distinctions of residence time and saltiness are matched by a third: the large contrast in the variety of life that lives in fresh waters and in

Table 2.1 Distribution of water on Earth, and the time taken for complete replacement (residence or renewal time) of each category.

Category	Water volume* (km^3)	Fraction of total (%)	Renewal time† (residence time)
Atmosphere	1.3×10^4	0.001	7–11 days
River channels	1.2×10^3	0.0001	7 days
Freshwater lakes	1.2×10^6	0.009	330 days
Saline lakes and inland seas	1.0×10^5	0.008	1–4 years
Soil water	6.6×10^4	0.005	?
Groundwater	8.2×10^6	0.62	60–300 years
Ice caps and glaciers	2.9×10^7	2.15	12 000 years
Ocean	1.3×10^9	97.2	300–11 000 years

*From Leopold [568].

†Estimates vary. Values are taken from Gregory and Walling [363] and Ward [996]. The residence time is defined as the volume of the water body concerned divided by the volume added to it in a given time. The dimensions of this are thus length^3 divided by length^3 time^{-1}.

oceans. Fifty-six phyla are present in the ocean and forty-one in fresh waters. No phyla are confined to fresh waters but fifteen phyla are found only in the ocean. This distinction probably applies even more to the total numbers of species present. Of the nearly 19 000 species of fish, just under 7000 are found in fresh waters; eighteen of the forty-six orders of fish are confined to the sea and only half as many to fresh water [718].

Nonetheless, water, fresh or salt, has many shared features as a medium in which to live. It is so different from air that its properties and the similar problems of living in it should be emphasized as much as the differences between the fresh water and the ocean. Chapter 2 explores first the similarities of living in and then the reasons behind the differences in species diversity between fresh and marine waters.

2.1 Properties of water

2.1.1 Physical properties

For all its familiarity, water (H_2O) is remarkable. It is the hydride of oxygen, an element that lies in the Periodic Table above sulphur and selenium and between fluorine and nitrogen. Elements within vertical groups and along the rows in the table usually have graded series of properties, but H_2O is clearly anomalous compared with the related hydrogen sulphide (H_2S) and hydrogen selenide (H_2Se) or ammonia (NH_3) and hydrogen fluoride (HF) (Table 2.2). It remains a liquid at earth-surface temperatures, whereas its position in the Periodic Table suggests that it should be a gas. It also freezes at a much higher temperature than would be expected.

It is not so easy to describe liquids as it is solids or gases. The molecules of solids are held in a more or less constant relationship with one another; individual molecules vibrate, but about a fixed position. On the other hand, the molecules of gases move fast and randomly and have no structural relationship with each other. Liquids can be seen either as highly condensed gases, in which the random molecular movements are dominant although the molecules are closer to one another, or as disturbed solids, in which the ordering of the molecules is much reduced but still present and important. In liquid water, in particular, some of the crystal structure of ice is

Table 2.2 Comparative properties of the hydrides of oxygen (water) and of those elements close to it in the Periodic Table. The upper value is the melting-point (°C) and the lower value the boiling-point (°C) in each case. (From Hutchinson [451].)

CH_4	NH_3	H_2O	H_2S	H_2Se	HF
−182.6	−77.7	0	−82.9	−64.0	−83.0
−161.4	−33.4	+100	−59.6	−42.0	−19.4

preserved in the liquid. It is this that makes water odd compared with hydrogen sulphide and hydrogen selenide, so what is the reason for it?

The water molecule has two hydrogen atoms held at an angle of 104° 27′ and at distances of nearly 0.1 nm from the oxygen atom. The hydrogen atoms are held at about 0.1 nm from each other. The molecule is covalently bonded, electrons being shared between the hydrogens and the oxygen, giving overall electrical neutrality, but the oxygen nucleus has greater affinity for electrons than that of the hydrogen. There is thus, on average, a slight displacement of negative charge towards the oxygen, which leaves a slight positive charge on the hydrogen atoms. Consequently there is an attraction between the slightly positive hydrogen on one molecule, and the slightly negative oxygen on the other, which links them together with a 'hydrogen bond'. The angles at which oxygen and hydrogen are held in the water molecule, coupled with this hydrogen bonding, result in a crystal structure for ice that is based on tetrahedra (Fig. 2.1).

Oxygen is at the centre of each tetrahedron, surrounded by four hydrogen atoms, two covalently and two hydrogen-bonded. Such a crystal is quite open, compared with those of other substances, which often have twelve neighbours (as opposed to four in ice) packed around each molecule. Ice consequently has a low density. It floats on liquid water and, by forming an insulating layer at the surface of water bodies, often prevents them from freezing solid and thus killing fish and other organisms.

The hydrogen bonding in ice is quite strong, because the displacement of negative charge towards the oxygen atom is powerful. The temperature at which melting of ice takes place – a measure of the energy needed to begin breaking down the hydrogen-bonded structure – is thus relatively high (see Table 2.2), compared with H_2S and H_2Se, where the charge displacement in the molecules is small.

The latent heat of fusion – the energy needed to convert the solid to the liquid – is, however, low for water compared with many other substances. This suggests that much of the orderly structure in ice is not destroyed on

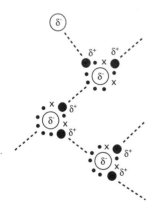

Fig. 2.1 The crystalline structure of ice is to some extent preserved in liquid water. Oxygen (large circles) and hydrogen (black circles) share electrons (dots from oxygen, crosses from hydrogen), but the distribution of the final charge is unequal. The slightly negative charge (δ^-) on the oxygen is attracted to the δ^+ charge on the hydrogen of a neighbouring molecule. Such hydrogen bonds bind the molecules into a structure in which molecules are bound in a tetrahedral structure, even in the liquid.

melting. More of the structure is destroyed as temperature increases but it is not until the relatively high boiling-point is reached (see Table 2.2) that all of it is lost. The high latent heat of evaporation – the energy needed to break down the crystal structure entirely – reflects this. It is over six times higher than the latent heat of fusion.

For organisms living in water, this tenacity of structure is important in giving water a high specific heat. Specific heat is defined as the amount of energy needed to raise 1 g of a substance through 1 °C. It takes much gain or loss of heat to change the temperature of liquid water, for temperature is a measure of the amount of free movement of molecules and hydrogen bonding partly locks water molecules together. Hence temperature ranges and fluctuations in natural waters are muted, perhaps 25 °C or less, whereas continental land habitats can have ranges of air temperature more than twice this.

Water has a maximum density at 3.94 °C (under standard pressure). This too depends on its retention of structure as a liquid. As ice melts, collapse of some molecules into the parts of the structure that are retained leads to a partial filling of the open crystal and an increase in density. This might be likened to a partly demolished building, where rubble fills the still-standing shell. A second process, of movement further apart of the released molecules, tends to decrease the density; the rubble is carted away. The former process dominates up to 3.94 °C and the latter at higher temperatures. The density differences around the peak are small but enough to layer the winter water in deeper basins. Colder layers (< 3.94 °C) float on warmer (> 3.94 °C) water at the bottom, which is insulated by the lighter surface layers from further cooling and freezing. At higher temperatures the less dense, warm water may float on cooler water, leading again to an isolation of the deeper layers from the atmosphere (see Chapter 6).

2.1.2 Water as a solvent

The charge displacement, which gives some properties of a charged ion to an essentially non-ionic substance, brings water its versatile properties as a solvent. The act of dissolving needs a chemical attraction between the solvent and the solute. A charged or ionic solvent cannot then dissolve an electrically completely neutral solute or a neutral one a charged solute. Because of the separation of charge in its molecule, water acts as a charged solvent and will attack ionic crystals, such as salts, and bring them into solution.

The extent to which such ions are dissolved depends on their attraction to water molecules. In turn, this depends on their own charge or valency and on the size of the ion. The attraction increases with the charge (either positive or negative) but decreases with the radius of the ion. This is because the charge is weakened by being spread over a greater surface area of the ion. If Z is the charge on and r the radius of the ion, the affinity for water can be measured as the ionic potential, Z/r.

The efficiency of water as a solvent does not simply increase with increasing ionic potential of the solute. Ions with Z/r less than 3.0 or greater than 12.0 are readily dissolved but those with Z/r between these values tend to be precipitated. Ions with an ionic potential less than 3.0 are cations derived from metals (Fig. 2.2). The charge is sufficient for attraction to the water molecules and to bind the ions with water (the results being called hydrated cations) in solution. The ions of sodium, potassium and calcium are good examples.

Some heavier elements have quite small ionic radii despite a high weight and a high charge. In these, which have an ionic potential between 3.0 and 12.0, the charge is sufficient to attract the oxygen atom in water so closely that the binding between the oxygen and one of the hydrogens in the water molecule is weakened. A hydrogen ion is then ejected into solution and a metal hydroxide is formed. Such hydroxides have little surplus charge left for the attraction of water molecules, and they precipitate. Aluminium and silica are good examples. So are iron and manganese, although these transition metals have several valency states and can also behave like the first group.

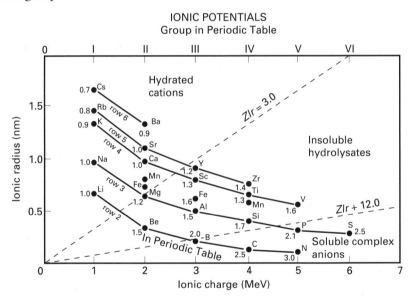

Fig. 2.2 Relationship between ionic charge (Z) and ionic radius (r) of several elements. Lines connect elements in the same row of the Periodic Table. Values of the ionic potential (Z/r) distinguish elements that behave as hydrated cations ($Z/r < 3.0$), insoluble hydroxides ($Z/r > 3.0 < 12.0$) and soluble complex ions ($Z/r > 12.0$). Cs, caesium; Rb, rubidium; K, potassium; Na, sodium; Li, lithium; Ba, barium; Sr, strontium; Ca, calcium; Mn, manganese; Mg, magnesium; Be, beryllium; Y, yttrium; Sc, scandium; Fe, iron; Al, aluminium; B, boron; Zr, zirconium; Ti, titanium; Si, silicon; C, carbon; V, vanadium; P, phosphorus; N, nitrogen; S, sulphur. (From Raiswell *et al.* [795].)

In the third case, elements with ionic potentials greater than 12.0, the charge attraction for the oxygen in water is very great. Both the hydrogens are ejected into solution as H^+ (where they bond with water molecules to form H_3O^+, hydrated hydrogen ions, and remain in solution) and an oxyanion is formed. Such oxyanions are called complex, for they involve two elements, and the ionic potential of the complex ion (as opposed to those of its elemental components) is reduced to a value at which it attracts water molecules and remains in solution. Nitrate, carbonate and sulphate are good examples, with the position of phosphate (Fig. 2.2) bringing it close to the borderline between the soluble oxyanion and the insoluble middle group.

The charge properties of water thus interact with those of other elements to bring ions into solution to varying extents, and the inorganic composition of natural waters, where the highly soluble sodium (Na^+), potassium (K^+), magnesium (Mg^{2+}), calcium (Ca^{2+}), bicarbonate (HCO_3^-), sulphate (SO_4^{2-}) and chloride (Cl^-) ions are normally major components, reflects this. The final composition, however, also reflects the availability in the surface of the Earth's crust of elements for potential solution (see Chapter 3). Because of this, almost none of the more soluble elements will approach saturation, except in conditions of very high evaporation. And in mixtures of ions the affinity of one to another may overcome the attraction of either to water and lead to precipitation. For example, both calcium and carbonate are highly soluble each in the absence of the other, but readily precipitate as calcium carbonate ($CaCO_3$) if mixed together.

2.1.3 Solubility of non-ionic compounds

Water is a charged compound and so substances whose molecules have no charge displacement will not dissolve in it. Many organic compounds, for example hydrocarbons, are thus not soluble. But a wide variety of organic compounds do have slightly charged (polar) groups in them, for example hydroxyl (OH^-), amino (NH_2^-) and sulphide (S^{2-}), and so have some affinity for and hence solubility in water.

In a sense, all substances will dissolve to some extent in that, if a source of them is held in contact with water, there will be some diffusion, although the equilibrium will lie heavily weighted towards the source for non-polar compounds, as opposed to the reverse for polar ones. Most atmospheric gases fall into the non-polar group, so solubilities of nitrogen (N_2), oxygen (O_2) and the inert gases are low. They vary with temperature and pressure and with their concentrations in the atmosphere (Table 2.3). Solubilities of gases decrease with temperature because the increased molecular movement increases the possibility of escape from the liquid to the overlying vapour, whereas solubility of ions increases with temperature. This is because ionic solubility is a chemical reaction, not a passive mixing. The

Table 2.3 Solubilities of atmospheric gases in pure water under conditions of 1 atmosphere pressure of each gas and in equilibrium with atmospheric air (21% O_2, 78% N_2, 0.03% CO_2) with a total pressure of 1 atmosphere. Both values are for a temperature of 15°C.

| Gas | Under 1 atmosphere | | With atmospheric air |
	(ml (STP) l^{-1})	(mg l^{-1})	(mg l^{-1})
Oxygen (O_2)	34.1	46.2	9.7
Nitrogen (N_2)	16.9	20.03	15.8
Carbon dioxide (CO_2)	1019	1897	0.57

STP, standard temperature and pressure.

exception to the general behaviour of atmospheric gases is carbon dioxide, which is relatively soluble. This is because it reacts with water, being itself polar, and exists in equilibrium with ions of carbon, such as bicarbonate and carbonate, in natural waters (see Chapter 3). A detailed treatment of the chemical composition of fresh waters is given in Chapter 3.

2.2 Land and water habitats and the evolution of aquatic organisms

Life evolved in water, but was it fresh or salt? The earliest fossils are of prokaryote microorganisms, which need access to small molecules of nutrients and gases and hence are also permeable to water. They must have at least a covering film of liquid water not to dry out. Representatives of all of the earliest phyla, however, are found in both the ocean and fresh waters. Some protistan families (Protista are eukaryote microorganisms, including algae and protozoa) are entirely marine, a comparable number entirely freshwater, and most have representatives, even fellow species in the same genus, in both habitats.

What is clear, however, is that multicellular animals evolved in the sea and that multicellular plants had at least three routes. One was within the ocean to brown seaweeds (Phaeophyceae), a second via fresh waters, or the sea, from the blue-green algae to the now mostly marine red algae (Rhodophycota) and a third through the freshwater green algae (Chlorophycota) to the freshwater stoneworts (Charophyceae) and the essentially land-living mosses and liverworts (Bryophyta) and vascular plants.

Subsequently, there was a secondary movement of some multicellular animal groups into fresh waters and on to the land and even a further movement of some land animals and land plants into fresh waters and back to the sea. This overall picture implies a considerable flux over time between the three main habitat types. The evidence depends on fossil and morphological comparisons and to a lesser extent on comparisons of the physiological modifications needed to exploit the three habitats. Physiological adjustment has been, on the whole, rather readily and quickly achieved, and movement among fresh waters, sea water and the land has not been hampered by insuperable physiological problems. The lower diversity of organisms in fresh waters compared with the other two habitats is probably thus not particularly a result of fresh water being a more 'difficult' medium to colonize, but some discussion of the problems faced in each habitat is needed.

2.2.1 Physiological problems of living in water

Organisms must be open systems, allowing movement of materials like oxygen and wastes between themselves and the environment. In satisfying this need, they are inevitably open also to movement of water and salts, whose amounts they must closely regulate, lest they dry out or burst or become too salty or too dilute for their enzymes to function.

Of the three main habitats, sea water is the most steady. Its salinity varies (except in very local instances) only between about 32 and 38 parts per thousand (g l^{-1}). The proportional composition of the salts in it varies even less from place to place. Marine invertebrates have internal solute concentrations close to those of sea water and thus spend little energy in regulating their internal environment. Fresh waters vary much more in concentration and composition (Table 2.4) and pose greater regulation problems.

Table 2.4 Concentrations of some of the most common ions in various natural waters. Values are given in millimoles per kilogram. A single set of values is representative for sea water, but fresh waters and inland saline waters vary greatly between themselves and even (for example, Lake Chilwa) between years, depending on water level.

Ion	Fresh waters		Inland saline waters		
	Amazon streams*	Barton Broad, UK†	Lake Chilwa, Malawi‡	Dead Sea, Israel§	Average sea water‖
Sodium	0.009	2.1	1.6–142	1955	475
Potassium	0.004	0.23		219	10.1
Calcium	0.001	2.8	0.24–0.95	481	10.3
Magnesium	0.002	0.46	0.04–0.74	2029	54.2
Chloride	0.031	2.4	0.74–88.0	7112	554
Sulphate	–	–	–	5.3	28.6
Bicarbonate	0.036	3.4	1.9–88.0	3.7	2.4

*Furch [306].
†Moss [677].
‡Morgan and Kalk [662].
§Steinhorn et al. [922].
‖Potts and Parry [786].

2.2.2 Brackish and freshwater invertebrates

Representatives of several phyla of marine invertebrates have adjusted to brackish conditions in estuaries and eventually to fresh waters. Estuarine brackish waters usually have a salt composition dominated by the seawater composition, but some inland brackish waters (Table 2.4), produced by evaporation of fresh water or salt springs in confined areas, may have a very different ionic concentration and an even higher total salt concentration than sea water. They are called athalassic (not sea-like) and, if appropriate, hypersaline or brine waters.

Colonization of estuaries, linking rivers and the sea, by marine invertebrates has meant coping with both reduced and variable salt concentrations, as the mixture of fresh and salt water varies with river flow and state of the tide. Some such colonizers (e.g. the lugworm, *Arenicola*) avoid the problem by burrowing in the sediment, which is generally infused with the denser, almost full-strength sea water and whose composition daily changes only a little. Others close up protective shells at low concentrations (e.g. mussels, barnacles) or move downstream towards the sea (some crabs). Many, at least in larger bodies of brackish water with some stability in salt concentration, such as the Baltic Sea, are osmoconformers.

Osmoconformity means they have reduced their internal osmoconcentration to close to that of the water by pumping out salts. This minimizes absorption of water by osmosis and consequent severe body swelling. As external salt concentrations are reduced further, such animals may begin to regulate their salt composition by active absorption and become osmoregulators. Such an ability was necessary for those animals which colonized fresh water, for an internal salt concentration as low as that of the external medium in fresh water would be insufficient to allow many enzymes to function. Freshwater invertebrates thus maintain, by active salt uptake, an internal concentration of between about 30 and 300 mosmol, compared with < 15 mosmol in the water outside. This means that, although most of the body may be relatively impermeable to both water and salt, they continually and inevitably take up water by osmosis, particularly through the gill surfaces, which must be large, because of the low oxygen concentration in water (see below), and permeable to small molecules.

In turn, they must eliminate this water as urine and, in doing so, inevitably lose some salts dissolved in the urine. The salt balance is maintained by active uptake through cells in the gills and, in some insects, through flat tail-fans or 'anal gills'. This uptake may be made less costly in energy by exchanging ammonium (NH_4^+ ions, produced as waste by food metabolism, for Na^+, and HCO_3^-, produced through respiration, for Cl^-.

The problems of coping with the more dilute fresh water (exclusion of water, maintenance of internal salt concentration) are certainly different from those of coping with sea water (regulating ionic composition at high salt level) but also not very costly in energy [785]. Representatives of many phyla occur in both the sea and fresh water (bivalve and gastropod molluscs, crustaceans and annelid worms being particularly good examples), but it is in the vertebrates, particularly fish, that evidence for a comparatively easy movement between the habitats comes most readily.

2.2.3 Osmotic relationships in vertebrates

Only one small group of vertebrates, the hagfishes, an early-evolved group of jawless fishes, has body fluids with both ionic and osmotic concentrations similar to those of sea water. The remaining jawless fish (the lampreys), the cartilaginous fish (sharks and rays) and the bony (or teleost) fish (Fig. 2.3), all have ionic concentrations and compositions in the body fluids much lower than and very different from those of sea water. The osmotic concentrations

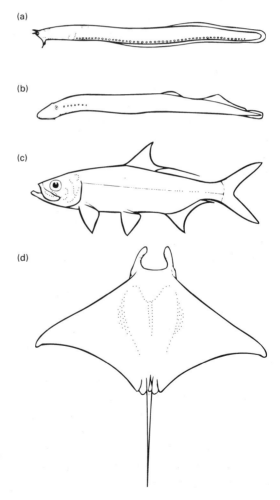

Fig. 2.3 Fish and their close relatives solve problems of osmoregulation in different ways. Shown are (a) hagfish, (b) a lamprey, a teleost fish, (c) the tarpon and (d) a ray, an elasmobranch fish.

of the freshwater fishes of all groups are similar to those of freshwater invertebrates and the fish cope with the same problems of high water uptake and salt loss by generally similar mechanisms.

The osmotic concentrations of the marine cartilaginous fish are about the same as those of sea water because large quantities of at least two organic compounds, urea and trimethylamine oxide, are stored in the body fluids to add to the low osmoconcentration provided by ions alone. Although jawed cartilaginous fish are now scarce in fresh waters, they (and the fish families that descended from them) may have had their evolutionary origin there. On moving back to the sea, they have found an alternative means of increasing their osmoconcentrations that is very different from that of the ultimate invertebrate ancestors. Amphibians are a largely freshwater group with low ionic osmoconcentration, but one species, a crab-eating frog of the mangrove swamps of South-east Asia, has colonized sea water, also by increasing the urea concentration in its blood to similar osmoconcentrations to those of sea water.

The bony fish appear also to have had their origin in fresh water, for the marine bony fish have a much lower osmoconcentration, as well as ionic concentration, than sea water. They thus tend to lose water by osmosis to the sea water and must compensate for this by drinking a large amount of sea water and then excreting the excess salt through glands in the gills (Na^+, Cl^{-1}) or in their urine (Mg^{2+}, SO_4^{2-}).

The low osmoconcentrations of the body fluids of both freshwater and marine bony fish allow a relatively ready movement of certain species between the sea and fresh waters. The migrations of lampreys, salmon and eels are well known, but many inshore marine fish penetrate estuaries to the lower reaches of lowland rivers. Some cartilaginous fish also move between fresh waters and the ocean by lowering their urea concentrations and quickening their urine production in fresh water.

Eels live most of their lives in fresh water and are called 'catadromic' because they move to the sea to breed. On reaching the sea, eels start to lose water by osmosis, and may lose about 4% of their body water before adjusting, by drinking sea water, in a few days. Young eels, moving to the rivers from the sea at first absorb water but soon compensate for this by increasing their urine flow.

2.2.4 Colonization of fresh waters from the land

Many freshwater organisms with close relatives in brackish water and the sea had a marine origin. Problems of adjusting to changed salinity were relatively easily overcome. The same is true of groups, including Protista and bony fish, which moved in the reverse direction. A second evolutionary pathway into fresh waters was, at least for the freshwater insects and higher (vascular) plants, from the land. Freshwater vascular plants often still retain terrestrial features, such as stomata and pollination by wind and terrestrial insects (Fig. 2.4). In both cases, the diversity of species in fresh waters is much lower than it is on land. Is this due to physiological difficulties posed by fresh water?

The main problem for land organisms is the threat of dehydration. In a sense, all land animals represent envoys moving out to foreign parts from a home base of fresh water, to which they must return to drink. Land plants, with their roots in moist soil, are effectively freshwater organisms in a freshwater habitat (damp soil) always perilously close to drying out. The colonization of more conventional freshwater habitats should thus have been relatively easy for land organisms.

Many other advantages are given by fresh water (Table 2.5). Water has a higher viscosity and density, eliminating the need for investment of much energy in supporting structures, such as wood or bone, and a high heat capacity, buffering temperature change. Carbon dioxide (CO_2) availability for photosynthesis is also much greater in water because of the dissolved reserve, as bicarbonate, which easily dissociates to form CO_2.

But there are problems also. The oxygen content of water is very low and the rate at which it diffuses to body surfaces is much lower in water than in air (Table 2.5). The great advantage of colonization of the land was access to a rich supply of atmospheric oxygen and the possibilities of greater

Fig. 2.4 Invasion of fresh waters has been relatively recent for certain groups, such as the aquatic plants and insects, which retain many land characteristics. These include aerial flowers and pollination (bladderwort, left) and life histories that depend on water for the juvenile but not the adult stages (damselfly, right).

Table 2.5 A comparison of certain characteristics of atmospheric air and fresh water. Values are given at standard temperature (273 K) and pressure (1 atmosphere (101.3 kPa)). (Modified from Schmidt-Nielsen [872].)

Characteristic	Water	Air	Ratio water : air
Oxygen concentration (ml l^{-1})	7.0	209.0	1 : 30
Density (kg l^{-1})	1.000	0.0013	800 : 1
Dynamic viscosity (cP)	1.0	0.02	50 : 1
Heat capacity (cal l^{-1} ($°C^{-1}$))	1000	0.31	3000 : 1
Diffusion coefficient ($cm^2 s^{-1}$)			
Oxygen	2.5×10^{-5}	0.198	1 : 8000
Carbon dioxide	1.8×10^{-5}	0.155	1 : 9000

manipulation of energy which this offered. A move to water was to accept the loss of this advantage. Respiratory mechanisms are thus important in understanding freshwater organisms.

2.2.5 Respiration in water

Aquatic animals must increase the surface area of the tissues (gills, which are evaginations and folds of the body surface, or lungs, which are similar invaginations) available for oxygen uptake (Fig. 2.5). Or they must maximize the rate of flow of oxygenated water past them, to maintain the highest possible concentration of oxygen at their surfaces. Alternatively, aquatic animals can continue to breathe air for all (birds and mammals) or part (amphibians, many insects) of their lives.

The smaller invertebrates may simply use their entire body surfaces as a gill as long as the path for diffusion into the centre of the body is no more than than about a millimetre or so (e.g. in flatworms, water-fleas, rotifers). Beyond that, more specilized folds of tissue, rich in a blood supply, are needed, and often some pumping mechanism which forces water past them.

Fig. 2.5 Freshwater animals include both gill-breathers and lung-breathers. The former, such as the Atlantic salmon (left), absorb oxygen from a stream of water pumped over a large surface area of highly vascularized, evaginated tissue. Lung-breathers, such as the European otter (right), use atmospheric air and must keep an equally large area of moist, invaginated tissue in the state of an internal freshwater environment.

For example, bivalve molluscs (mussels, clams, oysters) pump water in through tubes called siphons.

Fish force water over the gills by rhythmically filling the mouth with water and then reducing the mouth volume so that the water is forced over the gills and then out through a slit or slits behind the head. When great activity is needed, they may simply keep the mouth open and allow the surge of water created by their movement to flood the gills. The drawback to gill respiration is that large areas of tissue permeable to almost any small molecule must be exposed, so that the freshwater problems of osmotic water uptake and body salt dilution and the marine one of excess salt uptake are exacerbated.

2.2.6 Invertebrate air breathers in fresh waters

Insect groups have commonly invaded fresh waters. Often they have complex life histories, with juvenile larval or nymph stages. (Larvae look very different from the adult, whereas nymphs resemble the adults in many, though not all, respects (Fig. 2.6).) The aquatic juvenile stages may breathe water or air. The adults are usually air breathers and may live in water or

(a)   (b)

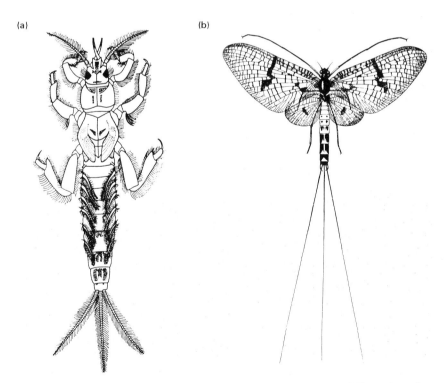

Fig. 2.6 In freshwater insects, the juvenile stages (larvae) may be very different from the adults, or they may preserve, as nymphs (a), the same basic pattern as the adult (b), as in the Ephemeroptera, or mayflies. This is *Ephemera danica*. (Based on Hynes [459] and Fryer [302].)

on land. Many nymphs of the mayflies (Ephemeroptera), for example, have gills on their backs, which they fan in the water, whereas mosquito larvae suspended from the surface-tension film at the surface absorb atmospheric air through a large pore or spiracle at their hind end (Fig. 2.7).

Water-beetle larvae have gills, but adult beetles and bugs (Fig. 2.7) acquire an air bubble at the water surface, which they hold under their wing covers when they dive. Through their spiracles they absorb oxygen from this bubble and, to maintain gaseous equilibrium, more oxygen diffuses into the bubble from the water, which is itself in equilibrium with the atmosphere. The bubble may thus give an oxygen supply for some time. It eventually collapses, because, as oxygen is absorbed by the animal, the percentage of nitrogen in the bubble temporarily increases. Nitrogen then diffuses to the water, which is in equilibrium with the lesser proportion of nitrogen in the air. The bubble thus progressively gets smaller and must eventually be renewed at the surface.

Water spiders maintain a similar temporary reserve of air under a 'bell' of silk strung between the stems and leavers of aquatic plants. They return to this bell to breathe between hunting forays. Some insects maintain a permanent air bubble close to their bodies by supporting it with a very dense mat of very fine, water-repellent, hairs called a plastron (Fig. 2.7). These prevent collapse, by temporary nitrogen loss, of the bubble and allow the animal to breathe atmospheric air indefinitely while underwater.

2.3 Time and fleetingness – a fundamental difference between freshwater and marine ecosystems

2.3.1 Low diversity and the colonization of fresh waters

Colonization of fresh waters from the land has thus not been difficult. The physiological problems of salinity tolerance have also been readily solved by organisms moving, both literally for some and in evolutionary terms for many, between fresh waters and the sea. What, then, is the reason for the relatively low diversity of freshwater communities?

Fig. 2.7 (*Oppposite*) Supply of sufficient oxygen is crucial to freshwater animals. Many beetles and bugs (*Sigara*, a) visit the water surface frequently to capture a bubble of air, which sticks to their undersides and supplies them for a time. Eventually, the bubble collapses and must be replaced. Others, such as the larva of the mosquito genus, *Culex* (d), attach to the surface-tension film and breathe atmospheric air through a large spiracle. Many simply absorb oxygen from the water over their whole bodies, or, as in the nymph of the mayfly, *Ecdyonurus* (b), have, on their abdomens, flat, paired gills, which provide large areas for absorption. The beetle, *Aphelocheirus* (c), has a plastron, or velvety mat of hairs on its underside, which supports a bubble of air for a long period. Oxygen dissolves from the water into the bubble and this obviates frequent visits to the surface. (Based on Fryer [302] and Hynes [459].)

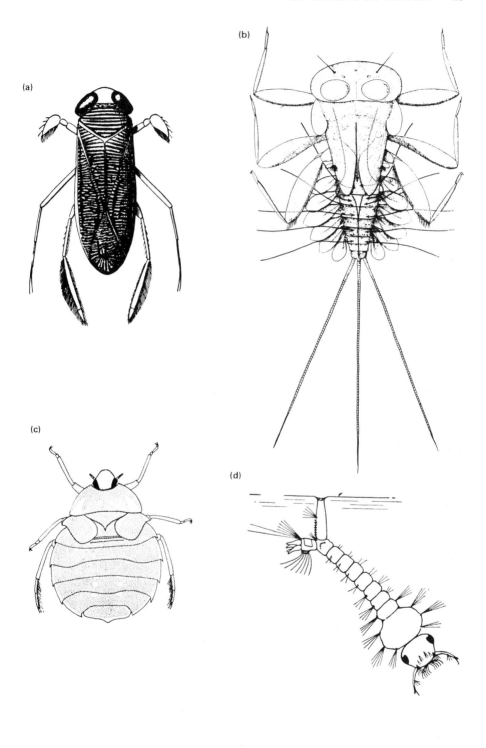

W.J. Sollas [905] thought that the flow of rivers might be a physical barrier to the delicate drifting larvae that characterize many marine animals; in general, freshwater animals do not have such larvae. But the copepods, widespread in fresh waters, do have larvae, called nauplii, to which the rivers cannot have been a barrier; and, in contrast, the vigorously moving marine squids seem never to have colonized fresh waters. Others [636] have proposed that the more widely fluctuating temperatures of fresh waters compared with those of the sea might also have posed problems. Yet the diverse intertidal zone is seasonally as variable in temperate regions as any lake, while the fluctuation in a large equatorial lake is no greater than that at the tropical sea surface.

The idea of change in the habitat extending to a longer time-scale may, however, be revealing. Fresh waters have small volumes and are liable to be disturbed by climatic changes which, for periods of thousands of years, bring drought to one area or ice to another (Plate 2). They are subject also to geological upheavals, which may alter drainage, emptying lakes and reversing river flows. The ocean has not been constant either. Continents have moved, altering the shapes of the ocean basins, and the polar glaciations have cooled the water. There has, nonetheless, always been an ocean. A continuous sequence of sedimentary rocks laid down in the sea testifies to at least one ever-present huge volume of water on a planet that for over 4 billion years has never been completely dry.

For over 100 000 years until only 10 000–15 000 years ago, most of the northern hemisphere, whose land areas now bear extensive river systems, and most of the world's natural freshwater lakes were covered with a polar ice-sheet several kilometres thick. The freshwater fauna previously present would have been crushed; some more active members doubtless moved towards the equator or were washed southwards in melt waters at the edge of the sheet. Further back, 250 million years ago, the present cool and damp northern England was an arid dune desert. In Africa, few areas of the continent have been undisturbed by volcanic and major geological move-ments in the last 300 million years, while whole mountain chains have risen over the whole of western North America in the last 200 million years. Between periods of change in one place, there may have been interludes – hundreds of thousands or even millions of years – of relative stability, but these are short compared with the continuity of the ocean.

The distinction between fresh waters and the ocean can perhaps be understood in terms of the different predictability of these habitats. The ocean has not been disrupted to the same extent as the fresh waters. Species have been able to differentiate without violent interruption. Assuredly, the rate of this process has been different in different places – coral reefs provide a more stable habitat than the polar seas at the edge of a fluctuating ice-sheet. But the existence still of species such as the coelocanth (*Latimeria chalumnae*), which has changed little from fossils recorded in the Devonian period, 300 million years ago, testifies to considerable stability.

The differentiation of species in fresh waters, however, may never have been able to proceed for long without disruption, and the discontinuity of freshwater systems, each separated in its own drainage basin, has hampered movement and escape when catastrophe has struck. Almost all natural temperate lakes were created about 10 000 years ago as the ice retreated, leaving a hummocky landscape and natural moraine dams across many valleys. These lakes have had to be colonized by freshwater organisms that had found refuge in the warmer areas towards the equator. There has been little time for more species to differentiate within the new freshwater habitats.

2.3.2 Evolution in freshwater habitats

That such an evolutionary process does go on, however, may be shown by a number of animal species that occur in lakes around the head of the Baltic Sea. These lakes, just after the ice retreated and sea level rose from the melt water, were continuous with the Baltic. Then the land, relieved of the pressure of the ice, rose relative to the sea level, and the lakes became isolated and freshened by the rivers. A group of animals, apparently originally brackish-water organisms, still present in the Baltic Sea, survived in them. In a series of lakes of different ages of freshening, some of these animals show alterations in body shape relative to their ancestors, which may indicate a gradual evolutionary change. An example is that of *Limnocalanus macrurus* (Fig. 2.8). That change will not proceed very far, however, for the area is likely to be glaciated again some tens of thousands of years hence.

2.3.3 Evolution in tropical African lakes

Away from the polar ice, fresh waters have still been disrupted so frequently as to prevent attainment of the diversity found in the ocean. In the Miocene period, some 25 million years ago, Africa was a continent of subdued relief with well-separated, shallow, swampy basins. Volcanic eruptions and movement of the crustal plates created a watershed that separates east from west and the rift valley, which houses the basins of many of the East African lakes. More recently, the wetter 'pluvial' periods, which bear some relationship to the warm, interglacial periods in higher latitudes, supported savannah and swamps in what is now the Sahara. There, rock paintings from only a few thousand years ago show hippopotamuses. A later drying out isolated Lake Turkana (Lake Rudolf) from its one-time connection with the Nile and left many lakes reduced to remnant salty basins.

In these lakes – for example, Lake Chilwa [515] in Malawi – the unpredictability from year to year (for occasionally they dry out completely, only to refill) perhaps mirrors the general unpredictability of inland waters (Plate 2). They support a limited biota of generalist species, unfussy in their

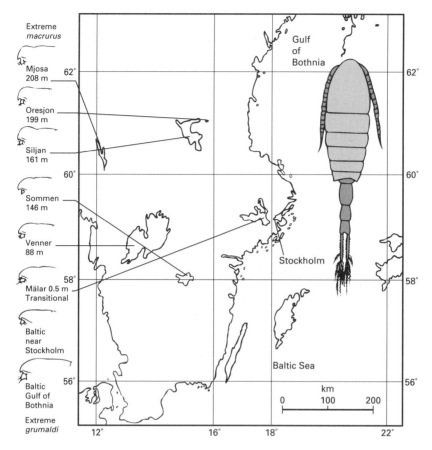

Fig. 2.8 Distribution of the copepod *Limnocalanus macrurus* in Sweden and Norway. This species shows a steady evolutionary change from the brackish *L. macrurus* stranded in freshwater lakes formed after the last retreat of the polar ice. The longer the lake has been isolated, the greater the change from the brackish *L. m. grumaldi* forms to the *L. m. macrurus* forms. Altitudes of the lake surfaces (a measure of the length of time that has elapsed since isolation from the Baltic Sea as the land has risen in the postglacial period) are shown under profiles of the head of the organism (seen from the side with the insertion of the antenna). An entire copepod, seen from the 'front' rather than 'side' view, is shown. Lake Malar has been cut off from the sea only since the early Middle Ages. (From Hutchinson [451].)

requirements for existance and breeding and taking a range of food, from algae to fish (Fig. 2.9). When Lake Chilwa dries out, for example, its fish move into the remaining trickling rivers or even into still wet, mud pools in the surrounding swamps.

In contrast, not far north of Lake Chilwa, Lake Malawi, in its relative longevity, echoes the ocean. Its basin is very deep and, although the water level has fluctuated as climate has changed, there has probably been a permanent water mass for a million years. In Lake Malawi and in the

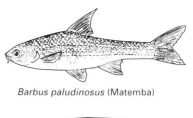

Barbus paludinosus (Matemba)

Clarias mossambicus (Mlamba)

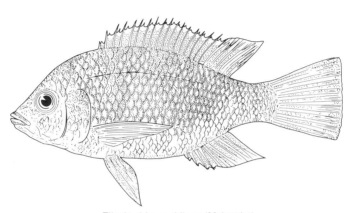

Tilapia shirana chilwae (Makumba)

Stomach contents (% of total items)

	Higher plants	Green algae	Blue-green algae	Diatoms	Crustacea	Snails	Insects	Rotifers	Fish
Barbus	14	22	1	1	56	0	3	1	1
Clarias	12	4	1	1	47	1	15	0	20
Tilapia	28	23	13	10	14	0	0	7	5

Fig. 2.9 The three main species of fish in Lake Chilwa, Malawi, and the average stomach contents of samples of several hundred fish of each species. Lake Chilwa undergoes irregular phases of drying out; its fish species have very broad diets and unspecialized habitat requirements, which allow them to cope with this variability. Each fish species, however, takes more of one particular item than the other two fish.

similarly old Lake Tanganyika and the Russian Lake Baikal, there has been a progressive differentiation of very many specialist species dividing the resources of the habitat finely among themselves. The fish form an excellent example (Fig. 2.10). Many species in these old lakes are endemic – they have evolved there and occur nowhere else.

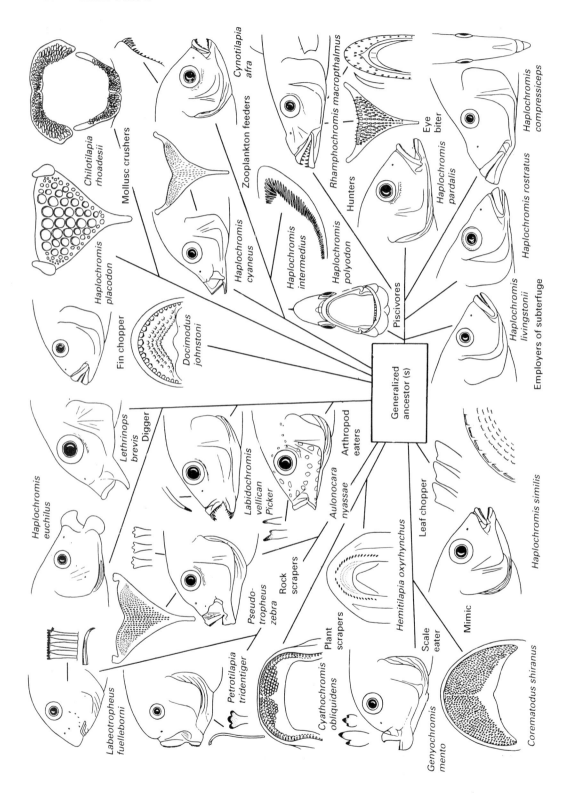

Chilotilapia rhoadesii

Mollusc crushers

Cynotilapia afra

Zooplankton feeders

Rhamphochromis macropthalmus

Hunters

Eye biter

Haplochromis pardalis

Haplochromis rostratus

Haplochromis compressiceps

Haplochromis placodon

Haplochromis cyaneus

Haplochromis intermedius

Haplochromis polyodon

Piscivores

Haplochromis livingstonii

Employers of subterfuge

Fin chopper

Docimodus johnstoni

Generalized ancestor (s)

Lethrinops brevis

Digger

Labidochromis vellican

Picker

Arthropod eaters

Aulonocara nyassae

Haplochromis similis

Haplochromis euchilus

Leaf chopper

Pseudotropheus zebra

Rock scrapers

Hemitilapia oxyrhynchus

Mimic

Labeotropheus fuelleborni

Petrotilapia tridentiger

Plant scrapers

Cyathochromis obliquidens

Scale eater

Genyochromis mento

Corematodus shiranus

The old lakes will not be permanent in geological terms – eventually they will disappear in some phase of earth movement – but for the moment their diverse communities support the idea that the fundamental distinction between fresh waters and the ocean lies in the general impermanence of the former and the stability of the latter. Fresh waters have long been subjected to disturbance; the resilience they still have in their abilities to accommodate the impacts of man owes much to this feature of their collective history.

Fig. 2.10 (*Opposite*) Specialization that has occurred in diet for a number of closely related small species of the genus *Haplochromis* in Lake Malawi. In some cases, whole heads and, in others, details of the pharyngeal bones (the triangular diagrams), upper part of the mouth (arch-shaped diagrams) or individual teeth are shown. Most descriptions are self-evident, but the mimic resembles a harmless (to other fish) plant-eater, although in fact it scrapes scales from the bodies of other fish species. The employers of subterfuge are fish-eaters (piscivores) that lie on the bottom, resembling rotting carcasses. Curious small fish come to inspect them and are soon eaten as the carcass 'comes to life'. Pickers delicately pick small animals from rock surfaces, while diggers use their sensitive lips to feel for animals while probing in sand or sediment. The eye biter is reported to remove eyes from larger fish, but also eats whole small ones. (From Fryer and Iles [304].)

3: From Atmosphere to Stream: the Chemical Birth of Fresh Waters

Only for an instant is liquid water ever naturally pure. It distils to the atmosphere, mostly from the ocean, and condenses to droplets or freezes to ice particles. Only at the moment of condensing or freezing is it ever likely to be pure water and probably not even then. Condensation requires nuclei of other substances – dust, salt, borne into the atmosphere from breaking sea spray, or single ions. Atmospheric gases quickly dissolve and, as the droplet is moved by the winds, it scavenges more salt and dust. The liquid droplet has become a very dilute fresh water. Ice crystals also pick up particles, ready to dissolve when the snow eventually melts.

The water that reaches the land is already complex, a dilute, weakly acidic, sea-water solution modified by dust, and varying (Table 3.1). Over cities it may become more strongly acid and near deserts it may have much more dust. It may acquire ammonium ions from the decomposition of animal waste near stock farms. Exhausts from vehicle engines or the burning of vegetation bring nitrate. The main sources of its composition, the atmosphere, the sea and human activities, will be discussed in turn.

3.1 Dissolving of atmospheric gases and the acidity of rain

3.1.1 Carbon dioxide and sulphur gases

Of the atmospheric gases that diffuse into water droplets, carbon dioxide, with its high solubility and reactivity, is the most important. Carbon dioxide (CO_2) is a major determinant of the acidity of the rain. It dissolves in water to form carbonic acid (H_2CO_3), a weak acid, which can dissociate as follows:

$$H_2CO_3\,(aq) \rightleftharpoons HCO_3^-\,(aq) + H^+\,(aq) \tag{a}$$

$$HCO_3^-\,(aq) \rightleftharpoons CO_3^{2-}\,(aq) + H^+\,(aq) \tag{b}$$

If CO_2 were the only substance whose dissolution resulted in the formation of hydrogen ions (H^+), the pH of rain would be 5.64, a value sometimes found in naturally forming rain. But there are other natural influences. Near active volcanoes, sulphur compounds, including sulphur dioxide (SO_2) and hydrogen sulphide (H_2S), are released into the atmosphere; sulphides of carbon, such as dimethylsulphide [17, 975] are released by bacteria and

Table 3.1 Concentrations (mg l^{-1}) of several major ions in rainwater, together with their percentages by moles (in parentheses) of the total sum of these ions compared with the percentage of the ions, on the same basis, in sea water.

Ion	Cumbrian Lake District, UK*	New Hampshire, USA†	Norwich, UK‡	Sea water
Sodium	3.1 (44.7)	0.12 (17.9)	1.2 (25.9)	(43)
Potassium	0.2 (1.66)	0.07 (6.43)	0.74 (9.5)	(0.9)
Calcium	0.2 (1.66)	0.16 (14.3)	3.7 (46.3)	(0.9)
Magnesium	0.3 (4.3)	0.04 (6.1)	0.21 (4.38)	(4.9)
Chloride	5.1 (47.7)	0.55 (55.4)	1.0 (13.9)	(50.2)

*Gorham [351].
†Likens *et al.* [573].
‡Edwards [236].

algae in the ocean. These sulphur gases react with water to produce hydrogen ions, and, because some, particularly sulphur dioxide, are very soluble, the effects of even low atmospheric concentrations may be large.

For an SO$_2$ concentration (partial pressure) in the atmosphere of 10^{-7}, the equilibrium pH is 4.9. Charlson and Rodhe [147] believe that natural rainwater is likely to have equilibrium pH values of 4.5–5.6 as a result of the combined action of carbon dioxide solution and sulphur compounds. Rainwater in remote regions, however, usually has a pH close to the upper limit of this range.

3.2 Contribution of sea spray to rain

In rain falling close to the coast (Table 3.1), the proportions of the most common ions often match those in sea water. There is also a general decrease in the concentrations of these ions in rain with distance inland (Fig. 3.1), where the influence of local dust may mask the influence of sea salt. Sea salts reach the atmosphere when bubbles, formed by wind at the sea surface, burst to fine droplets, which are readily carried by the wind. Dust is less important in determining rain composition than sea spray, because dusty land surfaces are those over which rain forms least readily. Proportionately, they are also much smaller than the area of the ocean. Dust, in reflecting local geology, will also have a much more variable effect. The high calcium concentration in Norwich rainfall (Table 3.1) probably reflects the abundance of chalky soils in the area.

3.3 Atmospheric pollution

3.3.1 Carbon and sulphur

The third major contributor to rain composition is human activity. This may

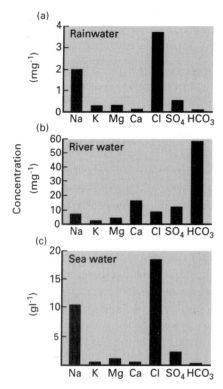

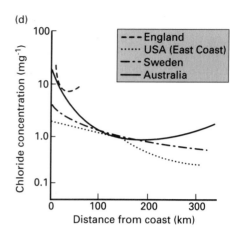

Fig. 3.1 The average major ion composition of rainwater is similar to that of sea water, because of the origin of these ions from sea spray (a–c). The effect is greatest closest to the coast (d; note logarithmic scale), which shows the concentration in rainwater of chloride, the predominant ion of sea water. Because of the varied additional effects of local geology, the composition of water that eventually drains to the rivers no longer reflects that of the ocean. The average total dissolved-solids content of (a) rain is about 7 mg l⁻¹, of (b) rivers 120 mg l⁻¹ and of (c) sea water 34.4 g l⁻¹. Na, sodium; K, potassium; Mg, magnesium; Ca, calcium; Cl, chloride; SO₄, sulphate; HCO₃, bicarbonate. (Based on Raisewell *et al.* [795].)

be smoke, with heavy metals, such as lead, chromium or zinc, released into the atmosphere from furnaces and smelters, but such effects are likely to be local. A more general influence comes from gases released by fuel burning. Sulphur dioxide comes particularly from power stations burning large quantities of sulphur-rich coal, and nitrogen oxides are released from vehicle engines and oil-burning power stations. These produce hydrogen ions when they are further oxidized and react with water, thus making rain more acid. Ammonia volatilized from farm-animal wastes is making an increasing contribution to the nitrogen content of rain.

The well-known increase in CO_2 concentration in the atmosphere, from the burning of coal and oil on a large scale in the last few decades, will have had some effect on the equilibrium pH of rain, but a comparatively slight one compared with other effects. The CO_2 content was about 0.029% in the nineteenth century [118], and 0.033% in 1976. It is now (1997) 0.036%, higher than at any time in the past 160 000 years. Predictions for the future suggest between 0.04% and 0.055% by AD 2030. The equilibrium pH value for the CO_2 concentration in the nineteenth century was 5.65 and that for 1988 5.6. Predicted pH values for 2030 range from 5.58 to 5.51.

The product of sulphur dioxide solution, HSO_3^-, may be oxidized by dissolved oxygen to form HSO_4^-, which dissociates to form the very strong sulphuric acid. This must also be formed from natural SO_2 (see above), but calculations, based on the amount of coal burned in the last few decades, suggest that the latter is a greater source (perhaps 65×10^6 t sulphur year^{-1} – mostly as SO_2) than natural ones (30×10^6 t year^{-1}, with about 7×10^6 as SO_2, the rest as carbon sulphides and other compounds). The burning is also concentrated into smaller areas.

Sulphur dioxide has been linked with lung disease and much effort is being made to reduce the release of sulphur dioxide into the atmosphere. Emissions have been reduced by 15% in the European Community (EC) since 1973 and are unlikely to increase in the next two decades, as a result of formerly voluntary conventions and latterly an EC Directive on large combustion plants. They may not decline greatly, however.

3.3.2 Nitrogen

The burning of oil and the production of nitrogen oxides (collectively called NO_x) have generally increased in parallel to carbon dioxide [285]. The primary oxide emitted is nitrogen monoxide (NO), which reacts in the atmosphere with ozone or hydroperoxyl radicals to form nitrogen dioxide (NO_2). Nitrogen dioxide is also produced naturally by reactions energized by lightning sparks. The reaction rate is very low and, although the supply of combined nitrogen from this source may be significant for the nitrogen supply to pristine land ecosystems, the 'natural' NO_2 has little overall effect on rain chemistry. Both NO and NO_2 may eventually oxidize to form nitrate and nitric acid.

Only half the global total NO_x production comes from human activities (cf. 90% of SO_2 production) but again it is concentrated in the more industrial areas. At present in the UK, just over half of the total consumption of petroleum products is by transport, of which the bulk (80%) is by private cars, with air travel contributing 15.3% [834]. This results in the release of about 1.57×10^6 t NO_x annually. There seems to be an increasing trend in nitrate concentrations in rainwater as a result [90]. Some NO_x is removed as nitrogen, ammonium and nitrous oxide (N_2O) by catalytic converters recently fitted to many new cars, but the latter two gases also have consequences for the atmosphere.

Nitrous oxide has increased by about 0.3% per year in the last few decades (Fig. 3.2) from about 290 ppb by volume in the nineteenth century

(a)

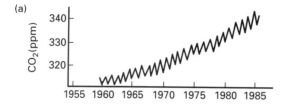

(b)

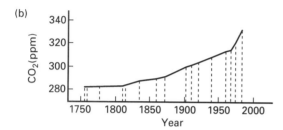

Fig. 3.2 Changes in carbon dioxide (CO_2) concentration in the atmosphere, from (a) direct records at Mauna Loa (Hawaii) and from (b) analyses of gas bubbles preserved in dated ice cores from Greenland (based on Cannell [118].)

to 318 by the early 1990s. It reacts in the atmosphere ultimately to give nitrate. There are natural sources of N_2O producing about 5–9 Tg year^{-1}. These sources include soil denitrification (reduction of nitrate) and marine nitrification (oxidation of ammonium) and they were formerly balanced by destruction of the gas in the atmosphere at similar rates to those with which it was formed. The present increase results from application of nitrogenous fertilizer, some of which is denitrified, direct combustion of fossil fuels and also the burning of vegetation in the tropics and subtropics.

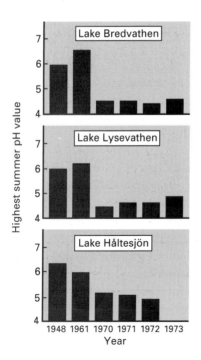

Fig. 3.3 Changes in the pH, between the 1940s and the 1970s, of three Swedish lakes, whose waters are little altered by the catchment from the composition of rain (based on data in Hultberg [446].)

3.3.3 Acid precipitation

The effects of solution and oxidation of SO_2 and NO_x has been to increase the H^+ concentration of rain and to reduce its pH to values below those naturally expected, in many parts of the world. Values down to 2.1–2.8 have been recorded in the USA, UK and Scandinavia, and averages are usually lower than 4.6 in these area [43, 154, 571]. These values are much lower than those obtained before 1960 (Fig. 3.3). Care is needed in making these comparisons, however, because techniques for determining pH, or bicarbonate concentration, which is often used as a surrogate for pH, have improved in recent decades. Older methods give slight overestimates [534].

Snow may also be very acid, for droplets of quite concentrated sulphuric and nitric acids (often called 'dry' fallout) may be picked up by the snowflakes as they fall. If it does not melt almost immediately, snow ablates – that is, vaporizes without liquid formation in very cold, dry air. The acid droplets may then become highly concentrated in long-lived snow banks, so that run-off water as the snow melts in spring may be extremely acid. The consequences of this are serious and are discussed in Chapter 4.

3.3.4 Nutrients delivered by the atmosphere

Nitrate increases in rain may have fertilizing (eutrophicating) as well as acidification effects, especially from volatilization of ammonia in animal wastes. Figure 3.4 shows the concentrations of ammonium ions in air around a large pig farm [465] and the total load of ammonium that fell at various points in the area. There were very high concentrations above the farm and the prevailing wind carried these to an area of particular impact to the north-east, with up to $275 \, kg \, ha^{-1}$ being annually deposited. This compares with around $120 \, kg \, ha^{-1} \, year^{-1}$ similarly deposited over large areas of Holland. Concentrations of ammonia in stock-raising areas may be between 20 and $30 \, mg \, m^{-3}$, compared with < 3 elsewhere.

The deposition in Britain is localized but compares in some instances with the same amount of fertilizer spread on land to grow arable crops. In the west of England and in countries like the Netherlands and Denmark, it is a widespread feature. Ammonia comes from the decay of urea in mammalian urine, uric acid in poultry excreta and the decomposition of proteins in faeces. It may be removed from the atmosphere as wet deposition by rain or by direct dissolution into soil and surface waters as a gas (dry deposition). The rates of volatilization depend on how manure is produced, stored and redistributed to the land. There are relatively greater losses from housed animals, uncovered slurry stores, poultry and manure spread on the land on windy days than when it is injected into the soil or spread on still days [753].

(a)   Concentration of NH$_3$ in air

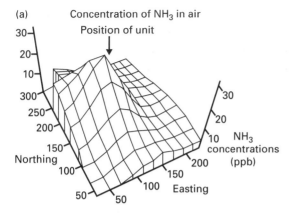

(b)   Ammonia deposition on surrounding land

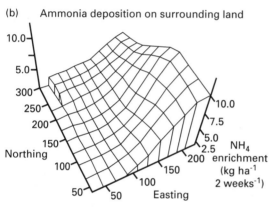

Fig. 3.4 Effects of a large UK pig unit on atmospheric ammonia (NH$_3$) concentrations (a) and deposition rates (b). The unit was located at coordinates 100 east and 230 north on the diagrams, with a drift of the gas towards the south-east before deposition as rain. Maximum deposition was around 500 kg ha^{-1} year^{-1}, which is around double the nitrogen fertilization rate used on intensively grown arable crops. NH$_4^+$, ammonium ions. (Based on Ineson [465].)

3.3.5 Models of nitrogen deposition

In Denmark [656], total nitrogen (N) deposition from the atmosphere has doubled between the 1950s and 1970s but has now levelled at about 14 kg ha^{-1} year^{-1} (8 kg as ammonium, 6 kg as nitrate). Most nitrate comes as wet deposition, and about equal parts of the ammonium deposition are wet and dry. Annual total N deposition in remote areas is about 10 kg N ha^{-1} year^{-1}, but not all of this is natural – some is blown in from elsewhere.

This raises two problems in understanding the contributions to rainfall of such substances: the ultimate source, especially where problems are being caused; and determination of the overall picture from the relatively few samples it is usually economic to take. Simulation models can be used to solve the latter problem [24].

Such models take available data on potential nitrate sources from the distribution of power stations and the density of road traffic and similar data on the distribution of farm stock. They use knowledge of the rates of emission from these sources and how they vary with the weather. Then, by incorporating data on wind patterns and rainfall, they predict the rates of deposition (Fig. 3.5). The results are then compared with observed data. If there is good correspondence, the model is seen to be successful. It can then be used to predict rates for regions lacking direct measurements and to determine the consequences of changes in agricultural policy disfavouring or encouraging stock raising, or in energy policy reducing the burning of fossil fuels. The current model has proved good at predicting dry deposition but underestimates wet deposition. Nonetheless, the patterns predicted are generally correct and such an approach, if increasingly refined, offers a powerful tool for environmental management.

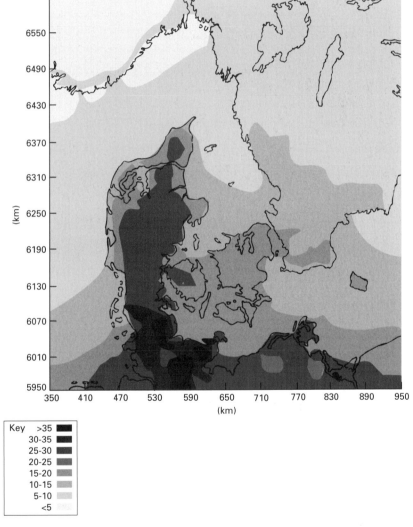

Fig. 3.5 Results of a model predicting total nitrogen (N) deposition from the Danish atmosphere. Rates are greatest towards the southern areas of more intensive farming and are given in $kg\ ha^{-1}\ year^{-1}$. (Based on Asman and Runge [24].)

3.4 The composition of water draining from the catchments

Rain and melted snow move through and over the land surfaces and become further modified before they pass to the streams that drain the catchments. Every stream water differs from the next and will itself vary with time – sometimes over minutes. This is because the final composition depends on the interplay of several variables.

First there is the initial composition and amount of rain and snow, and how the water passes through the catchment. It may run rapidly off hard rocks or soak slowly through permeable soil. Secondly, there are many thousands of substances, derived from the local geology, soils and ecosystems, which may become dissolved. And, lastly, the catchment may be altered by human populations: forest removal, afforestation, cultivation and fertilization of land, the building of settlements, industrial use and the disposal of wastes.

3.4.1 A chemical catalogue for runoff waters

Runoff waters are those at the point of delivery to a freshwater basin, usually a small stream. It is useful to look first at the final product – the general chemical composition of the runoff water – and then to derive this from the various sources listed above. A simple catalogue will include major ions, atmospheric gases, key nutrient ions, trace nutrient ions, other trace ions, refractory (difficult-to-decompose) and labile (very reactive) organic substances. There will also be suspended inorganic and organic particles, which may be able to interact chemically with the dissolved ones.

The order of this list roughly reflects the difficulties of analysis of substances within the groups and the historical sequence in which the groups have been investigated. Those most easily analysable were investigated first, and our view of the relative importance of the groups is to some extent conditioned by this historical baggage. Those which have been investigated for longest are often assumed to be of greatest importance. But no one has yet analysed completely a natural water sample and such an assumption may be very misleading.

Major ions, for example, are so called only because they are present in greater concentrations than others. The major ions include sodium (Na^+), potassium (K^+), magnesium (Mg^{2+}), calcium (Ca^{2+}), sulphate (SO_4^{2-}), chloride (Cl^-) and bicarbonate (HCO_3^-). They are generally dissolved in quantities of at least mg l^{-1} (parts per million) and usually vary only a little in concentration during the year (although this may not be true for bicarbonate, which may be absorbed by some plants and algae, and is not true of waters in endorheic areas). They are called 'conservative', their concentrations being not greatly changed by the activities of living organisms, which may require them but not in large quantities relative to those in which they are available.

In contrast, the key nutrients, which include phosphates (PO_4^{3-}, HPO_4^{2-}, $H_2PO_4^-$), nitrate, ammonium, sometimes silicate and occasionally iron, manganese, carbon dioxide plus bicarbonate and molybdenum, are not conservative. Their concentrations are generally lower than those of the major ions ($\mu g\, l^{-1}$ to $mg\, l^{-1}$), but, more importantly, the requirement for them by living organisms is high relative to the supply. Often they fluctuate greatly in concentration over the year.

The atmospheric gases were discussed in Chapter 2. Nitrogen is important because of the contrast between its relative abundance, as gaseous N_2, and the relative natural scarcity of those of its compounds (nitrate, ammonium) which are available to living organisms. Oxygen has a key role, particularly when absent, in setting the chemical scene for many important reactions. In the context of runoff water, however, it is generally present and of less importance. Under temporary circumstances, in some waters, CO_2 may be a key nutrient, but generally supplies are more than adequate because of its atmospheric reserve and high solubility.

With the trace-element ions, analytical problems become more important. Some, including copper (Cu), vanadium (V), zinc (Zn), boron (B), fluorine (F), bromine (Br), cobalt (Co) and molybdenum (Mo), are known to be needed by organisms (usually as enzyme cofactors) and are adequately present at $ng\, l^{-1}$ or $\mu g\, l^{-1}$ concentrations. Others (e.g. mercury (Hg), cadmium (Cd), silver (Ag), arsenic (As), antimony (Sb), tin (Sn)) are probably not required but are generally present at very low concentrations ($ng\, l^{-1}$ or lower). Probably almost all the natural elements not yet mentioned fall into this group. Both groups may be toxic if present at higher concentrations either through natural means (for example, in springs draining mineral lodes or percolating new volcanic lava) or discharged by industry.

The chemistry of these elements is often complex, with several oxidation states being available, or with different ions formed at high and low pH or in various possible combinations with organic matter – for example, cobalt in cyanocobalamin (vitamin B_{12}). Modest concentrations (say $100\,\mu g\, l^{-1}$) of iron, for example, could be toxic if it is present as ferrous ions (Fe^{2+}) in conditions of low oxygen concentration or ferric ions (Fe^{3+}) at low pH (3–4), but similar apparent concentrations at high pH (7–9) may be bound up in colloidal hydroxides largely inert to living organisms.

Analytical methods for these trace substances have often depended on conversion of the element to a highly soluble form for reaction and not been able to distinguish the several individual states in which the element is present in the natural water. These states are, however, distinguishable to living organisms and informative analysis may depend on a bioassay in which the reaction of a test organism is used rather than a chemical analysis. Analytical technology is, however, catching up.

The analytical problems become most difficult for organic compounds. The chemistry of most elements is sufficiently constrained for comprehensive knowledge to have been attained of the range of their possible compounds, and certainly of those that dissolve in water. This is not true of carbon, however; its ability to form straight and branched chains, regular and irregular rings, and to combine with other elements not only in one form but in a variety of isomers, means that an immense variety of organic compounds can exist. Organisms react to small differences, even isomers, which are difficult to detect in the laboratory.

Some of the larger molecules may be relatively stable, e.g. the residual products of decomposition in soils of organic matter, such as wood, and hence more readily analysable. Their very stability, however, means that, despite concentrations of the order of mg l^{-1} they are of least importance to the aquatic ecosystems. Other, smaller molecules – amino acids, alcohols, carboxylic acids, sugars, peptides – are more reactive, shorter-lived and present at low concentrations (ng l^{-1} to µg l^{-1}) but with a high turnover rate. They pose the greatest analytical problems but may be important as nutrients to many microorganisms.

The organic compounds that reach stream waters represent the stages, early and advanced, of decomposition of organic matter produced in the catchment (Table 3.2). The very earliest stages, in the form of leaf litter or smaller particles, may be present in suspension, as well as the colloidal and soluble later derivatives. In parallel, the inorganic component of the water composition represents the early (particulate) and later (dissolved) stages of decomposition of the rocks of the Earth's crust. From the mildly acid, very dilute and dust-changed sea water of rain, we need now to derive, through the catchment kitchen, the slightly saltier, more organic consommé of the stream waters.

Table 3.2 Origins and nature of organic substances washed into stream waters.

Origin in living organisms	Derivatives washed into draining waters
Proteins	Methane, peptides, amino acids, urea, phenols, indole, fatty acids, mercaptans, melanin*, melanoidin, yellow substances (gelbstoffe)*
Lipids (fats, waxes, oils, hydrocarbons)	Methane, aliphatic acids, acetate, lactate, citrate, glycolate, malate, palmitate, stearate, oleate, carbohydrates, hydrocarbons
Carbohydrates (cellulose, hemicellulose, lignin)	Methane, glucose, fructose, galactose, starch, arabinose, ribose, xylose, humic acids*, flavic acids*, tannins*
Porphyrins and plant pigments (chlorophylls, haemin, carotenoids)	Phytane*, pristane*, isoprenoid, alcohols, ketones, acids, porphyrins

*Generally refractory in water.

3.4.2 Rock weathering

Rocks and rain give most of the inorganic substances that reach fresh waters [324]. Rocks are made of minerals, most of which are based on crystalline combinations of silicon and oxygen, the two most common elements in the Earth's crust, with small additions of other elements. The silica minerals are very varied for, like that of carbon, the silicon atom can combine with up to four other atoms in a tetrahedral formation. The tetrahedra may then form lattices – chains, sheets and three-dimensional structures. As the structure becomes more complex, it becomes more stable and the number of available sites for the attachment of other elements is proportionately reduced. The Earth's crust is dominated by these minerals.

Following the formation of the Earth, heavy molten iron and nickel sank to the core, while the lighter silicon floated to cool and react within the surface layers to form the primary silica minerals in igneous rocks. Though stable under the pressure and high temperature of their birth and in the absence of oxygen and water deep in the crust, the minerals are reactive at the Earth's surface. Their weathering reactions release soluble ions and inorganic particles, whose properties determine the nature of later reactions and releases.

3.4.3 Weathering of igneous rocks

When exposed at the surface, igneous rocks are often massive. Cooling and crystallization during their formation lead to tight, intimate mixtures of mineral crystals. Released from overlying pressure, heated, cooled, frozen, thawed and penetrated by roots, the rocks crack, exposing more surfaces to water and oxygen. The process is slow and the small surface areas initially presented for access to water mean that relatively few soluble ions are initially released. The ones that are are those ionically (as opposed to covalently) held on the silicate lattices (Fig. 3.6). These can be attracted by the polar water and replaced by hydrogen in the lattice.

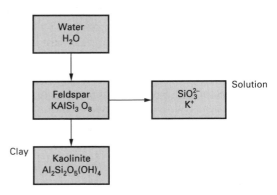

Fig. 3.6 Processes and products from the weathering of feldspar. SiO_3^{2-}, silicate; K^+, potassium ions.

The very dilute waters draining from these igneous rocks reflect the relative abundances of the more soluble metals bound in the minerals. Generally the abundance is in the order $Mg > Ca > Na > K$, but this is modified by the cations present already in the rain and derived from sea spray, so that the order is ultimately $Na > Mg > Ca > K$. Areas producing such waters include the Precambrian rocks of the Canadian Shield and the central part of the English Lake District.

Silicic acid ($HSiO_4$), is also released from hydrolysis of the lattice (Fig. 3.6) but often reprecipitates to form clay (see below) and appears in the drainage water in much lower quantities than those initially released. Lattice breakdown does not produce many soluble anions, therefore, and the anions in the water draining igneous rocks are those provided by sea spray and the reaction of carbon dioxide. The low degree of attack on these resistant rocks usually leaves sufficient H^+, ultimately derived from rain, to give an acid runoff water.

The weathering rate of minerals is increased by increasing temperature and acidity, a brisk percolation of water and the presence of oxygen. Increasing H^+ concentration not only increases the replacement of metal ions in the lattice but may also create conditions for dissolution of less soluble ions, such as those of aluminium, which is a very common component of lattices. Aluminium (Al^{3+}) is soluble below pH 4.5 and hence its concentration in drainage waters from igneous areas has been increased through the lower pH of rain acidified by the reactions of SO_2 and NO_x (see above).

Rapid percolation of water removes the soluble products of the lattice hydrolysis; it therefore promotes the reaction to replace these products, according to Le Chatelier's principle. Oxygen promotes the removal of elements like iron and manganese from lattices; these elements can exist in several oxidation states – for example, Fe(II), F(III) – and can be dislodged by oxidation.

Extensive breakdown of igneous rocks in the humid tropics can leave residues (ultimately called laterites) containing only the more resistant minerals, such as quartz $(SiO_2)_n$ and insoluble red or yellow oxides and hydroxides of Fe(III), while all the soluble ions are mobilized into streams. As the weathering goes on, the stock of such ions in the soil is progressively depleted, and the concentrations in the run-off water fall. In cooler and drier temperate regions, the breakdown is less rapid and some of the soluble products, such as silicic acid, are not immediately washed away but persist in the interstitial water, where they may re-form secondary silicate minerals by precipitation. These include the clay minerals – flat plates of silica tetrahedra interleaved with plates of aluminium oxides.

Clay-mineral particles generally have a net negative charge, which attracts H^+ ions, and are small (generally less than 2 μm) and present a large surface area per unit weight for further chemical reaction. Cations from rain may be adsorbed on to the clays, displacing H^+ into solution. The more acid the rainwater, the less likely it is that H^+ will be displaced from the clay and that metal ions will be leached out to the streams. Conversely, the higher the pH of the rain, the more likely it is that metal cations will be adsorbed by the clay and retained in the soil. In this way, through cation exchange, the clay minerals can begin to regulate the composition of the run-off water.

Waters draining igneous areas are thus dilute solutions of major cations, with suspended fragments of unweathered minerals and clays. There may also be low concentrations of silicate and other anions, such as phosphate, derived from scarce minerals present in the original rock. Phosphates, however, are only reasonably soluble at around neutral pH. At low pH they are precipitated as iron or manganese phosphate, so concentrations in the drainage water are generally very low (e.g. $< 1–5$ μg PO_4-phosphorus l^{-1}).

3.4.4. Sedimentary and metamorphic rocks

The weathered rock debris eventually comes to rest as an ocean sediment. To it may be added the remains of oceanic organisms, which have concentrated, from the water, essential elements, such as sulphur, phosphorus and nitrogen, and substances, such as calcium carbonate and polymerized silicate, which they had used to form cell coverings or skeletons. As the mass of sediment accumulates, it is compressed; water is squeezed out and calcium and other compounds may serve to bind it together into what will become a sedimentary rock. Eventually, this rock may be raised above sea level by earth movements and exposed to a new cycle of weathering. It may also be buried, compressed, reheated and melted to emerge, cooled, as a metamorphic rock, with silica minerals re-formed. A metamorphic rock will then weather again in much the same way as an igneous rock. The northern part of the English Lake District is of metamorphosed slates and produces waters similar to those of the Borrowdale volcanic rocks (Fig. 3.7).

3.4.5 Weathering of sedimentary rocks

Weathering of an unmetamorphosed sedimentary rock will be very different from that of an igneous rock. Despite the 95% dominance of igneous rocks by volume in the Earth's crust, the surface of the continental crust, where most of the weathering takes place, is 70% covered by sedimentary rocks. Overall, these have a much more widespread effect on freshwater chemistry.

Sedimentary rocks are built from jumbled particles; they are often porous, presenting a large surface area for water to percolate and many lines of weakness along which the rock may crack. The cements that bind

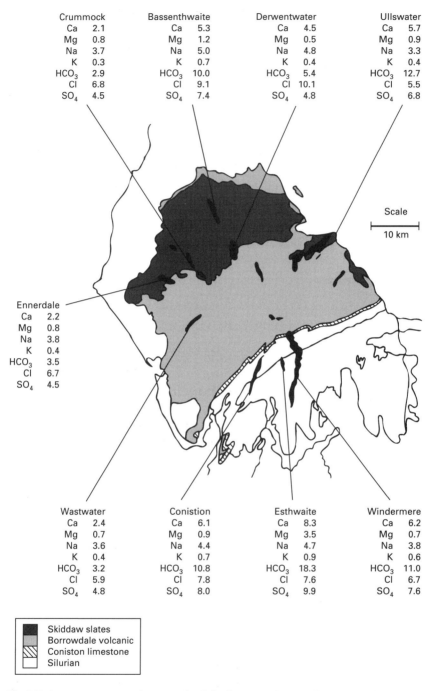

Fig. 3.7 Average major-ion chemistry (mg l^{-1}) of waters of some of the lakes of the Cumbrian Lake District (UK). The paucity of cations and bicarbonate in those to the west, drawing water from the igneous Borrowdale volcanic rocks, is well shown, whereas those to the south, served by sedimentary rocks, have much higher concentrations. Ca, calcium; Mg, magnesium; Na, sodium; K, potassium; HCO$_3$, bicarbonate; Cl, chloride; SO$_4$, sulphate. (Based on Macan [596].)

them are in general soluble and readily weather, releasing again the soluble ions and further exposing any primary minerals to decomposition. Their manner of formation, from inorganic and organic debris on the ocean floor, means also that anions, such as sulphate, carbonate and phosphate, are relatively abundant and this is reflected in the waters that drain from them.

There is a particular abundance of calcium carbonate, derived from the cell walls and shells of many marine organisms, in many sedimentary rocks (particularly chalks and limestones). This means that calcium and bicarbonate, released from the rock by the acids in rain, often dominate the major ions in such waters. Carbonates and other anions lead to a ready neutralization of the H^+ in rainwater, so that the drainage waters are neutral or even alkaline and ions such as aluminium are not so readily mobilized. The high pH of the soils that form from them means also that phosphates are again precipitated, though now as calcium phosphate, so that the run-off water is still poor in phosphate, though enriched in other ions.

3.4.6 Links between geology and water chemistry

The ready weathering of sedimentary rocks leads to deep soils, which are easily cultivable and hence quickly modified by human activity. It is more difficult, therefore, to isolate the effect of geology in determining the contents of drainage water from sedimentary rocks than it is for the igneous rocks, where landscapes, often mountainous and with thin soils, are least inhabited.

The English Lake District (Fig. 3.7), where the northern areas have igneous and metamorphic rocks and the southern parts sedimentary rocks, including bands of limestone, however, shows the distinction. The waters from the highest mountain lakes, the small tarns of the north and centre, are essentially accumulated rainwaters, little altered by rock weathering because rainfall is high and the flow-through is very rapid. The larger lakes, where the longer retention of water in the bigger, though still crystalline (igneous and metamorphic), rocky catchments has had some influence, have greater ionic concentrations, but calcium and bicarbonate are scarce and NH^+, Na^+, Mg^{2+}, Cl^- and SO_4^{2-} dominate the solution. The southern waters, on sedimentary Silurian rocks, have a dominance by calcium and bicarbonate at higher overall concentrations.

Further south in England, the subdued sedimentary landscape of the east Norfolk rivers (Fig. 3.8), draining chalk to the west and glacial deposits derived from chalk and other sedimentary rocks to the east, show the expected high major ionic concentrations. Closeness to the sea brings very high concentrations of sodium and chloride, through percolation near the coast and, to a lesser and decreasing extent with distance inland, through rain-borne spray to the rivers. The river concentrations of chloride are much higher than those in rain (which usually has $< 10\,mg\,l^{-1}$), because

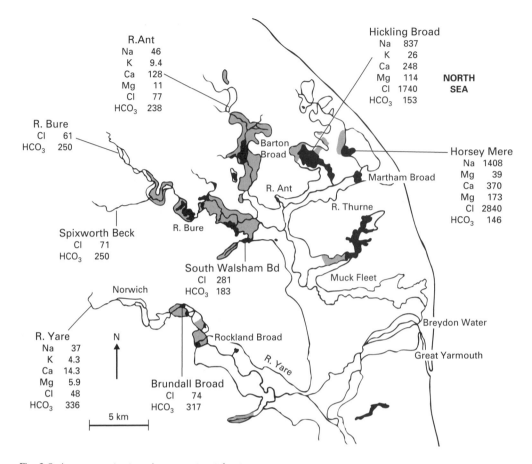

Fig. 3.8 Average major-ion chemistry (mg l^{-1}) of the Norfolk Broads, UK. The area is floored by soft sedimentary rocks and alluvium, and the effect of the chalk, to the west, on bicarbonate concentrations is well shown. Chloride concentrations tend to decrease westwards as the influence of the sea, either through spray, tidal movement upstream (South Walsham Broad) or percolation through nearby dunes (Horsey Mere, Hickling Broad) decreases. Na, sodium; K, potassium; Ca, calcium; Mg, magnesium; Cl, chloride; HCO$_3$, bicarbonate. (Based on Moss [677].)

much of the rain is evaporated in the catchment, effectively concentrating the salt in the eventual smaller volume of run-off water.

3.5 Effects of soil development and vegetation on the chemistry of drainage waters

Microorganisms, mosses, lichens and even higher plants are usually present when rocks weather. As the rock flakes accumulate, more organisms colonize and contribute litter to form a maturing soil. This development changes the nature of the runoff waters. It provides a source of combined nitrogen to supplement that brought in by rain; then the live biomass and undecomposed litter form a store into which soluble ions may be taken up, no longer being so vulnerable to leaching. Lastly, the organisms decomposing litter provide organic matter which may reach the streams.

3.5.1 Nitrogen fixation

There is little nitrogen in rocks. Atmospheric nitrogen gas is abundant but unavailable to most organisms. Nitrogen compounds ultimately reach living organisms mostly through the mediation of nitrogen-fixers in the catchment soils. All of these are prokaryotic, including some blue-green and other bacteria. Fixation means the reduction of nitrogen by the addition of hydrogen; it requires energy, is inhibited by oxygen and eventually produces amino groups ($-NH_2$), with which the fixers form their proteins. Nitrogen-fixers must protect their nitrogen-fixing enzymes (nitrogenases) from oxygen. Some live in waterlogged, anaerobic places; others place the enzymes in specialized cells, from which oxygen is excluded (Fig. 3.9), or in cells deeply encased in jelly; yet others associate intimately with oxygen-consuming bacteria, which create pockets of low oxygen tension in the soil. Like most prokaryotes, nitrogen-fixers grow best at around neutral or alkaline pH. Nitrogen fixation is thus most prolific in soils derived from sedimentary rocks, rather than those derived from igneous ones. The amino-nitrogen is released on the death of the fixers and decomposed to ammonium ions, which may be adsorbed by clays or taken up by other organisms. Ammonium is a reduced ion, from which energy may be obtained by oxidation; some bacteria, the nitrifiers, can oxidize it to nitrite and nitrate, in which latter form it remains available for plant uptake and, being very soluble, is easily leached to streams.

 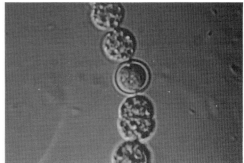

Fig. 3.9 Most nitrogen enters ecosystems from the atmosphere through biological nitrogen fixation. This is carried out by free-living soil bacteria, bacteria living symbiotically in the root nodules of certain higher plants, particularly legumes (left panel), and blue-green alga (right panel). Most of the fixation by the latter occurs in rounded, thick-walled cells, called heterocysts.

3.5.2 Storage in the plant biomass

The effects of vegetation in modifying drainage-water chemistry can be studied by measuring the income of particular elements to a land ecosystem and then the subsequent losses to the runoff waters to obtain element budgets. If the vegetation is then removed and the exercise repeated, more insight can be gained. Such experiments must use self-contained catchments which are small enough to be uniform and yet large enough to have general value. All the water entering as rain or snow and leaving by evaporation and stream flow must be accounted for; and the experiment must last several years to eliminate year-to-year effects of changing weather. Permeable sedimentary rocks, through which water may be lost to groundwaters, cannot easily be used. Reliable results are most likely to be obtained from well-sealed basins of igneous or metamorphic rock.

The Hubbard Brook study

Perhaps the fullest such study has been that at Hubbard Brook in New Hampshire, USA [573] (Fig. 3.10). The underlying rock is igneous granite with some metamorphic shales; the hilly land gives well-defined catchments, with forests of maple, beech and birch. Elements enter in rain and snow, as dust, by direct uptake (dry deposition) of gases, particularly SO_2, by fixation (of nitrogen) and by weathering of rock. All loss of elements, except some nitrogen by denitrification, is through stream flow. Atmospheric inputs (water and dry deposition (dust and gases)) can be collected and measured. Outputs can be determined by analysing the stream water and measuring its volume. Contributions from rock weathering of elements other than nitrogen are obtained by difference. Because the rocks contain almost no nitrogen, all nitrogen not entering in rain and dust is assumed to be fixed. Weathering (Table 3.3) supplies most of the Ca, Mg, Na, K and P, whereas rain and snow contribute significant amounts of Na, N and sulphur (S). Most of the nitrogen comes through fixation and nearly a third of the sulphur from dry deposition of SO_2 or as sulphuric acid droplets derived from it.

   Elements are incorporated into the ecosystem as biomass, litter or soil, or washed out in solution or suspension. In a mature forest, where growth and decomposition are balanced, most of the Ca, Mg, Na and S is washed out, largely in dissolved form, so that vegetation has little effect on the concentrations of these elements in the drainage water. Potassium and nitrogen are selectively retained by the ecosystem, although the high solubility of ions of these elements means that 20–30% is lost to the stream. Phosphorus is very strongly held, the stream losses being less than 1% of the annual supply.

   These data can also be looked at in terms of the net gain or loss of each element to the land ecosystem, by balancing the income from just the

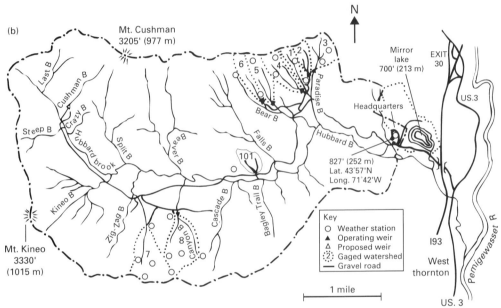

Fig. 3.10 Hubbard Brook is situated in the White Mountains of New Hampshire, USA (upper). Its catchment (lower) has been instrumented to provide data on weather and stream flows, and various subcatchments have been used to investigate nutrient flows and retention. (Based on Likens *et al.* [573].)

atmospheric sources (including nitrogen fixation) against the losses to the stream. This is a measure of the extent to which the rain chemistry is changed by the ecosystem. For some elements there is a net loss from the ecosystem (Na, K, Mg, Ca), but for others there is a net gain (N, P, S). For nitrogen this gain seems very substantial, but denitrification may balance some of this. This suggests an important role of vegetation – at least of forest vegetation – in affecting the concentrations of nitrogen and phosphorus compounds in the run-off water. These elements are particularly important in determining the productivity of fresh waters.

Table 3.3 Proportionate income (% of total) and fate (% of total) of various elements in the undisturbed forest ecosystem at Hubbard Brook, USA (based on Likens *et al.* [573]).

	Ca	K	Mg	Na	N	P	S
Income							
Rain and snow	9	11	15	22	31	1.4	65
Dry deposition (as gas)	–	–	–	–	–	–	31
Fixation	–	–	–	–	69	–	–
Rock weathering*	91	89	85	78	–	98.6	4
Fate							
Incorporated into vegetation	35	68	17	1	43	82	6
Incorporated into soil/litter	6	4	5	<1	37	18	4
Lost, dissolved, to stream	59	22	74	95	19	0.25	90
Lost, particulate, to stream	1	6	5	3	1	0.35	1

*Determined by the difference between atmospheric income and total losses to the stream.

This role of nutrient conservation is emphasized when the forest is removed. In a subcatchment of Hubbard Brook, the trees were cut and the wood left on site. Redevelopment of vegetation was prevented with herbicides. Some effects of felling, over the next 3 years, are shown in Fig. 3.11. They are quite dramatic, with increased gross losses in dissolved form of every element measured, especially nitrogen. In the undisturbed ecosystem, the fixed nitrogen released on decomposition in the soil is probably very quickly taken up as ammonium (NH_4^+) by the plants and not exposed to much risk of leaching. With the vegetation removed, the ammonium ion was oxidized to nitrate, whose high solubility, together with the lack of vegetation to take it up, led to the large losses. As well as the loss of dissolved substances, there was a fivefold increase in the particulate matter carried by the stream. The loss of phosphorus as soil particles was much greater than the dissolved loss of phosphorus and about 15-fold greater than that in the undisturbed forest.

The wider picture

The Hubbard Brook and similar studies have concentrated on forest ecosystems. There may be dangers in extending conclusions from these too widely, but some generalizations can be made. For example, the natural retention of N, P and, to a greater extent than other cations, K in natural systems suggests a general scarcity of these elements for plant growth. This is reflected in the use of N, P and K as the main agricultural fertilizers. In relation to plant need, rather more nitrogen fertilizer than phosphorus is used and this reflects the high solubility of nitrate and greater loss of it by leaching (see below) than of phosphorus. Of the elements required that are

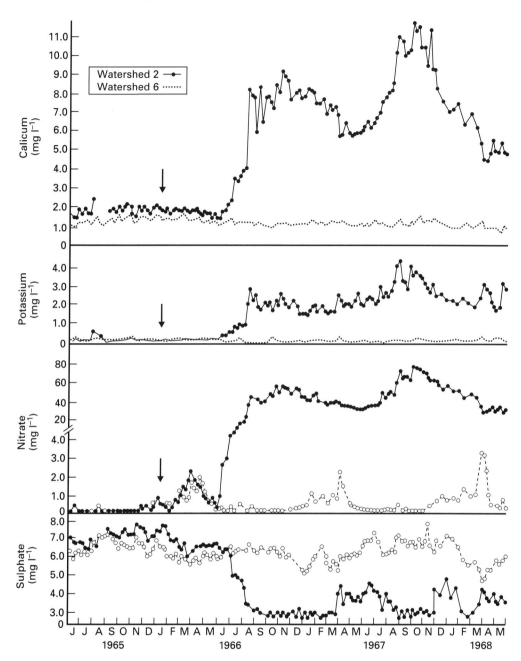

Fig. 3.11 Changes in the chemistry of stream water in two watersheds at Hubbard Brook in the White Mountains of New Hampshire, USA. The forest in watershed 2 was removed in early 1966 (arrow), while watershed 6 remained intact. Note change in scale for nitrate concentration. (Based on Likens and Bormann [572].)

derived largely from the Earth's crust rather than its atmosphere, the supply of phosphorus is smallest in relation to need (Table 3.4).

Much evidence now suggests that the felling of tropical forest for agriculture gives only a short period in which the soils are reasonably fertile [709]. Loss of the forest and the mechanisms by which it holds elements means that leaching losses, especially with very high rainfall, are greatly increased. Stream waters draining mature rain forest (Fig. 3.12) are very dilute indeed. For example, Amazonian forest streams average (in mg l^{-1}): Na, 0.22; K, 0.15; Mg, 0.04; Ca, 0.04; HCO$_3$, 5.5; Cl, 2.2 [306]. Most of the ions released by weathering are contained in the forest biomass. Destruction of the biomass, often with burning to produce a very soluble ash, leads to rapid loss of the nutrient stock, and little is left in the soil itself. A consequence of the small leaching losses from the mature forest on its anciently weathered, now depauperate soils, is that a major source of elements for the aquatic communities of Amazonian rivers is in the form of plant biomass falling into the water rather than leached ions (see Chapter 5). Such tropical waters may, on the other hand, have quite high concentrations of silicate and phosphorus, because the rock weathering has reached a very advanced stage.

A third generalization is that where landscapes lacking vegetation are created – for example, after glaciation (Fig. 3.12) – there appears to be a much greater run-off of ions and particles at first than subsequently when vegetation has developed. This change is a long-term one, so evidence for it comes from the analysis of sediments in lakes. In the English Lake District, sediments laid down just after the ice retreated and before vegetation developed are rich in cations (Fig. 3.13) [608] and also the easily eroded clay minerals. Those deposited after the colonization, initially of tundra and then of birch forest, are poorer in both.

Element	Supply ratio	Need ratio	Ratio of supply to need
Sodium	32.5	0.52	62.5
Magnesium	22.2	1.39	16.0
Silicon	268	0.65	413
Phosphorus	1	1	1
Potassium	19.9	6.1	3.3
Calcium	39.5	7.8	5.1
Manganese	0.9	0.27	3.3
Iron	53.6	0.06	893
Cobalt	0.02	0.002	100
Copper	0.05	0.006	8.3
Zinc	0.07	0.04	1.8
Molybdenum	0.0014	0.0004	3.5

Table 3.4 The relative supply of various elements, indicated by their abundance relative to phosphorus in the Earth's crust (supply ratio), and the relative requirements of these elements, again relative to phosphorus, by plants and algae (need ratio). The ratio of supply to need for all elements is greater than that for phosphorus. (Modified from Hutchinson [454].)

Fig. 3.12 Nutrient retention is greatest in systems with mature natural communities. Tropical forest (left) is the epitome of this. Retention is least in new, raw soils, such as those exposed by the recession of glaciers (right).

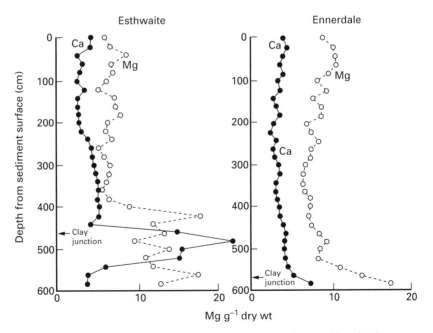

Fig. 3.13 Concentrations of calcium (Ca) and magnesium (Mg) in profiles of sediment from below two Cumbrian (UK) lakes, Esthwaite and Ennerdale. Both show increased values in the sediments laid down just after the glaciers melted back, exposing much mineral-rich rock debris to leaching. (Based on Mackereth [609].)

Long-term changes

The net retention of some elements by land ecosystems seems logical as long as the biomass is increasing and soil is accumulating. Once the balance between growth and decomposition, the 'climax' state, is reached, then as much of a given element that newly enters the biomass each year should leave it, and wash-out of elements from the land ecosystem to streams might increase again. However, the ultimate source of most elements is rock weathering and this does not go on indefinitely. The minerals become progressively leached in the upper and middle layers of soil where they are accessible to water and air. The provision of new minerals becomes less easy as the weathering proceeds deeper and deeper, insulating the fresh rock from sun, water and oxygen. The theoretically very large stock of elements in the Earth's crust is not all available. An undisturbed land ecosystem should thus, in time, deliver smaller quantities of dissolved ions to the streams draining it. Lack of any disturbance by geological or climatic processes or by man is, however, never an indefinite state for any ecosystem.

3.5.3 Vegetation and the supply of suspended silt and dissolved and suspended organic matter to drainage waters

Dissolved inorganic substances are not the only contributions made by catchment areas to the drainage waters. Eroded minerals and organic detritus are also important. So is dissolved organic matter. Silt loss tends to increase with run-off of water up to a point (about 25 cm year^{-1}) where it declines because the increasing precipitation supports a fuller vegetation cover, which reduces erosion. Forest is more effective (by about five times) at stopping erosion than grassland and other less bulky vegetation. Soft rocks and steep slopes will often offset this, however. Drainage waters of arid areas will thus naturally bear quite high concentrations of suspended solids, but those of wetter areas will not. Most inorganic particle loads result from disturbance of the landscape for agriculture. Up to 20 billion tonnes of suspended material is annually delivered to the oceans by rivers. Most comes from southern Asia and Oceania, where high rainfall, steep slopes, easily weathered rocks and highly intensive cultivation result in high loads. Overall, the inorganic particle load borne by streams and rivers is about five times that of the dissolved inorganic load [5].

3.5.4 Particulate organic matter

Organic matter may enter drainage waters as branches, leaf litter, bud scales, flowers, pollen and spores, the faeces and even carcasses of animals, and as finely divided soil humus. The amount of litter washed in will be high in dense, deciduous forest overhanging the stream valleys, and will enter mostly during autumn leaf fall. The abiding sound, during the dry season, in forests high in the Amazon basin is not the squawk of birds or the rasping of insects, but the steady, inexorable fall of leaves. There will be less export of solid organic matter where the vegetation is sparser, as in grasslands and savannahs.

3.5.5 Dissolved organic matter

Almost all natural waters have a dissolved organic concentration much greater than their suspended organic concentration. The total dissolved organic concentration (perhaps $1-20$ mg carbon l^{-1}) may rival that of total dissolved inorganic matter. Nonetheless, many substances are present and problems of analysis have prevented much generalization. The use of infrared absorption spectra now allows some distinction of compounds, but spectra of mixed substances are difficult to interpret. Gas chromatography, in which mixtures are separated, solves this problem and mass spectrometry can then be used to help identify the substances present and their amounts by comparison with standards of known compounds. Mass spectrometry bombards the sample with electrons and then characterizes the fragments by their weights. It can give an indication of the general formula of a compound but not its isomeric structure. Nuclear magnetic resonance spectroscopy can complete this task. Such analysis, however, is expensive and does not lend itself readily to routine monitoring of a group of substances some of whose members may be changing in concentration and composition very rapidly.

Much of the dissolved organic matter entering aquatic systems seems to be refractory. (The definition of this is pragmatic and depends on whether or not it is decomposable by high-intensity ultraviolet radiation in the laboratory.) Such substances persist in soils because they are difficult to decompose further. They come from the decomposition of lignin, have a structural core of benzene rings and polyphenols, are quite acid and have a high degree of oxidation and condensation. Those that are soluble over a wide range of pH values are known as fulvic acids and those insoluble at low pH are humic acids. They contain about 65% C and low amounts of nitrogen. The amounts dissolved may give an obvious brown stain to water from peaty areas, but all run-off waters will have some of these compounds. They are relatively inert, but may combine, for example, with easily precipitated metals, such as iron, and maintain these metals in solution as organic complexes. In this role, they are described as chelators. Some of these substances are formed in the aquatic system, as well as draining to it from the land.

The humic and fulvic acids formed in water tend to be richer in aliphatic groups than aromatic ones, with lower carbon and richer sulphur and nitrogen contents. They have lower acidity and a greater proportion of carbonyl groups than those derived from terrestrial sources, having been formed under reducing rather than predominantly oxidizing conditions.

Use of stable isotopes of carbon, nitrogen and hydrogen contained within the substances gives further insights [570, 797]. All organic substances contain a proportion of the heavy carbon isotope ^{13}C, derived from natural background sources. The proportion is very low relative to ^{12}C but detectable as a delta (δ) value, which compares the ratio (R) of the heavier to the lighter isotope in the compound, with an internationally agreed standard value:

$$\delta^{13}C(\permil) = \left(\frac{R\ sample\ -\ R\ standard}{R\ standard} \right) \times 1000$$

Heavy isotopes tend to react at different rates from lighter ones, so a negative value of $\delta^{13}C$ indicates discrimination against ^{13}C and a positive value selection for it. Similar principles apply to other stable isotopes, such as ^{15}N and deuterium (^2H).

Humic substances derived from terrestrial sources tend to have $\delta^{13}C$ values of -25 to $-28\permil$, δ^2H of -50 to $-100\permil$ and $\delta^{15}N$ of $+2\permil$. In contrast, those produced in aquatic systems have higher $\delta^{13}C$ values, between -20 and $-23\permil$, a narrower spread of δ^2H values, around $-105\permil$, and higher $\delta^{15}N$ values, about $9\permil$.

These isotopic characteristics, coupled with direct structural analysis can help characterize the sources of refractory organic substances and how they were produced. The difference in $\delta^{15}N$ values, for example, suggests that the source of nitrogen in terrestrial systems is fixation from the atmosphere (see above). That in aquatic systems comes mainly from combined nitrogen sources, such as nitrate, processed from the original fixed nitrogen with further isotope discrimination.

3.5.6 Labile organic compounds

Little of the soluble organic load delivered to streams is of labile compounds. Much of that produced in soils will have been rapidly decomposed, but some is leached and may be important, despite usually very low concentrations. Such compounds (amino acids, sugars, urea, indoles, phenols, amino sugars and polypeptides) may leach from insect-wounded leaves or carcasses or from decomposing litter or they may be actively secreted or exuded. Because of the high reactivity of these compounds, they are more difficult to study and little is known of production and transformation rates. Studies in lake water, using radioactively labelled compounds, suggest turnover times of only a few minutes or hours for the simple sugars and carboxylic acids so far investigated.

3.6 Effects of human activities on the composition of drainage waters

Rock weathering and sea spray still dominate the major ion composition of the world's fresh waters, but farming and settlement now probably have the greatest effect on key nutrient concentrations, while, increasingly, atmospheric pollutants and industry influence those of trace elements.

3.6.1 Agriculture

The greatest impact of human use of the land is the removal of the original vegetation cover and the destruction of mechanisms that conserved nitrogen and phosphorus in the land ecosystem. Thereafter, effects on the runoff water depend on whether there is continuous plant cover – in permanent grazing, for example – or whether the soil is left bare for part of the year. They depend also on how much fertilizer is used and when and on the number of stock kept and how the animals are managed. Loss of substances is greatest for a few years after active cultivation, but, if the fields are abandoned or come under less intensive management, there is less surplus to run off and export rates decline [429].

Export rates of nutrients

Rates of loss (export rates) of nitrogen and phosphorus from different sorts of land use and from stock are usually given as total P and total N (TP and TN), meaning the combined load of dissolved inorganic, dissolved organic and particulate compounds of these elements. Rates depend on climate and may be higher in arid than in moist regions because of soil erosion. For moist, northern temperate regions, Table 3.5 gives a summary. Some values are given as export coefficients, meaning the percentage of input (usually as fertilizer or stock feed) that is lost through leaching or wash-off of excreta. Data are not given for natural vegetation – work in this area has been mostly carried out for informing agricultural interests of potential losses of production – but, in general, natural rates are five- to 10-fold lower than those in agriculturally disturbed areas [736].

Disturbance of vegetation and soils by agriculture leads to rather greater losses of nitrogen than of phosphorus, although the long-held belief that phosphorus is held more or less indefinitely by absorption on to clay particles in soils is being challenged. When there has been a long history of fertilization, the soils can become saturated and phosphorus is then more easily leached. Most of the export of phosphorus probably still comes from human effluent and the excreta of stock.

A cow can produce as much phosphorus (up to $18\,kg\,year^{-1}$) as up to 1760 ha of forest or 300 ha of cropland. It will also excrete as much

Table 3.5 Examples of variation in export coefficients (% of input) and rates (kg ha^{-1}) for land use and rainfall or kg head^{-1} for stock and people) in northern temperate farming systems (from Johnes *et al.* [491]).

	Nitrogen (kg ha^{-1} or head^{-1})	Nitrogen (%)	Phosphorus (kg ha^{-1} or head^{-1})	Phosphorus (%)
Location				
Arable	2.8–120	–	0.06–5.77	–
Cereals	0.67–72.5	12	0.02–2.1	–
Grassland	0.15–30	1–40	0.02–4.9	–
Forest	0–13	–	0.01–0.88	–
Bare fallow	0.5–6.0	–	0.05–0.25	–
Rough grazing	3–6.4	–	–	–
Stock				
(1) Export from manure applied to land				
	–	1–25	–	3
(2) Manure voided				
Cattle	44.4–74.8	–	7.65–17.6	–
Pigs	6.6–18.8	–	1.4–5.63	–
Sheep	7.0–10.1	–	1.47–1.8	–
Poultry	0.2–0.9	–	0.1–0.3	–
People	1.86–8.6	–	0.3–3.9	–
Rainfall input	8.7–12.4	–	0.15–0.5	–

nitrogen (up to 70 kg year^{-1}) as exported from 70 ha of forest or 25 ha of arable land. In more traditional farming systems, where animals and crops were managed together, considerable reuse could be made, by manuring the fields, of the large amounts of phosphorus and nitrogen produced by the animals.

Recent trends in agriculture have separated stock-keeping from arable farming. Stock is often kept in intensive units, which produce a large quantity of wet 'slurry', which cannot often be disposed adequately to the land but penetrates eventually to streams and groundwater. Large areas of land are also devoted entirely to arable agriculture. Here the greatest profit is to be made from high-yielding cereal varieties, which demand heavy fertilization and the abandonment of traditional rotations, which left the land in pasture for at least 1 or 2 years in a cycle of 4 years.

The use of pesticides now allows annual cultivation for cereals. This leaves the ploughed soil bare for a part of each year, when nutrients are more vulnerable to leaching and soil to erosion. Just as the economic pressures for the intensive raising of stock (high production at low labour cost) have led to increased losses, particularly of phosphorus, to the drainage waters, so the intensive growing of cereals has increased the nitrogen losses and increased the concentrations in streams and rivers.

Nitrogen losses

Cultivation alone leads to a significant loss of nitrogen, but fertilization increases this. In England and Wales, the amount of nitrogen fertilizer (usually ammonium nitrate or ammonium phosphate) put on agricultural land increased from about $50\,000$ t in 1928 to nearly 1.3×10^6 t in the 1980s. Of the nitrogen reaching the fields, only about half is incorporated into crops, nearly a quarter is lost by volatilization of ammonia to the atmosphere and a seventh is lost by denitrification and incorporation into refractory compounds in the soil. Another seventh is lost to the run-off waters. The effect of these changes in farming practice has been to increase substantially the nitrate concentrations in lowland drainage (Fig. 3.14). This has followed particularly the 1947 Agriculture Act in Great Britain, which gave much support to farming through limited subsidies, and the EC's Common Agricultural Policy, which in 1977 gave effectively unlimited subsidy, in proportion to production, for most crops.

Such subsidies have now been reduced, though not discontinued, and there is a voluntary code of good agricultural practice for the protection of water [655], which represents a forward step. There is also a European Union directive allowing the designation of protection zones for ground-water used for domestic supply, where nitrogen fertilization is controlled, for there is evidence of potential health problems caused by nitrate (see Chapter 6). Nitrate concentrations still remain very high, however.

3.6.2 Catchment planning and export-coefficient models

Land use and practices change rapidly, often depending on the world economic situation and governments' use of incentives and subsidies in response to it. Tools are thus needed to predict the effects of such changes on the chemistry of water draining from the land. The ultimate tool would be a set of replicated, identical catchments where such changes could be directly measured. This is not possible, however, especially in a crowded country, and would produce results too slowly to be useful. The alternative is, again, to use a model that relates rainfall, land use, stock headage, human population and industry in a catchment to the concentrations of total nitrogen and total phosphorus in the run-off water.

Such a model is shown in Fig. 3.15. First, the land use of the catchment and the numbers of livestock and humans are determined from existing databases for a given year. Appropriate rates of N and P export from each type of land use (dependent on terrain and general fertilization rate) are then chosen from the available empirical literature (see Table 3.5). The rates are then applied to the actual area of each land use in the lake catchment to give the load exported from these sources.

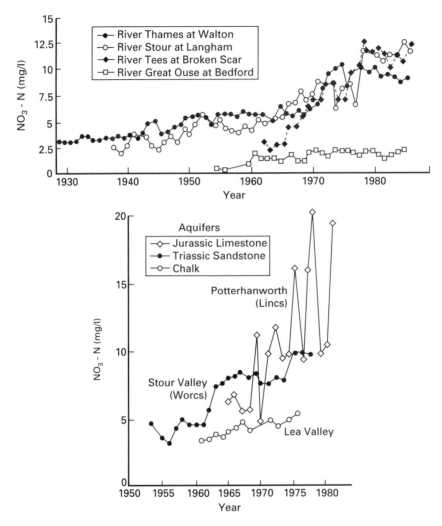

Fig. 3.14 Changes in nitrate-nitrogen (NO$_3$-N) concentration in four British rivers and three public water-supply boreholes. The World Health Organization recommended upper limit for water-supply is 10 mg l^{-1}. (From Heathwaite *et al.* [400].)

Similarly, appropriate annual rates of elimination of N and P from the major kinds of stock are determined from the literature with reference to major breed characteristics (largely upland versus lowland). Allowance is made for the proportions exported from different land types (e.g. tussocky grassland or short turf) and applied to the headage of stock in the catchment to give the loads exported. A similar calculation is made for humans and any industrial sources. Finally, the direct load from rain and snow is determined.

The total load is calculated and divided by the annual river discharge to give mean annual concentrations of total N and total P. These calculated

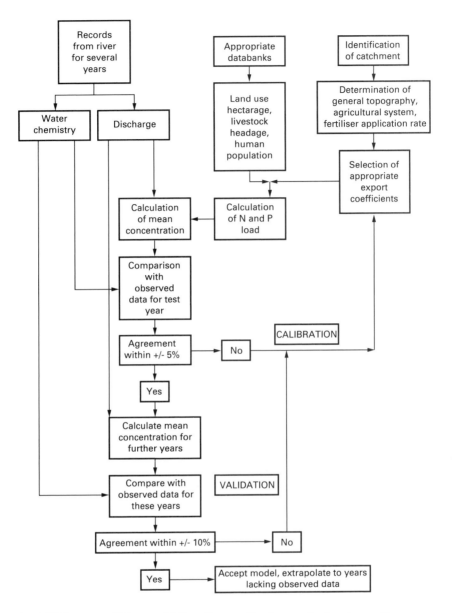

Fig. 3.15 Export-coefficient model flow diagram. N, nitrogen; P, phosphorus. (From Johnes *et al.* [491].)

concentrations are then compared with observed concentrations for the test year. If the fit between them is reasonably close ($< \pm 5\%$), no further adjustments are made. If there is a discrepancy, the export rates used are adjusted within the empirically determined range, until acceptable correspondence is obtained. During this calibration, experience is needed in selecting appropriate rates for the catchment and nutrient source concerned. This subjectivity is removed by the next stage (validation), in which the calibrated export rates are used to predict the concentrations for a different set of years for which independent records of land use, stock and people and observed TN and TP concentrations are available.

If, at this validation stage, the model calculates these concentrations with acceptable precision ($< \pm 10\%$), it is accepted and can be used to calculate concentrations for years for which no observed data are available. If not it must be recalibrated or abandoned. There can be an extremely close fit [490, 491], as shown in Fig. 3.16. It can then be used to predict the effects of proposed changes in land use or management or to determine historic water quality for comparison with the present. This can be used to set targets for water-quality improvement [692, 695].

The model can also give an overview of what particular activities contribute most or least to the total run-off of nitrogen or phosphorus. Figure 3.17 shows, for an average of 10 rural catchments in the UK, the trends between 1931 and 1988. Livestock have been the dominant source of both N and P, with increasing secondary contributions of N from cultivated land and P from people. These small rural catchments are not necessarily representative of the larger rivers to which sewage effluent is discharged from large lowland cities. In these, most N is contributed from arable agriculture and P from the human population [685].

3.6.3 Agricultural chemicals other than nutrients

Many pesticides are used to control insect, fungal and other diseases of crops, which may receive six or seven different pesticides in a growing season. Most of these substances are poorly soluble in water, being intended for uptake in the fatty membranes of their target organisms, but many are detectable in drainage waters at low concentrations. Pesticides are potentially dangerous and most developed countries screen new products and make arrangements to minimize dispersal of the pesticide to the environment in general. There is also strong pressure to develop products that are better retained on the land, less persistent in time and usable at low application rates. Nonetheless, the range of products and the inevitability that water drains from treated fields mean that pesticides will reach fresh waters. Other uses, including herbicide treatment of motorway central reservations and railway tracks, lead to prodigious run-off from hard surfaces. There are also accidental spills and irresponsible disposal of residual stocks.

There are 450 ingredients now approved in the UK for pesticide use. Data for 120 of them at 3500 sites in England and Wales [713] show that 100 were regularly detected in rivers and streams, but at low concentrations. There are severe problems in determining what concentrations are safe (see Chapter 4) and the only basis for assessing the concentrations

Fig. 3.16 Values predicted for total nitrogen (N) and total phosphorus (P) concentrations in the River Esk, in northern England, by an export-coefficient model, compared with actual observations. Two versions of the model were used: one specific to the catchment (triangular symbols) and one appropriate to upland regions in general (square symbols). Measured concentrations are shown as circular symbols. As in many rural English catchments, the P and N concentrations have about doubled since the pre-Second World War period, following postwar intensification of agriculture. (Based on Johnes et al. [491].)

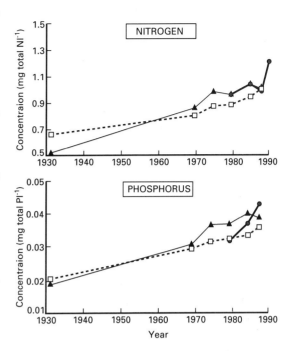

found is against more or less arbitrary environmental-quality standards (EQS). These are set for individual substances, based on their toxicity, persistence and potential for accumulation in non-target living organisms. Of the samples taken, 96% of concentrations were lower than the appropriate EQS. The most usual breaches were from pesticides (PCSD/ eulan and permethrin) used in protecting wool from moth infestations and

Fig. 3.17 Changes in percentage contribution from a variety of sources to nutrient loads exported from an aggregate of 10 British rural catchments between 1931 and 1988. Main diagram shows the percentage contributions and histograms show the total loads exported per unit area of catchment. (Based on Johnes et al. [491].)

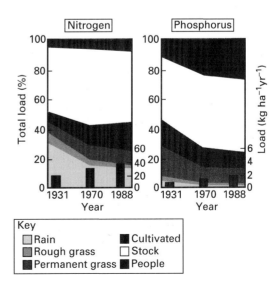

subsequently washing it for use in clothes and carpets and from veterinary medicines (diazinon) used for dipping sheep against scab. Groundwaters also sometimes exceeded a standard of $0.1 \, \mu g \, l^{-1}$ set by the EC for total pesticides in water to be used for human consumption. In these cases, herbicides (atrazine, diuron, bentazone, isoproturon and mecoprop) were usually responsible. The general presumption is that, for the time being, there will be continued and widespread use of a large range of substances in agriculture and other activities. Precautions against contamination (safe storage, no-spray zones close to streams, direct soil injection) are thus highly desirable.

3.6.4 Settlement

Expansion of villages towns and cities has led to increasing amounts of human sewage. Sewage is a mixture of wastes from laundry, bathing, cooking and the flushing of faeces and urine in lavatories. It is mostly water and to it may be added the rainwater collecting on the streets and pouring down the drains. Sewage contains much labile organic matter and, if released direct to streams, will cause severe deoxygenation.

The most common polluting discharge has been largely untreated sewage, as city populations have increased during the last two centuries in response to the availability of work. Many nineteenth-century rivers became completely deoxygenated for long stretches within and below towns, as did the newly dug transport canals. Discharge of raw sewage brought with it problems of disease, such as cholera, because drinking-water supplies became contaminated. Sir James Kay Shuttleworth, a physician, gives a graphic account of the awful horrors of cholera in Manchester in 1832:

A loop of the River Medlock swept round by a group of houses lying immediately below Oxford Road, and almost on the level of the black polluted stream . . . I was requested by one of the staff of the outpatients at the infirmary to visit a peculiar case in one of these cottages . . . I sat by the man's bed for an hour, during which the pulse became gradually weaker . . . as the evening approached I sent the young surgeon to have in readiness the cholera van not far away. We were surrounded by an excitable Irish population, and it was obviously desirable to remove the body as soon as possible. The wife had been soothed and she readily consented to be removed with her children to the hospital . . . None of them showed any sign of disease, and I left the ward to take some refreshment . . . On my return, the infant had been sick in its mother's lap, had made a faint cry and had died . . . When I returned about six o'clock in the morning, another child had severe cramps with some sickness, and while I stood at the bedside it died . . . Then later, the third and eldest child . . . the mother likewise suffered from a severe and rapid succession of the characteristic symptoms and died, so that within twenty-four hours the whole family was extinct [122].

Not surprisingly, from the middle of the nineteenth century, treatment works became common, but the problems of severe organic pollution were not substantially solved until well into the twentieth century for some city rivers, such as the Thames and the Rhine.

In isolated houses or hamlets, a very simple system of treatment, in which the sewage decomposes in an underground tank (septic tank) and the consequent effluent seeps away into the soil, may be adequate. This is not practicable for high population densities. The populations of many countries are increasing and aggregating; the problem of disposal of sewage is thus also increasing.

Conventional sewage treatment removes much organic matter from the sewage but leaves an effluent typically containing up to 5 mg l^{-1} NH_4-N, 40–50 mg l^{-1} nitrate (NO_3)-N and 10–20 mg l^{-1} PO_4-P from decomposition of the organic matter. In western countries, from the 1960s until the 1980s, about 40% of the phosphate content came from domestic detergents, in which sodium tripolyphosphate was used to remove calcium ions from solution and increase the efficiency of the surface-active cleaning agent, which would otherwise be precipitated by them. The phosphate content of many domestic detergents has been reduced in the last decade and accounts now for perhaps 20% of phosphorus in effluent.

Sewage effluent is released to streams usually at rates of only a few per cent of the total stream discharge. The stream community will normally decompose any remaining organic matter without severe problems, but the phosphate concentration will greatly increase compared with that in water draining from the land, by 10- or 100-fold. Increases in phosphate concentration from sewage effluent are now very prominent and in many lowland areas the bulk of the phosphorus comes from this source, just as the bulk of the nitrogen comes from the use of the catchment for agriculture.

This general picture is shown in Fig. 3.18, in which data have been collected from four European countries and the catchment of the River Po in Italy. In each case, agriculture includes both stock wastes and crop growing. The proportion of phosphorus due to sewage and its effluent rises with increasing population density, while that derived from natural background sources declines. Compare Sweden with the highly urbanized Po Valley, for example. Conversely, the highly agricultural lands of Latvia and Denmark contribute the bulk of nitrogen to their countries' budgets.

3.6.5 Industry

Many substances are released by industry to fresh waters. The nature and amount will vary locally but, for example, some 30 000 compounds of commercial significance are manufactured within the catchments of the North American Great Lakes and 2000–3000 new ones are added each

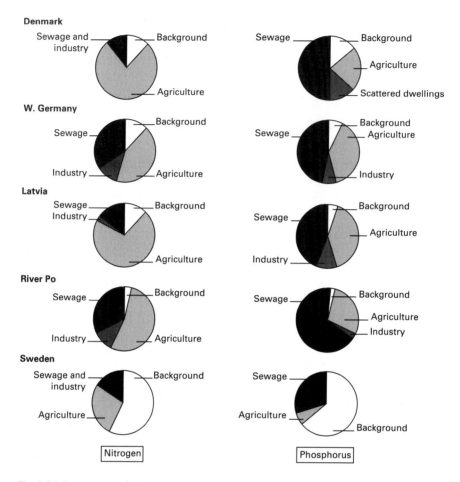

Fig. 3.18 Proportions of the total inputs of phosphorus and nitrogen into fresh waters attributable to industry, sewage, agriculture and background natural sources in several European countries and the River Po in Italy. Sweden has a lower population density and more remaining natural vegetation than the others. (From Kristensen and Ole Hansen [536].)

year. Some hundreds are potentially toxic. Such substances include acids, alkalis, anions (sulphide, cyanide and sulphite), detergents, labile organic waste (from food processing, for example), chlorine, ammonia, heavy metals (particularly Pb, Zn, Cd, Cu, Hg, Ni and Cr), oil, phenols, formaldehyde, polychlorinated biphenyls and radionuclides. The problem increased with the nineteenth-century industrial revolution, when the variety of industries was increasing and, for ambitious industrialists, the local stream was a cheap means of waste disposal. The rivers of the Mersey basin, for example, became and remain some of the worst-polluted waters in the British Isles.

The River Mersey

I remember well, as a child in the 1950s, watching from a bridge over the River Mersey, near my home in Stockport, a thick, brown, astringent-smelling water, eddying, rather fascinatingly, around the bridge stanchions. It was taking the effluent of tanneries, food processing, a gasworks, electroplating, oil and grease processing, a slaughterhouse, cotton factories, bleach works, dye works, lead–acid battery manufacture, paint and rubber works and the manufacturers of paper and glue. As small boys, we added our mite from the parapet with no guilt whatsoever.

Thirty-four per cent of the catchment of the Mersey is urban, with 828 000 people at a density of $1220\,km^{-2}$. Despite extensive improvements, it is still a severely polluted river [420, 712]. Heavy-metal concentrations are decreasing, but new organic pollutants are replacing them, such as pentachlorophenol, used as a rot proofer for textiles, carbon tetrachloride, used in the manufacture of plastics, various organic solvents and an increasing list of by-products of largely unknown environmental significance. Completely unsuspected new compounds are sometimes detected [620].

3.6.6 Disposal and consents

Many industrial discharges are now directed through a sewage (now called waste-water)-treatment works, where even highly toxic substances can be precipitated in a sludge and disposed in waste tips on land. The discharge of these substances to a river is then reduced. This is not always possible, because some wastes are so toxic that they would upset the operation, which is essentially biological, of the sewage-treatment works. In England and Wales, the Environment Agency issues 'consents to discharge', which limit the concentrations and amounts of particular substances that may be released direct from a factory or from a waste-water-treatment works to a river. This may require a factory to treat its own wastes on the site, but full treatment may be so costly that the factory might be put out of business if the standards were set too high. Large industrial companies, faced with increasingly stringent legislation, have seen the public-relations advantages of ready compliance. Smaller companies, however, may be unable to afford to be as environmentally responsible as would be desirable. Dilution of pollutants in the river thus remains a widespread solution, on the one hand, and problem, on the other. In old industrial areas, there are also many hidden discharge pipes, which have gone undocumented and therefore uncontrolled.

Trace organic compounds

As heavy-metal pollution is controlled, concern is passing to trace organic pollutants, including manufactured organochlorine compounds, such as polychlorinated biphenyls (PCBs). These comprise two linked benzene rings (biphenyl), in which chlorine replaces hydrogen in various combinations at up to 10 sites on the molecule. About 200 such compounds are possible and mixtures of them have been sold under the trade name Arochlor. They have useful properties – high dielectric constants, stability, non-flammability and cheapness – and were used as insulating fluids in electrical equipment, in cutting oils, plastics, paint, printing ink and carbonless copy paper. They are difficult to decompose, so they persist in the environment. Their use has now been restricted, but a large burden persists in waste dumps and derelict electrical machinery. Polychlorinated biphenyls are more soluble in fat than water, accumulate in organisms (see Chapter 4) and may sometimes be carcinogenic.

The advent of analytical techniques capable of detecting organic compounds at very low concentrations has revealed many additional trace organic problems in recent years. New organic compounds may cause physiological changes in aquatic organisms. They may, for example, compromise the immune systems of fish and, potentially, the human or other animal consumers of those fish. One particular problem, suspected earlier [618] but highlighted since 1990 [976], has been the occurrence of hormones and hormone analogues in rivers. These are not fully destroyed by waste-water treatment and may be derived ultimately from contraceptive pills and from paper and detergent manufacture. There have been cases of production of hermaphrodite fish in rivers with a high proportion of sewage effluent and at hormone levels barely detectable by conventional methods.

A common component of contraceptive pills, 17-α-ethinyloestradiol, stimulates vitellogenin production in male rainbow trout (*Onchorhyncus mykiss*) [259]. This protein is normally produced only by mature female fish in egg yolk. The extent of the problem is not clear, but there is justifiable concern, especially in view of evidence of declining sperm counts in the human population over the last 50 years. This may not necessarily be specifically linked with hormone derivatives in drinking-water. Sperm counts can be affected by diet, stress and potential toxins in spoiled food, fungi and plants, but increasingly a cautionary approach is needed. Many severe pollution problems have been denigrated as insignificant at first.

Non-hormones may also cause vitellogenin production in fish, though at concentrations far higher than those at which hormones act. A group of industrial-detergent surfactants, the alkylphenolpolyethoxylates (APEOs), nonylphenol and octylphenol, has come under particular scrutiny because about a third of the total production (180 000 t year^{-1}) enters aquatic environments [72]. Their concentrations in some rivers, especially those in textile-producing areas, where the detergents have been widely used, stimulate vitellogenin production in rainbow trout [72, 486]. The use of APEOs is now the subject of a voluntary ban in Britain.

3.6.7 Industrial atmospheric sources

Anthropogenic acidification of rain and snow began with the industrial revolution, when the burning of coal on a greatly increased scale released much SO_2 to the atmosphere. This caused bronchitis among townspeople and damage to vegetation within radii of a few hundred kilometres. The solution was partly to encourage the burning of fuels of lower sulphur content by householders and partly to build high chimneys at power stations and other factories. The SO_2 was thus released into higher and faster air streams, dispersed further and more diluted. It was not however, destroyed.

The local air-pollution problems of the 1950s in Britain had also been caused by smoke particles. Processes were introduced to remove these electrostatically before the effluent gas was released. The 'fly ash' so removed, however, had contained calcium and other cations capable of neutralizing some of the acid produced when SO_2 reacted with water in the air. Effectively, more acid was thus produced. Measures thus taken to reduce national air pollution turned the problem into an international one, with movements of sulphur dioxide in the prevailing winds for distances of thousands of kilometres, during which rain scavenged the acids. Meanwhile, the local production of NO_x by vehicles had also been increasing, as the number of cars increased.

The combined effects of acidic rain and snow on the drainage waters were seen markedly from the 1960s onwards, particularly in areas of igneous and metamorphic rocks, where there was little possibility of neutralization of the acid by carbonates in the soils. Such places have included the Laurentian Shield area of Canada, the New England states of North America [577] and hard-rock areas of Wales, Scotland, Norway and Sweden. Effects are less clear in sedimentary-rock areas, where the buffering capacity of the soils is great.

The drainage waters from the less well-buffered areas are not only decreased in pH but may also have high concentrations of aluminium ions, which, at concentrations as low as $100\,\mu g\,l^{-1}$, may be toxic to fish. Other metal ions may also be mobilized by the acid, but, in contrast to aluminium, the concentrations of those available for weathering in rocks is generally low (see Table 3.4). There is now, however, some suspicion that the bulk of the load of heavy metals reaching lakes such as the Laurentian Great Lakes may be derived from industrial smoke and may come via the atmosphere and rain [871]. The Great Lakes have proportionately small catchments and receive half of their water by direct rainfall.

3.7 The water rolls downhill

Of the four forces that physicists now recognize as governing the laws of nature (gravity, electromagnetism, the weak (radioactive) and the strong (nuclear)), it is the apparently humblest that now takes over the hydrological story. Beyond the intricacies of atomic and molecular structure that determine the nature of all the compounds that make natural waters so unexpectedly complex, it is gravity that transports it to the next stage, where we consider how this complex solution supports the increasingly complex systems of the streams and rivers, in Chapters 4 and 5.

4: Erosive Streams and Rivers

4.1 Introduction

Upland streams (Plate 3) have long excited casual visitors and anglers, as well as professional investigators. The water movement, the occasional glimpse of a trout and the relative naturalness of the surroundings all give a fascination, which is strengthened by a closer look.

Most rain does not immediately run off the land, but soaks into the soils. This store supports the dry-weather base flow of the upland streams. To this small volume, after heavy rain, is then added a flood flow (Fig. 4.1), dependent on the rainfall and the ability of the soils to soak it up. The chemistry of the base flow may be different (greater concentrations of ions) from the flood flow because of its longer contact with the soil and rock. The flood flow, in contrast, may carry organic debris, such as leaf litter, and more soil particles eroded from the land. The flood flow passes downstream as a pulse. It may move fast, carrying suspended clays and fine organic debris, gravels or sand with it. At times of very high flood, it may move large boulders, eroding the valley sides and the stream channel. The deep, narrow valleys of many upland streams have been cut in this way. The relatively meagre base flow moves fast enough only to carry very fine particles, so that, in dry weather, beds of sand and gravel are left among the larger rocks and boulders.

4.1.1 A model stream

We can imagine a model stream. Models allow general principles to be integrated for comparison with specific examples, none of which is like the model in all respects. The model stream lies in uniform geological terrain; it is most erosive in its headwaters, where the boulders left from a previous spate create an uneven or turbulent flow over a rough stream bottom. As the downward erosion of the stream bed continues, the slope is reduced and the stream's ability to erode its bed declines. The slope flattens until it becomes nearly horizontal, at its base level, when the stream (then usually called a river, but there is no absolute distinction) meets the sea.

The model stream changes steadily with increasing distance from the headwaters. Its catchment area increases, bringing a greater supply of water. The effects of local storms are evened out so that spates characteristic of

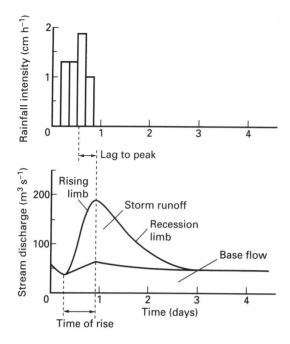

Fig. 4.1 Stream-flow hydrograph of the effects on stream discharge of a rainstorm (modified from Allan [5]).

the headwaters are less frequent. Downstream, the water movement is less turbulent so that pockets of finer sediment can be deposited among the rocks and persist even during floods. Periods when current speeds are capable of moving large stones become rare until eventually fine sediments dominate the bed, leaving bare stones only in the centre of the channel, where the current is greatest.

The size of the channel increases downstream as more water flows through. However, it does not usually increase in cross-sectional area at the same rate as does the area of catchment, for it is determined by the erosiveness of high floods, whose effects are muted downstream. It accommodates by meandering, thus increasing its length and capacity. The discharge (volume passing per unit time) and the cross-sectional area of the water passing in the channel are related by the average current speed:

Discharge $(m^3 s^{-1})$ = cross-sectional area $(m^2) \times$ speed $(m s^{-1})$

Paradoxically, in the model stream, the average current speed may increase downstream to accommodate the greater discharge. The less turbulent flow in the lower reaches, however, means a greater range of velocities over the cross-sectional area. These include very low speeds in contact with the bed and very high speeds in the midwater at the centre of the channel. The mean velocity is found at around 0.8 of the total depth in the centre of the channel. The turbulent, foaming water in the headwaters may have temporarily higher speeds, as it eddies in flood, but its average speed, largely a function of the base flow, is generally lower than that downstream.

Simple mathematical relationships describe changes in the channel with increasing discharge. Thus, if Q is discharge, W is width, D is mean depth and U is velocity:

$$W = aQ^b$$
$$D = cQ^f$$
$$U = kQ^m$$

These describe dramatic changes, as catchment area and its consequent discharge increase. In general, the rate of increase of width is greater than that of depth, while velocity increases least rapidly. The groups of constants, a, c and k and b, f and m, must each add up to 1, because $Q = WDU$. There is thus a degree of compensation among width, depth and velocity.

Accommodation of the channel to the moving water leads to patterns in the deposition of gravels and sands. Shallow riffles with coarse substrates and faster water alternate with deeper pools with finer deposits and slower water (Fig. 4.2). The riffles may form as bars at the outsides of bends, while the pools may be more or less continuous in the centre of the channel and on the insides of bends. These features provide different habitats for stream organisms.

In its lowest reaches, the channel may increase its storage capacity by winding across a flood-plain to accommodate the greater of its normal flows. The channel moves as the river erodes soil from the outside of its bends (meanders) and deposits it on the inside. This can be measured as a

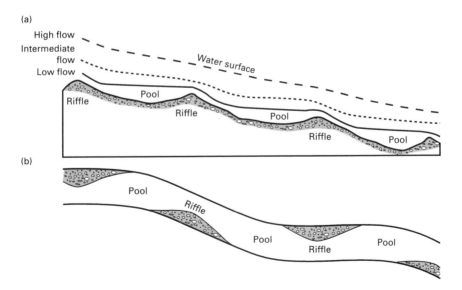

Fig. 4.2 (a) Longitudinal profile and (b) plan view of the sequence of shallow rocky riffles and the deeper pools, which often develops in naturally functioning streams, giving a variety of habitats for stream organisms (from Allan [5]).

sinuosity index – the actual channel length to the linear distance travelled downstream; values of up to 4 have been recorded. At very high flows, the channel may be unable to accommodate all of the water and water overtops its banks to cover its flood-plain. The flood-plain is not dry land, damaged, in our perception, by an abnormal feature we call floods, but a natural part of the river bed, which is used less frequently than the main channel to accommodate the highest river flows.

The model stream thus falls in orderly progression from uplands to the sea but does not exist! Changes in geology often prevent the establishment of a smoothly graded profile, because different rocks erode at different rates. Natural lake basins may interrupt the sequence. Uplifts of the Earth's crust may start lowland rivers cutting deeply into their beds as the land rises. And management by man – the creation of dams, the alteration of the channel itself in the interests of flood control – has upset any smooth change downstream. The model river, nonetheless, has value as a framework on which to base studies of river ecology. This is covered in two chapters, one about erosive streams and small rivers (this one) and the other about the larger rivers, where sediment deposition is predominant (Chapter 5).

4.2 Upland streams – three general questions

Many questions can be asked of any ecosystem. What organisms are found in it and how are they adapted to the particular conditions – in this case, the problems of coping with turbulent, erosive water? By what pathways does energy flow in the ecosystem? And what determines the distribution of different organisms and the compositions of their communities?

4.2.1 On the rocks

Stones, the pockets between them and the interstices of gravel and sand provide a complex architecture on stream bottoms. It is not stable, for high flows will change it, although boulders are only rarely moved. Given this, and a talent in microorganisms for producing glues, often as the outer layers of their slimy cell walls, permanently wet rocks soon become covered by an organic layer of bacteria, fungi, protozoa and algae. Some organic matter may also be deposited as a film by purely chemical means. Epilithic (Greek: *epi*, on, *lithos*, rock) organisms may be held flat to the rock or protrude from attachment pads (Fig. 4.3). Some may move through the organic film, which may include free enzymes, derived from bacteria, which continue to function and produce labile organic matter from polysaccharides and other polymers.

The film may be scoured by suspended sand and silt during floods, but this is unusual because of the way water moves over surfaces. The surface creates a frictional 'drag', which slows the flow to near zero at the surface

Fig. 4.3 Development of organic layers on stones in a New Zealand stream. (a)–(c) Development in darkness on a cleaned stone surface at the start (a) and after 1 month (b) and 3 months (c). Scattered slime can be seen in (b), with additional fungal hyphae and a diatom in (c). (d) Development in the light after 2 months, with the diatom *Cocconeis* (oval shape) and white structures, probably the cysts of protozoa. (From Rounick and Winterbourn [832].)

and reduces it for a short distance above. A viscous layer forms, a fraction of a millimetre thick, in which the flow is not turbulent and erosive, but smooth or laminated. Above this layer, for a few millimetres, until the water eventually reaches about 90% of the average velocity for the stream, is the 'boundary layer' where, as the current quickens with distance from the surface, the flow is increasingly turbulent. If the surface is rough, a laminar layer may be unable to form, turbulent water lies in contact with the surface and the boundary layer is thinned. Organic layers themselves make the rough smooth.

The nature of the organic layer depends on the water chemistry and the amount of light. Overhanging vegetation may limit the growth of algae and favour bacteria dependent on dissolved organic matter washed in from the catchment. A more open stream may have fewer heterotrophic bacteria and more photosynthetic algae. Some small plants – mosses (e.g. *Fontinalis*) and liverworts, a few genera of red algae (*Lemanea, Hildenbrandia, Bartrachosper-mum* (Plate 3) and, in the subtropics and tropics, one group of flowering plants, the Podostemonaceae (Fig. 4.4) – live on stones. These form flat crusts or small tufts, generally to the lee of the main current and sometimes enclosed within the boundary layer. Invertebrate animals are generally abundant, attached to rocks in the main flow (e.g. limpets) or in the gravel interstices, under the stones or among leaves packed into crevices (Table 4.1). Many organisms of turbulent streams persist by avoiding the turbulence, although they may often be torn loose. There are no specifically

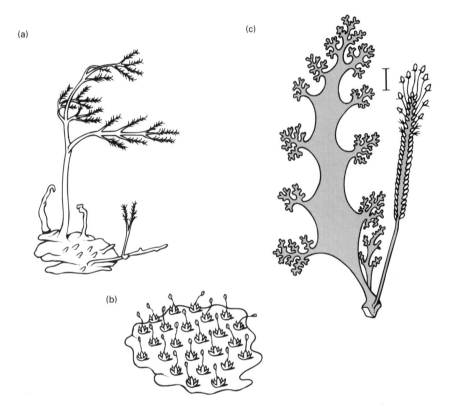

Fig. 4.4 Podostemonads are among the few groups of macroscopic plants that are able to survive in fast-flowing water. They have no roots but attach with holdfasts in the crevices of rocks, forming often-flattened green plates, a little like chewing-gum adpressed to the rock surface. They are flowering plants that produce small inflorescences and shown here are (a) *Wettsteiniola accorsii*, (b) *Zeylanidium johnsonii* and (c) *Mourera fluviatilis*. Each is only a few centimetres in size. The only other common macroscopic plants in such habitats are mosses, liverworts and red algae (Plate 3), but microscopic algae, such as diatoms, sheltering in the boundary layers, are very abundant if there is sufficient light.

Table 4.1 Number of genera of groups of invertebrates and mean biomasses found at sites in the Satilla River, Georgia (from Benke *et al.* [58]).

	Woody snags		Sand		Fine mud	
	No. of genera	Biomass (mg m^{-2})	No. of genera	Biomass (mg m^{-2})	No. of genera	Biomass (mg m^{-2})
Diptera	17	240–700	15	60–120	11	150–300
Trichoptera	9	1600–4200	0	–	3	20–30
Ephemeroptera	5	60–100	0	–	0	–
Plecoptera	2	100–140	0	–	0	–
Coleoptera	3	120–220	1	8–11	0	–
Megaloptera	1	260–380	0	–	0	–
Odonata	3	530–580	1	negligible	0	–
Oligochaeta	0	–	3	22	3	290–420
Totals	40	2900–6300	20	90–150	17	460–750

adapted, suspended organisms (plankton); there would be little time for their populations to grow before the water had moved far downstream.

4.2.2 Adaptation to moving water

The best way for an animal to avoid being washed downstream would seem to be permanent attachment, but most stream animals can move freely. Permanent attachment carries the risk of stranding during dry periods and, as a result, permanently attached organisms – for example, freshwater sponges – mostly survive on the undersides of submerged rocks. Motility gives flexibility, though with greater risk of displacement. Very small animals may avoid being sheared from the bottom by living within the boundary layer. In flows of $< 20 \, \text{cm} \, \text{s}^{-1}$, the layer may be several millimetres thick but often it is $< 1 \, \text{mm}$ and only the very smallest animals and microorganisms can take advantage of it. Most stream invertebrates are exposed to quite turbulent water and use streamlining to minimize the chance of dislodgement. Streamlining, in which the greatest body width lies about 36% along its length, confers least resistance to current. For the mayfly nymphs *Rithrogena* and *Epeorus*, the streamlining includes a flattened body (Fig. 4.5), which may allow the animal to crawl under stones. The long tails of mayfly and stonefly nymphs seem to act as fins, which turn the animal head on into the current, just as a small boat fares best if headed into the waves of a rough sea.

Leeches snails and freshwater limpets cling to rocks with suckers or suction-pad feet. Water penny beetles, *Psephenus* spp., have friction pads of small movable spines, which can be fitted into tiny irregularities of the rock surface. Hooks, grapples and claw-like legs have all been evolved, and some caddis-fly larvae make cases ballasted with heavy mineral particles. Some

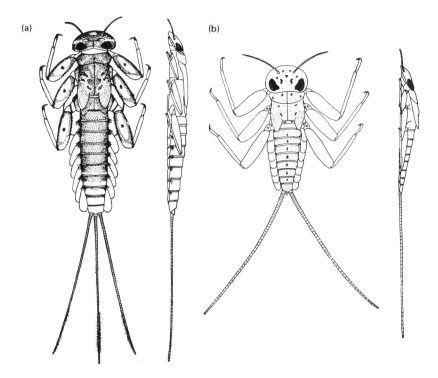

Fig. 4.5 Survival in fast-flowing waters is helped by several devices. One of them is evolution of a flattened body, as in the mayfly nymphs of *Rithrogena* (a) and *Epeorus* (b), which allow the animal to remain in the smooth-flowing boundary layer overlying stones and to avoid the turbulent water. (Based on Hynes [459].)

aquatic caterpillars (Lepidoptera) cocoon themselves under flat sheets of silks, like mountain tents in a windstorm. *Simulium* (blackfly) larvae (see Fig. 4.8) spin a pad of silk, which sticks to the rock and to which they attach with hooks. If the larvae are dislodged, they have a 'safety line' linking the body to the pad. By working it between the front proleg and rough spines on the head, the animal can 'climb' the line to regain the security of its silk pad on the rock. Other animals may shelter in the interstices of mosses [936, 937].

4.2.3 Drift

Despite these adaptations, displacement and drift downstream are common. Drift is complex [239, 459, 701, 968] and frequent among insect nymphs and larvae. Behavioural drift occurs when there is a consistent pattern and where such movements are probably adaptive; background drift is the continual accidental movement of a few organisms dislodged by the current; and catastrophic drift follows major disturbance by high floods or human interference.

Not all species or even size classes of the same species drift to the same degree. For some species, it is seasonal, peaks of travel coinciding with high current flow; for others, it may coincide with low current speeds. For *Gammarus* and *Baetis*, drift increases just after sunset, sometimes with peaks in the night, while in many chironomids there is no distinct diurnal pattern. Light appears to inhibit drift in some night-drifting species. This may be because they seek darker places within the streams as refuges from predation by day and emerge, to feed but also to be more vulnerable to dislodgement, at night. On the other hand, quite short experimental light periods (1 h in 24) may inhibit drift [699, 700].

Nets placed across streams passively catch huge numbers of animals. It might seem that upstream sections of streams would soon be completely depleted as animals drift and resettle downstream. Distances travelled in a single movement can be tens of metres. However, rarely more than 0.5% and usually fewer than 0.01% of the animals are moving at any time, although the cumulative drift over a given area over 24 h may be 10 to 100 times the density in that area [999].

Movements upstream along the bottom to some extent counteract the drift, although only to the extent of perhaps a few per cent replacement when drift rates are high [5]. There is also compensation by egg-laying adults that have emerged from drifted nymphs and larvae and flown upstream [623, 939]. Much movement of such adults is in other directions, but, since adults lay many eggs, only a few upstream movers may be able to compensate for quite large downstream drifting. For non-flying animals, the mechanisms of compensation are still obscure, but they must exist, because headwaters do not become progressively depauperate.

Active downstream movement has been found for some insect larvae just prior to pupation and emergence of the adults, which suggests that drift might not merely be a passive consequence of stream living but may have adaptive advantages. Müller [698] explained drift as the inevitable consequence of living in moving waters, to which natural selection had responded through evolution of upstream-return mechanisms. Waters [997, 998] thought it to be a consequence of 'excess production', removing surplus animals, perhaps through competition for space, that are eaten by fish. In support of this, Dimond [212] found that drift was low following denudation of a stream by insecticide contamination but increased as organisms recolonized to normal densities. More recent views see drift as potentially adaptive, there being benefits as well as costs to drifting [4, 275, 529].

The benefits are in rapid recolonization of newly wetted channels restored after drought or denudation by violent spates, or in acquisition of food when supplies are scarce. Drift of animals feeding on algae attached to stones was high when the algae were scarce [412]; this allowed dispersal to sites possibly richer in food. Drift rates of a net-spinning (see below) caddis-fly (*Plectrocnemia conspersa*) and a leaf-pack-inhabiting stonefly (*Nemurella picteti*) were exceptionally high (20% and 43% per day) in a southern English stream when densities of the former were so high (100 m^{-2}) that net-spinning sites and leaf packs were scarce. In contrast, drift rates were low in the same period for another stonefly nymph, *Leuctra nigra*, where its food supply, iron bacteria, was very abundant [968].

Drift may also help in avoiding predators. Larger animals, which are most at risk from fish predation, appear to drift at night more frequently than smaller ones [4]. In some Andean streams [275], drift was irregular or occurred by day in high-altitude sites that lacked drift-feeding fish. But, along a gradient of increasing vulnerability to such fish, the ratio of night to daytime drift for mayfly nymphs increased markedly. Where trout had established, following escape from local hatcheries to the high-altitude streams, the mayfly nymphs had started to drift at night.

4.3 Sources of food and energy flow in erosive streams

Many erosive streams receive most of their energy from organic matter washed into the stream, largely as leaf litter [84, 981, 1001]. The litter is processed to carbon dioxide by a succession of microorganisms and animals, which deal with successively smaller particles of it in a continuous sequence, like a factory production line, as it is moved downstream. This is common where deciduous forest covers the catchment or the stream is bordered by gallery forest, but not all streams depend so heavily on the processing of catchment-derived litter. In upland Britain, it is now more likely that the stream will be open to light and surrounded by grassland producing much less litter. Greater dependence on photosynthesis by epilithic algae might be expected. Here, several groups of streams will be compared.

4.3.1 Hot-spring streams

In volcanic areas, groundwater is superheated and forced through cracks to the surface, where it may emerge as very hot, often boiling, water or steam. Its contact with the underlying lava may have charged it with hydrogen sulphide, sulphuric acid or high concentrations of silicate. Even at boiling-point, living organisms are present, although they are absent in the superheated stream vents. As it emerges, the water may have some dissolved organic matter, ultimately derived from the land surface, and this may support mats or tufts of filamentous bacteria at the edge of the spring boil. The water has cooled a little there, but sterilized cotton or microscope slides placed at the boiling centre will also grow bacteria. If hydrogen sulphide (H_2S) is present, chemosynthetic bacteria, such as *Thiobacillus thiooxidans* and *Sulfolobus acidocaldarius* may grow attached to the rock at

85–90°C. They use H_2S to reduce carbon dioxide (CO_2) to produce organic compounds. As the water flows from the boil to form a stream, it acquires an often V-shaped set of coloured patterns on the channel floor. These are of organisms successively able to colonize the cooling but still hot water [94, 95]. The water cools faster at the shallow edge than the deeper middle, and the V pattern, pointing downstream, represents earlier colonization at the edge than the centre (Fig. 4.6).

The first obviously coloured V is usually a mat of *Synechococcus* sp., a photosynthetic unicellular blue-green alga, capable of growing at up to 75°C. It is not merely surviving high temperatures but, if cultured in the laboratory, grows best in a small range around 72°C. The *Synechococcus* mat often rests on an underlying one, sometimes several mm thick, of filamentous photosynthetic or heterotrophic bacteria. One of these, *Chloroflexus*, is capable both of photosynthesis, based on its orange pigments, and of feeding on organic matter produced by the *Synechococcus*. Further downstream, more complex blue-green algae occur, including *Mastigocladus laminosus*, which is filamentous and probably capable of fixing nitrogen.

All of the organisms so far mentioned have been prokaryotes. Eukaryotes, with the insides of their cells divided by membranes, cannot colonize very hot water. This may be because membranes, which can allow passage of large molecules, a necessity for proper functioning within the cell, are too 'holey' to survive much heating without breakage. The first eukaryotes to colonize are fungi, mingled in the bacterial mats, at about 62°C; eukaryotic algae follow at about 60°C and protozoa (*Cercosulcifer* and *Vahlkampfia*) at 57–60°C.

Thus far, only microorganisms are present, and one of the reasons for the prolific algal mats may be a lack of grazers, for no multicellular animals (or plants) can tolerate more than about 50°C. Vascular plants must wait until about 45°C and vertebrates 38°C. Only at 45–50°C are ostracods and the larvae of certain flies present and capable of chewing holes in the mats. The hottest springs, therefore, form very simple streams in their upper reaches, dominated by primary producers and heterotrophic microorganisms. Much of the biomass produced is washed downstream and not consumed *in situ*.

Stockner [927] studied a cooler hot spring (37°C), at Ohanapecosh in Washington, USA. He fitted a wooden trough, 12 cm wide, 3 m long, to constrain the water so that precise measurements could be made. The trough soon became colonized by communities previously present in the more irregular natural channel. Ohanapecosh springs are supersaturated with carbonate as they emerge and, on contact with air and through the, action of algae in removing CO_2 (see Chapter 6), a deposit of calcium carbonate (travertine) is formed and has built up to depths of several metres.

The stream was colonized by two filamentous blue-green algae, *Schizothrix calcicola* and *Phormidium* sp., and, among mats of these, the larvae of two flies, *Hedriodiscus trusquii* and *Caloparyphus* sp., were able to graze.

(a)

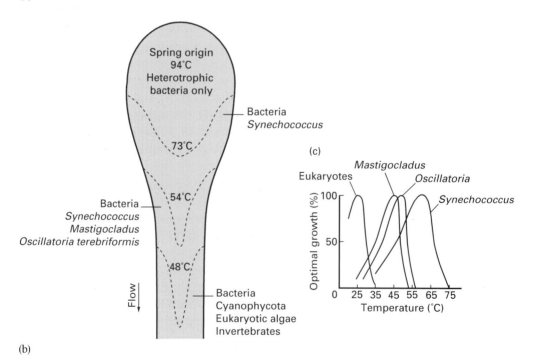

(b)

(c)

Stockner [927] measured the primary production of the stream by changes in its oxygen concentration as water flowed over the algal mats. He placed oxygen-detecting electrodes at each end of the channel and continually recorded the concentrations. From a baseline of the dawn concentration, he could then calculate the total net increase in oxygen during the day, after a correction for that which would have diffused into the atmosphere.

By repeating the measurements at night, he could similarly calculate the oxygen used up in respiration. The sum of the net uptake by day and the net loss at night gave the gross photosynthesis. This is because the measured production by day is less than the total photosynthetic oxygen production, as oxygen is simultaneously respired. To calculate this respiratory 'loss', it was assumed that the respiration rate of the community was similar by day and by night. Extrapolation of the night respiration rate to the full 24 h gave the total respiration of the community.

Some of the produced material was respired by the algae themselves, associated bacteria and the flies. These amounts cannot be separated, although the bulk of the biomass was algae and hence probably also the bulk of the respiration. Some material produced was dislodged by the current or the animals feeding. This was measured by sampling the water at the upper and lower ends of the stream, filtering it and weighing the material added over the stream stretch.

Some material was grazed; the amount was estimated from the number of larvae present and the amount they ate each day. Finally, travertine formation incorporates organic matter as well, so some of the production is deposited in the stream bed. This was determined from the difference between the gross production and the sum of its various fates described above. Table 4.2 gives the results of the study. Note the dependence on *in situ* photosynthesis (external (called allochthonous) sources of organic matter were negligible) and that little of the material (0.4%) was consumed by grazers. The bulk was apparently deposited or washed downstream.

4.3.2 Streams in wooded catchments – Bear Brook

It is not easy to account for all the sources of organic matter to a section of a stream and to balance these inputs with equally good measures of what

---

Fig. 4.6 (*Opposite*) (a) A hot-spring area in Yellowstone National Park, USA. The diagram (b) shows the occurrences of the major organisms of such hot springs as the boiling water emerges and cools. Successive organisms of decreasing thermal tolerance form V-shaped zones, because the water cools more rapidly at the edges than in the middle of the channel. *Synechococcus lividus* is a coccoid, and *Mastigocladus laminosus* and *Oscillatoria terebriformis* are filamentous cyanophytes. At the upper limit of its growth (74°C), *Synechococcus* is well beyond its optimum of about 60°C (upper panel), but heterotrophic bacteria can grow in the boiling water as it emerges.

Table 4.2 Energy budget for the stream flowing from one of the Ohanapecosh hot springs in 1966. Values have been converted to common energy units of kilocalories m^{-2} year^{-1}. (After Stockner [927].)

	Energy (kcal m^{-2} year^{-1})	Percentage of total
Input		
Gross primary production	4607	100
Output		
Community respiration	1158	25.1
Washed downstream	1461	31.7
Grazed	20	0.4
Deposited with travertine in stream bed	1968	42.8
Total	4607	100

happens to them. Such work has been carried out on a 1700 m section of Bear Brook, New Hampshire [269]. This stream was ideal because it lies on hard bedrock, which prevents deep seepage and directs all the precipitation that is not evapotranspired to the stream, allowing full accounting of the water movement. It is also a small stream uniformly surrounded by hardwood forest with a total area of about 0.6 ha.

The forest overhangs the stream and direct litter fall was measured by collection in boxes placed along the stream bank. Litter includes not only leaves, but also branches, bud scales, flowers, fruits and the exuviae and droppings (frass) of leaf-living animals. Although most litter fall is in autumn, it is not negligible at other times of the year, for there is a continual turnover of leaves even in summer; bud scales fall in spring and frass is a constant source in summer. Wind and the weight of snow break off tree branches in winter.

Litter is also blown sideways into the stream along the forest floor, and was collected in traps placed at right angles to the stream. Rain dripping directly into the stream from the overhanging leaf canopy in summer picks up dissolved organic matter exuded from the leaves and from leaf insects. This throughfall was collected and measured. At the top of the stream section, the entry of organic matter carried from upstream was measured as: coarse particulate organic matter (CPOM), greater than 1 mm in size; fine particulate organic matter (FPOM), less than 1 mm size, but collectable by filtration through a filter of pore size 1 μm; and dissolved organic matter (DOM), which passed through such a filter.

Dissolved organic matter presents few problems as its concentration is relatively steady and its total contribution can be calculated easily if the discharge of the stream is known. Fine particulate organic matter and, more so, CPOM present problems of measurement, because they enter the stream in pulses after thaws and rainstorms. Very frequent sampling was necessary to measure FPOM. Even this was unreliable for CPOM when it comprised merely the spreading of a 1 mm mesh net across the stream for a short period each week or two. Coarse particulate organic matter had to be continuously collected in a concrete ponding basin built into the stream. This collected all CPOM passing and gave a value 20 times greater than that obtained by regular discrete sampling.

Dissolved organic matter entered Bear Brook from soil seepage. Samples of this water were collected from seeps for analysis and its volume determined from the difference between the total amount of water entering the stream at the top of the stretch and that leaving it at the bottom. Finally, photosynthesis of the moss population of the stream (algae and higher plants were scarce) was estimated from the rate of production of oxygen by moss enclosed in glass bottles (Chapter 5).

Conversions of energy through animal consumption were estimated from biomass measurements and productivity data from other sites. Although these estimates were crude (the methodology is discussed later), the picture of energy flow in the stream would be little altered if they were even 10 times in error. Table 4.3 shows the energy budget that has been constructed. Most notable are the high contributions of litter (43.7%), particularly direct leaf fall, and of dissolved organic matter (46.3%). Autochthonous primary production was negligible in this stream.

The outputs of energy show some startling features. First, if all the measured outputs (CPOM, FPOM, DOM and utilization by animals) are added, a total of 4013 kcal m^{-2} year^{-1} was accounted for, whereas 6039 kcal m^{-2} year^{-1} entered. The difference, 2025 kcal m^{-2} year^{-1}, must be attributed to the respiration of microorganisms feeding on the organic matter entering

Table 4.3 Annual energy budget for Bear Brook, New Hampshire.

Inputs (kcal m^{-2} year^{-1}) [%]		Outputs (kcal m^{-2} year^{-1}) [%]	
Direct litter fall		Transport downstream	
Leaves	1307 [22.7]	CPOM	930 [15]
Branches	520 [8.6]	FPOM	274 [5]
Miscellaneous	370 [6.1]	DOM	2800 [46]
Side-blow litter	380 [6.3]	Respiration of microbes	2026 [34]
Throughfall (frass exudates)	31 [0.5]	Respiration of invertebrates*	9 [0.2]
Transport from upstream			
CPOM	430 [7.1]		
FPOM	128 [2.1]		
DOM	1300 [21.5]		
Groundwater DOM	1500 [24.8]		
Moss photosynthesis	10 [0.2]		
Total	6039 [99.9]		6039 [100.2]

*Obtained by difference.

the stream and thus processing about a third of it. The remaining two-thirds were exported for similar processing downstream. Such streams appear to be relatively efficient processing factories for the large amounts of organic matter they receive. The next section considers the mechanics of these processes.

4.3.3 Mechanics of processing of organic matter in woodland streams

Leaf litter is very different, chemically, from living vegetation. Labile, reusable substances have been translocated back into the perennating organs and what is left is largely cellulose and lignin, plus polyphenols, which may be metabolic waste products. Microorganisms have already colonized the litter before it has fallen and begun the decomposition, which, for most leaves, will be completed on the forest floor. Once litter has entered the water, most soluble matter is leached out within a few days or even hours [460]. The speed of this depends on the leaf species. Deciduous leaves are leached more rapidly than those of conifers. This DOM, plus that washed out of the catchment soils, may be taken up by microorganisms on the stream bed or washed downstream and used there. Some may be precipitated as part of the organic film on stones or aggregated into fine particles by apparently physicochemical processes, and may thus join the FPOM.

In the water, the litter will be colonized by aquatic microorganisms and, over a few weeks, an intricate succession of fungi appears in it. Bacteria are present but do not appear to be important. The fungi are largely from a group called the aquatic hyphomycetes. Their spores are often tetraradiate (Fig. 4.7), which favours their sticking to the litter like small grapnels when swept against it by the stream flow [468]. Leaf litter provides only part of the needs of the fungi. Although it is rich in carbon and energy, it contains little nitrogen and phosphorus and the fungi must obtain much of their supply from the water. The nitrogen-to-carbon ratio of the colonized litter increases as the fungi build up their biomass [347, 517, 518]. This is important for the next stage in decomposition by invertebrate animals, for the fungal biomass, rich in nitrogen, is a 'better' food for invertebrate animals than uncolonized litter. The speed with which different sorts of litter are consumed by invertebrates largely depends on their food contents.

4.3.4 The shredders

Mechanical abrasion breaks down some of the leaf litter to FPOM, but much of it is chewed by coarse particle-feeding invertebrates, the 'shredders', which bite out the softer parts – between leaf veins, for example – leaving the vascular skeleton for later abrasion or consumption [181, 182].

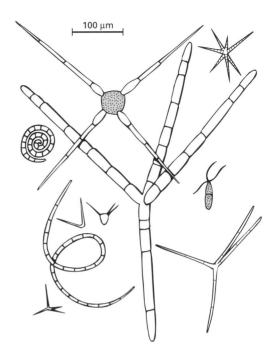

Fig. 4.7 Spores, mostly of aquatic Hyphomycete fungi, drawn from a sample taken below Sezibwa Falls, near Kampala, Uganda (modified from Ingold [478]).

Shredders include insect larvae and nymphs and Crustacea. Much work has been carried out on the crustacean *Gammarus*, which is easily maintained in the laboratory. When *Gammarus* were offered fungally colonized leaves of elm (*Ulmus americana*), alder (*Alnus rugosa*), white oak (*Quercus alba*), beech (*Fagus grandifolia*) and sugar maple (*Acer saccharum*), they showed a preference roughly in the same order as that in which the leaves support fungi – elm, maple, alder/oak, beech. This same order of preference appears to be shared by stonefly and mayfly nymphs [518]. The intrinsic properties of the leaves themselves help determine these preferences, for the order is also maintained if uncolonized litter is offered. The previous history of the leaves may also be important. Leaves from trees of paper birch (*Betula resinifera*) in Alaska that had been browsed by moose 2 years previously were found to decompose more rapidly than those from unbrowsed trees. They contained fewer tannins and had a greater nitrogen content [470].

Preference is always for colonized over uncolonized leaves, however, and can be influenced by artificial inoculation of particular fungi [40, 41]. Although the leaf packs also provide shelter from shearing currents, experiments using artificial plastic 'leaves' have shown that the prime role of the packs is as food rather than as microhabitat [817], and studies using radioactively labelled material have shown that much of the energy assimilated comes from the leaves rather than the microorganisms [553]. The role of fungi in the processing of leaf litter appears similar to that of the sandwich fillings commonly used by us to increase the palatability of bread.

There are usually several species of shredder present, and preference for different sorts of 'sandwiches' may explain why they can all coexist.

Wood may constitute 15–50% of the litter in streams surrounded by deciduous woodland, and even more in conifer forests. It takes years to decades to break down, compared with weeks to months for leaves, but it provides a habitat for specialist xylophages (wood lovers), including midges, beetles, caddis- and crane-flies, although at comparatively low biomass. The microorganisms rotting the wood are probably the main food sources for these invertebrates [13, 14, 918].

Microorganisms can degrade leaf litter alone, but the process is accelerated by shredder invertebrates [181], such that as much as 1.5% day^{-1} of the litter is converted to animal tissue, carbon dioxide or FPOM. Fine matter results from abrasion by water movement and from waste during feeding. It also includes the faeces of the shredders and material derived from the soils of the catchment. Fine particulate organic matter is additionally colonized by microorganisms as 'new' surfaces are exposed and forms a food source for a second series of invertebrates, the 'collectors', which collect the FPOM by filtration or deposit feeding. Sometimes half of the FPOM they ingest consists of the faeces of shredders.

4.3.5 Collectors, scrapers and carnivores

Finer debris is a rich food source for those invertebrates that can use it. Most are filter or deposit feeders, although some scrape the particles from surfaces, together with the attached algae growing there, where the current is slow enough to allow some deposition (Fig. 4.8). Typical scrapers include snails and freshwater limpets, which rasp at the rock surfaces with toothed organs called radulas, and some caddis-fly larvae and a few mayfly nymphs whose mouthparts have stiff bristles with which they scour the rock. Deposit feeders, which include burrowing dipteran fly larvae (particularly chironomids) and some mayfly nymphs, inhabit pockets of sediment in areas of slack flow.

Filter-feeders sometimes have fringes of fine hairs on the mouthparts (blackfly larvae) or legs (some mayfly nymphs), in which particles collect before transfer to the mouth. Others (some caddis-fly larvae) construct nets between stones and the bottom (Fig. 4.8). The nets may operate as filters, snares or deposition traps. A group of freshwater prawns in Dominican streams [299, 300] (Fig. 4.9) have long bristles on a pair of limbs, the chelipeds, near the front of the animal. These bristles are delicate but can be held together like the tip of an artist's wet paintbrush. They can then be used to sweep organic particles from surfaces, and other appendages transfer these to the mouth. Alternatively, the bristles can be expanded into a fan, which is held into the current and which, with the help of very fine setules on the bristles, acts as a filter. Apart from their intrinsic interest,

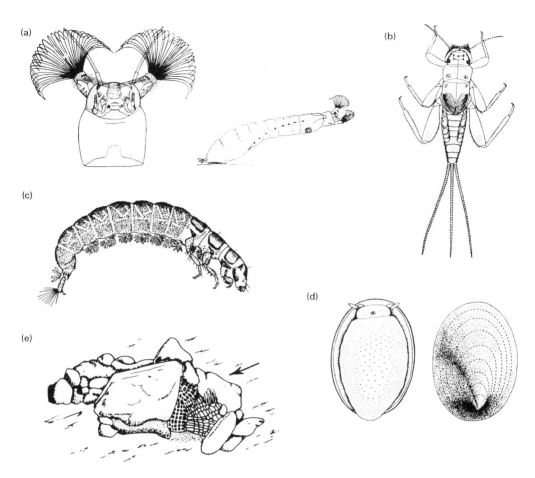

Fig. 4.8 Among the various feeding guilds of animals in streams are scrapers of rock and other surfaces, such as the freshwater limpet, *Ancylus fluviatilis* (d), and collectors of fine particles from the water flow. These include the blackfly larva, *Simulium* (a), which collects particles on its head fans, the African mayfly nymph, *Trichorythus* (b), which filters with brushes on its mouthparts and the caddis-fly larva, *Hydropsyche* (c), which spins a net across the entrance to a chamber of small stones (e), which it has constructed. (Based on Hynes [459] and Fryer [302].)

these prawns remind us that classification of stream (or any other) animals into feeding modes is but a convenience and that the categories are not always distinct. For example, shredders, such as *Gammarus*, thrive better in laboratory experiments if allowed access to the FPOM of their own faeces, as well as to fungally colonized leaf material, than if given the latter alone.

The food web of wooded upland streams (Fig. 4.10) is completed by carnivores, both invertebrate and vertebrate. The invertebrates include leeches, a variety of insect larvae and water-mites (Hydracarina), among others, and, just as shredders and collectors may ingest animal material with their predominantly detrital and fungal diet, the carnivores may also

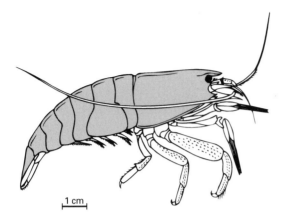

Fig. 4.9 *Atya innocous*, a stream-living prawn from Dominica. The two foremost appendages have brushes, which can be used to sweep deposited detritus from rocks or which can be held like reversed umbrellas against the current to filter particles from it. (From Fryer [300].)

1 cm

have a wider range of food than their name suggests. All the invertebrates are potential prey for fish in the stream. These may include small bottom-living fish that are permanent denizens and migratory fish, such as the salmonids, which spend only the early juvenile and final spawning parts of their life histories in these systems.

4.3.6 New Zealand streams

The idea of streams as factories for the orderly processing of allochthonous organic matter is an attractive one. Based on their experience with mostly wooded North American streams, Vannote *et al.* have created the 'River Continuum concept' [981] (Fig. 4.11) to summarize information about this idea. They see the life histories of shredders and collectors as being synchronized to take advantage of the annual pulses of incoming organic material – the collectors timed slightly later than the shredders. In the upstream sections, the ratio of *in situ* photosynthesis (P) to total community respiration (R) is found to be less than 1, indicating a dependence on the allochthonous matter. Further downstream, when the stream widens and is not so darkened by overhanging forest, so that the stone surfaces may be colonized by algae and mosses, the P : R ratio may be greater than 1. Later, when the river becomes large and is more likely to collect inorganic silt, which decreases light penetration, the P : R ratio may fall again, (Chapter 5).

The river continuum concept has attracted much support, and most stream systems probably do depend, to some extent, on allochthonous organic matter derived from the catchment. Whether the systems are quite so well ordered as the river continuum concept would suggest ought, however, to be questioned. Streams in New Zealand, for example, seem not to fit the continuum concept. Winterbourn *et al.* [1020] point out that New Zealand has a low timber-line, with many streams having great lengths above it, the slopes are steep, the rainfall is heavy and irregular and the

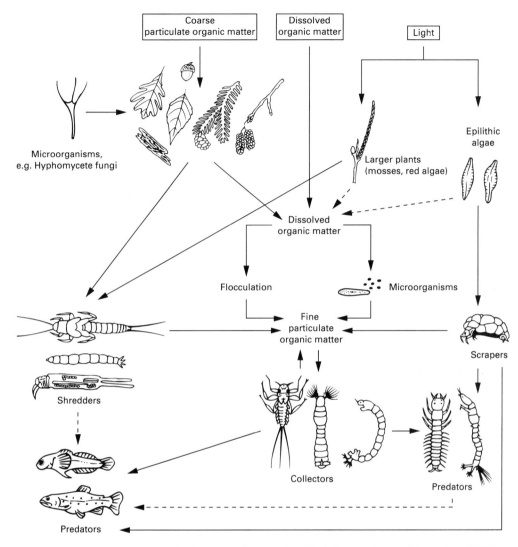

Fig. 4.10 Food web of an upland stream (modified from Mahan and Cummings [625]).

stream bottoms hence do not readily retain coarse litter. The forests of southern beech (*Nothofagus*) and podocarp hardwoods are evergreen, and the litter they produce is not concentrated into the autumn. It is produced continually in small amounts and is very woody and difficult to decompose.

The streams therefore lack a major source of CPOM and the regular changes that would favour an orderly processing of it. There are few shredders, so that the collectors feed on material eroded from the soils. In general, this is inorganic, so Winterbourn *et al.* [1021] emphasize the organic layer on the stones as a major food source, and they point out that many New Zealand stream invertebrates are opportunists, taking different sorts of food as they become available.

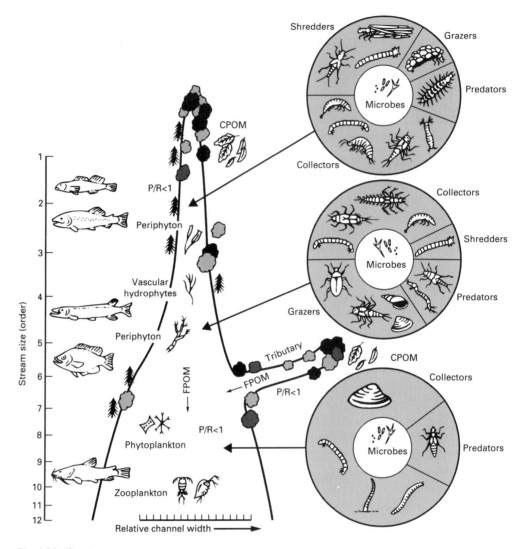

Fig. 4.11 The river-continuum concept of Vannote *et al.* [981]. Changes in the community are given in relation to the order of stream. In river systems, the smallest streams with no tributaries are called first-order streams. Second-order streams are formed by the joining of first-order streams, third-order by the joining of at most second-order streams and so on. The sequence of typical fish and the major primary producers is shown to the left of the diagram. P stands for autochthonous primary production, R for community respiration. CPOM and FPOM are coarse and fine particulate organic matter, respectively. Circles to the right show typical bottom-invertebrate communities.

Their life histories are not synchronized either, so that there is a further contrast with the North American streams. Winterbourn *et al.* [1020] do not deny the importance of allochthonous matter – the organic films on which the New Zealand stream animals depend are derived partly from absorption of DOM of soil origin – but they (and others [540]) question any global generality of the North American scheme. The uncertainty is compounded because we do not know what most streams would have been like under pristine conditions. It might seem certain that streams of arid areas were flanked by the scrub and grassland that characterize most of their catchments. However, had there not been overgrazing and stock trampling of banks, interruption and reduction of the flow by land uses leading to greater evaporation in the catchment or abstraction of upstream water for transfer elsewhere for irrigation, it is possible that most streams would have had wooded banks. Even in dry regimes, the stream environs have high water-tables capable of supporting a strip of woodland. The effects on a stream of a thin strip of riparian forest may not have been so very different from that of more extended woodland in the catchment. Our views may be strongly determined by the currently altered systems with which we have to work.

4.3.7 Fish in upland streams

To many people, fish are the most significant features of upland streams. Large amounts of money are spent in the maintenance of fisheries for salmonids, such as the Atlantic salmon (*Salmo salar*), brown trout (*Salmo trutta*), coho salmon (*Oncorhynchus kisutch*), chinook salmon (*Oncorhynchus tshawytscha*) rainbow trout (*Salmo gairdneri*) and brook trout (*Salvelinus fontinalis*). Some of these, particularly the *Salmo* species, are migratory fish, which make most of their growth in the sea but move into upland waters to breed. They are narrow-bodied, streamlined fishes, able to swim, sometimes for thousands of kilometres, up river and, in their juvenile stages, capable of bursts of high speed to catch drifting prey or flies dipping at the surface to lay eggs.

The requirements of these fishes (Table 4.4) are for highly oxygenated, cool water, especially for breeding. Fast flow, which maintains the bottom clear of silt, and rising waters stimulate the upstream movement of the adults. The eggs are usually laid in depressions excavated in the stream gravel and then covered to prevent their being washed away or eaten. Sites

Table 4.4 General habitat requirements of some adult salmonid fish species (based on Templeton [952]).

Fish	Optimum temp. for growth (°C)	Max. temp. tolerated (°C)	Spawning temp. (°C)	Min. O_2 (mg O_2 l^{-1})	pH
Atlantic salmon	13–15	16–17	0–8	7.5	5–9
Brown trout	12	19	2–10	5	5–9
Brook trout	12–14	19	2–10	4–4.5	4.5–9.5
Rainbow trout	14	20–21	4–10	4–4.5	5–9

are chosen where the water flow infiltrates the gravel, bringing well-oxygenated water continually to them. Silting is fatal, for it blocks this flow and the relatively large eggs asphyxiate; so also is any major reduction in flow that leads to the gravel beds being uncovered.

Salmonid fish are not, however, the only, or even the most numerous fish in the world's upland streams. Fish from many families (Fig. 4.12) [459] are well fitted for life close to the stream bottom. These include cyprinid fish, catfish, sculpins and gobies. In Europe, the bullhead (*Cottus gobio*) and loaches (*Nemacheilus* and *Cobitis* spp.) are found. In North America, a very large number of species of percids (e.g. *Percina*, *Ammocrypta* and *Etheostoma*),

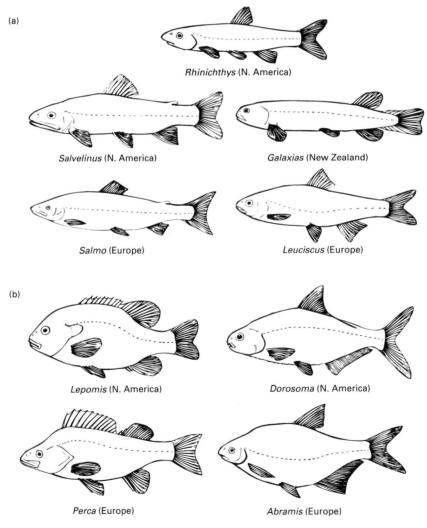

(a)

Rhinichthys (N. America)

Salvelinus (N. America)   *Galaxias* (New Zealand)

Salmo (Europe)   *Leuciscus* (Europe)

(b)

Lepomis (N. America)   *Dorosoma* (N. America)

Perca (Europe)   *Abramis* (Europe)

Fig. 4.12 Fish from relatively swift-flowing streams (a) and from more sluggish waters (b) (based on Hynes [459]).

often called darters, are typical. Bottom-living fish contrast with the salmonids, with their generally rounded body, a flattened underside and an arched back in cross-section. Their mouths are often turned downwards, which eases bottom-feeding, and their pectoral fins may be muscular and spiny and used to wedge them across crevices in the stream bottom.

Sometime, the fins move water from underneath the fish, so as to keep it on the bottom, or act as foils, against which the current presses, to the same end. The fins may be modified as suckers or friction pads. The swim-bladders – buoyancy devices in midwater fish – are much reduced and the skin is dark and mottled, perhaps to provide camouflage against larger, predatory, game fish. These small fishes also depend on swift currents for the successful production of young, for they lay eggs in small piles of gravel or stick them in groups under flat stones.

4.3.8 The Atlantic salmon

The Atlantic salmon is a well-known species. Some details of its life history will illuminate the problems of its fishery, to be discussed later. Atlantic salmon spawn in rivers in the USA, Canada, Iceland, Norway, Ireland, Great Britain and France and later mingle in the North Atlantic off Greenland. The eggs hatch in April–May after 70–200 days in their gravel nest, or redd.

On hatching, the larvae feed from their yolk sacs for about six weeks and then eat small invertebrates. They are then called parr and have a distinctive alternation of blue-grey 'finger marks' and red spots on their sides. The parr grow slowly and become silvery in colour, but, after 1–5 years, they move down to the river mouth and, now called smolts, feed on small fish and crustaceans. Eventually, they move into the ocean and grow rapidly on sand-eels, herring, sprat and other fish and crustaceans

Atlantic salmon spends 1–4 years in the sea and then migrates back, often to the river in which it was spawned. Subtle chemical differences in organic compounds released by related fish to the river water may allow it to recognize this river system, but there is some mixing of fish from different systems. Most adult salmon move into the rivers in late winter. They are fat but do not feed, although they retain for a time a reflex to bite at suddenly appearing prey; this is the basis for the usual recreational fishing method for them, in which an artificial fly is cast into the water for them to take.

Movement up river may mean their jumping waterfalls up to 3 m high and ascending longer rapids in stages. During the journey, the fat is used up, the sexual organs ripen and secondary sexual characters appear; the male develops a hook or kelp on the lower jaw. Mating pairs choose suitable patches of gravel in 0.5–3 m of water and the female excavates the redd by flapping her tail vigorously to form a hollow 10–30 cm deep and up to 30 cm long. The animals lie side by side, with violent trembling and jaw-gaping; the egg are shed and fertilized and then covered with gravel as the female excavates another redd upstream. The eggs are slightly sticky and denser than water. Spawning and migration make great energy demands on the salmon; the eggs may account for 25% of the body weight. After spawning most of the fish are very weak, only half their original weight and prone to fungal infection or stranding in shallow water. Most soon die as they drift downstream, although a few (about 5%) reach the sea, where they quickly recover and, called kelts, may return to spawn a second time in the next year.

4.3.9 Animal production in streams

Freshwater ecologists have spent much time measuring production, but such measurements, done well, are very time-consuming. They are needed for a better understanding of the ecosystem and for managing fisheries. Production measurement essentially involves counting the animals per unit area at successive times, measuring the growth the survivors have made between times and summing the product of these.

Sampling methods for invertebrates include nets, lift samplers and placement of trays of cleaned substrata. Most interest has centred on the macroinvertebrates, those more than about 3 mm in size when adult, and the net mesh most commonly used, about 1 mm, reflects this. The net is held against the bottom, facing upstream, and the bottom substratum is disturbed by feet or hands to dislodge animals into it. A known area may be sampled or, for relative measurements, the sampling goes on for a set time.

In sand and gravel, animals may penetrate to much greater depths (2–30 cm) than are sampled by net [156]. Usually, such deep-living animals are very small microcrustacea and nematodes and this has contributed to relative neglect of them. They can be sampled by first inserting a cylinder into the stream bottom to delineate the sampling area. Compressed air is then used to dislodge a mixture of water, air, sediment and animals through a series of coarse and fine nets. Some animals may be damaged by this.

Artificial substrata, usually comprising cleaned rocks fixed to a tray, may be left on the stream bed for several weeks for them to become colonized. Removal of trays, once colonized, allows a series of samples to be taken at intervals. The artificial substrata used should ideally closely reflect the natural substrata, but, for comparative purposes, the samplers may comprise simply piles of flat plates, spaced by washers and bolted together. Whatever the method used, a large number of separate samples must be randomly taken for numbers of an animal to be determined with reasonable precision, because stream animals are very heterogeneously distributed.

Fish are difficult to sample. For stony rivers, the best method uses electrofishing. An electric field of 200–300 V is created in the water between two electrodes, powered by a portable generator. With alternating current, the fish are temporarily stunned and removed from the water by a net for measurement and recovery before being replaced. With direct current, the fish swim towards the positive electrode and must be removed by net before they touch it. Sections of stream can be isolated by nets to prevent escape of fish. Most fish can be captured, but smaller ones respond less to the electric field than larger ones. The method is also potentially dangerous to the operators, who can be stunned or killed if they are not fully insulated.

If there were no losses, production could be measured simply by the change in weight of a population over a period. Much of the production may drift or be eaten, however, and a more sophisticated approach is needed. The basic data are numbers of an animal per unit area at a given time and the mean individual weight of the animals. It is tedious to weigh all the animals from a large enough sample to be statistically acceptable, so a previously determined relationship between length of animal or width of head and weight is often used. The simplest case is for populations that develop from eggs all laid at the same time. These are known as cohorts.

If the cohort begins with N_1 individuals of mean weight w_1 on hatching, the initial biomass will be $N_1 w_1$. At a future time, t, some animals will not have survived, but the remainder, N_2, will each be heavier, with a mean weight w_2. The total production will be that of survivors, $N_2(w_2 - w_1)$, plus that of those which died:

$$0.5[(N_1 - N_2)(w_2 - w_1)]$$

The assumptions are made that all those that died were lost halfway through the period and that weight per individual increased linearly during the period. Sampling at sufficiently frequent intervals reduces any error from this.

Expressed graphically, as it was first used (for fish) by Allen [10], the area under a graph of numbers versus mean weight, the Allen curve, gives total production for a cohort (Fig. 4.13). The sum of production of each cohort present gives total production. The method works well (given careful sampling) for organisms with synchronous reproduction or for those which can be easily aged – for example, by markings on the shell in bivalve molluscs or scales or the annual rings on the ear bones (otoliths) in the case of fish.

For animals which reproduce continuously and which cannot be readily aged, this method cannot be used. For these, determination of growth rates by serial weighing or measurement of captured individuals, in laboratory conditions resembling as closely as possible those of the habitat, may be used. Production is then given by multiplying the field biomass by the appropriate percentage increase in weight per day. Reviews of available methods are given in Benke [57] and Rigler and Downing [820].

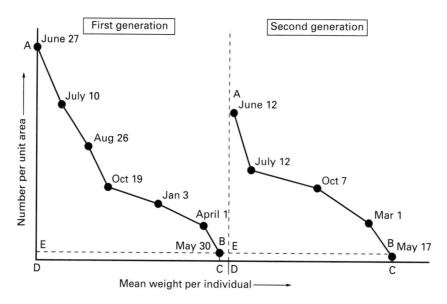

Fig. 4.13 Allen-curve method for determining the production of an animal population whose members can be recognized as distinct cohorts, each starting life more or less simultaneously. Numbers per unit area are counted on each of a series of sampling dates and plotted against mean weight of animals on these dates. Area ABCD represents total production; ABE the production of animals that were eaten or died before reproduction; BCDE the production of adults surviving to breed; and ABCD : BCDE the turnover ratio.

4.4 Stream communities

Studies of particular species show how a variety of complementary life histories and diets may allow coexistence of species in a given stream. This introduces the topic of the communities of organisms in streams and the factors which determine which particular species grow where. Most interest has centred on animal communities.

Some scientists believe that habitats are in general stable enough for their communities to be determined by competition between species, leading to a close packing of the small realized niches of many species. Such communities would be resistant to invasion by new species and similar in composition from place to place under the same general conditions. Such communities would also be rich in species, none of which would be extremely abundant, because the competition of others would prevent their unbridled expansion.

Others think that habitats vary a great deal, that disturbances of various kinds (e.g. unusual temperature, drought) frequently open up the habitat by eliminating particular organisms, which might or might not recolonize. Such communities, with organisms coping with wide ranges of conditions, would have fewer species. Each would occupy wide realized niches, with less competition between them because of the more or less continual availability of a temporarily unoccupied habitat. Such communities would have a random element in their composition and be less likely to be similar from place to place under the same general conditions. Differences between communities should also reflect differing physicochemical conditions to a greater extent than if competition shaped their composition. Real communities reflect both these viewpoints and the effects of predators are also important in structuring communities. The balance of importance of these factors may differ from time to time and from place to place in streams, as in all ecosystems.

4.4.1 Do distinct stream communities exist?

To answer this question needs a listing of the communities in a very large number of streams and a comparison between them. Wright *et al.* [1033, 1034] have analysed net samples of macroinvertebrates from 340 stony streams in Great Britain. The species found (total 587) included 29 snails, 20 bivalve molluscs, 54 oligochaete worms, 16 crustaceans, 34 mayflies, 26 stoneflies, 24 water-bugs, 88 water-beetles, 87 caddis-flies and 161 true flies (Diptera). Data were also collected for physical conditions (stone size, channel width and depth, distance from headwater) and water chemistry. All sites were unpolluted in a conventional sense. They did not receive discharge of significant amounts of toxic waste or sewage, but many were surrounded by land used agriculturally so that most had some influence of human activity.

The animal communities were analysed first by ordination, using detrended correspondence analysis (DCA), which compares each community with each of the others and then calculates a position for it relative to three graph axes. The position is determined by the community's similarity to each of the other communities. This creates a three-dimensional graph, with each community appearing like a star in a constellation. The constellation was a continuous one. Groups of communities did not separate out distinctly. There was continuous change, and this continuum could be related to the physical conditions. The trend was best correlated with the size of the bottom stones, the slope of the bed, concentrations of phosphate and nitrate and, to a lesser extent, distance from the source of the stream. The features best correlated with the change in community were ones that steadily change as the river moves from upland to lowland. This analysis suggests that the compositions of the communities owe most to the steadily changing physical conditions in the streams.

However, a separate analysis was also made, called two-way indicator species analysis (TWINSPAN), which attempts to classify the communities by progressively dividing the total into subgroups of two by means of the presence or absence of a particular (indicator) species in the two groups.

This can be continued through a series that creates first two, then four, then eight, then sixteen groups until ultimately each original community rests in a separate classification of its own. That is the point at which the analysis started!

The aim is to see if such classification is possible, so the division into groups is stopped at a point where there is a considerable imbalance in the numbers of communities placed in each of the two groups following a division. This occurred after the fourth level of division, so the analysis was interrupted when 16 groups had been determined. This is arbitrary, and groups so formed cannot be used as evidence that discrete communities, determined through competition mechanisms, exist, but it does confirm the relationships found by the ordination.

The 16 groups (Fig. 4.14) reflect, from left to right, a progression from upland to lowland sites. Stoneflies, many genera of mayflies, caddis-flies and some dipterans dominated the uplands, and worms, molluscs, flatworms, leeches, crustaceans, water-bugs and other Diptera the more lowland stony streams. The differential distribution of some of the insect groups and the Crustacea and molluscs might reflect the different pathways by which these animals have invaded fresh waters – in the former case by land and air and in the latter from the sea via the lowland estuarine stretches of the rivers (see Chapter 2).

A further test of the relationship was made by using multiple discriminant analysis, which tests whether the community composition can be predicted from a knowledge of the environmental variables alone. This was used to predict which group of the 16 generated by TWINSPAN ought to occur at a particular site and to compare the prediction with what was actually found. Predictions were correct in 76.1% of cases, with the correct group being given as second most probable in a further 15.3%. Again, this seems to confirm a strong relationship between the community and the physicochemical environment.

A study on streams in a heathy, sandstone area, in the Ashdown Forest of Sussex, has further illuminated this issue. The streams were acid (pH 4.8–6.5), with variable calcium and nitrate concentrations because of differences in local land use. They were also iron rich (0.2–2.8 mg l^{-1} of soluble Fe), and had stony bottoms. Townsend et al. [969] ordinated and classified the communities of thirty-four sites. They also used a simpler technique of stepwise multiple regression, which relates the degree to which one factor – a feature of the community, such as number of species in it – is correlated with another – say, pH or calcium concentration. It then goes on to include information on additional factors and calculates how much (as a percentage) of the variability in the first feature is explained by a lengthening list of these.

Again, a relationship was found with the physicochemical environment. The number of species was primarily related to pH, with 48% of variation

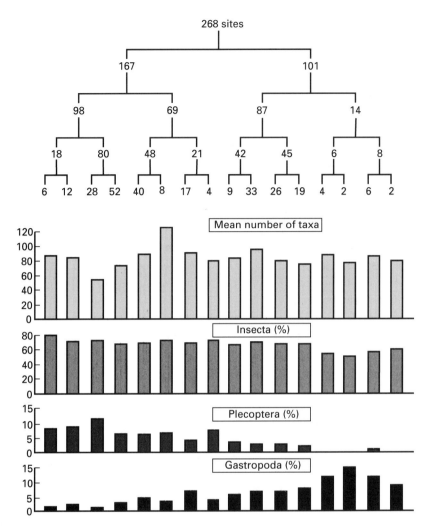

Fig. 4.14 Classification by two-way indicator species analysis (TWINSPAN) of invertebrate communities of rivers in Great Britain. Below the dendrogram, which shows the number of sites falling into each category, are the mean number of taxa and the percentage number of taxa (usually species) of certain invertebrate categories in each of the 16 groups in the lowest level of the dendrogram. (From Wright *et al.* [1034].)

in it accountable by the pH alone. For 20 species (out of 137), there were also significant correlations with physicochemical variables. The pH explained 49.3% of the variation in numbers of *Gammarus pulex*, the freshwater shrimp, pH and Ca together 56.2% and pH, Ca and nitrate 60.1%. For a stonefly, *L. nigra*, successively, pH, July temperature, iron concentration and maximum discharge explained 65.5%, 79.8%, 85.5% and 87.8% of the variation. Increasing numbers were positively related to iron concentration and negatively to the other three variables. Fish were absent from the more acid sites, but brown trout, bullhead and stone loach increased in numbers with rising pH.

The ordination and classification confirmed the importance of pH, with July temperature and discharge being secondarily important. A dominance of stoneflies in the more acid waters and *Gammarus*, *Simulium* and others at higher pH reflects the trends found by Wright *et al.* [1033]. These trends, determined from large numbers of data and very comprehensive computer-based analyses, it might, somewhat uncharitably, be pointed out, were recognized some time ago by naturalists working intuitively [595]. As often, Ecclesiastes 1:9 provides a succinct commentary.

The modern analyses add detail, though not necessarily increased comprehension. The existence of a correlation – with substratum type, in the one case, and with pH, in the other – does not explain the mechanism. In both cases, however, physicochemical factors explained substantial amounts of the variation in community composition, whatever the ultimate mechanism might be.

4.4.2 Competition in structuring stream communities

Some variation was not explained by the physicochemical environment, leaving the possibility that competition mechanisms might have some importance. Hildrew *et al.* [414] thought that, if competition is important, the sizes of niches should decrease with increasing species richness. The measurement of niche size is difficult, since the niche is an abstract concept with many dimensions. However, size of organism is generally related to the size of food that can be taken, and the greater the variation in size among the members of a species the greater the variety of food that can be taken. This might be used as a reflection of the overall 'niche width'.

If competition is important, it should lead to greater specialization between potentially competing species and thus to a more restricted size range of each. For a group of stoneflies feeding on fine detritus in streams in the Ashdown Forest, the variation in size was measured as the coefficient of variation (standard deviation as a percentage of the mean) of the width of the head at sites of different species richness. The variation was decreased as species richness increased. A further test was made by calculating the degree of overlap in diet (a measure of niche overlap), as expressed in the overlap in head sizes between species.

Complete overlap would suggest that the species are either competing so vigorously that one must soon be eliminated or that they are not competing at all – that food supplies are sufficient to support both – if they coexist indefinitely. However, if they are competing, natural selection should lead to specialization and a gradual decrease in overlap to zero. For a series of stream communities in the Ashdown Forest, the niche overlap did significantly decrease as the numbers of potentially competing species of detritus-eating stoneflies increased.

Erosive stream communities therefore seem to be structured primarily by physicochemical factors [261, 829] but secondarily by competitive mechanisms. There are also significant effects of predation by fish on invertebrates, which can result in maintenance of thick algal films on rocks in streams [788], and grazing by tadpoles can result in changes in the invertebrate community [542]. The problem will remain of explaining how the physicochemical competition and predation factors act in particular places. A fuller discussion is given in Allan [5].

4.5 Erosive streams and human activities

Erosive streams are often remote and, as a group, are perhaps affected least of the main freshwater ecosystems by human activity. However, they are not undisturbed. A simple classification into the effects their communities have on humans and vice versa includes, in the first category, a serious disease, river blindness, in the tropics, and the provision of recreational game fisheries. In the second are the effects of acid precipitation, changing agriculture and forestry and many engineering works.

4.5.1 River blindness and the Blandford fly

River blindness now infects about 40 million people. For some villages in Africa, it is a dominant part of their lives. The disease is caused by a nematode, *Onchocerca volvulus*, whose larvae (microfilariae) are carried by blackflies (buffalo flies), *Simulium* spp., which grow in swift-flowing rivers, where the larvae are attached to stones or sometimes freshwater crabs and collect fine particles from the flow.

The adult blackflies are blood-feeders and gouge a small hole in the human victim's skin from which to feed. They may then introduce microfilariae from a previous victim. The microfilariae develop in the human host to form long adult worms, the female up to 50 cm, the male 2–4 cm, which cause large fibrous swellings. Such swellings block lymph nodes and cause further swelling, as fluid collects around the groin, scrotum or hips. The adult worms live for up to 20 years, continually producing microfilariae, which migrate to the skin, making it dry and papery, and to the eyes. The skin population provides the source for infection of others via the blackflies, while blindness may result as microfilariae move into the cornea and die. This causes a reaction in which opaque protein is produced.

Control of *Simulium* with pesticides has generally been most effective when the pesticide is adsorbed on to the fine particles which the *Simulium* larvae collect. Control is inevitably short-term because particles are washed downstream so quickly. The disease can be treated surgically by removal of the adult worms and by drugs to kill the microfilariae. Medical sophistication is often unavailable in the remoter, poor countries, and it is unlikely that the disease will be controlled soon, if ever.

A local problem is caused by another simuliid, *Simulium posticatum*, the Blandford fly, in Dorset, UK. In taking its blood meal from human victims during May and June, it causes swellings, itchings, ulceration, dizziness and nausea, with more than a thousand people annually affected. The larvae occur mostly along 40 km or so of the River Stour, and attempts to control them, by removing the plants to which the larvae attach, have failed. Adding a bacterium, *Bacillus thuringesis*, to the river has been more useful. The larvae eat the bacteria, which contain a toxic crystalline body. The treatment is effective for about 1 km down river of the point of addition and kills over two-thirds of the larvae with apparently only minor side-effects [539].

4.5.2 Game fisheries

Angling is a second way in which upland rivers affect human societies. Fisheries for large and active salmonids support a substantial infrastructure of tackle dealers, bailiffs, river managers, hoteliers and others. There are also commercial fisheries. In the sea, these use drifting nets, in which the fish become held by their gill covers as they blunder into them at night. In estuaries, beach nets or fixed traps catch the fish migrating upstream.

The simplest fishery to manage is one that is discrete, with the fish confined to a known and accessible area. A river fishery for non-migratory brown trout is an example. A first essential is that good water flow and quality and gravel beds for spawning should be maintained. Secondly, the fishing mortality must be kept no greater than that of natural mortality in the absence of fishing. This mortality is often not known, nor is the extent to which an increased mortality of adults is compensated for by greater survival rates of the young. A conservative approach is desirable.

Fishing mortality can be controlled by issuing licences that limit the number of anglers, the size of fish they may remove, if they catch them, and the fishing season. The latter two ways help protect recruitment by preventing removal of fish that have not yet spawned and minimizing disturbance during the spawning season. Control of natural mortality poses greater problems. The fish are part of an ecosystem, which has values other than as a fishery. It may support otters, fish-eating birds, such as mergansers and cormorants, and piscivorous fish. A wise fisheries manager will accept that some of the fish will be taken by these predators. One of different bent will want to remove natural predators and will meet considerable opposition. Fishermen often believe that natural predators make huge inroads on the fish but, in general, this is not true and greater threats come from habitat deterioration.

The pressures for predator removal can nonetheless be strong. In

Alaskan rivers, the Dolly Varden charr (*Salvelinus malma*) [670] was believed to be a major predator on eggs, alevins (larvae) and smolts of the Pacific salmon, *Onchorhynchus nerka*, particularly when they moved downstream to the sea. Between 1920 and 1941 $300 000 were spent in western Alaska alone as bounty on the killing of the Dolly Varden. The Arctic charr (*Salvelinus alpinus*) was also incriminated. Examination of about 5000 charr stomachs, however, showed only 42 with any salmon remains in them. An expensive and pointless control programme had been based on belief rather than proper data. It was most unlikely that fish at the upper end of a food web would depend much on a similar predator, supplies of which, by the nature of energy loss along food chains, would be relatively scarce.

A fisheries manager may also be unsatisfied with the natural growth rates of the fish. There is little that can sensibly be done about this, for growth depends on many conditions over which the manager can have little control. Food supply and temperature are particularly important. For example, the attainable growth rates of brown trout in the UK are naturally [238] between 60 and 90% of the maximum determined under ideal conditions in a fish nursery [240, 241], because of too high or too low temperatures.

The management of fish that migrate is more difficult, for less control can be exerted in the open sea. When the major feeding area of Atlantic salmon was discovered off Greenland, there was an intense fishery and the numbers of fish returning to spawn in the rivers fell. The fishery has now been regulated by the extension of a restricted zone off the Greenland coast. Attention has passed to sea fisheries for the salmon in coastal waters. There are suggestions that seals remove many salmon off the Scottish coast and that too many fish are taken by inshore drift-net fishermen in England and Ireland before the fish enter the estuaries.

The Canadian government has, with difficulty, banned drift-net fishing off its Atlantic coast, so as to allow stocks to increase [822]. The fishermen came from small communities with a long history of such fishing, on which they largely depended. Although financial compensation was given, there was social discord. A drift-net fishery, though banned in Scottish waters since the 1960s, is maintained in the British Isles off the Northumberland coast, an area of high unemployment so that removal of it would have strong political overtones. A similar situation exists in Ireland. The extent to which the drift-net fishermen off the English and Irish coasts affect the runs of salmon in Scottish rivers is a controversial problem, not helped in its solution by an increase in illegal poaching both at sea and in the rivers.

Angling associations and fly-fishermen are a powerful and articulate group, so firmly convinced that drift-net fishing is responsible for reduced numbers of rod-caught fish that they have suggested a complete ban on Atlantic salmon fishing within 12 miles of the coast [1004]. The sea fishermen are politically much less influential. They are not helped by the fact that a single salmon is worth far more, in economic terms, as a river-caught fish, on which a whole economic infrastructure of management, equipment and accommodation depends, than as a sea-caught food fish [653]. The situation is further complicated by the increasing development of salmon farms, in which the fish can be reared relatively cheaply in floating pens in sheltered coastal waters. In the long run, changes in habitat are more likely to become the key issue for survival of salmon fisheries.

4.6 Alterations to upland streams by human activities

Streams have been altered by changes in their water chemistry (quality) and in the physical structure of the habitat. The former includes acidification [413] and eutrophication – the addition of acid and acid-mobilized metals and of nutrients, respectively. In the latter are the impacts of dam building and alteration of the river flow for the generation of hydroelectric power, the transport of water for industrial use and the lessening of flooding risk downstream. Linking these are effects, such as siltation and changes in land use, which influence both water quality and the physical habitat. Table 4.5 surveys the range of impacts that can affect native salmon and trout populations in UK fresh waters and categorizes most of the problems of erosive streams.

4.6.1 Acidification

Many rivers with catchments on thin, base-deficient soils, have become more acid in the last few decades or sometimes longer. Some neutralization of rain takes place in the catchment soils, but nonetheless there is evidence of rivers with pH values now below 5.0. A group of southern Norwegian rivers had pH values of 5.0–6.5 in 1940 but pH 4.6–5.0 in 1976–78. The pH values may fall to 3.0 after dry spells or after snow melts in spring.

The prime causes of the acidification are release of sulphur dioxide (SO_2) and nitrogen oxides (NO_x) into the atmosphere (see Chapter 3), but there are other possibilities. Catchments may become more acid as the bases in their soils are leached out (see Chapters 3 and 10), so that, without fertilizer and lime additions to soil for agriculture, there is an inevitable acidification over periods of thousands of years [762]. This, however, cannot explain the recent marked increase in acidification.

Conifer planting in the uplands may also lead to acidification. The trees take up and store, either in their foliage and wood or in the refractory litter that accumulates under them, much of the small stock of cations in upland soils. In their doing so, hydrogen ions (H^+) must be released to maintain electrical neutrality. The trees, however, also take up anions and release hydroxyl ions (OH^-), so the net effect is not clear. The large surface area of evergreen tree foliage may act as a collector of 'dry-deposition' particles (see

Table 4.5 Influences of man-induced changes on upland streams, either directly or consequent upon changes further downstream, with special reference to salmon and trout populations.

Activity	Disch	Vel	Temp	pH	DO	Bed	Silt	Tox	SS	Eutr	Shock	Obst	Gen	Overst	Dis	Exots	Overf	Mort
Fish cropping	–	–	–	–	–	–	–	–	–	–	–	–	–	–	–	–	–	+
Fish farms	+	+	–	+	+	–	+	+	+	+	–	–	+	–	+	+	–	–
Fish stocking	–	–	–	–	–	–	–	–	–	–	–	–	+	+	+	+	–	–
River impoundment	+	+	+	–	+	+	+	–	+	–	–	+	–	–	–	–	–	–
River regulation	+	+	+	–	+	+	+	–	+	–	–	+	–	–	–	–	–	–
River water transfers	+	+	+	–	+	+	+	–	+	–	–	+	–	–	–	–	–	–
Abstraction	+	+	+	–	+	+	+	+	+	–	–	+	–	–	–	–	–	–
Land drainage	+	+	–	+	–	+	+	+	+	+	–	–	–	–	–	–	–	–
Channel dredging	–	–	–	–	+	–	+	–	–	–	–	–	–	–	–	–	–	–
Farm and forest effluent	–	–	–	+	+	–	+	+	+	+	–	–	–	–	–	–	–	–
Sewage effluent	–	–	–	+	+	+	+	+	+	+	–	–	–	–	–	–	–	–
Industrial effluent	–	–	+	+	+	+	+	+	+	+	–	–	–	–	–	–	–	–
Extraction of sands and gravel	–	–	–	–	–	+	+	–	+	–	+	–	–	–	–	–	–	–
Change in bank vegetation	+	+	+	+	+	–	–	–	–	–	–	–	–	–	–	–	–	–
Change in stream vegetation	+	+	–	+	+	–	–	–	+	–	+	+	–	–	–	–	–	–
Roads	–	–	–	–	–	+	+	+	+	–	+	–	–	–	–	–	–	–
Vehicles	–	–	–	–	–	+	+	–	+	–	+	–	–	–	–	–	–	–
Entrapment of fish in turbines, etc.	–	–	–	–	–	–	–	–	–	–	–	–	–	–	–	–	–	+

Disch, discharge; Vel, water velocity; Temp, temperature; DO, dissolved oxygen concentration; Bed, bed movement; Silt, siltation; Tox, toxicity; SS, suspended solids; Eutr, Eutrophication; Shock, Mechanical shock (vehicles passing through fords or being washed in streams); Obst, obstruction of movement (dams, weirs); Gen, genetic change (escaped fish from farms); Overst, overstocking; Dis, diseases; Exots, exotic species introduction; Overf, overfishing; Mort, direct mortality.
+, indicates a deleterious effect of the activity expressed on a particular feature of the stream (e.g. discharge or oxygen concentration) or through a particular process (e.g. eutrophication, gentle disturbance).
–, indicates no apparent effect.

Chapter 3), so that, as rain washes these to the ground, the water may become more acid [380, 381]. Nonetheless, examination of changes in the diatom floras of lakes in the Galloway area of southern Scotland has shown that acidification has occurred whether or not the catchment was forested [277, 278].

Most concern for the consequences of acidification has been with the loss of fish stocks, but there are effects elsewhere on the stream communities. Most available information [291] is about physiological effects on individual organisms studied in the laboratory, rather than on whole ecosystems. The latter are more useful. Hall *et al.* [375] acidified part of the Norris Brook in the White Mountains of New Hampshire from April to September 1977 by steadily dripping in sulphuric acid. The pH fell from greater than 5.4 to 4.0. The immediate effect was mobilization of aluminium, calcium, magnesium and potassium from the channel sediments. There was an increase in invertebrate drift within 30 min of acidification, with the pattern and composition of the drift differing from an upstream control section.

The drift rate declined after a few days, largely because the populations of invertebrates had also declined, by as much as 75% for chironomids, tipulids, ceratopogonids and mayflies. Fewer of the latter and of stoneflies and dipteran flies emerged as adults during the summer. In the leaf litter, hyphomycete fungal populations fell and leaf decomposition rates were probably reduced, despite the abnormal spreading of a basidiomycete fungus. Algal growth on the stones increased, probably because of the loss of grazing and scraping invertebrates, and there was an overall decrease in diversity of the stream community. The major fish species, brook trout, migrated from the stream section, although a few, trapped in pools during the summer, survived the low pH with no apparent damage. Such fish were unusual, for acidification usually causes loss of fish, particularly salmonids. These losses come partly from a failure of reproduction and partly from death of the adult fish. The mechanisms involve both direct effects of hydrogen ions and secondary ones of the high aluminium concentrations that accompany pH values around 5.0.

Direct effects are felt particularly during episodes of great acidity following the melting of snow in spring, when salmonid eggs hatch. Although domesticated fish can be conditioned to low pH, wild fish eggs of Atlantic salmon fail to hatch if exposed to pH 4.0–5.5 at the stage where the embryo has developed eye pigment. This is because an enzyme, chorionase, secreted by the larva's snout, needs a pH of 8.5 to dissolve the egg membrane and allow the larva to escape. At pH 5.2, its efficiency is reduced to 10% [768]. Low pH also leads to inadequate storage of proteins in the egg yolk and reduces the number of eggs produced by the adult. Compared with controls at pH 6.7, flagfish (*Jordanella floridae*) produced only 2.1–8.2% of eggs at pH 4.5–5.0 [837]. The pH above which there was no effect on reproduction was 6.5 in this species.

Low pH can also upset oxygen uptake through the gills, and ion regulation. Values below about 5.0 cause an alteration in the permeability of the gills, which allows H^+ to move in and sodium ions (Na^+) to leak out; low blood pH reduces the efficiency of haemoglobin to combine with oxygen and leads to mucus clogging the gills and death from oxygen starvation [291]. In invertebrates, low pH may interfere with calcium uptake, preventing moulting of Crustacea, which require much calcium for their exoskeletons [628]. Crayfish disappear from streams at about pH 5.3 and *Asellus* at 5.2. *Gammarus* spp. need a pH of greater than 6.0. Birds, such as pied flycatchers, feeding on calcium-deficient and aluminium-rich aquatic insects may form thin-shelled eggs, which survive less well than normal eggs [724, 725]. Populations of dippers have declined in acidified streams [738–740] (Fig. 4.15).

(a)

Fig. 4.15 (a) The dipper (*Cinclus cinclus*) is a small bird that feeds on the invertebrates of upland streams. Its breeding success has been adversely affected by acidification, probably acting through the toxic effects of mobilized aluminium. (b) The graph shows the number of fledglings produced per pair compared with water pH in streams in Scotland. Pairs that lost their young through predation, flooding or desertion were excluded from the analysis, which shows a statistically significant relationship. (From Vickery [983].)

(b)

Aluminium ions appear to raise the pH thresholds below which the physiological effects discussed above occur. Toxic aluminium concentrations vary with pH and with species, but 200 µg l^{-1} appears lethal to brown trout at pH 5.0. Harriman and Morrison [381] found that forested streams in Scotland with aluminium concentrations up to 350 µg l^{-1} would not support fish or allow hatching of trout eggs placed in them, while in the River Tywi in Wales, Stoner *et al.* [931] found a correlation between the mortality of fish placed in cages in the river and the dissolved aluminium concentration. Aluminium salts were found deposited within the abundant mucus coating the gills.

The reversal of acidification effects in streams is difficult. The addition of lime (calcium carbonate) has been effective in lakes, but additions to streams are short-lived and widespread liming of acid snow banks would be impracticable and expensive. Restocking of fish is pointless until the acidity is removed, even if the restocked fish have been bred to live at low pH. The whole stream community – and hence the fishes' food supply – is affected. In any case, such responses to symptoms, not to causes, will almost certainly lead to further problems. The best way to solve the problem is to tackle it at source by reducing SO_2 concentrations from industry and NO_x emissions from vehicles. Progress has been made on the former, but the increasing numbers and use of motor vehicles have meant that acidification still remains a problem.

4.6.2 Changes in land use

Many upland streams in the UK may look attractive and 'natural' but nonetheless are severely degraded habitats. The stream in Fig. 4.16 (lower), for example, is much wider than it would be in the absence of overgrazing by sheep, which have removed overhanging vegetation at the edge, leading to erosion, widening and shallowing. The shallow water may become too hot for trout in summer and the lack of overhanging vegetation means no suitable habitat for the trout to lurk and rest.

Many such changes may be seen in upland landscapes. Where natural forest remains, it is increasingly used for timber, with ground disturbance by machinery; forest of exotic species is planted following drainage of peaty soils by the cutting of channels. Upland pastures are being fertilized to increase grass production. In the past 40 years or so, at least 150 000 ha, or 8% of upland moorland in England and Wales, has been variously reclaimed [716]. The ultimate effects on the streams are increases in nitrogen and sometimes phosphorus concentrations and in siltation. There may also be a decrease in discharge, when evapotranspiration of planted forest releases more water vapour to the air than open moorland or grassland. The distribution of flow during the year may be greatly altered, with reduced flows in summer [317]. Deforestation may also increase water

Fig. 4.16 Both of these streams appear 'natural', that in the lower picture perhaps more so than that to the right. However, the lower stream, in upland Britain, is greatly degraded. It is acidified and overgrazing by sheep has removed overhanging bankside vegetation, allowing bank erosion and widening of the channel. The shallowed water warms too much in summer and there is little cover for trout. There is no woody debris to hold litter and create varied habitats for invertebrates. The upper stream, set in a nature reserve in west Wales, is in better shape, at least above and below the ford. There is an abundance of leaf and woody debris available to fall into it and much overhanging vegetation in spring and summer.

temperature with removal of shading. The Atlantic salmon was excluded from sawdust-clogged streams flowing into Lake Ontario in the nineteenth century; reintroductions have failed because the waters are now too warm and the summer flows too low [900, 901].

Modern forestry may unexpectedly reduce the nutrient inputs to upland streams and the lakes they feed, through a mechanism concerning migrating salmon. After spawning, the fish often die and release phosphorus and nitrogen compounds to an environment generally devoid of them. The carcasses must not be washed away for this to happen. Woody debris – trunks, branches and twigs – will retain the bodies. If it is removed to ease the flow and minimize flooding or if it never reaches the stream because of intensive forestry practice, which converts even the brushwood eventually to chipboard, the carcasses are not retained. Cederholm and Peterson [145] found a significant correlation between retention of coho salmon carcasses in streams of the Olympic Peninsula, Washington, and the amount of large organic debris present. This comprised logs at least 3 m long and 10 cm in diameter lying in the water and within a metre of its edge. The carcasses were often taken to the edge by scavengers, such as racoons and black bear, and release of nutrients through mammal excretion could be as important as decomposition by microorganisms. The streams most productive of coho salmon were those with the most debris. The released nutrients support the litter/shredder system, on which the young salmon depend for invertebrate food. Many observations [370, 706] suggest an important role for woody debris and indicate that interference with the natural debris-strewn stream system is unwise.

Forest plantations, paradoxically, may also cause problems [652]. The dense shade cast by introduced conifers, such as Norway spruce, and the refractory litter they produce have reduced invertebrate populations [650]. Afforestation has diminished river flows, clogged spawning gravels and filled deeper pools, favoured by adult fish, with needles. After heavy rain, the drainage channels cut through the plantations allow rapid run-off, causing severe erosion of the bottom. The spates, however, are too brief to serve the upstream migration of salmon, which requires steadily rising waters.

4.6.3 Physical alteration of the streams

Upland areas are sources of water and hydroelectric power. The supply of both is irregular, for it depends on weather and season; the need for electricity by human societies also varies throughout the day. A reliable supply of water and power, recoverable at will, is ensured by damming upland rivers. The reservoirs can also be used to regulate the water flowing downstream, to control flooding in the lower reaches, while the river below the dam delivers the water cheaply from the reservoir to where it will be used. Dammed and regulated rivers are now more common than free-flowing ones.

Damming causes many changes to the river. It sets a blockage to fish movements, upstream or downstream; it may change the flow downstream by making it more irregular, with more extreme fluctuations, in response to the needs of power generation. It may make the flow more even if the reservoir is used for water storage or flood control. The nature of the river bottom will change in response to these changes in flow. The water quality may change too. If water is released to the river from the upper layers of the reservoir, where it has equilibrated with the atmosphere, it may be better oxygenated than if it seeped, only a short time before, from the catchment soils; conversely, if it has come from the deeper layers, it may be poorly oxygenated and contain suspended iron and manganese hydroxides or dissolved hydrogen sulphide (see Chapter 6). The nature of the suspended organic matter will have changed also. Leaf litter will have sedimented out in the reservoir, but planktonic (suspended) organisms may be washed into the river in large numbers.

The first problem, of fish migration, is theoretically easy to solve. Many fish can negotiate barriers of 1–1.5 m, and salmon species can often jump up to 3 m. Most dams are far higher, but fish ladders can break the ascent into manageable steps (Fig. 4.17). These might comprise a series of stepped pools on a gentle incline, beginning some distance below the dam, or climbing, as a spiral, within the dam itself. One problem is to encourage the fish to enter, because the flow down the ladders is often much less than that coming through the dam sluices. A series of diverting baffles may help. Some dams have powered lifts, in which the fish are transported from bottom to top. Scottish law has demanded provision of fish passes on all dams since 1860, but, even as late as 1933, the Grand Coulee Dam on the Columbia River was built without a fish pass and blocked 1600 km of river spawning sites to migrating salmon. If no provision has been made when the dam was constructed, fish can be netted at the foot of the dam and carried by road around it at the peak of the spawning season. Hatcheries may have to replace previous natural spawning areas, perhaps now covered by lake sediment.

Fig. 4.17 To allow migratory fish to move around obstacles such as dams or weirs, fish passes are necessary. These lessen the gradient and provide small steps, each of which is negotiable. More natural designs are preferable to one such as this, which is serviceable but ugly.

Passage downstream through dams may also be difficult. Fish moving into turbine intakes may be killed by the blades, and those taking the fall of perhaps tens of metres over the dam wall will be smashed on the rocks below. Turbine operation can be arranged to minimize damage, but death rates may reach 10–20% even so. A series of dams on a river may thus kill almost all the migrating smolts. The fish may not even reach the dam outlet because the reservoir lacks the strong currents that guide them down river. Again, they can be netted and moved by road, but this is expensive and increasingly recourse is made to a completely artificial system of stocking of the river with adults of non-migrating salmonid fish reared in hatcheries.

The fishes' problems will not be over even when they are in the river below the dam [98]. Although a minimum flow to be released from the dam is usually stipulated by law, the contrasts between low and high flows may be very great and change rapidly, especially with hydroelectric dams. The Kennebec River in Maine experiences flows of about $8.5 \, m^3 \, s^{-1}$ at night, when little power is being generated, but up to $170 \, m^3 \, s^{-1}$ by day, when the turbines are working fully. Twenty-five per cent of the river bed may be uncovered at night, stranding small fish, while the sudden high flows may damage them by abrasion against rocks.

4.6.4 River-regulating reservoirs

The problems of fish passage are very obvious but there are more subtle effects of dams on fish through food supply. For example, the Cow Green reservoir was built across the River Tees in the north-east of England to store water for industrial use. The estuary of the river is too polluted to encourage migration of salmon, but the upstream reaches, lying in open peat moorland, support populations of non-migrating brown trout and some smaller fish, including bullheads. The Cow Green dam was built just above a waterfall, Cauldron Snout, over which the Tees flows and then passes for several hundred metres before it is joined by a main tributary, the Maize Beck, which has not been dammed. This provided an opportunity [19, 176], through a study of the regulated Tees and the unregulated Maize Beck, to assess the effects of the dam.

Storage of water in the Cow Green reservoir causes changes in temperature and flow. The water discharged to the river is 1–2°C warmer in winter and cooler in summer, as a result of the large volume of water in the reservoir. Very high flows (more than eight times the mean) are eliminated, as are very low ones (< 0.1 of the mean), so that, overall, the river water varies much less in the River Tees than previously or than it still does in the Maize Beck.

The first effect of the dam was a great increase in the biomass of algae and mosses among the stones of the river below it. Previously, these had been scoured away by the higher flows. With the mosses came more and different bottom-living animals. The number of animals in a standard sample increased from 56 to 420 after regulation of the Tees, but remained about the same (99 to 77) in the Maize Beck. There were no changes in the diversity of the communities, but, while mayflies remained dominant in the Maize Beck and persisted in the Tees, the latter also supported increased numbers of *G. pulex*, the coelenterate *Hydra*, a group of oligochaete worms, the Naididae, and one of flies, the Orthocladinae. The nature of the invertebrate drift also changed. It became dominated by zooplankton organisms from the reservoir in the River Tees, and water-fleas (Cladocera) became prominent in the diet of brown trout.

Regulation of the Tees was reflected not so much in increased growth of the fish but in an increased population density. The growth rate of trout was about 80% of the theoretical maximum, probably because of the continuing low temperature, even though temperature fluctuations had been reduced. Population numbers of both trout and bullhead increased in the Tees and the bullhead bred earlier and more than tripled their net production. No very serious effects on the Tees can thus be attributed to the Cow Green dam. A diverse river community was maintained.

Other schemes may not be so fortunate. A comparison of the bottom communities of the River Elan below the Craig Goch dam and the nearby undammed Wye [877], in Wales, suggests some deleterious effects of damming. The bed of the Elan was clogged with iron and manganese deposits released from the reservoir. These are formed in the bottom waters by chemosynthetic bacteria and also discharged from the filter beds of a water-treatment works associated with the reservoir. Clogging results in reduced invertebrate diversity. Two solutions are possible. First, iron release could be prevented by artificially mixing the reservoir (see Chapter 7) and, secondly, water could be released to the river from the surface rather than the bottom waters. Surface water falls further and may more easily damage concrete structures at the foot of the dam, so release from the bottom is preferred.

4.6.5 Alterations of the fish community by man

The demand for salmonid fish by anglers and for the table has been growing. Native salmon and brown trout stocks in Britain and elsewhere are dwindling, because of changes in their habitat, and this trend is perhaps reinforced by the release of exotic, hatchery-bred species, which may compete with them [654].

The American brook trout (*S. fontinalis*) and the rainbow trout (*Oncorhynchus mykiss*) were introduced to Britain in the late nineteenth century. Initially, they were confined to hatcheries, for breeding was not successful in the wild. Hatchery fish have been used to stock fisheries, where the voraciousness of the rainbow trout and the ability of both to tolerate greater extremes of temperature, dissolved oxygen and, for brook trout, pH than the native brown trout make them especially favoured by fishery managers. Some populations have now bred in the wild and it is not clear whether stable coexistence of these more resilient fish with brown trout and salmon is possible.

There has also been a great increase in farming of rainbow trout for the food trade. From about 40 farms in 1970, producing less than 1000 t year^{-1}, the industry grew to over 400 farms, producing nearly 7500 t in the early 1980s. Fish farms take water from the rivers and return an effluent that may have a greater biological oxygen demand (from fish faeces). It may also contain antibiotics and disinfectants used to protect eggs against disease. Fish escape from the farms and may introduce diseases, always liable to be rife under hatchery conditions, to the wild stocks of other species. Viral haemorrhagic septicaemia (VHS) and infectious pancreatic necrosis (IPN) are examples.

The hatcheries favour genetically tailored strains, selected for high growth rates and disease resistance under hatchery conditions. Such fish may also be treated with steroids to produce only females, which grow fast, and to delay breeding so that spawning does not weaken fish, grown fat for angling, before they have been caught. Such fish may interbreed with native stocks. The progeny might survive in most years, but a severe year might result in widespread kills under conditions that naturally selected wild fish would survive.

The tendency to stock rivers with exotic fish for anglers has been widespread in the USA, where game-fishing commands large sums of money and where natural communities may have suffered considerably. Training of fishery managers in the past has emphasized the production of large fish for anglers, although a more sympathetic attitude to conservation of the smaller native species may now prevail. An article by E.P. Pister [775], a fisheries manager in the California Department of Fisheries and Game, describing his training and subsequent maturation to conserve small endemic fish of the desert springs and streams rather than removing them in favour of stocking exotic species for anglers, is enlightening reading. One of the introductions he formerly managed in the Owens River valley of California, to the near extinction of four native fishes, was the European brown trout.

5: Lowland Rivers, their Floodplains and Wetlands

There is a world of difference between the upland bubbling brook, with its rocky bottom and foaming-white water and the treacly floodplain river with its fringing swamps, bearing millions of tons of sediment to the sea each year (Plate 4). But this distinction develops gradually (Fig. 5.1). The river bed may at first develop pockets of finer sediments, which support submerged plants. The widening channel in the middle stages, even if overhung at the edges by trees, will become well lit over an increasing fraction of the bed, which will also allow more plants to grow in water still quite shallow.

Then, in the lower reaches, four gradual changes may take place (Fig. 5.1). Depth will increase, and the water may bear enough silt from erosion of the catchment to prevent much light penetration to the bottom, so the submerged plants may disappear. Secondly, at the edges, rooted or floating plants, which cope with the turbid water by emerging into the air above it, may start to form permanent swamps. In the bed, the animal community, still dependent on fine organic debris, will largely be deposit-feeding.

Thirdly, the river will meander more. Its channel size is determined by the average flow, so that at high flows it may overtop its banks and spread over its natural floodplain, taking fish and other animals with it to exploit the seasonal swamps and grasslands of the floodplain. Raised banks (levees) are created along its edges, as coarser material, borne by the flood, settles out first. Lastly, as the river becomes larger and water is retained in parts of the channel for longer periods, there may be time for a suspended community, the plankton, to develop before it can be washed down river to the sea. This community may at first be of animals, the zooplankton, feeding on suspended fine organic debris, but, if enough silt settles to allow enough light to penetrate, photosynthetic plankton (phytoplankton) may grow. Submerged plants may also recolonize the bottom. At this stage, the lowland river will be lake-like. This chapter begins with the colonization of middle-stage rivers by submerged plants and then looks at the development of swamps and floodplain ecosystems and the impact of man on them.

5.1 Submerged plants

Water differs greatly from air and yet the appearance of submerged plants is much like that of their land relatives. Most are angiosperms (with a few

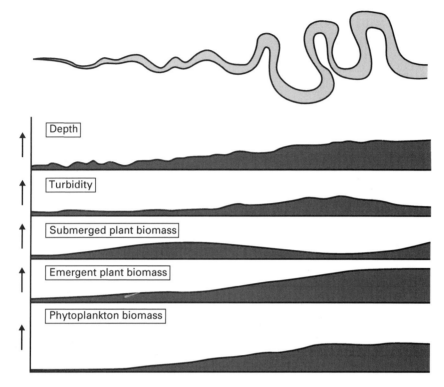

Fig. 5.1 Changes down a river system, as the erosive, often upland, system is replaced by the lowland floodplain system.

pteridophytes) and the basic pattern of roots, stems, leaves and flowers, adapted for wind or insect pollination above the water surface, is preserved. The picture of their evolution is one of sporadic and recent colonization of the water by a few of the available land families.

Submerged plants (Fig. 5.2) show many common features [158]. Often they have reduced root systems, little woody tissue and large internal air spaces (lacunae). Cuticles are spare, covering thin and sometimes dissected leaves, and there is an emphasis on vegetative spread by stolons, rhizomes or contracted shoots, called turions or winter-buds. The latter feature is often ascribed to difficulties of pollination under water; yet a few species have evolved thread-like pollen, which is transported by water currents. Structural characteristics reflect the advantages for physical support of the plant in a dense, wet medium.

Further features reflect the disadvantages of deoxygenation in sediment, low diffusion rates of gases in water and the shaded underwater environment (because of light absorption by the water itself). Reduced root systems are often related to the relatively high concentration of mineral nutrients in many sediments. Experiments with isotopes of nitrogen and phosphorus have shown ample uptake from the sediment by even small root systems

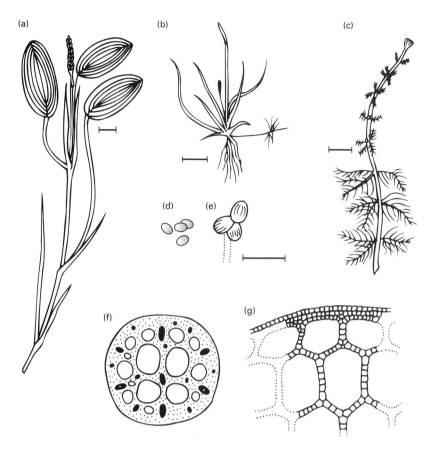

Fig. 5.2 Some features of aquatic plants. (a) *Potamogeton natans*, a submerged species with thin leaves which also has thicker, oval, floating leaves and emergent inflorescences; (b) a small submerged species, *Littorella uniflora*; (c) *Myriophyllum spicatum*, with much dissected leaves; (d) *Wolffia columbiana* and (e) *Spirodela polyrhiza*, two species of duckweed, with much reduced thalli, which float on the surfaces in quiet waters; (f) and (g) cross-sections of the petiole of white water lily, *Nymphaea alba*, and part of the stem of mare's-tail (*Hippuris vulgaris*), showing the abundant air spaces (lacunae). Scale bars represent 1 cm.

and translocation to the shoots [208]. This needs energy, but root respiration may be difficult in sediments that are deoxygenated by bacterial activity. The production of lacunae and air-tubes, through which oxygen may be moved to the roots, is one solution to the problem (see later).

The lack of woody supportive and water-conducting tissues is understandable in a wet viscous environment for plants that are buoyant from air spaces. The thinness and dissectedness of submerged plant leaves is not so straightforwardly explained. Turbulent water tends to pull plants from their anchorage and to batter them against adjacent rocks. Small species, such as the shore weed, *Littorella uniflora*, avoid this problem by growing tight mats of leaves close to the bottom, but risk being covered by shifting sand during spates. Taller plants can avoid burial but risk mechanical damage. Flexible, narrow leaves would seem to be well fitted to cope with turbulence, but are no less common in still lake waters than they are in swift streams.

Perhaps the main reason for selection of such leaves lies in the facilitation of photosynthesis. Diffusion of carbon dioxide (CO_2) into the leaves is slow and light is scarce. Submerged leaves show features like those of the 'shade' leaves deep in the canopy of woodland trees. Their thinness increases the ratio of outer photosynthetic tissue to inner tissue, where internal shading reduces photosynthesis but where respiration still continues. Chloroplasts are often present in the epidermes of shade and submerged leaves, but not in those of plants of well-lit places. This places photosynthetic tissue where it can best benefit from restricted lights and where the pathway for diffusion of CO_2 from the water is smallest.

The problem of obtaining CO_2 is acute in waters of low pH, where inorganic carbon concentration is low, with no reserve of bicarbonate ions. A group of submerged plants, called 'isoetid' forms, has features favouring use of sediment-derived CO_2 [1024]. Their root systems are well developed, their surfaces have thicker cuticles and they tend to be small and to live in clear, soft waters, where not only is bicarbonate content low, but so also is that of other nutrients, such as phosphorus and nitrogen compounds, so that competition for light with overlying phytoplankton is minimized. Some isoetids additionally use crassulacean acid metabolism (CAM) to store carbon dioxide temporarily as carboxylic acids at night, when they can absorb it but not use it in photosynthesis [81, 82, 521, 818, 847].

Some of the problems faced and overcome by plants in colonizing waters are shown by an elegant comparative study of plants in Danish streams [852]. Of 1265 species of herbaceous plant in Denmark, seventy-five are essentially terrestrial but can grow submerged, forty-five are amphibious and forty-one are obligately submerged. Among the amphibious forms, heterophyllous species produce some leaves resembling land plants and others of typically submerged form. Others (homophyllous) produce only one sort of (all-purpose) leaf. The photosynthetic rates of representatives of each of the four groups were measured under standard submerged conditions. Photosynthetic rates increased along the sequence, but the differences could be muted (Table 5.1) by enhancing the CO_2 content of the stream water. Only the obligate submerged group were able to use bicarbonate. The chlorophyll a concentrations in the plants also decreased along the series, suggesting a greater efficiency at photosynthesis underwater by the obligate submerged forms. The experiments suggest that obtaining sufficient carbon has been a key problem for aquatic plants to overcome. The 'terrestrial' plants that manage to survive underwater in streams can do so only because stream water is oversaturated with CO_2 from the groundwater, which passes into streams with high carbon dioxide concentrations derived from soil respiration.

Table 5.1 Comparative photosynthetic physiology of Danish stream plants. Plants were grown in natural stream water. Median values of the experimental results are shown. (From Sand-Jensen *et al.* [852]).

| Group | Photosynthesis: (mg O_2 g^{-1} dry wt h^{-1}) | | Final: | | |
	Ambient CO_2	Enhanced CO_2	pH (units)	CO_2 (μmol l^{-1})	Chlorophyll (mg g^{-1} dry wt)
Terrestrial	2	8.4	8.95	6.2	8.0
Amphibious (homophyllous)	4.3	13	8.95	6.0	7.5
Amphibious (heterophyllous)	10.8	27	9.05	4.8	7.1
Obligate submerged	14	34.7	10.48	0.04	6.1

5.2 Growth of submerged plants

Several factors may determine the growth of submerged plants in rivers. Plants need adequate light, an inorganic carbon source and other mineral nutrients, in addition to particular features of the environment that may favour a given species. Light is readily absorbed by water and its dissolved organic substances and suspended particles. The rate of light absorption in well-mixed water is exponential:

$$I = I_o e^{-kz}$$

where I is the light intensity (or, more correctly, the photon flux density) at a given depth, I_o is the light intensity in the water column z metres above it, e is the base of natural logarithms and k is the absorption coefficient, which defines the rate of absorption, in log units m^{-1}. Values of k express the proportion of the incoming light absorbed as it passes through a metre of water and vary with both wavelength (see Chapter 6) and the nature of the water. They may have values of < 0.01 in very clear waters but of as much as 20 in very turbid ones. Middle-stage rivers, with beds of aquatic plants, might have k values between 1 and 2 for the water, while the absorption by the plants themselves will increase these values within a plant bed.

Values of light intensity at which net photosynthesis (and hence growth) is just possible (gross photosynthesis > respiration) vary with plant species but have been determined in two English rivers [1008] to be between about 10 and 40 J m^{-2} s^{-1}. Maximum summer irradiance was about 320 J m^{-2} s^{-1} at the water surface, so a fraction of at least 3–13% of the available light was required.

By use of an expansion of the equation:

$$z = k^{-1} (\log_e I_o - \log_e I)$$

and substitution of values of 320 for I_o, 10 or 40 for I, and 1 or 2 for k, some idea can be obtained about the maximum depth at which net growth will still be possible in the rivers quoted above. The calculated depths range from 1 to 3.5 m, with three of the four values below 2.1 m. Many middle-stage rivers have depths much less than 3 m and, although the calculations must be made separately for each case, this example suggests that there will usually be enough light for plant growth to be possible on their beds. It also suggests that, in deeper rivers, the underwater light availability will be critical and that much reduction in incident light, for example by tree shading, could prevent growth, even in shallow water. The low rate of CO_2 diffusion into the bulky tissues of aquatic plants, as well as the relatively low concentrations of CO_2 found at high pH, could possibly lead to limitation of photosynthesis by carbon shortage.

This happens rarely, either because light is in even shorter supply or because the plants can use the often more abundant bicarbonate ions (HCO_3^-) directly as a carbon source. The latter is difficult to demonstrate unequivocally. At pH 4.5, essentially all the carbon is present as CO_2 or as carbonic acid (H_2CO_3), whereas at pH 9 the chemical equilibria markedly favour HCO_3^-. Experiments are set up in sealed containers, with no gas phase, where equal amounts of total inorganic carbon at pH 5 or pH 9 are supplied to replicate plants. If photosynthesis (measured as carbon-14 uptake, or sometimes oxygen production (see below)) is significantly greater at the higher pH than at the lower one, direct bicarbonate use is believed to have occurred [800]. This has been demonstrated in a number of species characteristically occurring in hard, bicarbonate-rich waters – for example, *Ceratophyllum demersum*, *Myriophyllum spicatum*, *Elodea canadensis*, *Potamogeton crispus*, *Lemna trisulca* and *Chara* spp. [455]. while aquatic mosses and plants of more acid, soft waters, particularly the 'isoetids', such as *Lobelia dortmanna* and *Isoetes lacustris*, seem to be confined to use of free CO_2.

Because of their bulkiness and hence long diffusion pathways, aquatic plants may nonetheless be at severe disadvantages compared with algae when CO_2 is scarce [7, 594, 895]. The rate of supply of either CO_2 or HCO_3 to the leaf surfaces will be influenced by flow rate, and photosynthesis should increase with current speed, up to some value where factors other than the rate of supply of carbon become important. Westlake [1006] found increasing net photosynthesis of aquatic plants with flow rate at velocities up to 0.5 cm s^{-1}. This is a very low flow for rivers and, although the flow within a plant bed will be much reduced, the flow outside the bed would need to fall well below 10 cm s^{-1} to have much effect. This might happen in dry periods, but the self-shading within a plant bed makes it more likely that light would become limiting long before carbon supply.

Nor is productivity likely to be limited by nutrient supply. High contents of nitrogen and phosphorus have been found in aquatic plant tissues [320]. In glasshouse-grown plants, growth of a variety of species increased with contents up to about 0.13% (as dry weight) of phosphorus and to 1.3% of nitrogen. Greater tissue concentrations gave no increase in growth rate. Wild-growing plants showed tissue concentrations greater than these critical values in most cases, except for plants rooted in sands [136, 455]. Many river species, however, will be rooted in fine gravel or sand, and nutrient availability may be low. Work on algae in upland streams [83] suggests a shortage of phosphorus, and Schmitt and Adams [873] and Christiansen *et al.* [149] found much higher phosphorus threshold concentrations for maximum photosynthesis of submerged plants than quoted above. The apparent increase in weed and algal biomass in lowland British rivers in recent years, where most rivers have been fertilized by agricultural run-off or sewage effluent, also suggests natural nutrient limitation.

The conditions that allow a particular plant to grow will vary greatly from species to species [390] and the plant community may also change in time because of the effects of one species on another. In chalk streams in Dorset, UK, good early growth of the water crowfoot, *Ranunculus penicillatus* var. *calcareus* (Fig. 5.3), accumulates silt, in which the watercress, *Rorippa nasturtium-aquaticum* can easily root. The *Rorippa* shades the *Ranunculus*, so

Fig. 5.3 *Ranunculus penicillatus* (formerly *Ranunculus pseudofluitans*) (based on drawings by Ross-Craig [830].)

that the biomass of it left to survive the winter is low. In turn, the early growth the following year is low and conditions for good *Rorippa* growth are not produced and its establishment is low. *Ranunculus* growth in late summer is thus high, as is the overwintering biomass; the cycle then starts again the following spring, if it is not interrupted by floods washing either species away at critical times [203].

5.3 Methods of measuring the primary productivity of submerged plants

It is useful to know the rate of production of aquatic plants, but the method chosen will depend on why the measurement is needed. The overall production of mixed-plant beds and their associated algae may be wanted or specific information may be sought, plant by plant, with or without the complications of the algal epiphytes that grow on them. A comparative view of the results of production estimates is shown in Table 5.2. Methods can involve the enclosure of plants in experimental containers (with consequent partial control of conditions) or can cope with the plants in their natural, but varying, environment.

5.3.1 Whole-community methods

The earliest 'whole-community' method was simple and involved the cropping, drying and weighing of the plant biomass at the time of its peak growth. Losses (grazing and mechanical damage) during the growing season were assumed negligible and the difference between biomasses at the start (negligible) and the end represented the net growth or net photosynthesis. A practical problem was that, if the natural variability of the habitat was to be allowed for, large numbers of bulky samples had to be taken. More important were the problems that underground parts (rhizomes, roots) were usually ignored and that losses of biomass during the season were often considerable. For six available estimates for submerged plants, a mean value of 0.65 has been obtained for the ratio of below-ground to above-ground biomass [1008]. A similar survey of losses of biomass during the growth season gave the turnover rate, a mean of twelve values of 1.9. Turnover rate is the ratio of total annual production (derived from methods discussed below) to maximum biomass and would be unity if there were no losses during the season. Values significantly greater than 1.0 suggest major losses. Values less than 1 are just as problematic, for they imply overwintering of considerable biomass, usually as roots and rhizomes.

A better method of determining overall community production is the upstream–downstream oxygen-change method (see Section 4.3.1) [372, 730]. A uniform stretch of river perhaps 100 m long is chosen and, over 24 h, the concentrations of oxygen are frequently (sometimes continuously)

Table 5.2 Annual productivity of various aquatic plant communities. Values are given in kg organic matter (ash-free dry wt) m^{-2} year^{-1}. (Compiled from Teal [951], Westlake [1008], Bradbury and Grace [86] and Woodwell [1028].)

Communities	Average	Range	Maximum
Freshwater phytoplankton		Negl.–3.0	
Submerged plants			
Temperate	0.65		1.3
Tropical			1.7
Floating plants			
Duckweed	0.15		1.5
Water hyacinth		4.0–6.0	
Papyrus		6.0–9.0	15.0
Reed swamps			
Typha (reed-mace, cattail)	2.7		3.7
Carex (sedge)		0.43–1.7	1.7
Phragmites	2.1		3.0
Tree swamps			
Alder/ash		0.57–0.64	
Spruce bog	0.5		
Cypress		0.7–4.0	
Hardwood	1.6		
Comparisons			
Tropical rain forest	2.3		
Boreal forest	0.9		
Savannah	0.8		
Temperate grassland	0.6		
Open-ocean phytoplankton	0.14		

Negl., negligible.

measured at its upstream and downstream limits. From the difference between the two oxygen curves so obtained, after rephasing to allow for the time it took for water to travel the length of the stretch, the net change in oxygen concentration over 24 h is calculated. It is taken to equal the gross photosynthesis minus the sum of respiration the net effects of diffusion between water and atmosphere and accrual, the addition of oxygen in seepage water along the banks. Accrual, which cannot easily be measured, is usually assumed to be negligible, while diffusion is estimated from the oxygen saturation levels and temperatures during the period; respiration is calculated from the oxygen changes during the night, extrapolated to the whole period.

5.3.2 Enclosure methods

The advantage of the above method is that it studies an unenclosed community; the disadvantages are that diffusion must be estimated rather than measured and that net primary production cannot be calculated. This is because the respiration estimate includes the activities of microorganisms and animals, as well as the plants. This also applies to methods of determining productivity of individual species by their oxygen production in closed containers, first developed for phytoplankton by Gaarder and Gran [307] in 1927. Samples of the plant material are placed in clear containers (usually bottles) and in containers made opaque with paint or black tape and are incubated in the natural habitat for several hours.

Oxygen is released in the clear container (light bottle, LB) by photosynthesis, but is simultaneously respired by the plants and associated heterotrophs. The change in oxygen concentration in the light bottle is thus:

$$\Delta O_2^{LB} = [O_2^{LB}] - [O_2^{I}]$$

where $[O_2^{I}]$ is the initial oxygen concentration and $[O_2^{LB}]$ the oxygen concentration measured in the light bottle after an exposure of t h.

In the dark bottle, respiration will have reduced the oxygen concentration, with a change:

$$\Delta O_2^{DB} = [O_2^{I}] - [O_2^{DB}]$$

where O_2^{DB} is the oxygen concentration after t h. Respiration rates in the dark and in the light bottles are taken as similar, although this may not always be so. It is assumed that oxygen produced or taken up by the plants is reflected in changes in the concentrations in the water. Air spaces in the plants may cause a lag in this [386], though not always [1007]. The gross oxygen production per hour and per unit dry weight of plant is calculated as the sum of the increase in oxygen concentration in the light bottle plus the decrease in the dark bottle, divided by the weight of plant used (w) and the time of incubation (t).

Gross photosynthesis = $([O_2^{LB}] - [O_2^{I}] + [O_2^{I}] - [O_2^{DB}])(tw)^{-1}$

Thus the initial oxygen concentration need not be known. However, if the amount of plant material is large compared with that of heterotrophs present, the change in the light bottle approximates to net plant production and that in the dark bottle to plant respiration, and a measure of $[\Delta O_2^{I}]$ will allow these to be separately calculated.

Enclosure in bottles is abnormal for plants of flowing waters, and photosynthetic rates could be altered [1014]; there are also difficulties in extrapolating the results of short experiments to entire-season production. The amount of photosynthetically useful radiation is continuously monitored and the ratio of whole-day radiation to that during the experiment is used to scale up the values. This assumes that photosynthesis is governed entirely by light, and this may not be true.

A second enclosure method measures the uptake rate of carbon as its radioactive isotope, ^{14}C, supplied as sodium bicarbonate (NaH^{14}CO$_3$) [920]. The method is sensitive and incubation periods can be much shorter than in the oxygen method. The plants are thus more likely to show rates close to their unconfined natural ones. The bicarbonate added rapidly equilibrates with the CO$_2$-bicarbonate system in the water and the addition is usually very small. Carbon-14 is taken up more slowly, because it is heavier, than the more abundant ^{12}C and corrections may be made for this. After a time, the plant is removed and exposed to fumes of concentrated hydrochloric acid to remove any adherent inorganic ^{14}C. Epiphytes can be removed and the amounts of ^{14}C incorporated into the epiphytes and plant separately measured. Calculation of primary production assumes that the ratio of ^{14}C uptake to total C uptake is in the same ratio as the supply of ^{14}C to the total amount of inorganic carbon present:

Total carbon uptake = (total inorganic carbon (^{14}C taken up))/(^{14}C supplied)

The total inorganic carbon present can be determined by routine analytical methods. The advantages of this method are offset by the problem that what is actually measured (gross or net production) is not known accurately. If ^{14}C taken up stayed in the cells and was not respired, gross production would be measured; if it was taken up in the ratio $^{14}C : {}^{12}C = x$ and respired in the same ratio, net production would be measured. In practice, it is probably taken up in ratio x but respired in some different ratio, y. It is assumed that respiration does occur, but the difference between x and y will determine how close to the true net rate of production the estimate is.

Paradoxically, approaches that use little equipment but more ingenuity may give the most reliable estimates of production for aquatic plants. They depend on a knowledge of plant structure and a willingness to carry out tedious but simple measurements. *Ranunculus penicillatus* var. *calcareus* (see Fig. 5.3) [198, 199] is common in lowland chalk rivers in southern England and grows stems up to 4 m long, bearing finely segmented leaves every 20 cm or so. Determining the biomass is not difficult with a sampler that cuts the weed from a known area, and the underground biomass is relatively small. Leaves continually break off or are cut to provide the cases of caddis-fly larvae, such as *Limnephilus*. Stems are broken by water-voles and moorhens to line their burrows and nests, respectively. Production was estimated from the total losses of leaves and stems from an area, plus the biomass left intact. Leaf loss was determined by counting the number of nodes that had lost their leaves along stems sampled at frequent intervals and was about 8.5% of the maximum biomass obtained. Whole-stem loss was determined by marking a sample of stems with coloured plastic rings. From the rate of loss of marked stems, an estimate of 77% of the maximum biomass was found for the stem loss.

Nets placed downstream of the areas under study showed remarkably little 'export' of plant material. The detached stems caught on other stems and often rerooted, so that the total net loss was small (about 30 g dry wt m^{-2} compared with a total annual production of 300–400 g m^{-2}). Most of the produced material, even at the end of summer, was retained and decomposed where it was produced, suggesting that downstream export was unimportant.

5.4 Submerged plants and the river ecosystem

Submerged plants in rivers provide microhabitats, spawning and egg-laying sites for invertebrates and fish, cover from predators for prey and lurking sites for predators. They also supply labile organic matter and entangle organic matter, such as leaf litter, washed from upstream. The beds may also affect water chemistry, particularly by the removal of nitrate ions.

Nitrate removal may be through growth, in which case it is only temporary, as the nitrogen (N) will be released on decomposition of the plants, or it may be by bacterial denitrification. Bacteria use nitrate as an oxidizing agent to release energy from organic matter, but commonly do so only at low oxygen concentrations. This need not mean that the entire water body need be hypoxic, but only parts of it, often sediment surfaces. The plant beds, in encouraging the accumulation of mud and in providing organic matter, may promote denitrification. The proportion of nitrate removed will vary with the size and discharge of the streams. In the Whangamata Stream in New Zealand, with discharges of about 0.1 $m^3 s^{-1}$, and with watercress beds, nitrate falls from about 0.6 mg NO_3-N l^{-1} to about 0.1 mg l^{-1} over 2 km. The plants take up about 560 mg N m^{-2} of bed day^{-1}, while denitrification accounts for about 56 mg N m^{-2} day^{-1} [436]. In another study [438], plant uptake was 1.14 g N m^{-2} day^{-1} and accounted for all of the nitrate loss from the stream. In larger rivers, with both discharges and nitrate concentrations an order of magnitude higher, the proportion of nitrogen taken up by the plants may be negligible [198], and any significant nitrate loss may be due mostly to denitrification. In the River Great Ouse, with losses of 274 g m^{-2} $year^{-1}$, 34% was due to plant uptake and 66% to denitrification.

Lastly, there is an aesthetic role of the river vegetation. Swards of water buttercups, brooklime and other flowers that characterize the edges of small rivers are attractive. In parts of Europe, favoured rivers for trout fishing may be carefully managed by the hand-cutting of parts of the plant beds and the leaving of others, to maintain ideal fishing habitats (Fig. 5.4). Elsewhere, plant beds in rivers are regarded with less favour and are extensively cut or removed with herbicides in the interest of flood management.

5.4.1 Plant-bed management in rivers

Extensive plant beds retard the flow and increase the depth of small rivers

Fig. 5.4 Management of stream vegetation is most sensitively done by hand, rather than machine cutting. This allows the encourage-ment of desirable species at the expense of silt-retaining plants and allows a good compromise between conservation management and flood control. Denmark is actively promoting renewal of such traditional techniques.

in spring and summer. This may flood riverside land now used for crops. Summer flooding leads to deoxygenation of soil and deaths of the crop roots. In the past, in Great Britain, such flooding was acceptable. The flooded 'water-meadows' were used for grazing of a flood-tolerant native flora, annually fertilized by the river silt. Many of these ancient habitats, very rich in plant species, have now been destroyed by ploughing [716, 891], following engineering operations to deepen and straighten the river channel (see later). The fields have been seeded with crop grasses or cereals and flooding is unacceptable to the farmers.

The frequency of summer flooding may have increased because of these changed farming operations. The removal of woodland and drainage of riverside land may have increased the summer flows through reduced water storage and evapotranspiration. Secondly, the increased run-off of agricul-tural nitrogen to the rivers has increased the growth of a filamentous alga, *Cladophora glomerata* (blanket weed) and other 'weedy' plants. Thirdly, cultivation to the river edge and fears of brushwood falling into the channel and accentuating the flood risk have meant removal of overhanging trees and bushes. Loss of this shade lifts any previous light limitation on the rooted plants and encourages their greater growth.

The plant beds are consequently cut in about a third of the main rivers and about 32 300 km of ditches and dykes. in Britain, usually by machine. Cutting is done two or three times a year and can be regulated to preserve some of the plants for fisheries and aesthetic purposes. The cut material must be removed to prevent its decomposition in the water, causing deoxygenation. Cutting, however, may increase the growth and biomass of the plants. Many are perennials, with a yearly rhythm of growth in spring and early summer, flowering in midsummer and senescence in late summer. Cutting just before flowering removes the suppressive effect on growth of the flowering hormones and encourages extension of side shoots. It also relieves the plant of the effects of its own self-shading, which also encourages further growth. Beds of *Ranunculus* in Dorset rivers [198] were reduced to half in biomass when cutting ceased. The biomass is closely related to the effectiveness of the beds in causing flooding.

Cutting is labour-intensive and costly. There has consequently been an increase in the use of cheaper herbicides in recent years, especially in slower-flowing waters. About eight compounds are officially cleared under the Pesticide Safety Precaution Scheme in the UK, although rather more compounds are actually used. One in particular, diquat alginate, is favoured for treating submerged vegetation. The safety scheme, in approving a compound, attends largely to the risks to stock and people and has little regard for the ultimate effects on natural communities. Indeed, it cannot, for data on such long-term effects are not available. Herbicides are largely unselective and kill most aquatic plants. In turn, this destroys the habitat for invertebrates and may cause deoxygenation as the plants rot. Herbicides may also be directly toxic to some animals. Repeated application can permanently destroy the habitat; recovery is often complete from light applications, and the disturbance caused by the creation of habitat 'space' during the effective period may allow additional species to colonize [705].

5.4.2 Ecotoxicology – the testing of potentially hazardous chemicals

The effects of chemicals such as herbicides are unpredictable in any precise way, despite much information on their toxicities to selected animals in the laboratory. Frequently, *Daphnia magna*, a water-flea, *Gammarus pulex*, the pond shrimp, or rainbow trout are used. Often the lethal concentration necessary to kill 50% of a batch of test organism (LC_{50}) in a given time (24–48 h) is measured and from such experiments decisions are made as to safe dosages [630]. Better, the no-observable-effect concentration (NOEC) is measured. However, the reactions of an organism may differ in its natural habitat [282], where it is in competition with others for food or space and where subtle effects on reproduction or behaviour may result at concentrations far below the NOEC in the laboratory.

Efforts have been made to standardize laboratory toxicity tests, and the precision obtainable lends authority to this approach among those who wish to use such chemicals in environmental management. Even then, different genotypes of the same species used in different laboratories give results varying over orders of magnitude [27–29]. Especially when used as a basis for legal controls [117], tests should be relevant to the environment, reliable (in the sense of being available on demand) and statistically repeatable. The problem is that the latter two requirements require standard, single-genotype systems, while the first demands complex multispecies systems that naturally vary with time. Calow [117] believes that the primary need is to be able to screen large numbers of new chemicals routinely and that therefore standardization, reliability and repeatability are the most important considerations. He argues that the gap in relating laboratory tests to natural situations (i.e. relevance) might be bridged by introducing safety factors (divide the LC_{50} or NOEC by 10 or 1000) and by choosing the most sensitive test organisms and ecologically important species. Unfortunately, sensitive species do not usually take well to laboratory conditions and, in multispecies communities, relative ecological importance is very difficult to ascertain.

The current approaches of ecotoxicology are technological conveniences which assume that commercial considerations should take priority and that producers should be given the benefit of the doubt. There are few chemicals that do not cause some environmental problem, and experience suggests that far more relevant testing should be done before any chemical is discharged to the environment. If such testing is impossible or too expensive, the chemical should not be used. Society can manage without most of the products of industry. The accuracy of almost all current ecotoxicological data in predicting the consequences of chemical use for natural environments is so low that this approach is useful only in screening out chemicals so overtly toxic that they should never be used.

A wiser approach to plant management in rivers than herbicide use has been proposed by Dawson [199, 200] and his collaborators [201, 202, 203], who suggest the replanting of bankside trees to reduce the water-plant biomass by shading. Trees on a northerly bank may cut down the incident light by about 20% in southern England and reduce the plant biomass by half. Trees and bushes planted on southerly banks will decrease light and biomass even more, but may reduce the organic matter available to the animal community very significantly. The leaving of gaps in the plantings of, say, 20 m in 100 m allows some compromise. The replacement of trees has advantages for riverine birds and otters, who make their nests and holts or dens among the roots of large trees, and as an attractive landscape feature. Furthermore, the costs of tree planting per unit length of river are about the same as the costs of a single year's cutting or herbicide treatment and additional maintenance over 10–20 years only doubles the initial cost.

As the river widens and begins to deposit more silt where its flow is reduced, emergent aquatic plants, with the bulk of their photosynthetic biomass above water, form swamps or marshes (Plate 4). Swamps normally have water standing above the peaty soil surface and are dominated by a few tall herb species, usually sedges or grasses, or trees, such as alder, willow or swamp cypress. Swamp grows closest to the main channel of a river. Marshes, on the other hand, form the more distant short grasslands on more mineral soils, flooded perhaps for only part of the year. The swamps and marshes in undisturbed river valleys may occupy a huge valley floor, the basis of a rich ecosystem involving indigenous peoples totally dependent on it. Alternatively, they may have been long drained to form rich agricultural soils, protected by embankments to prevent the river flood (Plate 4). The swamps and marshes act as huge filters for silt, change the chemical composition of water passing through them and are highly productive ecosystems.

5.5 Further downstream –swamps and floodplains

5.5.1 Productivity of swamps and floodplain marshes

Swamps and marshes are among the most productive of the world's ecosystems (see Table 5.2). They are dominated over great tracts by single, large, vigorous species (Fig. 5.5). In northern temperate regions, reed, *Phragmites australis*, reedmace or cattails, *Typha* spp., bulrushes, *Scirpus* spp. and other monocotyledons often predominate, with tree swamps of alder (*Alnus*) and willow (*Salix*). In warm temperate climates, members of the Restiaceae (South Africa, New Zealand), the saw-sedge (*Cladium jamaicensis*) or swamp trees – the swamp cypresses, *Taxodium* and tupelo, *Nyssa* (southern USA and Caribbean) – are found. In the tropics, the igapo is a more diverse swamp forest in parts of the valleys of the black-water Amazon tributaries, while papyrus (*Cyperus papyrus*) characterizes many

a

b

c

Fig. 5.5 Some typical swamp plants. (a) *Taxodium* sp. (swamp cypress); (b) *Cyperus papyrus* (papyrus); (c) *Phragmites australis* (common reed).

Central African swamps. A full list would be much more extensive and the floodplain marsh grasslands may be very diverse. Submerged and floating plants will be present in channels in the swamp, although the emergent vegetation will provide most of the biomass.

The high productivity stems from the favourable environment. Only in extreme drought is water short; a continuous supply of river-borne silt brings abundant nutrients; carbon dioxide is readily available from the atmosphere to the emergent parts and, through the water, to the submerged parts; and light probably becomes limiting during the growing season only when the stands become dense enough for self-shading to occur. We measure the productivity of swamps usually by methods that depend on the harvesting of biomass, partly because of a belief that little of the production is directly grazed, most entering the food webs as detritus following the productive period. The few estimates made of turnover of the biomass, however, suggest that losses during the growth season may be quite high. Mathews and Westlake [639] applied the Allen-curve technique (see Chapter 4) to cohorts of emerging shoots of a grass, *Glyceria maxima*, and found an annual ratio of above-ground production to maximum above-ground biomass of > 1.5. Values of this ratio for other temperate emergent plants (*Phragmites, Typha, Scirpus*) [1008] are close to 1 but for papyrus in African swamps were 1.8–3.6 [957]. A comparison [1008], using the data of Schierup [862], among different methods of calculating the productivity of a stand of *P. australis* showed that the lowest estimate (measurement of maximum above-ground biomass) gave a value of 1143 g dry wt m^{-2} year^{-1}, which was only 55% of the best estimate (2085 g dry wt m^{-2} year^{-1}) obtained when turnover of shoots, roots and rhizomes was measured. Many estimates of swamp production are probably underestimates.

5.5.2 Swamp soils and the fate of the high primary production

Dead organic material falls continually to the swamp floor. The plants impede water movement, so relatively little is washed downstream and much of the production must be oxidized by secondary producers or stored as peat. Processing through the food webs in the oxygen-rich environment of the emergent shoots appears negligible. Budgets (see, for example, Table 5.3) suggest that only a small fraction of green tissue is eaten by insects, such as grasshoppers and leaf-miners, and that, although birds, such as geese, may eat both green tissue and seeds and rodents, such as muskrat and coypu, will excavate rhizomes, the impact of these activities is small.

Perhaps this is because the plant biomass is bulky, because of the air spaces in it (see below), and has a high water content, and hence a low energy content per unit volume. If direct grazing is a negligible pathway by

Primary producers	Net primary production
Phragmites communis (australis)	15 000
Utricularia vulgaris	150
Planktonic algae	500
Periphyton	1 000
Total	16 650
Herbivores	Energy uptake
Plant-eating insects	40–60
Muskrats	< 10
Total	< 70
Algal feeders	Net production
Chironomidae	10
Snails	30
Asellus aquaticus	10
Total	50

Table 5.3 Energy budget for a freshwater swamp. Values are in kcal m^{-2} year^{-1}. (Compiled by Howard-Williams [432] from Dokulil [217], Imhof [463] and Imhof and Burian [464].)

which the organic matter is oxidized, an alternative is fire. Tropical and subtropical swamps, such as those of the Florida Everglades, may lose much organic matter through lightning-induced fires in the dry season [826] and this may be why swamps in seasonally dry and hot regions do not accumulate much organic matter on the swamp floor. In cool and wet regions, storage of partly decomposed organic matter (peat) is prominent. The primary production also enters detritus pathways, with microorganisms and animals living in the water and the upper layers of peaty soil.

5.5.3 Oxygen supply and soil chemistry in swamps

Oxygen diffuses slowly into water and sediment and, because of microbial and animal respiration, swamp sediments and, in warm periods, the overlying water are near-anaerobic. Deoxygenation limits the rate of decomposition and favours rapid peat build-up, so that, if the flow of water is low, the soil may have little inorganic content. Waterlogged soils pose problems for plants, for their reducing nature changes the chemistry of the overlying water [250, 894]. The degree of reduction is measured as the redox potential. A calomel (Hg$_2$Cl$_2$)–platinum (Pt) electrode is immersed in water or wet soil, while connected into a circuit that includes a standard hydrogen electrode. Free electrons move to the calomel–Pt electrode to an extent reflecting the number available and are measured on a galvanometer as a relative negative potential, in millivolts. A large number of free electrons available for donation indicates a reducing environment. In an oxidizing medium, electrons move away from the calomel–Pt electrode, giving a relative positive electrode potential difference. Well-oxygenated waters may have redox potentials greater than $+500\,mV$.

As heterotrophs decompose organic matter in sediments and the oxygen concentration falls, so does the redox potential, to values of about $+200\,mV$ when anoxic conditions are reached. Decomposition will have produced inorganic ions, such as phosphate and ammonium to add to the ions already present in the interstitial water. Bacterial respiration, becoming anaerobic, may produce organic acids (butyric and acetic). The carbon dioxide content will have risen and the pH will consequently have fallen.

As oxygen is depleted, oxidized ions may be used by microorganisms to process organic matter. Some protozoa and many bacteria (e.g. *Achromatium*, *Bacillus*, *Pseudomonas*) reduce nitrate to nitrous oxide (N_2O) or molecular nitrogen (N_2) through denitrification. Deoxygenation also favours nitrogen fixation by free-living bacteria (*Clostridium pasteurianum*) and bacteria in nodules on the roots of swamp trees, such as alder (*Alnus*). Eventual decomposition of these organisms gives a supply of ammonium ions to the soil.

Denitrification is generally complete as the redox potential falls to $+100\,mV$, when anaerobic bacteria, unable to oxidize carbon compounds further, may release ethylene, and bacteria or perhaps inorganic chemical processes begin the reduction of manganese of (Mn^{3+}) ions to the more soluble (and toxic) Mn^{2+}. At slightly lower redox potentials, *Clostridium* and other bacteria reduce the orange-red ferric ion (Fe^{3+}) to green-grey, also more soluble, ferrous (Fe^{2+}) ions. If water is moving through the soil, these soluble ions may be washed out leaving a pale green or grey appearance (gleying).

Just below a redox potential of $0\,mV$, bacteria produce methane (*Methanobacterium*, *Methanomonas*), hydrogen (*Clostridium* spp.) or phosphine (PH_3) (marsh gases), which may ignite as 'wills-o'-the-wisp'. *Desulphovibrio delsulphuricans* and *Desulphomaculatum* oxidize molecular hydrogen (H_2) or organic matter by reducing sulphate in the interstitial water to sulphide. In turn the sulphide may precipitate Fe^{2+} to form black, iron sulphide in the most intensely reducing soils. These processes produce soils with potentially toxic ions (Fe^{2+}, Mn^{2+}, hydrogen sulphide (HS^-)) and organic acids, no oxygen and ammonium rather than nitrate as a nitrogen source for roots. An example – a mild one, for there is some water flow under the mats of papyrus – is shown in Fig. 5.6 of how these processes may in turn affect the composition of the overlying water.

The chemical changes that occur on waterlogging are reversed if the flooding is seasonal and alternates with drying. Carbon dioxide concentration may then decrease, pH may rise and oxygen may diffuse into greater depths. The reduced iron may be oxidized to Fe^{3+} to release energy by iron bacteria of the genera *Gallionella* and *Sphaerotilus*, giving deposits of rust-coloured ochre. Oxidation of sulphides to sulphate by sulphur bacteria (e.g.

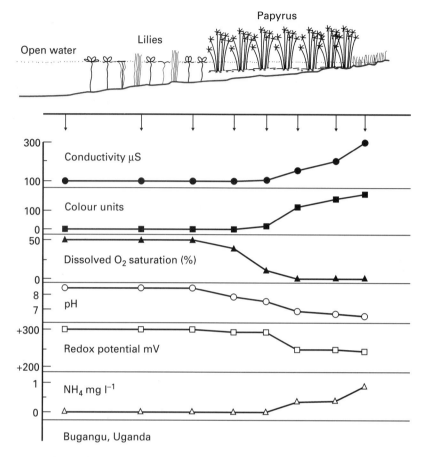

Fig. 5.6 Chemical processes in waterlogged soils at the fringe of a tropical papyrus swamp. Decomposition in the stagnant swamp water leads to increased dissolved ion levels, high humic acid levels, giving brown-stained water, and high ammonium (NH_4) levels. Oxygen (O_2) is used up rapidly, redox potentials are decreased and carbon dioxide (CO_2) production decreases pH. (Redrawn from Carter [132].)

Beggiatoa) may produce insoluble gypsum if calcium is abundant, or sulphuric acid if it is scarce, as in many peats. This reduces the pH to 2 or 3 and forms acid sulphate soils. Ammonium ions may be oxidized by nitrifying bacteria (*Nitrosomonas*) to nitrite and then (*Nitrobacter*) to nitrate. On reflooding, the nitrate will be denitrified, so a swamp with a seasonally fluctuating water level may ultimately remove much combined nitrogen from the water. Fenchel and Finlay [263] give a more detailed account of anaerobic bacteria.

This interlinked set of inorganic chemical and bacterially mediated reactions thus produces a swamp soil and an overlying water that are problematic in more ways, for the higher organisms that colonize them, than oxygen lack. Ways in which plant roots cope with conditions in the sediment and animals with those in the water will now be reviewed.

5.5.4 Emergent plants and flooded soils

Flooding kills many plants, because their roots are unable to respire aerobically and the toxic products of anaerobic metabolism (acetaldehyde or ethanol) reach lethal concentrations. Swamp plants often grow better in drained soils than in waterlogged ones, if freed of competition with other plants. They persist in swamps by either avoiding deoxygenation or tolerating it by supplying oxygen from the atmosphere to the roots via the emergent leaves and stems [172].

Avoiders may produce shallow roots (pine (*Pinus*) and rushes (*Juncus*)) or may continually replace damaged roots (some grasses). In estuarine mangrove swamps, the trees may produce aerial roots, which protrude above the soil surface and absorb oxygen through pores (lenticels). Tolerators limit the formation of toxic anaerobic products, excrete them to the water or form non-toxic ones instead of ethanol. They may also be able to detoxify ions, such as Fe^{2+} and Mn^{2+} which diffuse into the roots from the soil.

Most swamp plants have systems of internal air spaces and it is often assumed that these act as channels for the diffusion of oxygen to the roots and rhizomes. Sometimes they do [849], although diffusion alone may be too slow to give a ready supply. In water lilies, there is mass flow of air from the emergent leaves to the rhizome [188, 189]. If polyethylene bags containing air are sealed over the young leaves by day, the bags collapse as air is withdrawn from them. Internal pressure, slightly greater than atmospheric, is created in the air spaces of the leaf by absorption of heat. The inner parts of the leaf are in contact with the atmosphere through stomata on the upper surfaces, but the rate of diffusion of air through these is insufficient to prevent build-up of pressure in the young leaves. The pressure in the young leaves forces air down through the petioles and into the rhizome, from which air is exhaled via the older leaves. As the leaves age, the stomata enlarge and diffusion through them is much faster. In rice, there is a mass flow of air through the continuous bubbles that coat the outside surfaces of the submerged parts of the leaves and stems [798].

Not all emergent plants, however, have such bubbles or unoccluded air spaces. Sometimes, plates of cells interrupt the passages. Diffusion through these is slow and mass flow is prevented [455, 878]. Sometimes, closed lacunae provide internal surfaces in contact with air, on which toxic reduced ions (e.g. Mn^{2+}, Fe^{2+}, HS^-), diffusing into the root can be oxidized. Diffusion of oxygen from some roots may have this effect at the outer surface in contact with the soil, where iron compounds may be precipitated [1025].

Some marsh plants – for example, *Filipendula ulmaria* and *Phalalaris arundinacea*, do not have any air spaces. In these, metabolic mechanisms [170, 171] may sometimes better explain flood tolerance. Finely divided adventitious roots (those arising along a buried or submerged stem) may be produced, providing a large surface area, through which ethanol and acetaldehyde can diffuse out. This happens in several flood-tolerant trees, especially where there is moving water to remove the excreted compound. Secondly, the rate of anaerobic metabolism can be controlled to minimize ethanol production [174]. Lastly, the end-products of anaerobic metabolism can be diverted to less toxic compounds – pyruvic acid and glycolic acid (willow), malic acid, shikimic acid (iris and water lilies) [173] or amino acids. Eventually, these compounds must be oxidized, if root function is to continue, either by translocating them to better aerated parts of the plant or when oxygen movement to the roots is greatest. Not all flood tolerant plants produce such compounds [899]. When they are produced, they may not entirely replace ethanol [520] and may have other roles, such as in combining with toxic reduced ions to form less toxic compounds [251].

5.6 Swamp and marsh animals

Floodplains often have spectacularly large populations of birds and mammals; fish may be equally abundant. The reasons are the great diversity of habitat to be found on the floodplain (Fig. 5.7) and the high productivity of the floodplain vegetation.

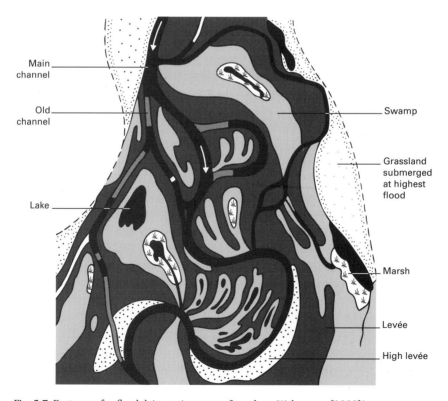

Fig. 5.7 Features of a floodplain environment (based on Welcomme [1003].)

(a)

(b)

(c)

(d)

Plate 1 The Pantanal in Brazil (Chapter 1) is one of the largest freshwater systems in the world and yet one of the least intensively investigated. It includes rivers and the shallow wetlands of their flood-plains and lakes, a rich flora and fauna and spectacular displays of reptiles, birds and mammals. It provides services to people in terms of water storage and flood control, grazing and tourism. Characteristic of the Pantanal are the jabiru storks, the crocodilian jacare, the capybara – a large herbivorous rodent – and seasonal cattle herding.

[facing page 144]

(a)

(b)

(c)

(d)

Plate 2 Fresh waters are fleeting, in geological terms (Chapter 2). Many lakes have been formed from glacial action around 10 000–15 000 years ago. As the present climate warms (Chapter 11), glaciers recede, leaving moraines to act as dams, and new glacial lakes are being formed. The fine clay, initially suspended in the waters, scatters blue light, giving a turquoise appearance (lake in the Canadian Rocky Mountains, (a). The basins of some lakes in the tropics are much older, but still geologically young. A few have had permanent water for some hundreds of thousands of years (Lake Victoria, (b) (Chapters 6 and 9); others may dry out and refill in cycles of only decades (Lake Chilwa, Malawi, (c) (Chapter 5). Many small lakes have been created very recently to serve human purposes. An Indian dam (d) serves for domestic washing, watering of water buffalo and fishing for food.

Plate 3 Upland or bed-eroding streams in their natural state draw much energy to support their ecosystems from washed-in leaf litter (Chapter 4). Their accessibility (a), the rich animal communities among their stones (b) and the possibilities of catching a salmon or trout (c) fascinate people of all ages. A detailed examination of their structure reveals intricate organisms, such as the freshwater red alga, *Bartrachospermum* (d).

(a)

(b)

(c)

(d)

(a)

(b)

(c)

(d)

Plate 4 Lowland rivers have wide beds, which they fully occupy for only part of the year. In drier periods, they retreat to their central channel, leaving the flood-plain to support wet grasslands and swamps (Chapter 5). In undisturbed systems, such as the Hayden Valley in Yellowstone National Park, USA, these support grazing mammals, such as bison (a). Alas, the fertile soils that river flood-plains provide when drained and a general tendency for human beings to interfere with and change what they do not fully understand have led to drainage of many flood-plains and engineering of their rivers into soulless conduits (b). Elsewhere, the reed swamps provide valuable commodities, such as thatch for roofing (c). However, swamps are favourably perceived in the literature favoured by young children, although Nungu, here (d), might be advised to make less free with hippopotamuses. Only after the age of about 7 or 8 do they learn prejudice from their elders [15].

Swamp invertebrates are remarkable. Groups of air breathers (see Chapter 2) – beetles, some Diptera, pulmonate snails – are predominant, together with specialized members of some otherwise water-breathing groups. An African oligochaete worm, *Alma*, has developed a deeply grooved tail, which is richly supplied with blood-vessels. The tail is extended to the surface of a waterlogged, deoxygenated floating mat of vegetation and the groove flattened out in contact with the air. Some snails – for example, *Biomphalaria sudanica* – have both gills and lungs.

Among the fish, air breathing is crucial for those which stay in the swamp all year. At low water, they are confined to the deeper channels and pools, where the stagnation of the water and the large numbers of animals in a small volume make deoxygenation more severe than in the flood season. It may also make, for the observer, a spectacular collection of fish, turtles, water snakes and alligators in, for example, the sloughs (pools) of the Florida Everglades (Fig. 5.8).

Modifications for air breathing in swamp fish (Fig. 5.9) range from a flattening of the head, allowing the fish to remain in a thin layer of water at the surface, where oxygen concentrations are greatest, to a development of lungs (the bichir, *Polypterus*, and the lungfish, *Protopterus*, *Lepidosira* and *Neoceratodus*). These fish will drown if denied access to the atmosphere at the water surface. Between these extremes are modifications of existing organs for air breathing. Some catfish (*Clarias*, *Anabas*) and the electric eel (*Electrophorus*) have bony supports, which minimize the collapse of the gill filaments (a collapse would reduce their surface area for absorption) and allow the gills to be used in both air and water. The swim-bladders, normally buoyancy regulators, and parts of the gut have, in some species, become richly provided with blood-vessels to allow swallowed air to be absorbed. The Amazonian *Arapaima* and the North American bowfin (*Amia*) and garpike (*Lepisosteus*) have modified swims-bladders; and the catfish *Plecostomus* and *Hoplosternum* have vascularized stomachs and intestines, respectively.

These air-breathing fish normally remain within the swamp. Despite its hypoxia, the swamp is a permanent and predictable habitat, compared with the temporary aquatic conditions of the floodplain grasslands. This favours economy in reproduction, with relatively few, but large, eggs, each with a high chance of survival, being produced and often guarded by the parents in floating nests of vegetation or mucus froth. Where the swamp dries out seasonally, some annual species may survive as eggs buried in damp mud (e.g. some Cyprinodont fishes [581, 582] or even (some lungfish) as adults cocooned in a muddy chamber lined with body slime. The fish breathe through a tube emerging at the mud surface.

5.6.1 Whitefish and blackfish

In the Mekong River floodplain, the swamp-tolerant fish are referred to as

Fig. 5.8 Pools excavated or extended by the alligator (*Alligator mississippiensis*) are extremely important during the dry season in the Everglades, when they form refuges for spectacular concentrations of fish, turtles, snakes, birds and other animals.

'blackfish'. Often they are dark in colour, with small or few scales, and come from a group of families that include the siluroids and anabantids (catfish), channids (ophiocephalids), osteoglossids, bichir and lungfish. In contrast are the silvery, scaled 'whitefish' of other families, with very different characteristics. The terms 'blackfish' and 'whitefish' conveniently describe similar functional groups in many tropical floodplains.

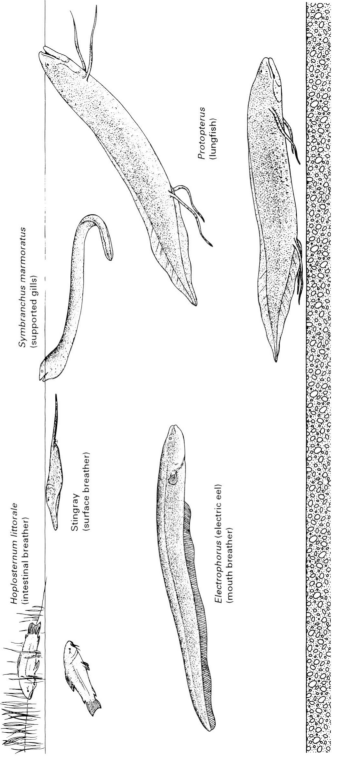

Fig. 5.9 Air-breathing fish from South America and (*Protopterus*) Africa.

Fig. 5.10 Sitatunga, an antelope well adapted, with wide-spreading feet, for swamp life.

The whitefish are migratory. They move upstream from the dwindling channel of the river in the dry season and breed in the headwaters or an upstream floodplain in high flood, where they move to the fringes, where oxygenation is highest. Later, they feed on the floodplain and then retreat to the main river as the flood goes down. The whitefish produce many small eggs, which are scattered and each of which stands only a small chance of survival. This strategy is appropriate where the conditions for breeding (well-oxygenated water and often some form of structural support for the eggs) may be short-lived and where rapid reproduction is needed in an environment likely to change rapidly and unpredictably.

The contrast between the tolerators of the permanent swamp, such as the blackfish, and the migratory whitefish, who take advantage of it seasonally, is reflected to some extent in the mammals and birds. Among the former, the sitatunga (Fig. 5.10), a Central African antelope, is a more or less permanent swamp-dweller, with feet splayed to allow it to walk over soft, often floating, beds of vegetation, whereas another antelope, the red lechwe, migrates on to the floodplain grasslands as the water retreats and fresh growth begins. Something of the same pattern can also be seen in the indigenous human societies associated with floodplain wetlands.

Fig. 5.11 Part of a Madan village, Iraq (from Thesiger [956].)

5.7 Human societies of floodplains

5.7.1 Tolerators – the marsh Arabs

Regarding the Madan, the marsh Arabs of Iraq (Fig. 5.11), Wilfred Thesiger [955, 956], Gavin Maxwell [642] and Gavin Young and Niall Wheeler [1039] paint a picture of a people permanently dwelling in tall reed swamps. These span the 6000 square miles around the confluence of the Tigris and Euphrates near the Arabian Gulf. *Phragmites australis* grows to 4 m in height and dominates the swamp; it is flanked by seasonal swamps of *Typha angustata* and a sedge, *Scirpus brachyceras*. The Madan are descended, in part, from the Sumerians, who founded the great Mesopotamian civilization at the edge of the marshes 6000 years ago; a Sumerian legend has it that Marduk, the Great God, built a platform of reeds on the surface of the waters and created the world.

Perhaps this was how the ancient marsh Arabs built their houses, as it still is in those few who have survived drainage of the marshes in recent years. On the platform, simple houses of reed bundles are built and also elaborate *mudhifs*, or guest-houses, which may be 20 m long and 6 m wide. The bases of their construction are pillared arches of reed 2 m in girth and tightly bound. Inside these structures, an elaborate social life is possible in comfortable carpeted surroundings.

Outside the buildings, the platform houses the three to eight water-buffalo, which are the mainstays of a marsh family. These placid animals are taken to graze where the water is shallow enough for them to touch bottom, on sedge or water plants (*Polygonum senegalense, Jussiaea diffusa, Potamogeton lucens, Cyperus rotundus*) and, on return at nightfall, are fed the young shoots of sedge or reed cut during the day. Buffalo provide milk, meat and dung for fuel, while birds are shot for meat or, in the case of pelicans, for the soft pouch skin, which, once cured, forms excellent drum and tambourine skins.

Fish, particularly *Barbus sharpeyi*, are speared with forks from lines of boats set across channels, or may be poisoned with flour and dung bait laced with extracts of the toxic plants, *Digitalis* or *Datura*. Increasingly, gill nets are laid and the fish sold to the cities. Boats are usually of wood, brought in from the coast, but, as elsewhere, simple canoes are built from *Typha* bundles bound together. On the swamp edges, rice is cultivated and provides a staple diet; the rice-fields also attract wild boar. A metre or more at the shoulder and aggressive, the boar account for many deaths and injuries among Madan, who surprise them while punting through the channels or cutting reed.

The Madan were originally largely self-contained; visitors comment on their hospitality and the complexity of clan and tribal relationships. They also suffered from poor health – high infant mortality, yaws, schistosomiasis, dysentery, tuberculosis. Increasingly, health care has been made available and there has been trade in woven reed mats and fish with areas outside the marsh. The pattern was set for maintenance of a traditional culture benefiting in a sympathetic way from other influences. The current situation of most Madan is probably far different from this description, however. For reasons purportedly of agricultural development, the Iraqi government has dug canals, which are draining the marshes. Denied the traditional marsh resources and influenced by the brutalization inevitably following the Gulf War in the area, the Madan seem likely to be forced to join the mainstream of westernized ways. Commentators who have observed this [179, 213, 722, 1027] share the regret that Thesiger and Maxwell would have articulated no less sadly.

5.7.2 Migrators – the Nuer of southern Sudan

The Nuer, for comparison, are migrators, rather than permanent swamp-dwellers. They are a Nilotic people living in the valleys of the White Nile (Bahr el Jebel), Bahr el Ghazal and River Sobat around their confluences in the Sudd swamp in Southern Sudan (Fig. 5.12). Like the Dinka, who occupy the eastern part of this region to the Nuer's western, they have a mixed

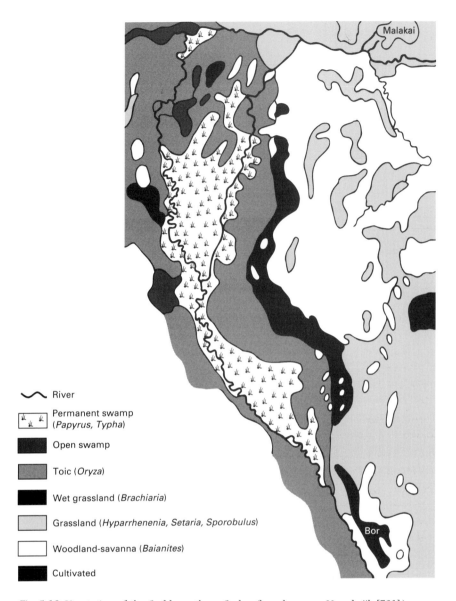

Fig. 5.12 Vegetation of the Sudd, southern Sudan (based on van Noordwijk [721].)

economy and a social system that rests heavily on the flood cycle [148, 254] (Fig. 5.13).

Heavy rain falls, between April and September, on to the flat plain, floored by clay, lacking trees and crossed by large rivers. Close to the rivers is a permanent swamp of papyrus, reed, *Phragmites mauritianus* and floating or part-floating beds of water hyacinth (*Eichhornia crassipes*) and hippo grass

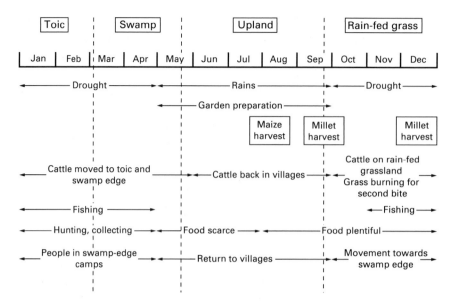

Fig. 5.13 Annual cycle of the Nuer (based on Evans-Pritchard [254].)

(*Vossia cuspidata*). At the edge of the permanent swamp are regularly flooded swamps, with *Typha australis* and grasslands, called toic. The toic has grasses, such as *Echinocloa pyramidalis*, *Echinocloa stagnina* and wild rice, *Oryza barthis*, and is annually fertilized by silt-laden river water.

Further distant from the rivers, the toic is replaced by grasslands, which flood (to depths of up to a metre and largely with rainwater) in the wet season and are less fertile. The upland, which is not flooded, is reached often some tens of kilometres from the rivers. The area is rich in both large aquatic animals and many bird species, which move on to the grasslands as the water recedes. These include most of the world population of the Nile lechwe and several hundred thousand tiang, an antelope subspecies of the tsessabe (*Damaliscus korrigonus*).

The Nuer have permanent villages at the edge of the upland, close to the plain. They keep cattle, and movement further into the upland is discouraged by the risk of nagana (cattle trypanosomiasis), a protozoan blood disease carried by the tsetse fly *Glossina morsitans*, characteristic of shady woodland. The floodplain is clearly unsuitable for permanent villages, but the Nuer must stay close to it, for it provides grazing for the cattle. The swamps beyond it are a source of fish when milk yields are low in the dry season.

When the rains begin, in May, the village gardens are sown with millet (*Sorghum*), some maize, beans, gourds and tobacco, which will be harvested in July and August (Fig. 5.13). This work is done mostly by older people, the younger being still in temporary camps near the swamp, where, at the end of the dry season, there is still some grazing and where the lagoons and diminished stream channels are a concentrated source of fish. Everyone returns to the village by June and the wet season provides an ample food supply, based on milk and its soured products and meat from sheep and goats. Blood is drawn from the necks of the cattle, coagulated and roasted.

Meanwhile, fishing is difficult, for the fish are dispersed in the flooded grasslands, which are difficult to cross. The lack of trees means that boats are scarce. When the millet is harvested, there are wedding, initiation and other ceremonies and the social organization is at its most complex. A second sowing of millet in August is harvestable by December, when the rains have stopped and when, particularly if the rains have been poor, food may be scarce.

The cattle are taken by the younger people on to the floodplain in August, while the older people remain to harvest the millet. The grazing produced on the rain-flooded grassland is of low quality for cattle and soon dries. Some areas may be burned to bring new shoots, but the main grazing is in the toic. Camps are made on mounds at the swamp edge, to which the older people also move after the millet harvest. These camps simply comprise windscreens of swamp grass, and the social life is family-based and geared to the business of survival. Milk yields fall as the toic grasses shrivel and the millet store dwindles.

To the rescue comes fishing, the third focus of the Nuer economy, together with collection of wild dates (*Balanites aegyptica*), wild rice seeds and water-lily rhizome (*Nymphaea lotus*) to eat. The swamp fish include catfish (*Bagrus, Clarias*) and lungfish (*Protopterus*) which are speared behind temporary grass dams as they attempt to move back to the river. A line of withies, thin branches of ambatch (*Aeschynomene*), may be pushed into the mud in front of the dam. The quivering of these as the fish move against them signals the target for the spearsman. Lungfish aestivating in the mud may be sounded out by tapping the surface with a pole or by recognizing sounds the buried fish make when a finger is scraped over a gourd. They can then be dug for. There is some hunting, largely for products such as skins of waterbuck for bedding or of hippo to make shields. Cattle stealing, especially by the Nuer from the Dinka, is not unknown.

When the rains return, the older people move back to the villages to prepare the gardens and the cycle is completed (Fig. 5.13). Each of its three parts is essential; diseases, such as rinderpest, prevent any great expansion of the cattle population; the climate precludes total dependence on grain, the crops of which often fail; and the dispersion of the fish in the wet season means that fishing cannot be the sole source of food. Since 1961, a rise in the discharge of the Nile, linked with a 1–2 m rise in the levels of Lake Victoria at the head of the river, has led to an extension of the permanent swamp and a marked reduction in the grassland. The Nuer have turned more to fishing, although they still regard themselves primarily as cattle keepers.

5.8 Floodplain fisheries

For many people, inland fisheries are their major source of protein and, among fresh waters, floodplains provide the most productive and diverse fisheries. Part of the Amazon floodplain illustrates this. In the Amazon basin, the rains are seasonal and water levels change greatly between the wetter season (December to June) and the drier (July to November). Differences between high and low water may be 16–20 m. This means that the forests are flooded, sometimes to the treetops, and the lagoons and lakes alongside the rivers may vary in depth from 16 m to less than 1 m.

The fish fauna is diverse and productive, with strong tendencies for the fish to migrate. One of its particular characteristics is dependence on the food supplied by the forests at high water. Among the characins, which, with the catfish, constitute 80% of the list of at least 1300 species, are many species feeding on seeds, fruits, flowers, insects and monkey dung falling from the forest canopy. Fish returning to the main channels as the water recedes are fat from this source. Fruit and seed eating appears to benefit the trees also, because often only the fleshy fruit is eaten and the seed is not digested but is defaecated to germinate elsewhere [354, 355].

The epiphytic microorganisms (periphyton) that grow on the trees and on tree debris are another important food, especially for some large catfish species. Around Porto Velho, on the Rio Madeira, 87% of the commercial catch comes from nine genera, of which a third of the species depend directly on forest seeds and a quarter on forest detritus and periphyton; the rest are predators on the seed and detritus eaters. Reptiles and mammals are also integral members of the system. Where their numbers have been reduced by hunting, it has sometimes been noticed that fish (production has fallen [270, 271].

These animals may mobilize nutrients from the sediments and soils to the water when they feed on plants (manatee, capybara, tapir) or from larger fish (dolphins, turtles, caimans, anacondas). They move, at high water, into the swamps or flooded forest, where most of the fish spawn, producing young that require small planktonic organisms as their first food. The pulse of nutrients excreted by the pursuing predators may support planktonic production at the appropriate time.

A varied fish fauna leads to a varied fishery, and methods [354, 355] vary from the simple and cunning, but dependent on a deep knowledge of the habits of the fish, to the use of modern monofilament gill nets, backed by refrigerator ships. The former support subsistence and the latter commercial fisheries. The methods can be grouped into those exploiting fish in the rapids, the main channel, the swamps and the flooded forest.

Twenty kilometres above Porto Velho on the Rio Madeira lie the Teotonio rapids, with a complex of fast and quieter water and pools as the water moves between the boulders. Many catfish species migrate up and down these rapids prior to spawning and a long oral tradition exists to advise fishermen of where to catch particular species at particular times as they negotiate the cataract. Wooden platforms are built over a particular stretch of quieter water through which a big catfish, the dourada (*Brachyplatystoma flavicans*), moves. The water is stroked with curved hooks bound to a 5–8 foot pole. Once a fish is gaffed, the hook comes free on a line attached to the pole and the fish is played until it tires and can be pulled out.

In the main channel, use of the gill nets also depends on knowledge of seasonal migrations. Some characin genera, *Brycon* and *Mylossoma* live in a clear-water tributary, the Rio Machado, during the low-water season, but, as the flood rises, move downstream into the more turbid Rio Madeira to exploit the detritus and periphyton on the flooded forest trees in the main river valley. They move in a 10–14-day period when the water is a few metres below the peak flood. The fishermen watch for dolphin activity, which indicates a characin shoal, in the tributary. They then manoeuvre upstream of the shoal and place the 100–200 m gill net in a horseshoe pattern before beating the water with paddles to scare the fish, causing them to reverse direction and move into the net, which is then drawn tight with a rope threaded through the bottom of it. In May to July, the same method is used for other species as they return from feeding in the Rio Machado forests to move upstream in the main river.

In the swamps and the flooded forests, the methods depend on individual skills. The swamps harbour cichlid fish, which move along the edges of the floating mats of vegetation. At night, they can be paralysed with a light and stabbed with a 1.5 m, pronged spear. One voracious cichlid predator, *Cichla ocellaris*, is lured by movement through the water of a tassle of strips of red cloth or birds' feathers, in which are embedded hooks. It mistakes the lure for its prey. The picarucu (*Arapaima*), which must return to the surface to breathe every few minutes, is harpooned when it does so.

In the flooded forest (a refuge against large-scale fisheries, where nets would be tangled), individuals can be caught from a knowledge of their diet. Seeds and fruit are released from the trees sporadically and are scarce. They represent large but infrequent meals and are usually snapped up as they fall (their characteristic shapes and sizes making specific 'plops' thought to be recognized by the fish). These noises, particularly those of the jauri, a palm tree, (*Astrocaryum jauri*), can be simulated with metal ball-bearings or nuts cast on lines. The fish are harpooned as they dart in to feed. Seeds can also be used as bait in simple rod-and-line fishing, with palm-fruits or those of rubber trees (*Hevea*) or a cucurbit (*Cuffa*) being the most successful baits.

5.9 Modification of floodplain ecosystems

5.9.1 Wetland values

River floodplains are notable parts of landscapes. To some peoples, they constitute home. To others (those who live on the surrounding uplands), they may represent a threat – from disease or the fear of being drowned in them. Or they may be seen as 'wasteland', useless but potentially convertible by drainage to fertile agricultural land. The former fears are reflected in the association of swamps with malevolent trolls, water witches and the like in traditional children's literature. Swamps have acquired a poor image, and nations have encouraged the drainage of them with alacrity (Table 5.4), until the recent realization of their value as wetlands [223, 224, 629]. This value, reflected in increasing legislation to protect swamps, where before there was only legislation to finance the draining of them, rests in three areas: flood control, sediment and nutrient retention and wildlife [432, 433] (Table 5.5).

Swamplands are large sponges, with often great surface area. A small increase in water level results in the temporary storage of water, which

Table 5.4 Threats to the maintenance of particular groups of functioning wetlands. Freshwater marshes for this purpose are the extensive wetlands often associated with lakes, peatlands include large areas of muskeg, blanket bog and wet tundra and swamp forests include boreal forest wetland. (From Dugan [224].)

	Floodplains	Freshwater marsh	Peatland	Swamp forest
Drainage (agriculture, forestry, mosquito control)	+ +	+ +	+ +	+ +
Dredging and stream channelization	+ +	+	–	–
Filling in (waste disposal and industrial development)	+ +	+ +	–	–
Conversion for aquaculture	+	+	–	–
Construction of dams, walls for flood control, irrigation	+ +	+ +	–	–
Pollutant and sediment discharge	+ +	+ +	–	–
Mining for peat, coal, gravel, etc.	+	–	+ +	+ +
Groundwater abstraction	+	+ +	–	–

+ +, Major threat, +, significant threat, –, no or minor threat, on a world basis.

Table 5.5 Values provided to human societies by particular sorts of wetlands. For definitions see Table 5.4. (From Dugan [224].)

	Floodplain	Freshwater marsh	Peatland	Swamp forest
Functions				
Groundwater	+ +	+ +	+	+
Groundwater discharge	+	+ +	+	+ +
Flood control	+ +	+ +	+	+ +
Erosion control	+	+ +	−	−
Sediment/toxicant retention	+ +	+ +	+ +	+ +
Nutrient retention	+ +	+ +	+ +	+ +
Biomass export	+ +	+	−	+
Storm/wind protection	−	−	−	+
Recreation/tourism	+	+	+	+
Products				
Forest products	+	−	−	+ +
Wildlife	+ +	+ +	+	+
Fisheries	+ +	+ +	−	+
Grazing	+ +	+ +	−	−
Agriculture	+ +	+	+	−
Water-supply	+	+	+	+
Attributes				
Biological diversity	+ +	+	+	+
Cultural	+	+	+	+

+ +, Major importance, +, importance, −, lesser or negligible importance, on a world basis.

would otherwise rush downstream after a heavy storm to damage human settlements. The effect is to spread out the flood peak over time, reduce its height and minimize erosion of the downstream banks. It is estimated that, for the Charles River catchment in Massachusetts, the loss of the 8422 acres of wetland (a small proportion of the total catchment) would result in average annual flood damage downstream of $17 million (1985 value) [731]. In the spreading out of the water, some is also lost by evapotranspiration, and this may also help alleviate downstream damage, although it also means a loss of water that might have been used for irrigation or domestic or industrial water.

There is some controversy over the amounts lost [433, 574, 898]. Evapotranspiration from the swamp vegetation may actually be less per unit area than from an open-water surface, by as much as 40%. The vegetation reduces its own loss by the closing of stomata in hot, windy conditions, whilst also shading and cooling the water surface. A wetland may thus save water. However, it will only save it if it would otherwise be spread out in a sheet as a lake behind a dam, for example. More water will be lost in total if the water spreads out in a swamp than if the water is allowed to move downstream in a direct channel.

5.9.2 Swamps and nutrient retention – reed-bed treatment

Sediments and nutrients are retained in swamps [435, 967], the former by the effects of reduced current [723] and the latter if the swamp is laying down peat, which inevitably retains some mineral nutrients. Nitrogen will also be removed from the water by denitrification. These properties may improve the quality of the downstream water. Sediment might otherwise block irrigation channels or the main channel if used for navigation, or might need to be expensively filtered out if the water is used for industry or domestic supply. Excessive nitrogen and phosphorus can cause problems for lakes (see Chapter 6) and increasing nitrate concentrations may be a threat to human health.

Mineral retention in swamps has been seen by civil engineers as a treatment for domestic and industrial effluents, which is much cheaper than building and operating works for chemical treatment. Small 'artificial' wetlands of reed, *Phragmites*, are being established at a number of sewage-treatment works [321, 828]. Large natural swamps are also being viewed as potential treatment works [209, 512]. What are the limitations of such treatment?

The capacity of a swamp to cope with sediment and chemicals is not infinite; only for denitrification does the adding of a substance not cause changes in the swamp, because the ultimate destination of the nitrogen is the atmosphere and not the swamp itself. Adding sediment to a natural wetland will cause successional changes. The swamp surface will rise relative to that of the water until a drier system is produced, which floods for only part of the year and eventually not at all. Such successional changes, which eventually preclude the use of the ecosystem as a nutrient retainer, may be slow, however, and retention in the short term may be very successful. Succession might be prevented by harvesting the vegetation and thus removing some of the accumulated nutrients. There is a fallacy in this, however. Cutting of the above-ground plants at the height of their growth, when they contain most nutrients, will eventually kill them, while removal when they are senescing will result in little harvest of nutrients, which will, by then, have been translocated to the rhizome. In any case, the major sink for the nutrients (other than the atmosphere for nitrogen) is the sediment. The plants create the optimal environment for sedimentation but do not, when the vegetation is fully established, have any net uptake from year to year.

Swamp soils will retain metals, including toxic ones from industrial processes, and burial in the sediments may seem attractive. However, plant roots may absorb such elements and mobilize them into food webs, or the build-up of such substances may ultimately kill the swamp plants or interfere with important bacteria in the sediments. Experiments with such disposal should be encouraged, but only with specially established artificial wetlands, not with natural ones where the properties of flood control, denitrification and natural sediment retention are too valuable to be jeopardized.

Recent surveys [161, 828] of such constructed wetland systems have tempered earlier enthusiasms and increased our understanding of how they might best be used. Many such systems have now been constructed, usually on a small scale. In Europe, they usually use common reed (*P. australis*) but sometimes other species (*Typha, Phalaris*). They most effectively remove sediment and biological oxygen demand (a measure of the labile organic content of the water) and a varying proportion of the dissolved nitrogen compounds. They are much less reliable at removing phosphorus, and may become sources to the outflow water after a year or so.

Reeds nearly double the rate of oxygen input into the sediments compared with simple diffusion from the atmosphere [21], but respiratory demands of the plant biomass use most of the additional oxygen [92]. Some reed-beds have been used for the treatment of raw sewage and have functioned poorly; the best results have been obtained where sewage effluent has been run into them for final 'polishing', where areas have been large in relation to the population served, where the growth season of the plants is long, and where a variety of plant types has been used in a mosaic of habitats through which the water flows. Original estimates of a requirement of about $1 \, m^2$ of reed-bed per person have been found to be too low and $2–5 \, m^2$ may be more appropriate. This means that space requirements are high, with a town of 100 000 people needing up to 50 ha to produce an effluent that has had most of its added N removed. Constructed reed-beds may be best used for hamlets, small villages or farms to treat the effluent from septic tanks or other pretreatment processes that have already removed much of the organic matter.

Constructed wetlands offer some advantages to wildlife conservation and they use no energy, pumps or mechanical devices (until they need to be dug out and regenerated). There can be no objection to further investigation to improve their performance in aspects for which they are well fitted. Natural floodplain swamplands, however, have aesthetic and wildlife values, which are important and much greater than those of any constructed wetland. In the USA, 20% of threatened or endangered plants and animals are associated with wetlands. Floodplains figure prominently in lists of national parks and wildlife reserves. The cheap disposal of domestic or industrial waste cannot be justified at the expense of long-term damage to a habitat that is being increasingly lost through drainage.

5.9.3 Floodplain swamps and human diseases

The close links between cycles of flooding, high fish production and movements of people and stock that support human settlement in floodplains are not without problems. Some of the most devastating human disease organisms have also fitted their life histories to this productive system (Fig. 5.14). Hundreds of millions of people are exposed to malaria, schistosomiasis and filariasis, tens of millions to yellow fever, encephalitis, trypanosomiasis and lung and liver flukes. The vectors of dispersal are often wetland animals. Such vectors occur elsewhere – at lake margins and in irrigation ditches, for example – but the extensiveness of floodplain swamps and the large associated human populations make such areas particular foci. Destruction of these areas is thus often advocated on health grounds.

Malaria is perhaps the most familiar. It was not confined to the tropics until relatively recently, for 'paludism' or 'marsh miasma' was a risk of the

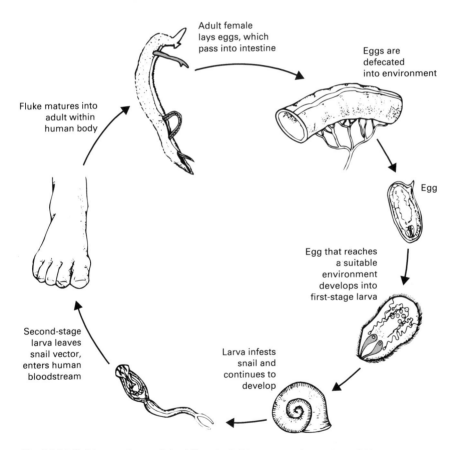

Fig. 5.14 Life history of one of the bilharzia (schistosomiasis) parasites, *Schistosoma haematobium.* The scales vary. The adult fluke is a few millimetres long, the eggs and larvae are fractions of a millimetre and the snail about 1–2 cm.

European fenlands until the nineteenth century. As late as 1827, people were fearful to enter the fens around the Wash in England because of the disease. In the eighteenth century, parishes in Kent associated with marshlands, such as Romney Marsh (Fig. 5.15), had much higher infant mortality than those of the uplands, and clerics declined to occupy these parishes for fear for their health [214]. Similar incidences of the disease occurred in the coastal marshes around the Thames and the Essex coast.

Malaria is a protozoan parasite of red blood cells and is carried by about 60 of the 400 or so species of the mosquito genus *Anopheles*. The female needs several blood meals to complete development of her eggs. In feeding, she may transfer cells of the malaria parasite *Plasmodium* between human hosts. There are at least four important *Plasmodium* species; some (*P. falciparum*) are more dangerous than others (*P. vivax*), but all are fatal if not treated. *Anopheles* lay their eggs as floating rafts on almost any still water suface. Water among aquatic plants is ideal, as long as it is not completely deoxygenated, for there may be some cover from egg predators.

Control of malaria has long been sought, because over 2 billion people are exposed to the disease; about 250 million suffer from it at any one time, with 1.3 million deaths each year. In 1956, the World Health Organization began a major campaign of spraying settlements with the inexpensive insecticide dichlorodiphenyltrichloroethane (DDT) to kill adult mosquitoes.

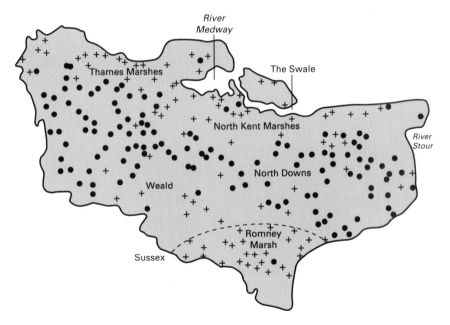

Fig. 5.15 Births and deaths in Kent between 1661 and 1681. Filled circles show an excess of baptisms over burials in the parish reservoirs and crosses an excess of burials over baptisms. The crosses coincide with marshland parishes, where ague (malaria) was prevalent. (Based on Dobson [214].)

This was very successful, but plans for spraying have often fallen into disarray for social or financial reasons or because war has intervened. Resistance to DDT has developed in some *Anopheles* species. Drugs are available, but resistance of the parasite to many of these has developed and costs often preclude their widespread use. A third line of control might be immunization against the parasite or genetic engineering of the host or parasite so as to decrease its transfer or virulence. Much research now concentrates on this.

If the latter is successful, the fourth line – destruction of the swampland habitat by drainage, a method used in the nineteenth century in Europe – might be avoided. Indeed, drainage and replacement of the swamp with irrigated agriculture and a plethora of canals may sometimes provide a better breeding habitat for the mosquitoes than the original wetland. In the UK, the marsh fever was probably *P. vivax*, carried by *Anopheles atroparvus*, which breeds in slightly saline water. From before the sixteenth until the early twentieth century, it was a debilitating disease of the poor, who tended cattle on the rich marshland pastures. The mosquito overwinters in dark, warm places, such as those provided in the hovels of the marshmen, who developed resistance to the disease, helped by opium, alcohol and, more effectively, quinine from the Peruvian cinchona bark, which they bought from the local apothecaries. Incidence of the disease declined from the seventeenth century onwards and this has been attributed to drainage. However, the mosquito breeds in small stagnant pools and prevalence of the disease was greater in warm, dry years, when such pools were not readily flushed out. The mosquito is still common and supported an outbreak of malaria in north Kent during the 1914–18 war, based on reintroduction of the parasite by troops returning from Greece and India. It is more likely that decline of indigenous malaria was due to generally improving health and housing conditions and increased herds of farmstock, which form alternative blood sources for the mosquitoes [214].

Mosquitoes carry other diseases. *Aedes* and *Culex* transfer arbor and other viruses, among which yellow fever and Japanese B encephalitis are well known, with the former being preventable by immunization. Others, such as Marburg virus, are at present incurable, though fortunately rare. More widespread (about 300 million current cases) is filariasis, often called elephantiasis, after a characteristic symptom. Several mosquito genera (*Mansonia*, *Culex* and *Aedes*) carry the microfilariae, the larval dispersal stages, of the nematode worms *Wuchereria bancrofti* and *Brugia malayi*. The microfilariae are released into the human bloodstream from large (often several centimetres) adult worms, which grow in the lymph tracts, causing blockage and swelling of the tissues and eventually death. The microfilariae are produced rhythmically in large numbers each day and are present in the bloodstream at the time when the local mosquitoes are most active.

Mansonia, an important vector of *Brugia*, has larvae which, like those of other mosquito larvae, breathe air, but not by spiracles held in the surface-tension film at the water surface. *Mansonia* larvae have a saw-edged siphon bearing a spiracle and cut their way into the roots of floating aquatic plants, such as the water cabbage, *Pistia stratiotes*. They use the air supply contained in the lacunae of the plant. In this way, they may be less exposed to predation than they would be at the water surface.

No less important as disease vectors in floodplains, are snails. Many carry stages of flatworms (flukes, Trematoda), which cause debilitating and often fatal disease. Most widespread are *Schistosoma* species, which cause bilharzia or schistosomiasis. The adult flukes (Fig. 5.14) occur in male–female pairs in the veins around the intestine or bladder of the human host. An individual may carry only a few pairs or very many, each producing up to several hundred eggs per day. The eggs have spines, the position and number varying with species, and pierce through the gut or bladder wall, eventually to be voided with the faeces or urine. If, as often, this is in fresh water, the eggs will hatch to miracidia, which infect snails of particular genera. Later, cercariae are released from the snails to reinfect humans when bathing or wading. Schistosomiasis can be controlled by provision of organized sanitation, although finance and custom may prevent this. As with many European public lavatories, the great outdoors may personally be far preferable to an overused pit. Also, the main source of schistosome eggs, the 10–24-year age-group and more particularly the 10–14-year-olds, is the one most likely to make free in the open air.

Schistosomiasis is a debilitating disease, affecting whole villages and undermining much will for an active life. It can be treated with drugs, metrifonate, oxamniquine and praziquantel being safe, effective but relatively expensive. Control of the disease in the past has concentrated on killing the snails, using copper compounds, which proved relatively ineffective, or the synthetic Bayluscide.

Such approaches do not work well on large swamps, although they are locally effective – in village ponds, for example. Draining or filling in of swamps has been recommended to remove the snails' habitat, but this may ultimately increase the incidence of the disease. The irrigation schemes established in the Nile valley at Gezira following the damming of the floodplain by the Sennar Dam in 1924 led to an increase from 1% and 5% to 21% and 80% in the incidence of *Schistosoma haematobium* and *Schistosoma mansoni*, respectively, in the local populations. Most of the 74 countries and 600 million people exposed to the disease will receive only local relief in the foreseeable future.

Schistosomiasis is only one, though the most widespread, of the floodplain fluke diseases. Others include Busk's fluke (*Fasciolopsis buski*), affecting fifteen million people, lung flukes (*Paragonimus*) and liver flukes, affecting five and thirty million people, respectively. All have snail vectors and also a second swampland host. For example, Busk's fluke – a large animal, 7–8 cm long, present as adults in human blood-vessels – is carried in China and Thailand by the snail genera *Segmentina* and *Hippeutis*. The snails release not cercariae, but metacercariae, which attach to water-chestnut plants, where they change to cercariae. Water chestnuts (*Trapa natans*) are floating rosette plants with a flower which, although initially above water, droops as the fruit forms and dips underwater. The metacercariae attach to the fruits, which are often collected and eaten uncooked by children, who become infected. Reservoirs of infection in the Far East are domestic pigs, in which the adult flukes also live. The pigs may be penned over ditches so that their excreta fertilize the water in which fish are cultured for food. The fluke eggs passing in the faeces then have a very good chance of reaching the snail host.

Liver flukes include *Clonorchis sinensis*, the Chinese liver fluke, whose vectors are snails of the genus *Bithynia* and a fish, the Chinese grass carp (*Ctenopharyngodon idella*), under whose scales the cercariae develop. In Taiwan, *Opisthorchis felinus* similarly uses the common carp, *Cyprinus carpio*. In the Far East, fish is frequently eaten raw and the parasite is transferred. In Europe, the sheep liver fluke, *Fasciola hepatica*, may also infect humans, though not seriously, through their eating raw watercress, its intermediary host between snail and mammal. Finally, in West Africa and Asia, the usually fatal lung flukes are transferred to humans by the eating of raw freshwater crabs, on whose gills the metacercariae encyst.

5.10 Drainage and other alterations to floodplain ecosystems

There have been considerable changes to floodplain ecosystems, mostly for drainage and flood control. Drainage brings into cultivation the accumulated silts and peats, which make extremely fertile soils. Flood control may allow settlement in the outer reaches of the river bed by confining it to its main channel. Often both of these aims are pursued together. Some of their consequences are illustrated by the Florida Everglades.

5.10.1 The Florida Everglades

The Everglades is a complex of tree and sedge swamps based on a shallow river almost as wide, up to 100 km, as long, 180 km. It flows slowly from the lowlands around Lake Okeechobee to the sea at the south-western tip of Florida, USA (Fig. 5.16). The flow is only a few hundred metres per day and the river is at most 30 cm deep over much of the area. The main freshwater community is of emergent aquatic plants, dominated by the saw-sedge, *C. jamaicensis*, which grows several metres tall, so that the description 'river of grass' is embodied in the title of a well-known book on the area [218].

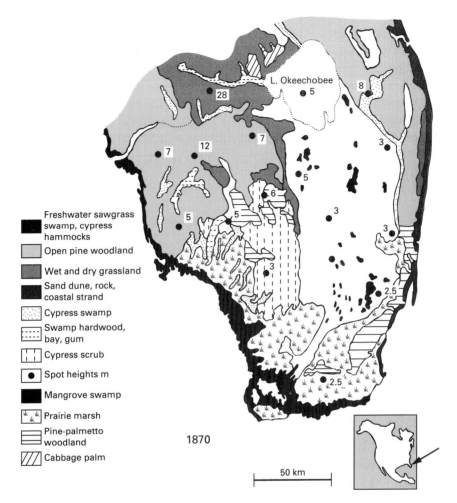

Fig. 5.16 Natural vegetation of the Everglades as it was in 1870 (based on Caulfield [144].)

Southern Florida is floored by a porous limestone (Miami oolite) and is very flat. Running along the east coast is a ridge of rock some 6 m above mean sea level, and the west coast also bears a slightly wider but still narrow and subdued 'upland'. In the basin between the two lie the Everglades, with the land dipping only 7 m from near Lake Okeechobee to the sea. Most of the 200 cm or so of rain falls in June and July, sometimes associated with hurricanes, which also occur in September. The winter is dry and the natural water supply to the Everglades is then at its least.

South of Lake Okeechobee is an area of several thousand hectares of peat, up to 4 m deep, laid down over several thousand years in swamps associated with the lake. This acts as a sponge, taking up water in wet periods and releasing it steadily all the year round. The flow penetrates the Everglades along three main watercourses or sloughs, running south or

south-westwards. When the water reaches the sea, the saw-grass community is replaced by a tangled intertidal mangrove forest up to 30 m high
(Fig. 5.17). The steady flow of fresh water confines the mangrove to the
coast by stopping inland penetration of salt water.

Dotted among the saw-grass, on islands of oolite standing a few centimetres above the general basin level, are woods of slash pine (*Pinus caribaea*)
and palmetto (*Serenoa serrulata*) (Fig. 5.17) and sometimes clumps of hardwood trees. Small groves of other trees, swamp and pond cypresses (*Taxodium* spp.), occupy depressions in the oolite where peat has accumulated.
The saw-grass does not succeed to drier forest, because fires, begun by
lightning at the start of the rainy season, have occasionally burnt the surface
vegetation and litter, though left undamaged the deeper peat and rhizomes
of the saw-grass [825]. The pine woodland is also fire-resistant and is
replaced by hardwood only after a long period, when fires have not
intervened.

(a)

(b)

(c)

(d)

Fig. 5.17 The richness of the Florida Everglades depends on the variety of habitats
created by differences in water-table and the balance between fresh and saline water.
The driest areas are the pine–palmetto woodlands (a). The sawgrass marsh (b) depends
on a slow flow of shallow fresh water and periodic light fires. The cypress swamps
(c) grow in deeper, permanently flooded areas. The forest floor with 30-cm high
pneumatophores (aeration roots) of the cypress is shown. At the coast, the mangrove
forest (d) replaces the freshwater swamps.

This complex of plant communities supports a rich fauna, of which the birds, some 250 species of them, are best known, for the area is at the junction of several migration flyways [439]. Ducks, ibis, spoonbills, herons, pelicans, coots, plovers, gulls, terns, storks and cranes are abundant. Other vertebrates are no less exciting. The alligator is most famous, but the American crocodile lives in the mangrove swamps, and snakes and turtles are common. There are fifty-seven species of reptile and seventeen of amphibia. Some twenty-five mammal species – opossums, racoons, wildcat, otter, white-tailed deer, mountain lion, black bear and, offshore, the manatee and dolphin – have been recorded, and half of these depend significantly on the freshwater communities. The fish (240 species) and invertebrate faunas are even richer.

Not surprisingly, the first threat, now much reduced, to the Everglades came from poaching of the rich fauna. In the 1890s, millions of feathers of egrets and other birds formed the raw materials of a thriving millinery trade, and, by 1930, 100 000 alligator hides were being processed into leather each year in Florida tanneries. The alligator population was reduced to only 1% of its original level by the 1960s, but it is now recovering owing to stringent protection [314]. The real danger for the Everglades' ecosystem, and also for the whole of southern Florida, now comes from interference with the natural drainage patterns.

Around 1882, it was realized that the peatlands around Lake Okeechobee would be extremely fertile if they could be drained of the standing water that covered them for eight months of the year. A canal was dredged between the Caloosahatchee River and Lake Okeechobee (Fig. 5.18). Sugar cane and winter vegetables thrived on the drained areas. In 1925 and 1928, hurricanes were severe enough to cause flooding of water from Lake Okeechobee into adjacent drained areas, which had then been settled. These areas had naturally accommodated floods, but, in 1928, between 1500 and 2500 settlers were drowned. Embankments were made around the lake and more drainage canals were dug, so that future flood water could be drained rapidly to the sea. The canals (Fig. 5.18) emerge at the coast among the built-up areas of the eastern oolite ridge and divert the flow from its former southward progress through the Everglades. There are now some 2400 km of canals and 250 powerful pumps. Half the former freshwater area has been lost and remaining flows are only a tenth of their natural volume. Pressure for drainage has been stimulated not only by farming interests but also by the warm Florida climate. For decades, the area has been promoted as a retirement and holiday haven, and the demand for building land, otherwise confined to the conastal ridges, has been aggressive. The resident population of southern Florida has increased from 3 to 12 million in the past forty years.

The canal system incorporates large, shallow reservoirs, called water-conservation areas. These can be used for temporary storage of water released from Lake Okeechobee and are needed to delay the flow to the sea

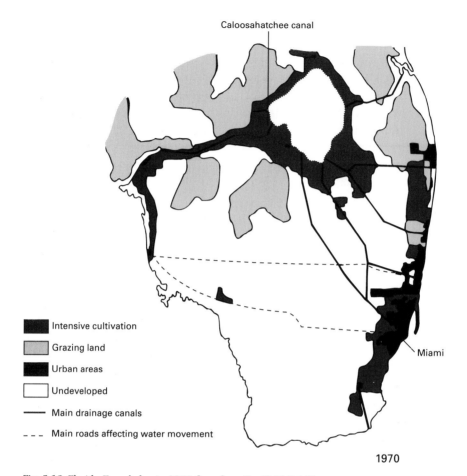

Caloosahatchee canal

Intensive cultivation

Grazing land

Urban areas

Undeveloped

——— Main drainage canals

‒ ‒ ‒ Main roads affecting water movement

Miami

1970

Fig. 5.18 Florida Everglades in 1970 (based on Caulfield [144]).

sufficiently for the groundwater aquifer, from which the coastal cities derive their freshwater supply, to be recharged. They had to be built because the natural peat sponge south of Lake Okeechobee, which bore exactly these functions, has been much reduced. Drainage results in rapid oxidation of the peat, which, once 4 m deep, now disappears at the rate of 2 cm year^{-1}.

A main road, the Tamiami Trail, skirts the southernmost of the reservoirs, and water may be released through culverts under it to the remaining southern portion of the Everglades. The supply is now insufficient and the natural seasonal rhythm of flow is not always maintained. Water is released to the Everglades largely when it is convenient for the drainage system. Between 1962 and 1965, none was released at all, so that, in 1968, it was necessary to pass legislation to guarantee at least a minimal supply for the Everglades National Park, a 500 km^2 remnant of the original 10 000 km^2. In 1970, plans were blocked just in time to prevent a large part of the adjoining Big Cypress swamp from being developed as an airport [916].

In the pristine Everglades, animals congregated in the deeper parts of the sloughs in the dry season and breeding cycles were related to this. The concentration of invertebrates at the edge of the receding water provided the wood ibis, a wading bird, with a rich food supply during the fledgling season. Food and fledging are now often out of phase. Nests built on the swamp floor in the dry season may be destroyed by unseasonal inputs of water. The wood stork, which depends on a predictable supply of fish and Crustacea for its fledglings, has been reduced in numbers by 73% since the 1930s.

The overall lack of water is probably most crucial, however [225]. The Everglades have always been subject to light fires; indeed, the diversity of their vegetation depends on fire to prevent succession in the *Cladium* swamps. But the extreme drought now caused by diversion of the water-supply, particularly in years of low rainfall, has led to especially destructive fires biting deep into the peat. Uncontrollable fires also threaten adjacent urban areas, as well as causing smoke pollution for long periods. In the dry season, fewer pools persist as refuges for fish and reptiles. This had led to heavier than normal predation and mass fish deaths due to deoxygenation. The lack of water is affecting the cities also. The ground-water aquifer is not being recharged rapidly enough to prevent sea water moving into the oolite and contaminating the drinking-water wells. New wells have had to be drilled further inland. A fishery worth $20 million year^{-1} for estuarine shrimps, which move into the freshwater Everglades for part of their life history, is at risk.

The estuary of Florida Bay adds a final twist [461]. Enclosed partly by the islands of the Florida Keys, the Bay formerly supported vast meadows of sea grasses, a community of completely submerged angiosperms. To the ocean side of the Keys is North America's only major coral reef. In the late 1980s, the sea grasses began to disappear and more than 20 000 ha have now gone. Phytoplankton populations in the water have increased and there is greater turbidity. More recently, there has been widespread death of coral species on the reef [995]. The latter may be due to disease, but this is unlikely to be the entire cause. It is more likely that declining water quality in the Bay is responsible. The reduced flows from the Everglades have led to increased salinity in the Bay. Increased intensification of agriculture and rapid movement of water through the canals to the sea, along the coast and perhaps into the Bay may also have increased the nutrient, particularly nitrogen, load. Silt loads may also have increased through this route, and together these factors are almost certainly linked with the changes in the sea-grass meadows and coral reef.

Ironically, the problems in the Bay have emphasized the need to increase the flows through the Everglades and, in 1996–97, work began on a major project to pump more water southwards through the system. Constructed wetlands will be used in an attempt to limit movements of nutrients from the sugar-cane fields south of Lake Okeechobee. Some experts believe that the increased flow will decrease the salinity of the Bay and restore the sea-grass meadows; others argue that unless the nutrient and silt contents of this water are also markedly improved, pumping of more low-quality water to the sea will simply worsen the problem. It may be a classic, but not unfamiliar, case of a positive feedback in the creation of problems, once the functioning of a natural system is disturbed.

5.10.2 Drainage and river management in temperate regions

The floodplains of most temperate rivers have been greatly altered. Rivers now rarely expand over the plain, but are confined in deepened and often embanked channels, sometimes perched above the surrounding land. The surrounding soils, especially if peaty, have oxidized and shrunk on drainage, sometimes by a metre or two. Even with embankment, there may still be some risk of flooding of the now agricultural or urban land of the plain, so the channel may be managed.

Its capacity to carry water may have been increased through dredging and deepening; it may have been straightened (canalization) to minimize siltation on bends, and its cross-section, particularly in urban reaches, may have been formed, sometimes with concrete, into a trapezoidal shape, which least impedes flow. Bank vegetation, overhanging trees, debris and aquatic plants may be regularly cut, because these may also impede the flow and increase the risk of flooding [102, 719]. All such measures decrease the variety of habitat for the channel ecosystems, while the floodplain swamps may be reduced to communities capable of growing in the ditches draining the land. Urban, industrial and agricultural pollution may reduce water quality, so that the river-channel ecosystem is altered in so many ways that it may be difficult to link cause and effect.

The extent of river alteration is now very extensive. In the USA, over 26 500 km of rivers are so managed, while in Britain a quarter (8500 km) of all main river lengths in the lowlands have been severely altered (canalized, dredged, piled) and virtually all the remainder is managed in a lesser way (removal of aquatic plants and bankside trees and shrubs). Management is equally extensive in mainland Europe.

The effects are often to reduce the numbers, production and diversity of the bottom communities [46]. In the upper Rhine, the turtle *Emys orbicularis* has probably been eliminated, and, in one Swedish river, the smooth flow brought about by channelization has increased the incidence of biting blackflies (*Simulium*). A main reason for the decline of otters in British lowland rivers is loss of habitat, for cavities between bankside tree roots are preferred sites for nest building.

A managed and canalized river is not pretty (Plate 4) and there is doubt that the costs to the community of the management, even if aesthetic and conservation considerations are ignored, are matched by even equal benefits. Important functions, such as the nitrogen removal brought about by floodplain wetlands and swamps and the flood-storage capacity of the swamps, are clearly lost. The consequences of removal of such capacity upstream are that more flood-defence works must be installed downstream. The water must go somewhere.

Current opinion, alarmed at the consequences of intensive agriculture, has forced a more sensitive approach [719]. The drainage engineer's ideal of a straight, smooth, trapezoidal channel may be softened by the provision of lengths of near-natural channel, by the keeping of bends as relief channels and by limited planting of trees at the edge of a shallower berm adjacent to the main dredged channel (Fig. 5.19). This change is welcome but only cosmetic. When it becomes necessary to provide submerged lengths of pipe to provide enough cover for fish, then it seems clear that our practice of floodplain drainage has been unbalanced [244].

It is poignant that past civilizations in South and Central America before the Spanish conquest found ways of cultivating swamps by building raised platforms of soil in them, on which a range of crops were grown [18, 190, 191, 530, 742]. The platforms were fertilized by recycling the silt accumu-

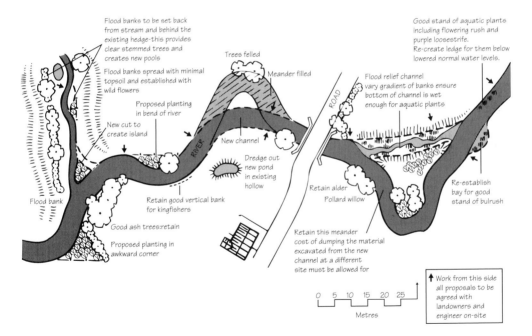

Fig. 5.19 Proposals for the engineering of drainage schemes have become more sympathetic, though still highly damaging, to rivers. In this example of a proposed design, one meander has been straightened but another kept by provision of a flood-relief channel. It represents the designer's ideal. The proposals may not prove acceptable to the landowners and engineer working on the site. (From Newbold *et al.* [719].)

lating in the channels around the platforms and by growing nitrogen-fixing legumes with the other crops. The method allowed much of the swamp to be preserved without drainage. These methods are currently being revived in Mexico [191] and around Lake Titicaca [125].

But against too sweeping a condemnation of the civil-engineering approach to river management must be set two considerations. First, such management has its origins, on a small scale, several centuries in the past, when it must have appeared entirely desirable. Once drainage and flood control have begun, they cannot easily be ended without abandoning farms or moving people. There is understandable resistance to this; maintenance of cultivated land represents for many people a coping with the vagaries of nature and perhaps subconsciously of themselves. And, secondly, in the developing world, people moving from the countryside to the cities to find work may frequently find the only places to build their shanties are on the edges of floodplains. Canalization may provide a less risky home for them. Problems of the floodplain wetlands ultimately arise from those of population, poverty and inequitable social and economic systems, as much as economic greed.

5.10.3 River restoration

In some countries, a realization of the benefits of more than cosmetic restoration of rivers and their floodplains has led to some reversal of the damage caused by unsympathetic river engineering in the past [473]. The key to successful restoration is recognition that a river is not just a series of separate lengths, each of which can be treated alone, but a connected system, in which the upstream lengths and the catchment all combine to determine its overall nature. Sear [879] points out that this system takes a long time to adjust to changing conditions. Sediment loads and deposition, channel plan and form (meanders), depth and width, riffles and deeper pools are not random features but respond to the overall as well as the local conditions.

Many restoration attempts have been disappointments for lack of proper geomorphological understanding, for, unless all the former conditions of a river length are restored, including apppropriate land use, it is rarely possible to reinstate the particular, pre-engineering features. A new design, a compromise that adds function, aesthetics, diversity or some combination of these and yet reflects the new constraints, is needed for a successful project [99–101]. We thus rehabilitate most rivers, rather than restoring them to some previous pristine state.

Before restoration work is carried out, it is useful to know how the existing system falls short of a fully functioning one. A number of diagnostic tools exist. The system for evaluation of river conservation value (SERCON) [78] assesses the river on many criteria, including biological diversity, and has been devised to identify relatively undamaged river stretches for designation as conservation sites. The river, channel and environmental system (RCE), devised by Petersen [767], is an effective way of assessing more quickly the state of the river stretch and has the advantage of comparing the stretch against a notional baseline of what it would have been like in an undisturbed state. It pays most attention to the extent to which the channel has been modified and hence gives a rapid index of the potential loss of functional values. It applies, however, only to northern temperate streams up to about 3 m wide in lowland agricultural landscapes. These, however, are systems that have suffered considerable damage.

The RCE takes about 20 min for assessment of a 100 m stretch of river. Points are allocated under 16 headings (Table 5.6), with most points being given to states of the river closest to the notional baseline pristine state (a heavily shaded stream, with a wide wooded riparian zone, from which numerous debris dams of twigs and timber are supplied, and to whose floodplain nutrients and water are supplied from the main channel at regular intervals during the flood season). The greatest emphasis is given to major structural characteristics of the channel and bank (riparian) zone that can be reliably assessed, and the least to those features (e.g. fish community) which may be more difficult to determine and hence subject to the greatest risk of error. The scheme works well and its results correlate satisfactorily with other, more time-consuming, schemes, which use assessments of the benthic invertebrate community (see below).

5.10.4 Approaches and methods of river and floodplain restoration

The lead in work to reinstate a fully functioning river and floodplain system has been taken by Denmark [473, 622]. Previous legislation encouraged severe management of rivers and floodplains for drainage or power generation. From 1973 (the Environmental Protection Act) and particularly from 1982 (the Watercourse Act), new laws promoted gentler techniques, concentrating on diverse communities, fisheries, amenity and aesthetics. The results have been dramatic, although it will be many years before the new techniques have been applied to all the relevant watercourses.

The first approach is that of gentle management. Previously, water plants were vigorously cleared, often by machine. This wrecked trout and invertebrate habitat, led to movements of sand and silt and smothered spawning beds. The new (or rather readopted approach, for the new techniques are old, traditional ones) is to cut only when absolutely necessary for flood control. The water-carrying capacity of a channel is only marginally increased when all the plants are cut. Cutting by hand is encouraged to favour plants like water crowfoot (*Ranunculus baudotii*), which block the channel less than those like bur reed (*Sparganium*), which accumulate too much sediment. No more than two-thirds of the channel

Table 5.6 The riparian, channel and environmental inventory for small, streams in agricultural landscapes. The scheme is essentially that of Petersen [767], but small modifications of terminology have been made following trials under UK conditions. For a 100 m stretch, one of the four possible scores is selected for each characteristic. Finally, the scores are added and the total interpreted for recommended action.

Score	Recommended action
293–360	Monitoring and protection of existing status
224–292	Selected alterations needed for restoration of full functional value
154–223	Minor restoration needed
86–153	Major restoration needed
16–85	Complete structural reorganization is required

Characteristics	Points
1 Land-use pattern beyond immediate riparian zone	
(a) Undisturbed forest, natural wetlands, bogs or mires	30
(b) Permanent pasture with some woodlands and swamps and little arable agriculture	20
(c) Mixed arable and pasture agriculture	10
(d) Mostly arable agriculture	1
2 Width of riparian zone from stream edge to field	
(a) > 30 m	30
(b) 5–30 m	20
(c) 1–5 m	5
(d) No riparian zone	1
3 Completeness of riparian zone	
(a) Intact, without breaks in vegetation	30
(b) Breaks at intervals of > 50 m	20
(c) Frequent breaks, perhaps with erosion gullies and scars every 50 m	10
(d) Deeply scarred and disrupted or stream is channelized	1
4 Vegetation of riparian zone	
(a) > 90% plant cover of well-established trees/shrubs or native marsh plants	25
(b) Newly established species along channel, mature trees behind	15
(c) Mixed grasses and sparse trees/shrubs	5
(d) Mostly grasses, few or no trees/shrubs	1
5 Retention devices	
(a) Channel with rocks and old logs set firmly in place	15
(b) Rocks and logs present but back-filled with sediment	10
(c) Retention devices loose; moving with floods	5
(d) Channel of loose sandy silt; few channel obstructions	1
6 Channel geometry	
(a) Ample for present peak flows, width to depth (W/D) ratio < 7	15
(b) Overbank flows rare, W/D 8–15	10
(c) Barely contains present flows, W/D 15–25	5
(d) Flows frequently spill out, W/D > 25, for stream is channelized	1

[As sediment collects in a stream exposed to erosion in the catchment, its bed may fill with sediment and it may become relatively shallow. Hence the emphasis on W/D ratio. It may then heat up excessively in summer and will be unable to contain even moderate discharges unless it widens its bed, causing severe bank erosion in the process.]

7 Channel sediments	
(a) Little or no channel enlargement from sediment accumulation	15
(b) Some gravel bars of coarse stones but little silt	10
(c) Sediment bars of stones, sand and silt common	5
(d) Channel divided into braids or stream is channelized	1

Table 5.6 (*Continued*)

Characteristics	Points
8 Stream-bank structure	
(a) Banks stable, of rock and soil firmly held by plant roots	25
(b) Banks firm but only loosely held by roots	15
(c) Banks of loose soil with sparse plants	5
(d) Banks loose and easily disturbed	1
9 Bank undercutting	
(a) Little evident or confined to areas with tree-root support	20
(b) Cutting only on curves and at constrictions	15
(c) Frequent undercutting	5
(d) Severe, banks falling in	1
10 Feel and appearance of stony substratum	
(a) Stones clean of silt, rounded, may have blackened patina of iron and manganese salts	25
(b) Stones without sharp edges but with slight gritty feel	15
(c) Stones with sharp edges or obvious gritty cover	5
(d) Sharp edges and thick silt cover	1
11 Stream bottom	
(a) Several sizes of stones packed together, with obvious interstices	25
(b) Stony bottom easily moved but with little silt	15
(c) Bottom of silt, gravel or sand, stable in places	5
(d) Loose bottom of sand and silt, but no stones	1
12 Riffles and pools, or meanders	
(a) Distinct, occurring at intervals of $5-7 \times$ stream width	25
(b) Irregularly spaced	20
(c) Long pools, separated by short riffles, meanders absent	10
(d) Meanders and riffles/pools absent or stream channelized	1
13 Aquatic vegetation	
(a) Absence of mosses and patches of algae	15
(b) Algae (filaments or films on stones) dominant in pools, vascular plants at edge	10
(c) Mats of filamentous algae present, some vascular plants, few mosses	5
(d) Algal mats cover bottom, vascular plants dominate channel	1

[This scoring may seem counterintuitive. However, in a wooded stream, shading prevents much plant or algal growth, and nutrient availability is low where an extensive littoral zone acts as a nutrient filter. A damaged system is thus one that encourages plant growth by access of light and nutrients.]

14 Fish	
(a) Native stream fish (trout, dace, minnow) obvious	20
(b) Native stream fish scarce and difficult to locate	15
(c) Fish of sluggish waters (cyprinids, such as carp or roach, or, in North America, centrarchids, such as bluegills) present but native stream fish absent	10
(d) Fish absent or very scarce	1
15 Detritus	
(a) Leaves and wood without sediment	25
(b) Leaves and wood scarce; fine flocculent debris but without sediment	10
(c) No leaves or woody debris; coarse and fine organic matter with sediment	5
(d) Fine sediment, anaerobic below surface, no coarse debris	1
16 Macrobenthos	
(a) Many species present on all types of substrate	20
(b) Many species but only in well-aerated habitats	15
(c) Few species but found in most habitats	5
(d) Few, if any, species and then only in well-aerated habitats	1

width is cut and then not before July or August, so as to allow time for development of trout fry. Sinuous pathways are created for the water among the plants, thus increasing habitat complexity and giving a variety of current flows. There is no mechanical digging into banks, so that they collapse and blanket the channel with soil. Where banks need attention, a graded, not precipitous, slope is left, with its vegetation as intact as possible. Grazing meadows to the sides of the streams are fenced to prevent cattle destroying the banks and watering-places are carefully controlled by directing the cattle to places provided with gravel hard standing, again to prevent mobilization of sediment.

The next step is to improve the habitat by structural modification, such as provision of new gravel beds for invertebrate colonization and trout spawning, where these have been smothered by sediment. Sediment traps – essentially upstream ponds in the line of the channel – may be needed to maintain these new beds. Banks may be stabilized to prevent further erosion, overhanging structures provided for fish shelter and large stones to deflect currents and provide areas of faster flow. Where the migrations of animals have been blocked by weirs and dams, these may be replaced by stepped riffles, rather than vertical walls, or bypasses that allow movement in a channel around the dam.

Both gentle management and structural improvement require continued attention. Unless the sources of sediment, in erosion of the surrounding agricultural land, are removed, new gravel beds will become smothered or will eventually be moved downstream by currents. They are rehabilitation measures, not permanent restoration. Provision is made, however, in the Danish Acts, for full restoration and there are now a large number of examples where straightened channels have been remeandered (Fig. 5.20)

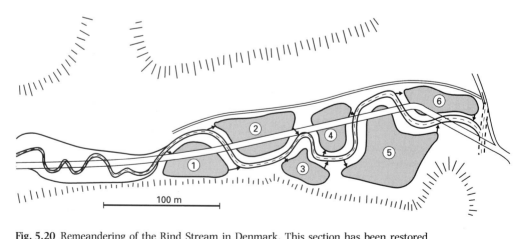

Fig. 5.20 Remeandering of the Rind Stream in Denmark. This section has been restored for nature-conservation purposes and includes six shallow basins (numbered) for colonization of wetland communities. The former straight engineered channel is shown, as well as the restored meandering channel.

and the previously deepened channel bottom raised so that water again spills out during heavy rain on to the floodplain. Streams that had previously been directed underground in culverts have been brought back to the surface and the fully functioning system thus restored. Such restoration requires that land previously used for agriculture is given back to the river system. Financial compensation is not normally paid for this; the landowner gains in having a pleasanter environment, a potential trout fishery and public approbation in a society in which environmental reponsibility is increasingly valued.

The UK has been much tardier in such restoration, lacking the necessary encouragement by government and hindered by an increasingly profit-motivated agribusiness community. There is some softening of engineering schemes by cosmetic solutions, perhaps a slightly gentler approach to river management than in the past and a very few and small excursions into true restoration, which have been undoubted successes and which might encourage further action. Land and river management in the UK is now motivated mostly by economics and the exploitation of extensive subsidy systems [388]. There has been progress in the management of existing floodplains in the UK to encourage waterfowl and wading-bird populations and general landscape amenity, while permitting cattle and sheep grazing. But it has been gained in reserves owned by conservation organizations, and in Environmentally Sensitive Areas (Fig. 5.21), covering a minority of the land, where payments are made to landowners for management practices that meet environmental objectives.

Fig. 5.21 The Broads (UK) Environmentally Sensitive Area includes the Halvergate marshes, part of which is now managed as a bird reserve. The marshes are former floodplain and intertidal marshes, which were partly drained by wind-pumps from the nineteenth century onwards. More recently, they were endangered by more powerful electrical pumps but some compromise has now been reached over maintenance of relatively high water-tables.

In such floodplains (Table 5.7), the water-table is raised to the surface of the land or above in winter to force soil invertebrates to the surface as prey for the birds. Lower water-tables are kept in summer, though sufficient to maintain ponds on the floodplain to encourage ducks and other waterfowl to remain and breed, while allowing soil invertebrates to multiply in reasonably aerobic soil conditions. Cattle grazing is allowed at levels that do minimal damage to nests on the ground and encourage a structured tussocky sward in some areas and a closer sward (higher stocking levels) in others for geese.

Table 5.7 Typical regime for management of a floodplain for the encouragement of bird communities in the UK (based on Armstrong *et al.* [20]).

Month	Management	Reason
November	Keep water-table at surface Flood some tussocky fields to 5–50 cm Leave most fields unflooded but with some pools in low-lying areas	Forces invertebrates to surface for wader food Frees seeds for food Attracts waders and grazing waterfowl
December–March	Drain flooded tussocky fields. Flood some other fields. Keep surface wet in others	Maintains cycle of conditions and exploits subtleties of habitat present in different places
April	Lower water-table to –20 cm Maintain shallow floods in some fields	Prevents soil anoxia and invertebrate death Allows early start to wader breeding Encourages breeding in some fields and pairs of waterfowl to remain
May	Keep water-table at –20 cm Maintain pools and water-filled ditches	Keeps invertebrate food close to soft surface for probing waders Feeding and breeding sites for waterfowl
June	Hydrological regime as in May. Put cattle on land early in month (< 2 beasts ha^{-1} to give 250–300 livestock units ha^{-1} between June and October) Graze some areas more (> 300 livestock unit days ha^{-1} from June to October)	Creates tussocky grassland for nesting but trampling regime that does not put nests at high risk Produces even, short turf for grazing waterfowl
July–September	Lower water-table to –50 cm	Exposes fresh mud in pools for wader feeding and minimizes cattle damage to surface
October	Allow water-table to rise naturally. Remove stock	Prevents competition for grazing with geese

5.11 Lowland river channels

Whatever happens to the floodplain wetland, the river channel remains, except where the river has been dammed to form a lake. A pristine lowland river channel is a varied habitat, with bends, fringing and some submerged vegetation, patches of silt and gravel. The animal communities are modifications of those further upstream (see Chapter 4), with organisms of rocks and gravel, but many more that feed on the silt deposits and can tolerate lower oxygen concentrations in the warmer, more organic-laden water. The deposit feeders include oligochaete worms, chironomid larvae and bivalve molluscs; insect larvae and Crustacea are less diverse than upstream.

Even without severe canalization, the channel is likely now to be much modified by human activities, particularly pollution. As a river valley has been developed, it has been organic matter from sewage which has been the initial pollutant, together with readily noticeable inorganic sediment from the washing of coal or mineral ores. Many subtropical rivers clearly show the effects of these, particularly in the dry season, when flows are low and dilution ineffective. I have stood on a river bridge over a heavily polluted river in Puna in India and watched what my Indian colleagues called the 'black flowers' (Fig. 5.22): circular clouds of ferrous sulphide, borne to the water surface by methane from the anaerobic sediment, and then gently subsiding to be replaced in a continual blooming.

Fig. 5.22 'Black flowers' in a river receiving raw sewage in Puna, India. The 'flowers' are flocs of ferrous-sulphide-rich sediment, which rise to the surface, with bubbles of methane generated anaerobically from the sewage by bacteria in the mud.

There may have been added variety of other obvious pollutants – for example, dyes, bleach and heat, if industrialization has developed. A public outcry may have followed, either from the epidemics of cholera or typhoid that have accompanied sewage pollution of the water-supply or due to the demise of the fish population. Legislation has usually been passed to curb the worst of this obvious pollution and river quality has improved. Next, as industry and agriculture have become more sophisticated, the variety of pollutants has increased, although controls have usually been set to limit the amounts of such substances that can be discharged. The developed world's rivers are generally at this last stage, and the developing world's at the first or second. Only rivers in remote regions are generally unpolluted. In the final sections of this chapter, I shall look in more detail at what happens following development of a river floodplain.

5.11.1 Pollution by organic matter

Input of organic matter is, of course, a normal feature of streams and rivers. The detritivores of streams (Chapter 4) and the sediment communities (Chapter 8) of slow-flowing rivers depend upon it for most of their energy. The main differences between natural organic input and gross organic pollution are that the former tends to be in large packets, such as leaves, with a low surface-to-volume ratio, or is relatively refractory when finely divided, while the latter is usually soluble or finely divided and very labile. Organically polluted water rapidly becomes deoxygenated. In rivers, loss of most invertebrates and fish follows and the remaining 'pollution community' comprises a mass of filamentous bacteria (including 'sewage fungus', *Sphaerotilus natans* and others), colourless flagellates, ciliate protozoons and anaerobic chemoautotrophic bacteria, such as *Beggiatoa* (Fig. 5.23).

As the organic matter is decomposed, this community may be replaced by one in which filamentous algae, such as *Cladophora*, predominate. These are stimulated by the release of ammonium and phosphate from the decomposition, and support numerous chironomid larvae, which may cause a nuisance when they emerge as flies. Myriads of oligochaete worms may develop. These further process the organic matter and, as oxygen levels begin to rise again, a crustacean, *Asellus*, and other moderately tolerant invertebrates become abundant. Eventually, as the water again becomes fully oxygenated, the 'clean-water fauna', with a high species diversity, is able to return [458].

Organic pollution in the developed world is controlled through use of the same biological processes that break down organic matter in a river, but they are concentrated into the small area of a waste-water-treatment works rather than along several kilometres of waterway. The deoxygenating ability, or oxygen demand, of a sewage effluent, the product of a waste-water-treatment works, may be crudely assessed by measuring the

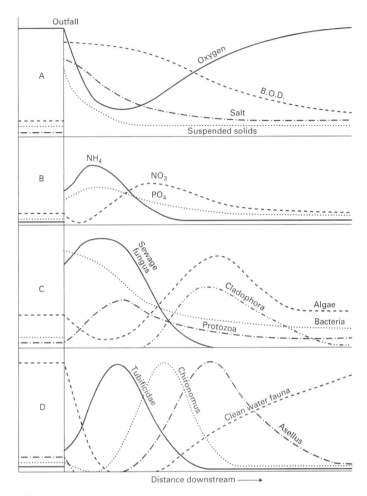

Fig. 5.23 Effects of discharging substantial amounts of decomposable organic matter on a river. Values are relative. BOD, biochemical (biological) oxygen demand; NH_4, ammonium; NO_3, nitrate; PO_4, phosphate. (From Hynes [458].)

rate at which oxygen disappears from it (diluted, if necessary) in a sealed bottle kept for 5 days in darkness at $20°C$. The biological oxygen demand (BOD) of crude sewage is around 600 mg O_2 l^{-1} (5 days)$^{-1}$ and of unpolluted river water less than 5 mg O_2 l^{-1} (5 days)$^{-1}$. Good sewage treatment reduces the BOD of the discharged effluent to, at most, 30 mg O_2 l^{-1} (5 days)$^{-1}$ and usually less than this.

5.11.2 Sewage treatment

The raw material pumped to a treatment works is mostly water, with 1–2% of dissolved, particulate and colloidal organic matter. Domestic waste is only one component, the bulk of it being bath and kitchen water, rather than faeces and urine. In urban areas, there are drainage water from the streets and some factory effluents, although these are usually controlled so that metal and other poisons do not inhibit the organisms harnessed at the works to remove organic matter.

After screening for objects like dead cats, the sewage has few particles larger than a few millimetres – passage through the sewerage pipes has generally broken up large lumps – but a macerator now completes the process. The 'foul water' is now greyish brown and turbid and still has a BOD of 600 mg O_2 l^{-1} (5 days)$^{-1}$. It is led into large tanks, the primary sedimentation tanks, where a sludge settles out, and the BOD of the overlying water decreases by 30–40%. The primary sludge is pumped to digesters, while the supernatant water enters the secondary stage.

The digesters are closed tanks, in which the sludge is fermented by *Methanomonas* spp to produce methane, which may be used to heat the tanks to 30–35°C to promote the process. After several weeks' digestion, the residual sludge should have an almost pleasant earthy smell. Its disposal is a problem, solved by composting it with urban refuse on rubbish tips or using it to fertilize land if it does not contain large quantities of heavy metals. Previously, it could be dumped at sea, but this is now prohibited in Europe and much of it will have to be incinerated in the future and the ash disposed to waste tips.

Meanwhile, the supernatant water from primary sedimentation is usually treated in one of two ways – the trickling filter or the activated-sludge processes. The former is the older method and removes a slightly lower proportion of the residual BOD than the latter, which, however, requires more control and is therefore more expensive. Both depend initially on the same principle – breakdown of organic matter by a similar community of bacteria and protozoa to that which responds to gross organic pollution of a river.

Trickling filters (Fig. 5.24) are typically circular beds, up to 50 m in diameter and about 2–2.5 m deep, of small rock fragments. Usually they are set on a valley slope below the primary sedimentation tanks so that water

Fig. 5.24 Trickling filters are a widely used method for treating waste water. Bacteria coating the stones in the beds decompose organic matter in waste water as it trickles through the beds. Oligochaetes and fly larvae graze the bacterial film and prevent clogging. Insectivorous birds are attracted to the area to feed on the emerging flies.

moves under gravity through pipes to the centre of the bed and out along radiating arms. Gushing through holes, the water jets move the arms so that the bed is evenly sprayed. A film of bacteria and grazing protozoa develops on the rock fragments and converts the organic load into bacterial and protozoon cells and CO_2. The process may be aided by aerating the water before it enters the filter. Growth of the bacteria would soon clog, waterlog and deoxygenate the filter, were it not for grazers on the bacterial and protozoon film. These are a few species of oligochaetes and fly larvae, whose respiration removes yet more organic matter as CO_2 and a small proportion on emergence of the flies, sometimes to cause a nuisance in neighbouring housing areas.

The water trickles through the bed and out into channels. Its BOD has been reduced to about 60 mg O_2 l^{-1} (5 days)$^{-1}$, and it is a solution rich in phosphate, ammonium and nitrate ions, with sloughed-off bacteria, invertebrate faeces and residual refractory organic matter. The BOD is reduced to 20 mg O_2 l^{-1} (5 days)$^{-1}$ by settling of these solids in tanks to form a secondary sludge. This may ultimately be dried and disposed of to farmers or dumps or be digested with primary sludge.

The alternative activated-sludge process is carried out in large tanks. Primary effluent is led into the tanks, seeded with a floc of particles of bacteria and protozoa, largely ciliates, kept back from the previous batch and vigorously aerated. This promotes rapid growth of the floc (called activated sludge), and some rotifers and nematodes may graze it. The plant is usually run on a carefully regulated and monitored continuous-flow system to maintain a high ratio of floc to incoming organic matter. The effluent is taken to ponds, where the floc is settled, and then pumped to sludge digesters, while the supernatant may be the final effluent or may be further 'polished' by filtration or percolation through adsorptive activated-carbon columns to remove dissolved organic matter. In tropical regions, the sewage may simply be led into shallow lagoons, where it is bacterially oxidized, aided by the photosynthetic oxygen production of dense populations of phytoplankton (usually Chlorophyta, including *Chlorella*, *Scenedesmus* and *Chlamydomonas*), which develop in the nutrient-rich medium.

From the point of view of BOD and turbidity, the standards of 20 mg O_2 l^{-1} (5 days)$^{-1}$ and 30 mg l^{-1} suspended solids, set by a Royal Commission in the UK in 1913 for final effluents, are easily achieved. Even good effluents, however, may still contain pathogenic gut bacteria and viruses, such as those of poliomyelitis and infectious hepatitis, although these should be greatly diluted by the river flow and undetectable a little way downstream. Nonetheless, where water is removed from the river for treatment to produce drinking-water, the effluent may be treated with ozone before it leaves the works. In many countries, a further treatment is given to remove phosphorus and sometimes nitrogen compounds. This may be by chemical precipitation or passage through wetlands (see previous).

5.11.3 Pollution monitoring

The extent of organic-pollution damage to the river community is usually assessed by the presence or absence of indicator species, for example, sewage fungus, or groups of sensitive organisms in what are generally called biotic indices or by some mathematical measure of community diversity. River systems are so extensive and change so much that a simple, rapidly applicable measure is needed. Measures of community diversity need no taxonomic expertise, but may be less sensitive than those which use indicator species. There is much debate about which method is most informative [404, 405].

The communities examined are usually those of invertebrates, where a large sample can be easily obtained (compared with fish) and easily examined (compared with microorganisms). In North America, a method based on diversity indices and using the diatoms that will grow on glass slides suspended in the water has been used. Very large numbers of cells are counted on the slides and a graph plotted of number of species against their frequencies in the community. A truncated normal curve (Fig. 5.25) is obtained, with the rarest species usually not detected and hence truncating the left-hand end of the curve. The peak of the curve and the spread of its right-hand tail are used to interpret the overall effects of the pollution. A low peak (few species of intermediate frequency) and a long tail (a small number of abundant tolerant species) are considered to indicate a high degree of pollution, whereas a high peak and short tail indicate a pristine community. The method uses simple apparatus, integrates effects over a period (that of exposure of the slide) and can be standardized to a high degree. Counting of the slides is tedious, however.

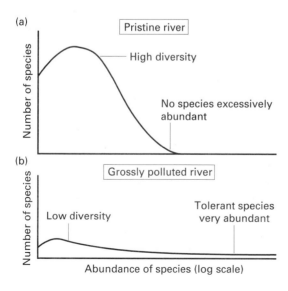

Fig. 5.25 Distributions of the frequencies of diatom species grown on slides left in (a) pristine and in (b) grossly polluted waters. The severity of the pollution can be assessed from the height of the peak and the extent of the tail towards the right.

In England and Wales, chemical measures and biotic indices using invertebrates find greater favour. A survey of rivers was made in 1980 [715] and subsequently every five years, based on some simple chemical criteria and the National Water Council's biological score system. River lengths were classified from class 1a – water of high quality – to class 4 – grossly polluted – on the basis of their dissolved oxygen concentration, BOD, ammonia concentration and ability to support fish. At the same time, the invertebrates were sampled and the communities scored according to the system in Table 5.8.

This allocates points to particular families of invertebrates present, with the highest points given to families least pollution-tolerant (e.g. certain mayfly, stonefly, caddis-fly and beetle families) and the lowest to oligochaetes and chironomids, which normally live in organic sediments and hence can most easily tolerate an increased organic load. The points are added for each family present to give the biotic score. Scores for lightly polluted rivers may be greater than 100 and, for heavily polluted ones, less than 10 or even zero.

In general, there is a reasonable correspondence between the chemical classification and biological score, but the latter, in measuring the effect of pollutants over the period in which the community has developed, offers an advantage. Chemical methods may give different results, depending on the time and date of sampling. A short discharge of pollutant may be missed by chemical sampling, for it will soon be washed down river. But, in killing organisms, its effect on the community will be detectable for some time.

There are some problems in using biological schemes. Prime among them is that the score given to each family is a subjective assesssment of that family's tolerance to predominantly gross organic pollution. Many other sorts of disturbance can influence the invertebrate community. Thus acidification may remove many families and yet leave the high-scoring stoneflies intact to give an impression of a high-quality river. The scheme also takes no account of natural variability. A low score may not necessarily mean a polluted stream if the stream is a sluggish lowland one where the high-scoring stonefly, mayfly and caddis-fly families would not be expected to occur under normal conditions. And, thirdly, the score for a given site tends to increase with increasing sampling and sorting effort.

This latter issue can be met by calculating the average score per taxon (ASPT), which is less sensitive to sampling effort. It is the total biotic score divided by the number of families found. The first two drawbacks can be accommodated by using a scheme that compares the current community, reflected in its specific composition or its biotic score, with a baseline determined for physically comparable pristine sites and assessing the degree to which the site falls short of expectations set by the baseline. The development of the river invertebrate prediction and classification system (RIVPACS) offers such an approach.

Table 5.8 National Water Council biotic scores system. (Also called the Biological Monitoring Working Party (BMWP) score) Scores are summed for each family present to give the biotic score.

Families scoring 10
(Mayflies) Siphlonuridae, Heptageniidae, Leptophlebiidae, Ephemerellidae, Potamanthidae, Ephemeridae
(Stoneflies) Taeniopterygidae, Leuctridae, Capniidae, Perlodidae, Perlidae, Chloroperlidae
(Beetles) Aphelocheiridae
(Caddis-flies) Phryganeidae, Molannidae, Beraeidae, Odontoceridae, Leptoceridae, Goeridae, Lepidostomatidae, Brachycentridae, Sericostomatidae

Families scoring 8
(Crayfish) Astacidae
(Dragonflies) Lestidae, Agriidae, Gomphidae, Cordulegasteridae, Aeshnidae, Corduliidae, Libellulidae
(Net-spinning caddis-flies) Psychomyiidae, Philopotamidae

Families scoring 7
(Mayflies) Caenidae
(Stoneflies) Nemouridae
(Net-spinning caddis-flies) Rhyacophilidae, Polycentropodidae, Limnephilidae

Families scoring 6
(Snails) Neritidae, Viviparidae, Ancylidae
(Caddis-flies) Hydroptilidae
(Bivalve molluscs) Unionidae
(Crustacea) Corophiidae, Gammaridae
(Dragonflies) Platycnemididae, Coenagriidae

Families scoring 5
(Water-bugs) Mesovelidae, Hydrometridae, Gerridae, Nepidae, Naucoridae, Notonectidae, Pleidae, Corixidae
(Beetles) Haliplidae, Hygrobiidae, Dytiscidae, Gyrinidae, Hydrophilidae, Clambidae, Helodidae, Dryopidae, Elminthidae, Crysomelidae, Curculionidae
(Caddis-flies) Hydropsychidae
(Dipteran flies) Tipulidae, Simulidae
(Triclads) Planariidae, Dendrocoelidae

Families scoring 4
(Mayflies) Baetidae
(Alderfly) Sialidae
(Fish leeches) Piscicolidae

Families scoring 3
(Snails, bivalves) Valvatidae, Hydrobiidae, Lymnaeidae, Physidae, Planorbidae, Sphaeriidae
(Leeches) Glossiphoniidae, Hirudidae, Erpobdellidae
(Crustacea (Water-hog louse)) Asellidae

Family scoring 2
(Diptera (midge larvae)) Chironomidae

Families scoring 1
(Bloodworms) Oligochaeta (whole class)

5.11.4 The river invertebrate prediction and classification system

The RIVPACS scheme [1032] is based on samples of the macroinvertebrates in 438 sites in just over seventy near-pristine rivers in the British Isles. No river is entirely pristine in a country as populated as Britain, but these rivers are among the least affected by human activities. Some 603 taxa were found, with 459 insects and 144 other species.

Some key environmental features were also recorded for each site (distance from the river source, slope, mean particle size of the bottom materials, altitude, discharge, width and mean depth of stream, latitude and longitude, alkalinity, mean and range of air temperature) and multiple discriminant analysis was used to relate these factors to the invertebrates at each site. Given a knowledge of the physical environment of a river, the invertebrate species expected there in the absence of major human impacts could be predicted. This then allows the currently determined community (the observed, O) for any site, pristine or impacted, to be compared with that predicted for the site under pristine conditions (the expected, E). The ratio of O to E, measured as biotic score, number of species or families or ASPT, can then be used as a measure of the status or quality of the site.

Table 5.9 shows the use of the scheme for a site on the River Parrett in Somerset, located just below a sewage-treatment works. A number of families (Gammaridae, Simulidae, Hydropsychidae, for example) expected to occur with high probability were absent, reflecting a degree of pollutant stress from the works. The biotic score was much lower than expected, as was the number of families found and the ASPT. Observed/expected values can be used as a basis for a classification system, with values around 1 representing undisturbed sites and lower values being used to characterize increasingly poor sites. The RIVPACS scheme thus measures the change of state of a site and therefore gives a much more valuable assessment than the biotic score scheme alone, which merely measures current state without establishing any significance for that state.

5.11.5 Current problems of river pollution

Whereas water pollution in the past often involved very obvious changes, it now more often involves the release of small amounts of substances that may have more subtle effects on behaviour or reproduction of organisms. These may be organic, such as pesticides or herbicides, or inorganic, such as some of the heavy metals (see Chapter 3). The latter are now usually directed through a waste-water-treatment works, where they can be precipitated out in the sludge. The treatment company can regulate the amount delivered to the works, so that it is insufficient to poison the bacteria in the digestion plant, but the final effluent often contains more metal than natural drainage water, except in areas of ore-bearing rock.

Table 5.9 Biotic scores and RIVPACS predictions for a site (Lower Severalls) on the River Parrett in Somerset (from Wright [1032]).

(a) Environmental data

Water width 3.6 m; mean depth 35 cm; substratum composition: boulders and cobbles 33%, pebbles and gravel 36%, sand 22%, silt and clay 9% (mean size on a logarithmic (phi) scale: −2.57); altitude 35 m; distance from source 7.2 km; slope 5.2 m km^{-1}; discharge category 1; mean air temp. 10.5 °C; air temp. range 11.8 °C; latitude 50°53′N; longitude 2°46′W; alkalinity 4.7 mequiv l^{-1}

(b) Families found (asterisked) and families predicted, with their percentage probability of occurrence, by RIVPACS for this site

Oligochaeta (100)*	Chironomidae (100)*	Elmidae (100)*
Baetidae (100)*	Sphaeridae (99.9)*	Gammaridae (99.7)
Glossiphoniidae (96.5)*	Simulidae (95.4)	Limnephilidae (94.2)*
Hydrobiidae (90.7)*	Hydropsychidae (89.6)	Tipulidae (89.3)*
Erpobdellidae (89.1)*	Ephemerellidae (88.4)*	Asellidae (86.7)*
Dytiscidae (84.2)*	Hydroptilidae (83.7)	Ancylidae (79.3)*
Lymnaeidae (78.6)*	Haliplidae (75.6)*	Caenidae (71.2)
Planorbidae (69.1)*	Planariidae (68.7)*	Rhyacophilidae (67.2)
Polycentropodidae (66.6)	Leptoceridae (65)	Sialidae (64)*
Sericostomatidae (54.4)	Leuctridae (54.4)	Ephemeridae (54.4)
Goeridae (52.5)	Leptophlebiidae (51.6)	Physidae (44.1)*
Hydrophilidae (38)*	Dendrocoelidae (33.9)*	Notonectidae (6)*

(c) Scores

	Biotic	No. of taxa	ASPT
Observed (O)	98	23	4.26
Expected (E)	178	32	5.6
O/E	0.55	0.72	0.76

ASPT, average score per taxon.

5.11.6 Heavy metals

Heavy metals are those with a specific gravity of five more. Manganese, iron, copper, zinc, molybdenum, cadmium, mercury and lead have attracted most notice. Some are required as trace elements by living organisms, although others (Cd, Hg, Pb) are not. At more than 'normal' concentrations, even the trace elements are frequently toxic, although the mode of action is often obscure. Selective inhibition of particular enzymes seems one likely reason. Normal concentrations of these elements in unpolluted waters are around 1 µg l^{-1}, with zinc concentrations around 10 µg l^{-1}. Effluents from metal industries and from the exposure of metal sulphides to air and water in mine-waste dumps may increase these concentrations by several orders of magnitude. The bacteria and blue-green algae that colonize hot, volcanic springs (see Chapter 4) may tolerate extremely high levels of these metals and also of fluoride and arsenic. In 'normal' streams polluted by heavy metals, highly resistant strains of algae may be selected for study [1015]. Dense populations of green algae (Chlorophyta) may frequently be seen in the water of recirculating fountains playing over the bronze statues of long-dead worthies in European cities!

Animals fare less well, however, and, in Welsh mountain streams polluted by the waste from old lead mines, both zinc and lead levels may lead to a depauperate invertebrate fauna of insect larvae, *Tanypus nebulosus* and *Simulium latipes*, and some flatworms. Crustaceans and oligochaetes seem to be particularly susceptible, and fish are absent from such streams [506]. Salmonid fish tend to die at lower concentrations than coarse fish [507].

Death of specific organisms is not the only consequence of heavy-metal pollution. There may be reduced growth rates of those that survive and accumulation of the metals in their bodies by factors of many thousands over the concentrations found in the environment. There may therefore be effects on amphibious but unresistant predators, such as water birds.

It would be convenient to be able to establish the concentrations at which each heavy metal causes physiological or behavioural changes or death in each species likely to be subjected to heavy-metal. As for herbicides (see Section 5.4.1), LC_{50} values can be established. Unfortunately the LC_{50} varies not only with species, but with life-cycle stage, concentrations in the water of other heavy metals, pH, oxygen, bicarbonate, organic matter and temperature. The situation is similar for other pollutants that may have subtle effects at very low concentrations, such as pesticides and organic waste products, such as polychlorinated biphenyls (PCBs) (see Chapter 3).

A very large literature exists giving LC_{50} values (or other related measures of toxicity, such as the highest concentration giving no measurable effect or the lowest concentration giving a significant effect) for a limited range of test organisms. As suggested above for herbicides, the values may be worthless as a means of deciding what amounts of a substance can safely be released to a natural community. At best, they allow a regulatory authority, working mostly on the basis of experience and inspired guesswork, to set a consent either for direct discharge to a river or through a waste-water-treatment works.

The consent takes into account not only the concentration but also the total volume permitted to be released to a river. The consent must also allow for the period when natural flow in the river will be lowest (and concentration of pollutant therefore highest), for other effluents upstream and downstream and, in deference to political realities, the economic state of the industry and area. Too stringent a standard may cost the industry so much to remove the pollutant that it becomes unprofitable, and an appeals procedure discourages a pollution officer from allowing too great a safety margin. In the final analysis, officers hope to maintain at least some fish populations and to minimize obvious visual effects of pollutants. Regulatory authorities in the UK have powers to prosecute (after warnings) if the industry consistently exceeds the consent, once the period in which it may appeal is over.

5.11.7 Problems in pollution management

There are still problems. First, under the British Control of Pollution Act (1974), discharges that are in accord with good agricultural practice, defined by the Ministry of Agriculture, Fisheries and Food and thus, through lobbying, by the industry itself, are specifically excluded. There can be no prosecution, therefore, for any effects of run-off of agricultural chemicals from field spraying or of soil particles, even if these cause measurable deterioration in the river community.

Secondly, there is as yet no absolute standard to which a river ecosystem must be held. At present, with one of the periodic reorganizations of the regulatory authorities in the UK relatively recently, the new organizations, the Environment Agency and the Scottish Environment Protection Agency, are still considering how to determine standards. The uses of the river are surveyed and the standards are likely to be set chemically (not biologically), dependent on whether the water is largely for, for example, fishery or industrial or amenity use. Previously set standards were flimsy, in terms of understanding their effect on natural ecosystems, and there were some notable omissions – for example, standards for phosphate concentrations, which reflected a concern among the water industry at the cost of meeting any standard set of phosphate discharge from its own sewage-treatment works. The general principle in the UK, however, is likely to be that standards will reflect use and that much lower standards will be accepted for waters used for, say, industrial or irrigation abstraction, compared with those used for salmonid fisheries.

Other countries – for example, most of the European Union – prefer the setting of absolute standards of concentration in the final river water for all such waters, irrespective of use. Such a system has the advantage that it cannot be easily manipulated to suit local political concerns and emphases on particular uses, but it might also mean that controls for high-quality rivers, such as those in which salmon and trout breed, could be lessened. At present, the British authorities can insist on higher standards than those prescribed by the European Economic Community (EEC), as well as permit much lower ones, for example, in areas of heavy industry.

5.12 Interbasin transfers and water needs

Finally, there are problems of removal elsewhere of the water itself. Water is essential to human societies, but their sizes may be determined by factors other than the availability of the local water-supply. Conurbations arise, for example, because raw materials for industry are locally available, because of the siting of transport and communications or because of an equable (often warm and dry) climate. Soon, they may outstrip the local water-supply and water must then be imported from areas of apparently greater supply and lower population. The consequences, both to the donor river system and to the natural and human communities dependent on it, have not been primary considerations for the usually aggressive recipient societies demanding the water as of right.

This is a particular problem of arid areas – Australia, South Africa, Central Asia – but is not confined to them. In 1984, the total of such transfers in Canada was $14 \, km^3 \, year^{-1}$ and there is a proposal to divert $50–60 \, km^3$ from rivers flowing into the Arctic seas of Russia to the former Russian republics. Such transfers [192] take little account of when and how much water is moved and the seasonal needs of organisms. Exotic fish, plants and invertebrates may incidentally be moved with the water; reduced flows in the donor river may lead to a reduction in water quality, or the recipient river may be worsened and water-borne diseases may be spread. There will inevitably be a need to transfer water to a greater extent as world populations expand, particularly in urban areas, but the process needs more care than in the past. The Jonglei Canal illustrates some of the potential pitfalls.

5.12.1 The Jonglei Canal

The Sudd has long been regarded as a sink for evaporative losses of water by distant populations in northern Sudan and Egypt, lower down the Nile, where water is needed for irrigation and industralization. A scheme was proposed in the 1970s to bypass the Sudd by a canal, the Jonglei Canal, passing from Bor to Malakal (see Fig. 5.12). The canal was to be 360 km long, 60 m wide and 6 m deep. It would disrupt the lifestyle of the Dinka, Shilluk and Nuer peoples (see Section 5.7.2), but it was argued that their traditional way of life would be improved by the provision of organized cattle ranches, veterinary services, irrigated agriculture and generally improved social services that the wealth to be provided by the canal would provide. There was considerable concern in the southern Sudan, whose local authorities were not included in the planning consortia behind the project.

Excavation began in 1978 with the world's largest mechanical excavator (Fig. 5.26) and had covered 267 km by November 1983, when work was suspended by civil war in Sudan. Costs by then were more than double those projected and work has not yet resumed. Planned crossing points for the canal have not been built and the structure now forms a barrier to the movements of cattle and wildlife. There has been little sign of provision of alternative social organization for the Nilotic peoples. If and when the canal is completed, it seems, from independent studies [440], that the proposed

Fig. 5.26 Part of the excavator used to dig the Jonglei Canal to divert the Nile away from the Sudd swamps. The dredging buckets can be seen to the left. Long arms (not shown) extend to the right to remove the dredged material. The size of the machine can be judged from the human figures below it. (Based on a photograph in Charnock [148].)

replacement lifestyle, even if provided to the extent promised, will be an inappropriate alternative for the conditions and that the Nilotic peoples will be disadvantaged to a large extent.

5.12.2 Assessment of the water needs of donor rivers when transfers are proposed

Assessment of the minimal amount of water needed to maintain a complex river and floodplain system in the face of demands for transfer of water elsewhere is difficult. Components of such systems have evolved to cope with the natural water regime, with its annual cycles and year-to-year variations. The system is best left alone. However, our society has not yet developed to the stage where it can accept this and adjust its own lifestyles to the limitations of the Earth's natural resources. Techniques are thus being demanded which provide the impossible – estimates of how much water is essential for environmental and conservation needs *vis-à-vis* industrial and agricultural needs where development of the catchment is proposed.

There are some schemes available which attempt to estimate the water needs for maintenance of particular target species, usually salmonid fish in temperate regions, but these fail to consider the support of other, more important functional values of the river system. They are also dangerously seductive in their implication that such needs can be simply calculated. Engineers, in particular, have a touching faith that, where a number is quoted for a biological system, it must mean something and absolves them of further worry over conservation issues.

The best examples of the emergence of potential techniques (under protest from professional ecologists) come from arid countries, where availability of water resources is already a major problem [734]. Rivers feeding the Kruger National Park in South Africa rise in lands under development for agriculture and forestry and they flow, in their lower courses, into a wild savannah with a spectacular assemblage of game. As well as being among the best managed of the world's wildlife areas, the Park is a major tourist attraction and, as such, a large source of income for the country. Some of the rivers (Fig. 5.27) are already damaged. The Letaba, Olifants and Crocodile Rivers are polluted and colonized by alien weeds; the Letaba and Luvuhu once flowed all year but now dry up in summer; the Crocodile is regulated to give a constant flow and no longer preserves its natural seasonal cycle. These problems have been largely caused by increasing agricultural abstraction and clearance of land, leading to decreased run-off and fast responses to flood and drought. The Sabie River, in a currently largely undeveloped catchment, is the least affected, but there are now plans to farm the catchment, leading to demands for an estimate of how much water is needed to maintain the system within the Park.

There are two approaches to determining this. The first estimates the needs for consumptive (water-supply for humans and for animal drinking, natural evaporation) and non-consumptive uses (self-purification following a natural (say, a drowned elephant) or human pollution impact, sediment transport, prevention of reed encroachment, sufficient water depths for maintenance of pools for hippopotamuses and crocodiles and for fish movements over riffle areas, flooding of the floodplain and maintenance of its vegetation, appropriate temperature and water quality (especially salinity).

An estimate of the environmental needs of the already developed Luvuhu River is given in Table 5.10, where a value of 76.1 million m^3 $year^{-1}$ is given. This is about a quarter of the mean annual run-off. However, flows are frequently well below the long-term average. The value quoted is also greater than the total flow found under natural conditions during a degree of drought likely to recur once every 50 years and about the same as that found once every 20 years. A value given as a proportion of the mean flow is thus misleading. More significantly, the minimal required flow is greater than the total flow currently found in most years under present conditions of development of the catchment. Further development is thus not advisable and, indeed, current projects should be

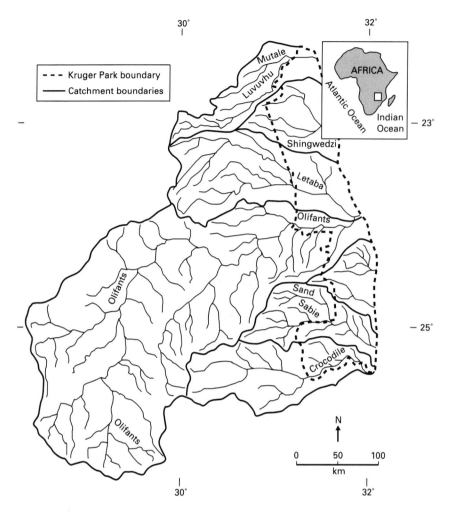

Fig. 5.27 Map of the rivers flowing into the Kruger National Park, South Africa. The park is shown by a hatched line. Most of the catchments lie above the park and hence land use outside the park is crucial to the quantity and quality of water that enters the park. (From O'Keefe and Davies [734].)

curtailed if the functions of the river in the Park are to be maintained.

The second approach has been applied to the higher-quality Sabie River, for which an additional seven irrigation dams have been proposed, including a very large one, the Madras Dam, on the borders of the Park. This approach uses an assessment of the conservation effects under particular flow regimes. Scores are accorded to particular aspects (habitat diversity, number of endemic species that could be supported, invertebrate diversity), according to an agreed scheme, and the scores added to produce a total that represents the conservation status under a given regime. There is some subjectivity in this, but it indicates the relative impacts of a series of

Table 5.10 Water needs of the Luvuhu River in the Kruger National Park in relation to current mean annual run-off. Values are given in millions of m³ year⁻¹. (From O'Keefe and Davies [734].)

Use	Need	Subtotal
Consumptive use		29.9
Animal drinking	0.2	
Human use	0	
River evaporation	4.7	
Riparian evapotranspiration	25.0	
Non-consumptive use		46.2
Maintenance of fish habitat	31.2	
Additional need for sediment flushing (1 in 10 years flood)	15.0	
Grand total		76.1
Mean annual run-off		328
Percentage needed in average year		24%
Present human use in upstream activities (forestry, irrigation)		21%
Recent minimum flow		28
Needs for Kruger Park under minimum flow		> 100%

progressive changes to the water regime (Fig. 5.28). The assessment shows that there has already been a nearly 20% reduction in status, that the operation of one dam, the Injaka Dam, would reduce this by a further 10% and that the operation of the Madras Dam would account for a further 30%, resulting in a system reduced to only 40% of its value under natural conditions.

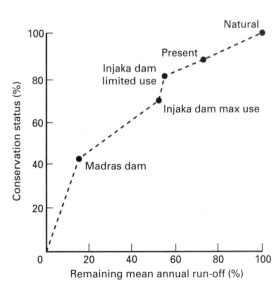

Fig. 5.28 Changes in conservation status of the Sabie River, South Africa, that would follow the introduction of various developments for irrigation [734]. The present, relatively high, conservation status is shown.

6: Lakes, Pools and Other Standing Waters: Some Basic Features of their Productivity

From the floodplain ecosystem of Chapter 5, it is a small step to imagine a basin containing a volume of water relatively large compared with its annual inflow and outflow. Small and shallow lakes are indeed formed on floodplains by masses of emergent or semifloating plants blocking the drainage of floodwater back to the river. Nonetheless, there is likely to be a great influx and efflux of water and the nature of the lake will be greatly determined by that of the river, with much movement of suspended matter and relatively large changes in water level.

The main river channel itself may begin to show lake-like (lacustrine) features in its floodplain stages. Theoretically, such rivers should not support a plankton community – one in which the organisms are suspended or weakly swim in the water mass. The rate of replacement of the water is generally too high for the organisms to reproduce before being washed away to the sea. Natural river channels, however, are not smoothly flowing conduits. The irregularities caused by tributary deltas, shallow bays and gravel bars cause parcels of water to be retained long enough for several

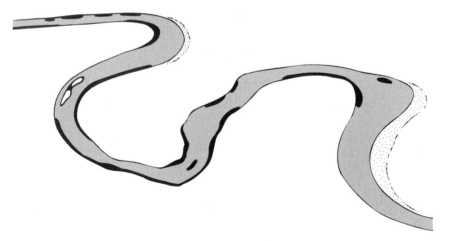

Fig. 6.1 Rivers are not uniformly mixed. Irregularities in the bank and bed create zones where water is retained for longer than average. These show up in satellite images as slightly warmer areas. This diagram shows, in black, areas with greater than 18°C in a several-hundred-metre stretch of the River Severn. They are also associated with low velocities (< 0.03 cm s^{-1}), deposition of fine silt and build-up of algal populations. (Based on Carling [124]).

generations of microscopic phytoplankton to be produced. As the parcels move downstream, they may be retained again and again in such 'dead' or storage zones for varying periods and soon a substantial community may have built up [807, 808] (Fig. 6.1).

If a substantial basin is created, perhaps by the movements of ice or rock or by deliberate damming, the nature of the water mass will change (Fig. 6.2). Water moving in will have a proportionately smaller effect on the mass stored in the basin, so levels will change slowly and with smaller amplitude. Suspended matter will be sedimented out as water movement is slowed by the mass of that in the lake. The water, cleared of river-borne particulate matter, will support much more growth of the suspended photosynthetic community, the phytoplankton, and a submerged plant community to greater depths than in the turbid river. The longevity of the water mass in the lake may allow a physical and chemical structure to develop, with horizontal and vertical patterns, in contrast to the simpler, constantly changing patterns in a river system. Sediment may accumulate on the bottom in layers, containing a fossil history of the lake over time.

The nature of lake ecosystems in large basins will depend increasingly on water chemistry. For, denied the dominance of energy washed in as suspended material from catchment or swamps, the lake ecosystem must function on what is synthesized within the water and hence on the dissolved nutrient supply. Eventually, the lake basin may fill with sediment and succeed to a swamp, while its inflow rivers find renewed channels

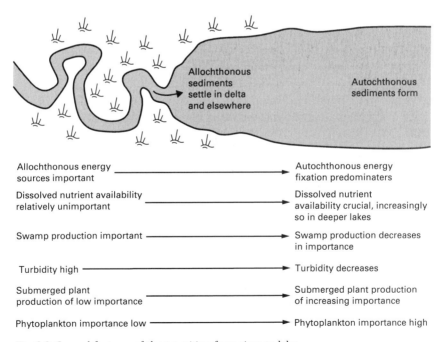

Fig. 6.2 General features of the transition from river to lake.

across the original basin. Lakes, for this reason, are often less permanent features of the landscape than river systems.

Variants on the lake theme are many [111]. The water-supply may be large in relation to the basin volume, as in a riverine lake, and yet create a standing-water ecosystem in temporary rock pools fed by rainwater. These are washed out repeatedly in the rainy season but may persist for weeks or months over the dry season until they finally disappear. In other cases, the basin may be large but the water-supply so low that there is no outflow and water leaves only by evaporation. The catchment-derived water-supply, in evaporating, leaves salts and the basin becomes an endorheic (internally drained) salt lake. Most lakes are exorheic, with some flow-through and no long-term salt accumulation, but the distribution of climates over the world's land surfaces is such that endorheic lakes are the 'normal' lakes over about the same total area as the more familiar exorheic ones (see Fig. 1.3).

6.1 Exorheic lakes

The longest-lived lakes are those formed in continental trenches by separation of the plates of the Earth's crust. Two series of lakes in the Rift Valley of East Africa (Fig. 6.3) are examples. Such basins include the deepest in the world (Lake Baikal 1741 m, Lake Tanganyika 1435 m), frequently with steep sides and such masses of water that, if emptied, they might take tens of thousands of years to fill again at present rates. A useful statistic is the theoretical replacement time or hydraulic residence time – the mean volume (km^3) divided by the annual inflow rate (km^3 year^{-1}) – or its reciprocal, the flushing or washout rate per year. For the upper layers of Lake Tanganyika, the residence time is probably several thousand years. Such lakes are old (see Chapter 2), several hundreds of thousands of years, although, as climate has changed over this period, so have the depth and volume of the lake.

More gentle geological (tectonic) movements may create shallower basins as land subsides. Between the west and east arms of the East African Rift Valley lies the world's largest lake, Lake Victoria (Fig. 6.3), covering 69 000 km^2. It is about 80 m at its greatest depth and, by the standards of smaller lakes formed in other ways, is not shallow. However, in comparison with the larger Rift Valley lakes, it has extensive areas of fringing shallows, covered by papyrus swamp. Lake basins formed, in the manner of Lake Victoria, by gentle subsidence are probably not uncommon close to areas of major earth movements associated with ancient earthquakes and volcanic activity; Lough Neagh in Northern Ireland, Great Britain's largest lake, is also an example.

Volcanic lava flows may dam rivers. Lake Kivu (Fig. 6.3) was formed by the damming of the upper Rutshuru River by seven lava flows from the Virunga volcanoes within the last 20 000 years. Extinct volcanic craters, their vents plugged by solidified lava, may harbour mountain-top lakes of

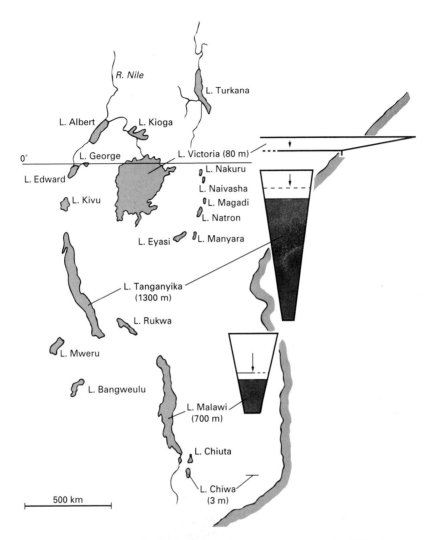

Fig. 6.3 Some major East African lakes. For Lakes Victoria, Tanganyika, Malawi and Chilwa, scaled profiles are given, with arrows indicating the maximum depth to which the water column is mixed during the year. The darkened parts of the profiles indicate permanently unmixed and deoxygenated water. (Based partly on Beauchamp [48].)

singular beauty in near-circular basins (Fig. 6.4). Their small catchments, the crater rims, mean that little particulate matter is washed in and that a shortage of key nutrients may severely limit phytoplankton growth in their very clear waters.

Although tectonically produced basins may be old, the majority of the world's lake basins are relatively young. Most (Fig. 6.5) were produced in the recent glaciation and date back only to 10 000 or 20 000 years ago. As the ice melted from the land masses of North America and Eurasia, an uneven surface of hollows scraped by its movement was left in the rocks

Fig. 6.4 Crater lake, Oregon, USA. The lake has formed in a volcanic crater at the top of a mountain. Wizard Island, in the foreground, is an old volcanic cone.

and debris. Such hollows might be large – for example, those forming the North American St Lawrence Great Lakes (Superior, Huron, Michigan, Erie and Ontario), which cover a total of nearly a quarter of a million km² – or only a few hectares in size. Many of the lakes of Canada, the North

Fig. 6.5 Lakes formed by glacial activity. Pools can be created by simple scraping of ice over the rock surfaces (Cumbrian Lake District, upper left); corrie or cwm lakes are formed in basins, where, at the head of a glacier, rock is plucked by freezing and thawing from the surrounding rock walls (Goats Tarn, upper right); moraines may dam a valley deepened by the movement of the glacier (Buttermere, lower left); and melting of icebergs tumbling from a retreating glacier and buried in drift material washed out from it creates basins called kettle holes, in which lakes may form (Quoisley Mere, lower right).

American Midwest, Scandinavia and Russia were formed like this or through the melting of ice blocks carved from the retreating glacier and partly buried in washed-out rock and gravel debris. Such basins are called kettle holes; good examples are some of the Shropshire and Cheshire meres in England (Fig. 6.5).

Glaciers form steep-sided lake basins in upland areas. At the head of the glacier, towards the mountain top, contact of the ice with the backing rock wall and seasonal freezing and thawing may pluck off rock fragments, which grind out a basin under the ice as it moves downhill. Such are the corries, cwms and tarns (Fig. 6.5) of the Scottish, Welsh and English mountains. The ice moved rock fragments down pre-existing river valleys and, when it melted, left rock debris as a pile or moraine, eventually to form a dam in the deepened valley. The main English Lake District lakes (see Chapter 3), many Scottish lochs, the larger lakes of the European Alps and many in Norway were formed in this way.

There are many ways of forming lake basins other than by ice action (Fig. 6.6). Landslides, sand-dunes and shingle ridges may block streams;

Fig. 6.6 Lakes may be formed in many ways other than glaciation. Upper left, dam created by beavers, Alberta, Canada; upper right, shallow pan filled seasonally by the rising waters of the Pongolo River, South Africa; lower left, lake created by the digging of peat in the River Bure valley, Norfolk, UK, between the ninth and thirteenth centuries; lower right, Clywedog dam built in the twentieth century for regulating river flow and storing water, in mid-Wales.

dissolving of limestone by rainwater may end in the collapse of underground caverns to give a basin on the land surface. Beavers or the
movements of great semifloating rafts of papyrus may block side streams on
a floodplain. Especially there are the vast numbers of small ponds and
smaller numbers of large man-made reservoirs. Humans are probably now
as important as ice was in forming lake basins.

6.2 The essential features and parts of a lake

Light penetration is an important feature of a lake or pond water [527]. The
body of water must persist long enough for silt to settle out and allow a
suspended community of primary producers, the phytoplankton, to photosynthesize. The dimensions of the water mass may be too big for winds to
mix it completely at all times; it may thus acquire a structure. And, because
the phytoplankton needs dissolved nutrients, the nature of the catchment
(see Chapter 3) is important. The rate of supply of nutrients may determine
the plankton production. Finally, the relative clarity of the water may allow
submerged plant and bottom-associated algal communities to develop over
parts of the bottom, dependent on the shape of the basin. These and any
fringing swamp communities may contribute to the productivity of the lake
by providing organic detritus for the sediment communities (benthos) or the
plankton. The aquatic plants may also be important as habitats for larger
animals (the nekton), which move between the fringes and the open water.
These four aspects will be examined in more detail.

6.2.1 Light availability

Radiation beams from the Sun (Fig. 6.7). Much is dissipated in space, but a
still large residue of about $1350 \, J \, s^{-1} \, m^{-2}$ (W) reaches the top of the Earth's
atmosphere. This radiation has wavelengths from about 0.15 to 3.2 μm

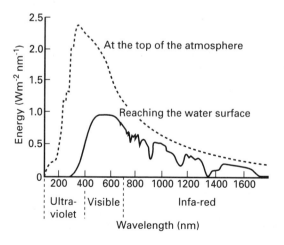

Fig. 6.7 Spectrum of energy
from the Sun. Much of the
ultraviolet radiation is
removed by ozone absorption
in the atmosphere, and large
amounts of infrared radiation
are absorbed by oxygen,
water vapour and, at the
longest wavelengths
(> 2400 nm, not shown),
carbon dioxide.

(150 to 3200 nm), with the peak of energy at around 480 nm. The range between 400 and 700 nm ('light') is usable in photosynthesis (photosynthetically active radiation (PAR)). The rest is either very short-wave ultraviolet (UV) radiation or longer-wave infrared and other radiation. Molecules of oxygen, ozone (O_3), water and carbon dioxide in the atmosphere absorb some radiation, and the energy is dissipated in heat.

Some, largely infrared, radiation may be scattered from molecules, dust particles or cloud and be lost to space. Shorter wavelengths penetrate the atmosphere almost completely, except for some UV radiation, which is absorbed high in the atmosphere by ozone, so that, of the radiation received at the ground, PAR constitutes about 50% compared with about 38% in extraterrestrial radiation. The absolute value of the amount at the ground varies with latitude, time of day, cloud cover and dustiness of the local atmosphere and is potentially 420 J s^{-1} m^{-2} (annual average) at the equator and half as much (197) at the polar circles, with considerable seasonal change. Cloud cover may reduce these maxima by as much as 50%.

Further losses may then occur through reflection at a lake surface and this is particularly important above about 40° latitude, where losses begin to rise from about 2–3% to 6% at 60° and 35% at 80°. Wind roiling may reduce the reflection losses, particularly at very high latitudes. The net effect is that the amount of PAR available to the aquatic community is much reduced from its potential maximum at all latitudes and the annual average received at a lake surface generally declines from equator to poles.

Water is much more efficient at absorbing and scattering light than the atmosphere. There are four main components: the water itself, dissolved yellow organic substances, commonly called *gelbstoff*, suspended phytoplankton and suspended detritus, both inorganic and organic. The latter two categories may scatter as well as absorb the light and, because scatter takes many directions, there is an upward flux of light (a few per cent of the total) as well as a downward flux. The net effect is a downward flux, the amount of which dwindles as the light is attenuated (attenuation = absorption plus scattering). The behaviour of the dominant downward flux has proved adequate for a reasonable understanding of the behaviour of light in water but the absolute amount of energy received at a given point (the scalar irradiance) may be higher by a factor of up to 2 than the downward flux.

For a given wavelength of light in a uniform mass of water, attenuation will be exponential and will follow the equation:

$$I = I_o e^{-kz}$$

(see Chapter 5) where I is the irradiance (amount of light energy received per unit area) at a depth z from the surface, I_o is the irradiance just under the surface of the water (i.e. after reflection losses, e is the base of natural logarithms and k is an exponential coefficient, which can be calculated from the gradient of the straight-line expansion of the equation:

$$k = z^{-1} \log_e(I_o/I)$$

The units of k are depth^{-1} and it is correctly called the net downward attentuation coefficient, although it is usually called the extinction coefficient. It describes the fraction of light energy that is converted to other forms per unit depth and its value depends on the wavelength and the nature of the water. In very clear waters, values of less than 0.01 m^{-1} may be found (less than 1% of the energy absorbed per metre), while in the milky clay-laden waters of lakes at the feet of glaciers, values may be of 10 or more (the energy being almost totally converted within only a few centimetres).

Values of k may increase with depth, even in completely mixed water, because of the increased scattering that occurs with increased path length of the light in the water, but the deviation is usually not great. Where the water mass is not well mixed and layers of different character are present, k will also vary with layer, and calculations must be made using values of I, I_o and z appropriate to each layer.

Figure 6.8 shows attenuation of several wavebands (narrow ranges of wavelength) in a lake and illustrates how some wavebands are more rapidly attenuated than others. The reasons for this are the differential capacities of water, *gelbstoff*, phytoplankton and detritus to absorb at different wave-

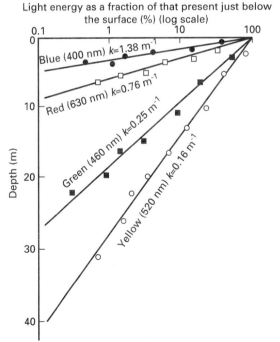

Fig. 6.8 Absorption of light of various wavebands in a typical lake.

lengths. Water itself is a blue substance, absorbing strongly above 550 nm (orange and red light) and below 400 nm (UV) but barely at all in the blue and green wavelengths (400–500 nm). For this reason, waters with very little dissolved organic material or suspended matter – the midoceans and very infertile large lakes – look blue. In most lakes, this blue light, before it can be scattered out of the water to the eye, is absorbed by the yellow substances, derived mostly from the soils of the catchment and sometimes very abundant – for example, in peaty areas. Such lakes look brownish because these substances absorb least in the yellow and orange parts of the spectrum.

Phytoplankton comprises many groups of algae, which have combinations of pigments absorbing particular wavebands. Chlorophylls absorb in the blue and red regions, carotenoids in the blue and green, and the blue and red biliprotein pigments (phycoerythrin, phycocyanin, allophycocyanin) in the yellow and green regions. In a mixed phytoplankton community, the yellow and green light is usually least absorbed. Finally, inorganic suspended matter and detritus generally scatter light to a greater extent than they absorb it, although some detritus is brown in colour and absorbs in the blue wavebands. Scattering is greatest at shorter wavelengths, so the blue end of the spectrum is most affected.

Each lake water comprises a unique mixture (or set of mixtures, for the composition varies with season) of these four components, so there is no single picture of how the spectrum of irradiance will change as the light is progressively attenuated with depth. However, in waters dominated by neither silt, *gelbstoff* nor algae, but having some of each, the usual pattern is of a background light that becomes yellower and greener as light fades with depth. In peaty waters, red and orange light come to dominate with depth, although all wavelengths will be rapidly extinguished, while in the clearest waters, bluish-green light will characterize the final fade-out. Ultraviolet light is absorbed by the water very rapidly in all cases.

6.2.2 The euphotic zone

Photosynthesis can use all wavelengths in the spectrum of PAR and each quantum of light (photon) is approximately equal in value, independent of wavelength. A measure of the total photon flux per unit area passing down the water column is valuable. The extinction coefficient for photon flux is greater in the surface waters as the more readily absorbed wavebands are removed, and declines with depth as the more penetrative bands are left. However, the effect of increased scattering with depth may cancel this out, giving a net effect that corresponds well with the exponential absorption equation.

The photon-flux density is measured in $\mu mol\, m^{-2}\, s^{-1}$ or μEinstein $m^{-2}\, s^{-1}$. One μEinstein is about 1.5 μmol or 0.24 J. Surface values may vary from zero in the polar winter to about $2400 \, \mu mol \, m^{-2} \, s^{-1}$ at the equator in full sun. A temperate cloudy day might have $300 \, \mu mol \, m^{-2} \, s^{-1}$. At values around $10 \, \mu mol \, m^{-2} \, s^{-1}$ (very approximately), the energy absorbed in photosynthesis (gross photosynthesis (P_g)) by an algal cell just balances the maintenance energy needs of the cell (respiration, R). Net photosynthesis $(P_g - R)$ is then zero and no new production can occur. The depth in the water column at which photon-flux density falls to around $10 \, \mu mol \, m^{-2} \, s^{-1}$ is called the euphotic depth (Z_{eu}) and the layer of water above it, in which net photosynthesis is theoretically possible, the euphotic zone.

As a rule of thumb, the euphotic depth for phytoplankton is that at which about 1% of the surface light still remains. It lies higher in the water column – say, 5% of surface light – for bulkier aquatic plants, whose respiratory needs per unit weight are higher than those of microscopic algae. It can be calculated from the extinction coefficient of the photon attenuation (or, more crudely, that of 'white light', measured with a light meter sensitive to all wavelengths of PAR):

$$Z_{eu} = k^{-1} \log_e(100/1)$$

which simplifies to $4.6/k$.

Alternatively, experiments in which the attenuation of different wavebands have been measured have shown that the equation:

$$Z_{eu} = 3.7/k_{min}$$

usually holds, where k_{min} is the extinction coefficient of the most penetrative waveband – green in most lakes, red in peaty waters. Values of k for photon extinction may be as low as $0.06 \, m^{-1}$ for very clear water, e.g. Crater Lake, between 0.15 and $2.3 \, m^{-1}$ for many lakes, and as much as $15 \, m^{-1}$ for highly turbid, silt-laden lakes, such as those in semiarid regions like South Africa and parts of Australia. Respectively, these values would suggest euphotic zone depths of 77 m, 31–32 m and 0.3 m. In the middle and latter ranges, Z_{eu} may be much less than the mean or maximum depths of the lake and net photosynthesis may thus be confined to a relatively thin surface layer. A theoretical maximum value for the depth of the euphotic zone of around 200 m is imposed by the absorption properties of water itself. Figure 6.9 shows the consequences of differences in the euphotic depth and the depth of mixing for the balance between respiration and photosynthesis in phytoplankton communities.

A crude measure of the depth of the euphotic zone may be obtained by dangling a weighted, flat, white disc (called a Secci disc, after its inventor) into the water and recording the depth (Z_s) at which it just disappears to an observer at the surface. This depth is a measure of the transparency of the water and, although it varies with observer and surface conditions, it does bear an approximate relationship to the depth of the euphotic zone:

$$Z_{eu} = 1.7 \, Z_s$$

Fig. 6.9 Relationships between euphotic depths and mixing depths in lakes. If the phytoplankton is mixed so deeply (to Z_m) that their respiration (R) in the water column is greater than the photosynthesis (P) it can carry out in the illuminated part of it (to Z_{eu}), the population will decline. Where mixing is less deep compared with the euphotic depth, respiration and photosynthesis may balance and, at lesser mixing depths, net growth will be possible. Shaded areas show the extents of photosynthesis (curvilinear area) and respiration (rectangular area).

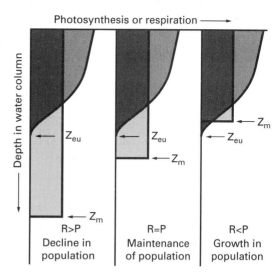

The value 1.7, however, is a mean and values from 1.2 to 2.7 have been recorded.

The concept of the euphotic zone leads to the convenient division of a lake into littoral and open water (pelagial). The littoral is that part of the lake where the euphotic zone extends to the bottom, so that net photosynthesis by either algae or larger plants is possible on the bottom. The zone includes both the bottom and the overlying water. The open-water zone is the remainder of the lake, where the bottom lies below the euphotic zone and whose communities are dependent on the import of organic matter from the overlying water or the littoral zone. There is no distinct boundary between the zones, because the depth of the euphotic layer will vary seasonally to some extent.

6.2.3 Thermal stratification and the structure of water masses

A lake surface churned by wind suggests a well-mixed environment, but this is rarely true. A shallow lake may be relatively uniform throughout the year and a deeper one also in a windy winter, but in many cases temperature differences may develop down the water column. These may divide (stratify) the lake into layers, which may not readily mix. Stratification may be diurnal, forming by day in shallow lakes in the tropics and breaking down as the water cools each night, or it may be seasonal or even permanent. It may cause major differences in water chemistry with depth. In a large lake, the patterns of stratification may differ from one part to another.

Stratification is driven by absorption of solar radiation. The longer wavelengths (above 700 nm), which impart most heat, are absorbed readily.

The attenuation coefficients for such wavelengths are high (10–100 times greater than those for PAR) and heat radiation is mostly absorbed in the top metre or two of water. Theoretically, this should create an exponential fall in temperature from the surface in a still water mass, but there is generally some wind disturbance. The wind mixes the heated surface layers downwards but may be insufficient to mix them to the bottom. This results in an upper, warmer, isothermal layer, a few metres or tens of metres deep, which, because water decreases in density with temperature above about 3.94°C (see Chapter 2), floats on cooler, denser, usually also isothermal water. Traces of the idealized exponential temperature curve may be detectable in the lower layer, but deep currents usually destroy them.

Between the upper layer, the epilimnion, and the lower layer, the hypolimnion, is a transitional zone, called the metalimnion, where a temperature gradient, the thermocline, of as much or more than $1°C\ m^{-1}$ may be detected. This structure is called direct stratification (Fig. 6.10). It is found for most or all of the year in lakes in warmer climates, in temperate

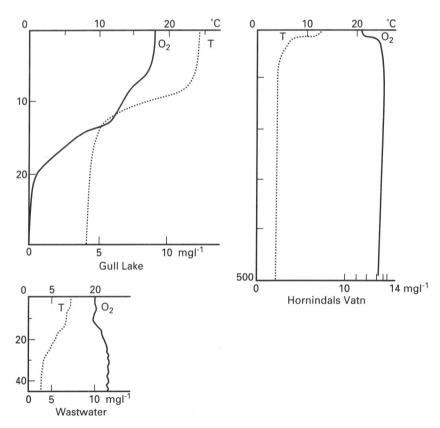

Fig. 6.10 Oxygen (O_2) and temperature (T) profiles in midsummer from three lakes. Gull Lake is relatively fertile, but the other two are infertile. Vertical depth scales are in metres. (Based on Moss [672], Macan [596] and Hutchinson [451].)

lakes from spring to autumn or for a few days in calm periods in lakes so shallow that a slight increase in wind will mix the water completely again. Even in lakes as shallow as 3–4 m, it may be semipermanent if the lake is very well sheltered by dense forest. Conversely, in somewhat deeper lakes in open landscapes swept by the wind, no persistent stratification may ever form.

The temperature range between epilimnion and hypolimnion may be from 20°C to 4°C in temperature midcontinental lakes (e.g. those of Finland or the American Midwest) with severe winters and hot summers. It may be from 18°C to 10°C or a little below, in maritime temperate lakes (e.g. those of the UK), where the water mass does not cool so much in winter or warm so much in summer. In the tropics, the difference may be only a few degrees – say, 29°C to 25°C – but the stratification may be stable, because the change in water density per degree change in temperature is much greater at high water temperatures than it is at lower ones (see Chapter 2).

The annual course of stratification is illustrated for three lakes – Lake Victoria (East Africa), Lake Windermere (UK) and Gull Lake (Michigan, USA) – in Figs 6.11–6.13 by means of depth–time diagrams. Temperatures obtained from a succession of depths are graphed for the dates on which they were obtained. Isotherms are then drawn connecting points of equal temperature, much as contours are drawn along equal heights in making a map. The greater the vertical slope of the lines, the less is the temperature gradient, so that vertical lines indicate completely mixed (isothermal) water masses. The more the lines tend to the horizontal, the greater the temperature gradient in the water column and the stronger the stratification.

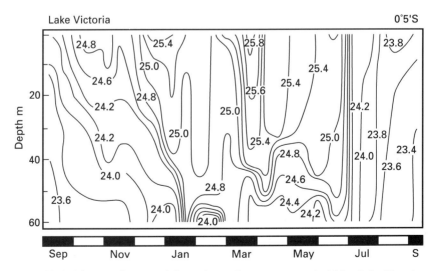

Fig. 6.11 Depth–time diagram of thermal stratification in a tropical lake, Lake Victoria, Uganda. Isotherms are in °C. (Based on Talling [941].)

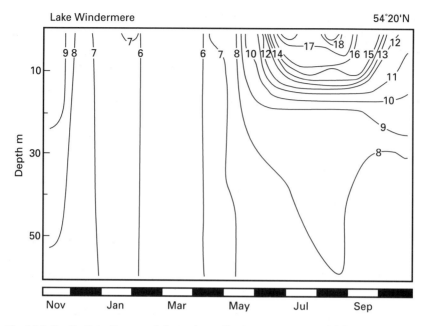

Fig. 6.12 Depth–time diagram of thermal stratification in a temperate lake experiencing a maritime climate, Lake Windermere, English Lake District. Isotherms are in °C. (Based on Jenkin [480].)

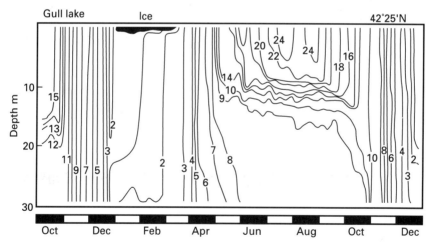

Fig. 6.13 Depth–time diagram of thermal stratification in a temperate lake experiencing a continental climate, Gull Lake, Michigan. Isotherms are in °C. (Based on Moss [672].)

A period of direct stratification ends when the water column is mixed from top to bottom. In temperate lakes this is usually caused by windier weather and cooling from the surface in autumn. Eventually, the density gradient becomes too small to be stable under the prevailing wind conditions. In Lake Victoria (Fig. 6.11), astride the equator, the surface fall in temperature is provided by the evaporative cooling of the trade winds, which begin to blow just after the middle of the year. The two shorter periods of mixing in Lake Victoria are probably not features of the main body of the lake but represent movements of permanently isothermal shallow water from the margins to displace the 'usual' water mass. This emphasizes that there may be horizontal as well as vertical differences in the structure of water masses.

The season of complete mixing in temperate lakes may last from one period of direct summer stratification to the next (Fig. 6.12) – a condition known as monomixis (single mixing). This also occurs in many tropical lakes, but not in temperature continental lakes with intensely cold winters. The maximum density of water at around 3.94°C (in practical terms, 4°C) leads, in these, to inverse stratification during winter with the warmest and densest water usually at 4°C at the bottom (although it may be cooler in lakes towards the poles) and with an upward gradient of colder and less dense water to the surface, where ice forms at 0°C. The consequences are important, for ice and the snow that may accumulate on it act as insulators to further heat loss, and it is rare for a lake to freeze to the bottom. Fish and other organisms are therefore protected from being frozen solid.

Inverse stratification (Fig. 6.13) is ended by the melting of the ice and warming of the surface water in spring. Spring mixing, at first at 4°C and then at higher temperatures, precedes sufficient warming to create direct stratification in most sufficiently temperate lakes. This condition, with two mixing and two intervening stratified periods, is called dimixis. In polar regions, the summer is too short for this and, after a period of mixing at a maximum temperature perhaps around 10°C, inverse stratification sets in again for the autumn and winter (Fig. 6.14), giving again a monomictic state, distinguished as cold monomixis from the temperate conditions of warm monomixis.

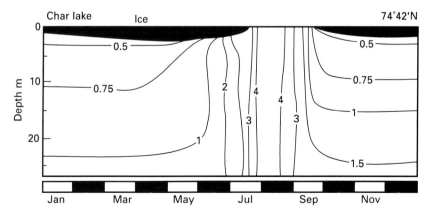

Fig. 6.14 Depth–time diagram of thermal stratification in a polar lake, Char Lake, Canadian Arctic. Isotherms are in °C. (Based on Schindler *et al.* [867].)

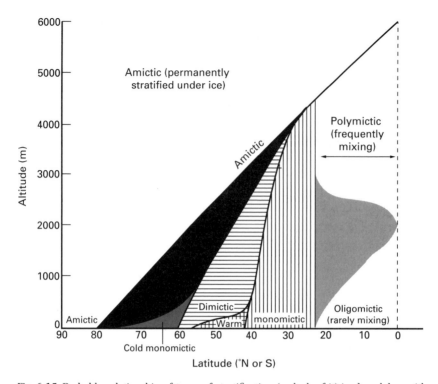

Fig. 6.15 Probable relationship of types of stratification (or lack of it) in deep lakes with altitude and latitude. Transitional regions between warm monomictic and dimictic types are shown. Also, in the tropics, a region of mixed types, mainly variants of the warm monomictic type, is indicated at midaltitudes. (Based on Hutchinson and Löffler [456].)

At the highest latitudes and altitudes, there are lakes which are permanently covered by ice and never mix (amictic). At the lowest latitudes, in equatorial forests, are lakes which probably mix very rarely [360], in unusually cold spells. At high altitudes in the tropics, there are deep lakes which may mix very frequently, with only brief periods of direct or indirect stratification. Figure 6.15 [456] summarizes the most common stratification patterns with latitude and altitude.

6.2.4 Key nutrients

Take a sample of water from almost any unpolluted lake and add, in some convenient experimental design, a range of ions alone and in combination at concentrations of a few tens of $\mu g\ l^{-1}$. Leave the flasks in good light and the chances are very high that you will soon notice much greater growth of algae in flasks to which phosphate has been added than in those to which it has not. For some waters, it may have been necessary to add both phosphate and nitrate or ammonium ions and, in others, nitrogen compounds alone may suffice [647, 671, 685, 809, 984, 985].

The experiment is even more convincing if done in larger containers set in a lake [880] or even on whole lakes [866]. In Canada, a lake shaped roughly like a violin, though with a much narrower waist, was divided with a vinyl-reinforced nylon curtain sealed into the mud and to the banks at the waist [863]. Phosphate was added to one side but not to the other. Only on the side to which phosphate was added was there a dramatic increase in phytoplankton growth. Phosphorus is, on average, the scarcest element in the earth's crust of those required absolutely for algal and higher-plant growth (see Chapter 3). Its compounds are also relatively insoluble and there is no reservoir of gaseous phosphorus compounds available in the atmosphere, as there is of carbon and nitrogen.

The case of nitrogen is more complex, since, despite the huge supply (nearly 80% of the atmosphere), elemental nitrogen is relatively unreactive and available only to a few nitrogen fixers. Nitrogen is converted to available compounds by lightning sparks and by nitrogen-fixers (see Chapter 3). The ultimately available nitrate and ammonium ions are very soluble and hence readily transported into waterways. Nevertheless, in some areas, where rocks are particularly rich in phosphate or where mechanisms capable of concentrating phosphorus occur [691] the local supply of nitrogen may be less than that of phosphorus and nitrogen may be the nutrient that sets the maximum biomass of algae that can be sustained.

6.2.5 Nutrient 'limitation'

Phosphorus and nitrogen where appropriate, have been referred to as 'limiting' nutrients. This graphic term can cause confusion. The stock of such nutrients in a water body does seem to set an upper limit to the average total algal crop that can exist at any one time (Fig. 6.16). But this upper limit may not be reached, because washout or grazers may be removing the crop as fast as it is being produced. Figure 6.16(b) shows some of the data in Fig. 6.16(a) plotted on linear scales and illustrates the great range of actual crop found at a given phosphorus concentration as a result of such processes.

Although the total potential crop may be set (limited) by the stock of P or N, the contributions of individual species within it may be set by other factors. Diatom crops in early spring may be set by silicate, a particular flagellate species by stocks of a vitamin, or most of the phytoplankton community by nitrogen, while potential crops of nitrogen-fixers are set by phosphorus or a trace element essential to N fixation, such as molybdenum.

In another sense, none of the individual cells of the crop may be physiologically short of ('limited' by) N or P. Continuous regeneration of N and P from bacterial decomposition of detritus and from grazing, coupled

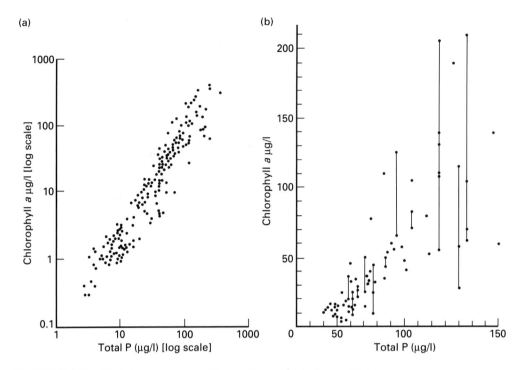

Fig. 6.16 Relationship between summer chlorophyll *a* concentration and total phosphorus concentration in lakes, with data plotted (a) on log scales and (b) on linear scales. (Based on Shapiro [884].)

with supplies of these elements from the inflows, may be just sufficient to meet the needs of the growing cells. Growth may be balanced by loss through grazing and nutrient generation from the consumed cells. The algae may then show no evidence of individual nutrient deficiency (growth-rate limitation), while the total amount of algal biomass at any one time is still set by the amounts of P or N in circulation (biomass or crop limitation).

Further confusion may come from attempts to understand the role of light. In winter, light availability may set the total amount of biomass sustainable in the water, and no amount of fertilization with N and P will increase the crop. In summer, light may still be limiting if the algal crop becomes so dense (because of a very large availability of N and P) that it starts to shade itself. Light may limit the growth rate of suspended phyto-plankton if the wind-mixer layer is very deep so that phytoplankton cells spend much time in water layers where they cannot obtain sufficient light [510].

Although the N or P supplies are low and the mixed layer is shallower than or about the same as the euphotic zone, the rapid attenuation of radiation with depth might suggest a key limiting role for light. However, in the epilimnion or throughout the water column of an unstratified lake, cells are circulating through the light gradient and thus experience the higher surface-light intensities for some part of their day. Enough light energy is usually received for full use of the nutrient supply. This generalization is for the crop as a whole and may not apply to individual species or where suspended silt or clay accounts for much of the light absorption, as in lakes of arid regions.

6.2.6 How the total phosphorus or nitrogen concentrations of a lake come about

Phosphorus enters water bodies not only as inorganic phosphate ions (PO_4^{3-}, $H_2PO_4^-$, HPO_4^{2-}) but also in inorganic polymers, organic phosphorus compounds, living microorganisms and dead detritus. Nitrogen has a similar array. Only some of these forms are immediately available for plant and algal growth, but others may become so through microbial activity. The sum of all the forms, total phosphorus, P_{tot}, or total nitrogen, N_{tot}, is a measure of the potential fertility of a water. A rather infertile lake may have only about 1 µg P_{tot} l^{-1}, or 50 µg N_{tot} l^{-1}, while the most fertile may have up to a milligram or more of P or 10–20 mg N l^{-1}, with an unbroken continuum between.

The concentration of a substance, usually expressed as its key element, is determined by a number of factors, which can be quantitatively related in a general equation [987]. The amount of a substance entering a water body per unit time and per unit area of a lake is called the areal loading, L. There are usually several sources of loadings. They include subdivisions of the catchment area, if it is not uniform (its land use or geology may vary), and the atmosphere, including dry fallout and rainfall and N fixation. There may be loadings from the excretion and defaecation of visiting birds, such as gulls, and other vertebrates, such as hippopotamuses, and release of the substance from the underlying sediment into the water. The latter is called an internal loading and the others are external. There will also be losses to the overflow, as particles sedimenting to the bottom and through denitrification.

The incoming load is diluted into the water entering the lake and this in turn is mixed into the existing lake volume. Thus the final concentration achieved is related to the size of the lake (related to z, the mean depth) and the washout or flushing rate (ρ), as well as the incoming load (L). The net exchange rate with the bottom (loss or release) (σ) also influences the concentration in the lake. In most lakes, the longer the water is retained, the greater is the opportunity for phosphorus to be lost to the bottom, and concentrations in the lake are lower than those in the inflow waters. If the units of L are g m^{-2} year^{-1}, z is in m, and ρ and σ are in year^{-1} (that is, the fraction of the amount in the lake lost to the overflow or the net amount lost or gained from the sediment per year), a simple model for a well-mixed lake, which gives the concentration (m) in mg l^{-1}, is:

$$m = L/z(\rho + \sigma)$$

This is known at the Vollenweider equation. The term L can be expanded to equal the sum of the separate loadings from different sources; each can usually be measured by straightforward techniques. The values of z and ρ can be readily obtained from survey and standard hydrological techniques, but σ must be measured and this is difficult. It is perhaps better to replace σ in the model with R, the fraction of the total load that is retained in the lake water and not sedimented. Using R, the equation becomes:

$$m = LR/z\rho$$

The value of R can be obtained as the ratio of the amount of substance leaving through the outflows divided by the total influx. If measurements of the components of this model are used to calculate m and the calculations are then compared with measures of m and found to give reasonable correspondence, the model can be used to understand how the concentration of substance has been established and to investigate how changes in loading or hydrology might affect the concentration. Applied to phosphorus or nitrogen, this can be a useful tool in managing the lake where concentrations have become a problem and need to be reduced (see below).

The value of R is often much less than 1, because phosphorus is easily locked permanently into sediment and nitrogen componds are denitrified. Maintenance of phytoplankton production in a lake is then dependent on a continued supply of nutrients from the catchment area or atmosphere. In contrast, from the rich source in the sediment, sufficient nutrients may be returned, especially in shallow and in tropical lakes, to the water to relieve the lake of dependence on the catchment. At present, and possibly because most research has been conducted on relatively deep, temperate lakes, catchment dependence is regarded as the general case and independence as a special case, but future work may alter this perspective.

6.2.7 Demonstration of catchment dependence

Catchment dependence can be readily shown by use of Lund tubes (Fig. 6.17) [589, 590]. These are cylinders of butyl rubber 20 or 40 m in diameter, floating at the rim and sealed into the sediments at the bottom so that they isolate a body of water within a lake from the catchment. When this was done in Blelham Tarn, in the English Lake District, a reduction in phytoplankton crop followed very rapidly in the tubes (Fig. 6.18). Available nutrients, particularly phosphorus, were taken up by the algae and found their way quickly into the sediment. Isolation meant that they could not be replaced from the catchment. The lake did not accumulate nutrients in its

Fig. 6.17 Lund tubes being used in an experiment on the shallow Hickling Broad, Norfolk. The floating rim can be seen, surmounted by a line of fluttering plastic strips designed to discourage the perching of birds. This proved a failure, as demonstrated by the white droppings. The tubes had eventually to be covered by a tent of netting to prevent inadvertent guanotrophication.

water, but was dependent for maintenance of its production on continued external supplies.

6.2.8 Consequences of thermal stratification

Isolation of the hypolimnion from the epilimnion is often not complete, because water movements between the two layers may cause mixing upwards in the region of the thermocline. When wind blows strongly over the lake surface, it creates a surface current, which is balanced by a deeper current in a reverse direction along the bottom of the epilimnion. As this current rubs along the top of the hypolimnion, it may cause eddies that pare portions of hypolimnion water and mix them with the epilimnion. The effects of strong winds are to deepen the epilimnion by erosion of the hypolimnion – as happens completely at overturn.

Strong winds blowing in narrow basins may also set up internal waves along the interface between epilimnion and hypolimnion, which continue long after the wind has abated. Initially, the wind may pile up epilimnion water at the windward end, as well as creating a return current (Fig. 6.19). This pile of water deepens the windward thermocline by displacing an equivalent volume of hypolimnion water to the leeward end, where the thermocline is brought nearer the surface. When the wind ceases, the two

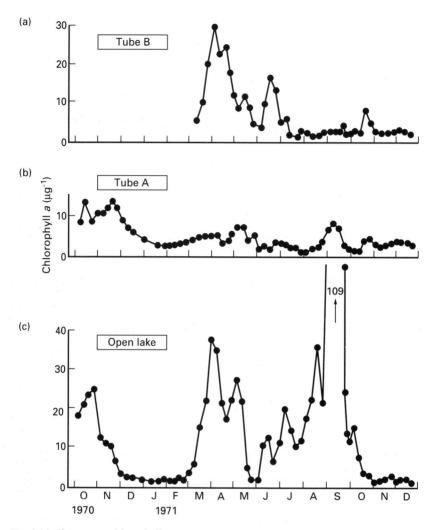

Fig. 6.18 Changes in chlorophyll *a* concentrations in Blelham Tarn (a) and in two Lund tubes, (b) A and (a) B, placed in the lake. The water in (a) tube B was isolated from March 1971 onwards and that in (b) tube A from mid-1970. (Based on Lund [589].)

layers may rock, or oscillate, as the displaced waters in the two layers run back, overshoot and then run in the original direction. Such oscillations may take several days to die down completely and, as the two layers move against one another, there may be mixing at their interface. Several such waves (seiches) may travel in different directions in a basin of complex shape, and waves travelling in a circular motion around the edge of the lake may be imposed by the earth's rotation. For all this complex activity, however, it is still true to say that the epilimnion and hypolimnion of a lake may be essentially unmixing for long periods, particularly in continental climatic regimes.

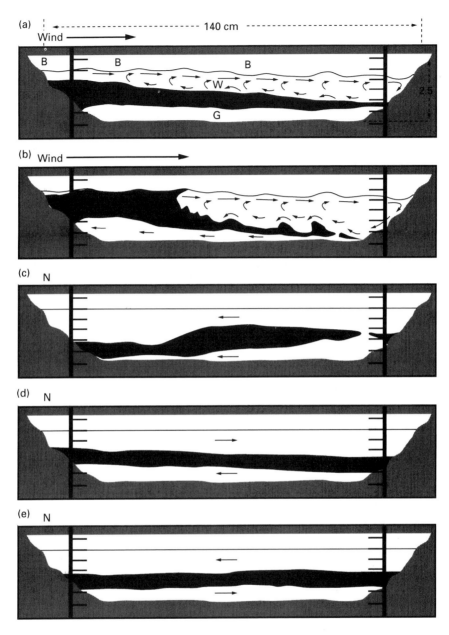

Fig. 6.19 Seiche formation demonstrated in a model tank, with stratification created by layers of (W) water (epilimnion), (P) phenol (metalimnion) and (G) glycerine (hypolimnion). In (a) wind from a blower piles epilimnion water at the leeward end and, if wind strength is increased (b), there is some mixing of epilimnion and hypolimnion water. When the wind drops (c, d, e), the epilimnion water flows back and piles up at the former windward end. Hypolimnion water moves in compensation to leeward, and an oscillation between the two layers (a seiche) begins and continues, with progressively decreasing amplitude, for some time. (Based on Mortimer [668].)

6.2.9 Consequences for water chemistry

Gases in the atmosphere maintain dynamic equilibria with the same gases dissolved in epilimnion water, such that the concentrations of all but CO_2 are predictable almost entirely from the water temperature (see Chapter 3). Respiration and photosynthesis may lead to temporary departures from the equilibrium of oxygen and carbon dioxide over a day, even in well-mixed, open water, but there will always be a tendency to return to equilibrium. Because of its isolation from the atmosphere, the water of the hypolimnion (and that under ice) deviates in its dissolved-gas chemistry.

In such water, photosynthetic oxygen production is small or zero, and yet respiration of bacteria, associated with falling detritus or sediments, and bottom-living animals continues. Oxygen concentrations decrease and carbon dioxide concentrations increase. In particular, a continual rain of phytoplankton cells, detritus, animal faeces and corpses and associated bacteria falls through the metalimnion to the hypolimnion. The greater the epilimnetic production, the greater the supply to the hypolimnion and the greater the demand on the irreplaceable (until overturn) oxygen reserves there. The effect on the concentration of dissolved oxygen in the water is determined not only by the epilimnetic production, but also by the hypolimnion volume and temperature.

A given epilimnetic production may cause complete anaerobiosis of the hypolimnion in a lake 10 m deep with half its volume below the thermocline and at a temperature of 10°C, but have a negligible effect on oxygen concentrations in a 100 m-deep lake with 90% of its water at 4°C in the hypolimnion. In both cases, however, the rates of hypolimnetic oxygen depletion per square metre of water (called the areal hypolimnion oxygen depletion rate) could be similar and determined by the rate of supply of decomposable organic matter.

Hypolimnetic oxygen concentration, therefore, is not a reliable guide to the productivity of a lake. In practice, however, many less productive lakes are situated in deep basins in rocky, upland catchments and for all these reasons have high hypolimnetic oxygen concentrations. In contrast, many productive lakes have shallow basins in the subdued relief of fertile lowland catchments and greatly deoxygenated hypolimnia. Profiles of oxygen con- centrations for three lakes in summer are shown in Fig. 6.10. In the two lakes that are infertile, the oxygen concentrations increase in the hypo- limnion. This is because very little oxygen is being used up in decomposition, while the amount of oxygen capable of dissolving in the very cold hypo- limnion water is much greater than that in the warm epilimnion.

Particles falling into the hypolimnion are decomposed by bacteria and protozoa, with release of ammonium, nitrate (if some oxygen persists), phosphate, silicate and other ions. These substances may be returned to the

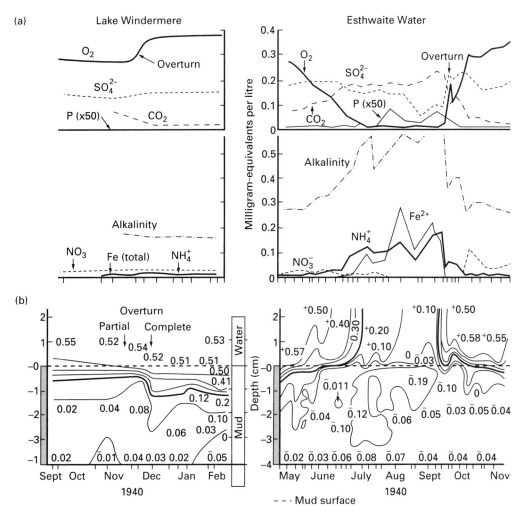

Fig. 6.20 (a) Changes in the chemistry of the water just (10–20 cm) above the sediment surface in Lake Windermere, which had an aerobic hypolimnion, and Esthwaite water, whose hypolimnion became anaerobic from July to September. (b) Redox potential (volts) measured in undisturbed cores taken from the sediment surface during the same period in the two lakes. The approximate lower limit of the oxidized microzone is at about + 0.2 volts. The oxidized microzone remains intact in Windermere, but is destroyed in Esthwaite. (Based on Mortimer [667].)

surface waters at overturn. In a tropical lake, they may then stimulate increased production. In a temperate lake, the overturn usually occurs at a time when production is becoming limited by light and when the increased winter flow from the catchment area is restocking the lake with nutrients anyway.

6.2.10 Sediment and the oxidized microzone

Bacteria in the surface sediments below the thermocline absorb oxygen, which is replaced by diffusion from the overlying water so long as supplies last. Below this surface layer, called the oxidized microzone (Fig. 6.20), bacterial activity uses up oxygen faster than it can be replaced and the sediment becomes anaerobic millimetres below the surface. The oxidized microzone is usually brown-red in colour, because it includes oxidized iron compounds, largely oxides and hydroxides. Other substances are present, especially ions, such as phosphate, which are adsorbed and immobilized within the layer and largely prevented from diffusion into the overlying water.

Below the oxidized microzone, bacterial activity continues for some centimetres. It is largely anaerobic, as different groups of bacteria, lacking access to oxygen, use other electron acceptors to oxidize organic matter [502, 667]. The reactions are those that also go on in deoxygenated swamp soils (see Chapter 5). Some of the products of these bacterial activities may diffuse through an oxidized microzone and accumulate in an oxygen-depleted but still aerobic hypolimnion – for example, CO_2, ammonium, manganese (Mn^{2+}); others – ferrous ions (Fe^{2+}), hydrogen sulphide – may be reoxidized by aerobic bacteria or inorganic processes in the oxidized microzone.

As the hypolimnion becomes completely deoxygenated, the oxidized microzone disappears [667]. The redox potential (see Chapter 5) falls at the sediment surface (Fig. 6.20) and redox-sensitive substances begin to escape from the sediment into the hypolimnion. Ferric ions (Fe^{3+}) are reduced to Fe^{2+} between about +300 and +400 mV, when there may still be 1–2 mg l^{-1} of oxygen dissolved in the adjacent water. Ferrous and phosphate ions then accumulate in the hypolimnion. At lower redox potentials, perhaps −100 mV, release of soluble sulphide ions occurs and hydrogen sulphide starts to accumulate. The hypolimnetic water thus becomes greatly changed as it becomes deoxygenated. It steadily accumulates CO_2 (with a decrease in pH), Mn^{2+} and NH_4^+ then Fe^{2+}, phosphate and sulphide. Any nitrate present from previous aerobic decomposition is denitrified and, as bacteria previously confined to the anaerobic sediment are able to colonize the water, methane may also accumulate.

Such hypolimnetic waters may become hostile to many eukaryotic organisms and fish are excluded at quite modest oxygen concentrations. At the concentrations found, Fe^{2+}, NH_4 and H_2S may be toxic. Few invertebrates can survive in anaerobic sediment overlain by anaerobic water. Nonetheless, the anaerobic sediment surface and the hypolimnion may have a relatively rich community of bacteria and of anaerobic Protozoa [263], which feed on the bacteria. These include flagellates (diplomonads and retortamonads), amoebae (rhizomastigids) and ciliates (trichostomatids

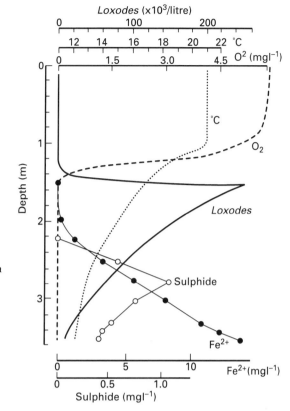

Fig 6.21 The distribution of a ciliate protozoon genus, *Loxodes*, in the water column, in August, of a small lake, Priest Pot, in the UK. Peak abundance is in anaerobic water, but in the absence of the highly toxic sulphide ions. (Based on Finlay *et al.* [268].)

and heterotrichids). There may also be the large amoeba, *Pelomyxa palustris*, which lacks mitochondria but derives energy from the activity of its symbiotic anaerobic methane-generating bacteria. As deoxygenation develops above the sediment, there may be a distinctive layering of different protozoan taxa [267], dependent on their tolerance to anaerobiosis and the availability of bacterial food (Fig. 6.21).

Overturn results, of course, in a reversal of these changes, as the hypolimnion water is reoxygenated. Under inverse stratification, similar deoxygenation and associated chemical changes may occur, particularly if ice cover is prolonged. Winter kills of fish are then not uncommon in shallow lakes.

6.2.11 Shallow lakes, tropical lakes and other scenarios

Much research has been carried out on the moderately or very deep, stratifying lakes of the northern temperate zone. The model discussed above works very well for such lakes. Its key features are that: the catchment provides nutrients, whose rate of supply determines primary production within a lake; phosphorus is of primary importance; nutrients do not accumulate in the water, but are washed out or accumulate in the sediment, where they are more or less permanently stored. The model, however, does not work quite so well for other water bodies, particularly shallow and wetland lakes, where the littoral zone dominates and aquatic plants, rather than phytoplankton, dominate the primary productivity, or for tropical lakes. There is a tendency to regard such lakes as 'exceptions', but this is to underestimate their extent and importance. Shallow lakes form a particularly complex case.

For many tropical and warm temperate lakes, the climate is one of dry and wet seasons, both being warm with long day lengths. They do not have the winter pause in growth of cold temperate lakes, when the lake is also usually well flushed out by incoming water from the catchment. In the wet season, nutrients enter from the catchment but supplies are usually low, because the soils of warm regions are often ancient and have long ago been leached of many of their nutrients. In contrast, the catchment soils of cold temperate lakes are young, derived from rock debris freshly exposed by glaciation only a few thousand years ago, and provide relatively abundant nutrients. The productivities of tropical lakes, however, are not necessarily low, because the high temperatures promote a more rapid mineralization and recycling of nutrients. The lack of washout during the dry season must also lead to greater retention of nutrients, so that, although the catchment

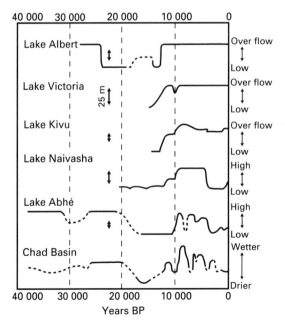

Fig. 6.22 Water-level changes in some African lakes over the last 40 000 years. High levels generally coincide with an increase in numbers of the diatom genus *Stephanodiscus* in the plankton. *Stephanodiscus* competes best at low silicon (Si)-to-phosphorus (P) ratios in the water, which are characteristic of wet periods. BP, before present. (From Kilham and Kilham [526].)

annually provides only low loadings, the lake system is able to use a much larger supply, conserved over several years and kept in circulation.

An insight into this comes from Kilham and Kilham [526], who point out that, in palaeoecological studies (see Chapter 10) of African lakes, one particular diatom genus, *Stephanodiscus*, dominates when water levels have been high (Fig. 6.22). *Stephanodiscus*, at least in laboratory culture, is associated with high phosphorus concentrations and low silica-to-phosphorus ratios in the water. At high water levels in a lake, there is a greater opportunity for nutrients to be recycled (because of the greater water-column depth and the time it takes for particles to reach the sediment) than at low water levels.

They point out that the concentrations of nutrients found in inflow waters to tropical lakes do not vary greatly with discharge. This contrasts with temperate lakes, where much higher loads are often carried at high discharge. In the tropical case, this means that the higher discharges at high water levels are not associated with higher nutrient loads. Therefore, the increased production at high water levels, indicated by *Stephanodiscus*, is not a function of increased nutrient loads from the catchment but of internal processes within the lakes, which are likely to be more important for the maintenance of production than new catchment supplies.

Internal recycling of phosphorus, coupled with an emphasis on nitrogen

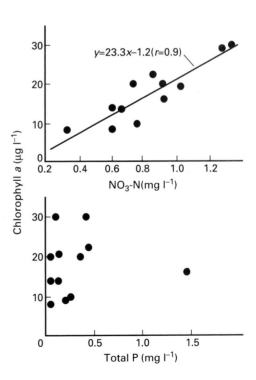

Fig. 6.23 Relationship between growth-season phytoplankton chlorophyll *a* concentration and phosphorus (P) loading (as mean annual total phosphorus concentration) (upper panel) and nitrogen (N) loading (as mean winter nitrate-N concentration) (lower panel) in a series of deep meres in north-west England. The relationship with nitrogen is statistically significant, but not that with phosphorus. (Based on Moss *et al.* [690].)

as the key limiting nutrient, may also be important in some, perhaps many, cold temperate lakes. A series of moderately deep, naturally formed, glacial lakes, called the Meres, lies in north-west England. They were formed by the melting of icebergs buried in glacial-drift deposits about 12 000 years ago and have some apparently peculiar characteristics for north temperate lakes. Their total phosphorus concentrations are absolutely high (several hundred µg l^{-1}), compared with their nitrogen concentrations, on which their phytoplankton crops most closely depend (Fig. 6.23) [690, 691]. The lakes are fed largely by groundwater, but the phosphorus concentrations in the supply are much lower than those in the lakes and there are no major pollutant sources that might explain the high lake values. This suggests that phosphorus has been concentrated in the water.

The retention times are long, because there are no main outflow streams, and this suggests a situation like that in tropical lakes in arid regions, where little phosphorus is washed out annually. The absorptive capacity of the surface sediments may also be saturated, because of a relatively long exposure to high concentrations, giving an additional factor maintaining high phosphorus concentrations in the water. The situation in shallow lakes (see below) adds further complications to the general assumption of phosphorus limitation in freshwater lakes.

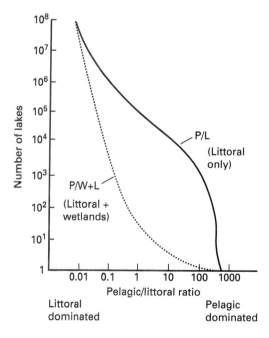

Fig. 6.24 Ratios of the area of littoral to the area of deeper open water (pelagic) in the lakes of the world. Most of the world's lakes are small and dominated by shallow water. Only a few are comparatively large and dominated by their pelagial communities. (From Wetzel [1012].)

6.2.12 Aquatic plant communities and the morphometry of basins

Thus far, light, nutrients and the structure of the water mass have been considered as important features of standing waters. A fourth aspect is the balance between open water and fringing and bottom communities in shallow waters. There are vastly more shallow lakes and wetlands than there are deep lakes (Fig. 6.24). Shallow waters are also likely to be more productive than deep ones, because of the possibilities of recycling of regenerated nutrients from the bottom waters for continued phytoplankton growth in summer and because of the contributions made by extensive beds of aquatic plants and their attached epiphyte (periphyton) communities.

Emergent, floating and submerged aquatic plants all contribute to the aquatic-plant production of lakes. The emergent swamps are confined to water less than about 2 m deep and the effects of wind on floating plants will confine populations of these to sheltered areas. Submerged plants penetrate from only a few decimetres to perhaps 100 m or more, depending on the water clarity. They include vascular species, usually not found below about 10–11 m, perhaps because of intolerance to pressure [322, 912], and bryophytes and charophytes, which, lacking pressure-sensitive lacunae, may go much deeper. Wave action may prevent colonization of the most shallow water, although communities of microscopic algae will generally be found attached to the rocks laid bare of sediment by the waves (see Chapter 8).

Submerged plants have problems in obtaining sufficient light and enough carbon dioxide from the water, because of slow diffusion into bulky tissues. They nevertheless have an advantage, in that they are rooted (or, for bryophytes and charophytes, have rhizoids) in a sediment generally much richer in nutrients, especially phosphorus, than the overlying water, from which they can often also absorb nutrients, though slowly, via their leaves [208].

Sands and gravels are nutrient-poor, but finer sediments generally are not, and, in lakes fed by catchments that provide few dissolved nutrients to the open water, the rooted plants may be greatly advantaged. Data on plant production (and that of associated epiphytes) are much scarcer than those on phytoplankton production, and few studies have considered all communities in the same lake (Table 6.1). Some general hypotheses might be expected to be true, however.

First, aquatic-plant production should be less dependent on the nature of the pristine catchment than that of phytoplankton. Even infertile catchments contribute enough sediment to supply a quite dense crop if the nutrients are annually recycled between the growing parts and storage tissues in roots and rhizomes. This does not mean that aquatic plant growth might not be nutrient-limited; they may be restricted by nitrogen availabil-

Table 6.1 Relative importance of different producing communities in lakes (from Westlake *et al.* [1009]).

Lake	Depth (m) Mean	Max	Area (km^2)	Photosynthesis (% total) Aquatic plants	Periphyton	Phytoplankton
Lawrence (USA)	5.9	13	0.005	51	22	26
Marion (Canada)	2.4	7	0.13	27	61	12
Borax (USA)	0.7	1.5	0.43	0.7	43	57
Latnajaure (Sweden)	16.5	43	0.73	20	20	60
Mikolajskie (Poland)	11	28	5.0	12	6.5	82
Myastro (USSR)	5.4	11	13	7.4	5.5	87
Chilwa (Malawi)	2	7	1400	63.6		36.4
Kiev reservoir (USSR)	4	19	990	6.4	500	44

ity, because nitrogen compounds are denitrified rather than stored (as in the case of phosphorus) in sediments.

Secondly, the contribution made by aquatic plants should be greater in shallowly sloping basins than in steep ones. The ratio of mean depth (z) to maximum depth (z_{max}) may give a measure of this. Thirdly, a high degree of indentedness of the shoreline should favour aquatic plants by providing sheltered water, where sediment can accumulate. The shoreline development, the ratio of the shoreline length (l) of a lake to that (l_o) of a circle of the same area as the lake, is a measure of this.

And, fourthly, the proportion of the total production contributed by aquatic plants might be expected to decrease with fertilization of the lake by human activities, owing to competition for light (or CO_2) with the phytoplankton. Initially, the absolute production of aquatic plants might increase with fertilization, owing to leaf uptake. In pristine lakes, therefore, the proportion of the total production contributed by submerged aquatic plants might then be directly proportional to the relative shallowness and the shoreline development and inversely to the nutrient concentration in the water [903].

6.2.13 Alternative stable states in shallow lakes

Even in shallow lakes with high nutrient loadings, it is common to find low phytoplankton production, especially during summer, provided that there are substantial beds of aquatic plants. Conversely, it is also possible to find such lakes, under similar loading rates, that have dense phytoplankton growths and few or no aquatic plants. Different basins of the same lake may show these alternative states. In one such case, Hoveton Great Broad (Fig. 6.25) in Norfolk, UK, one basin, Hudson's Bay, is dominated by water lilies.

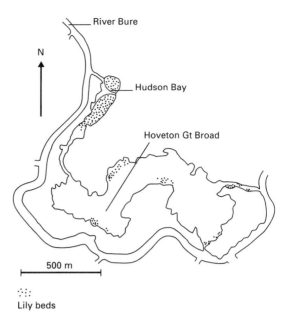

Fig. 6.25 Hudson's Bay and Hoveton Great Broad, Norfolk. Most of Hudson's Bay is dominated by water lilies, with only a small area of open water.

Lily beds

In open water adjacent to the lilies, phytoplankton crops are low in summer (Fig. 6.26). They rise in spring and autumn, in response to the high nutrient loadings from the associated river. The linked, larger basin has few lilies and extremely turbid water during all of the growth season (Fig. 6.26). Nutrients are abundant in the clear summer water in Hudson's Bay and water removed to the laboratory will support large algal crops. However, there is a major difference between the basins, in that the turbid basin has relatively few grazing zooplankters, while such animals are plentiful in Hudson's Bay (Fig. 6.26), especially within the lily beds, from which the animals drift out at night to graze on any developing phytoplankton. This maintains the clear water and a good light climate for the submerged leaves of the lilies. It appears that the lilies provide refuges for the animals against fish predation and allow large and effective grazer communities to coexist with the fish. Where such refuges are absent, as in the turbid basin, fish are able to remove most of the grazers and phytoplankton algae can use the available nutrients, thus depriving any developing plants of sufficient light.

An established aquatic plant community may compete in other ways with phytoplankton and these are discussed in Chapter 8. They constitute a series of mechanisms that stabilize the clear water and plant community against changing circumstances, such as an increase in nutrient loading. In experimental ponds (Fig. 6.27), Balls *et al.* [33] attempted to displace the plant community by adding nutrients, but failed to do so (Fig. 6.28). Only when the plants were manually raked out was an increased phytoplankton

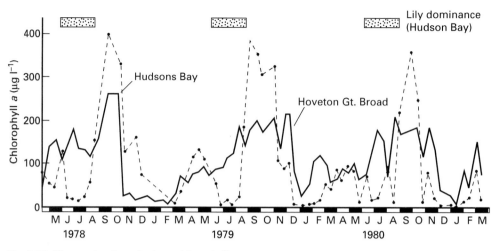

Fig. 6.26 Changes in phytoplankton chlorophyll *a* in Hoveton Great Broad and Hudson's Bay.

crop able to take advantage of the added nutrients. Conversely, in many shallow lakes that are turbid with algae, substantially reducing the nutrient supply has little effect.

Over a wide range of nutrient loadings, plant-dominated, clear water and turbid, algal-dominated states [859] can exist as alternatives, each

Fig. 6.27 Experimental ponds used at Woodbastwick, UK, for studies on the switch between plant-dominated and phytoplankton-dominated systems in shallow lakes.

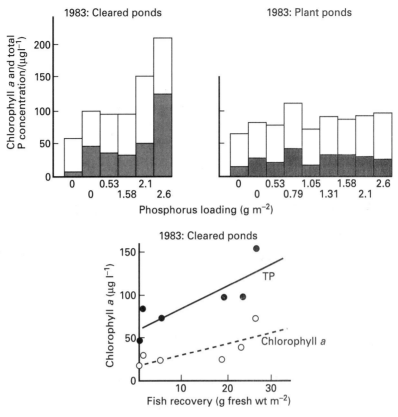

Fig. 6.28 Effects of nutrient loading on the total phosphorus (P) concentrations established in experimental ponds (Fig. 6.27). Only when plants were manually cleared were concentrations able to build up and support a substantial phytoplankton community. The ponds were stocked with fish and the extent to which the phytoplankton community was able to take advantage of the added nutrients depended on the surviving fish stock (lower panel). Where this was high, zooplankton communities were low and chlorophyll concentrations increased. (Based on Balls *et al.* [33].)

stabilized by buffer mechanisms (Fig. 6.29). At very low nutrient loadings, probably only plant dominance can occur, because the algae are severely nutrient-limited while plants can use sediment sources of nutrients. It may also be that, at artificially very high loadings, phytoplankton have such an advantage that plants are permanently excluded. The range where the two communities can be alternatives, however, is very great (perhaps from 25 µg l^{-1} total P to several mg l^{-1}) (Fig. 6.29). To move from one state to the other requires the operation of switches, which are discussed in detail in Chapter 8. Manual raking out, as in the experiment described above, however, is an example of a forward switch.

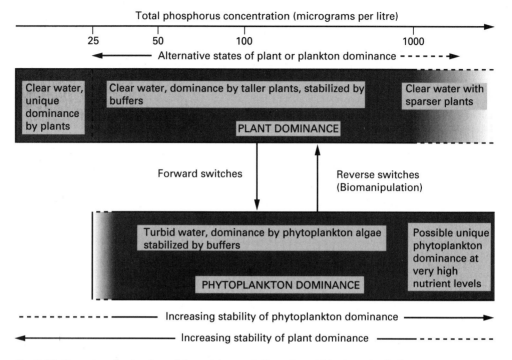

Fig. 6.29 Current understanding of the existence of alternative stable states in shallow-lake ecosystems.

6.3 General models of lake production

Four main features of lakes that contribute to their productivity have now been discussed. Can they be built into an overall model explaining lake productivity? Because many lakes have become more productive recently, posing problems for their management (see below), the matter is important. Unfortunately, most of the syntheses have related features of the lakes only to phytoplankton production, and the fringing and bottom plant communities have been ignored.

Between 1964 and 1974, data on lake productivity were collected from all over the world. Brylinsky and Mann [110] and Brylinsky [109] attempted to draw generalities from these data by making correlations among lake variables and photosynthetic production. Latitude was closely correlated, but nutrients had minor effects. The data available on nutrients, however, were relatively few and did not include the total amounts (e.g. P_{tot}, N_{tot}) in the water, only the immediately available forms. At a given time there may be little available, because it has already been taken up into the plankton crop. The important effect of latitude was expressed not so much through light but through some other correlate, such as temperature or precipitation. This result was surprising, in that the conventional wisdom was, and is, that nutrient supply is the most important determinant of production, as measured by the accumulation of biomass. Experiments, either deliberate or unintentional, where lakes have become more productive due to nutrient addition in sewage effluent or agricultural drainage, support the crucial role of key nutrients.

When mean chlorophyll *a* concentration, a measure of phytoplankton biomass, was included in the correlations with propductivity, it was found to be very significant, overshadowing the effect of latitude. Because algal biomass is an indirect measure of the total amount of potentially limiting (to crop size) nutrient in circulation, this may indirectly indicate a key role for nutrients. The pre-eminence of latitude in correlations using abiotic variables alone may then rest solely on the lack of an adequate measure of nutrient availability in the calculations. On the other hand, because algal biomass and algal productivity are not independent of one another, a circularity is introduced, which technically invalidates the statistical techniques used.

Following the work of Brylinsky [109] and Brylinsky and Mann [110], D.W. Schindler [865] repeated the exercise, using data gleaned from equally wide sources. Schindler's data included the energy-related variables used by Brylinsky, but also incorporated better measures of nutrient availability. Variables tested included latitude, mean chlorophyll *a* concentration, mean total phosphorus concentration, P loading, N loading, mean depth and residence time of the water. Schindler found no significant correlation between phytoplankton productivity and latitude or between phosphorus loading and productivity. However, when he adjusted the loading for the effects of water residence time (see Section 6.2.6) to give an effective measure of total P concentration, he found highly significant correlations with productivity and with chlorophyll *a* concentration. Schindler did not use data from lakes where he expected the key nutrient to be nitrogen (those with input ratios of N : P less than 5 : 1 by weight) and did not have sufficient sites to make separate correlations with nitrogen concentration. Again, conventional wisdom suggested then that such sites were scarce.

Schindler argues cogently elsewhere [864] that, because of the availability of atmospheric nitrogen and the widespread occurrence of N-fixing blue-green algae, nitrogen limitation can only be temporary. It should be countermanded sooner or later by the development of crops of such algae to the potential set by phosphorus in the water. This is theoretically true but assumes that nitrogen-fixing algae more or less instantaneously develop in a nitrogen-limited state. In practice, they may take several weeks to build up substantial crops, by which time other environmental factors, such as decreasing seasonal temperature or increased flushing rate may have begun to disfavour their growth.

The analyses of Schindler [865] and of Brylinsky [109] and Brylinsky and Mann [110] are clearly at odds. To reconcile them, Schindler suggests that the effects of latitude found by Brylinsky cannot be due to light. For a 50-fold range of light energy available from the polar regions to the equator, a 1000-fold range in production is found. He suggests that the latitude effect may be related to greater speed of recycling (see Chapter 7) of nutrients with increasing temperature at decreasing latitudes. This is perhaps supported by findings [646] that photosynthetic rates in pristine, saline East African lakes were around $30 \, g \, O_2 \, m^{-2} \, day^{-1}$ ($11 \, g \, C \, m^{-2} \, day^{-1}$) a value seldom found elsewhere, except when large quantities of nutrients are added by human activities. It is also consistent with the idea that internal cycling of nutrients in warm lakes is relatively more important and external loading relatively less important than in cold temperate lakes.

A more recent analysis (by Kalff [514]) correlated data for total phosphorus concentration, phytoplankton chlorophyll a, zooplankton biomass, phytoplankton primary production and fish yield for up to 183 lakes across a latitude range running from the poles to the tropics. Latitude was adjusted to normalize for differences in altitude. There were no latitudinal trends in phytoplankton or zooplankton biomass, but weak tendencies for total phosphorus to decrease and phytoplankton primary productivity to increase with decreasing latitude. The increasing productivity into the tropics is consistent with the conclusions of Brylinsky [109] and Brylinsky and Mann [110] and is explained by higher temperature, as discussed above. The reduction in average total phosphorus concentration is consistent with the increasing dependence on in-lake nutrient cycling in warmer, drier regions and with the ideas of Kilham and Kilham [526].

However, none of the variables was closely correlated with latitude. Kalff [514] concludes that attempts to find strong global patterns in lake function are probably futile and that relationships are more likely to be demonstrated on a more restricted regional basis. Essentially, conditions in lakes are much more likely to be affected by local conditions than by planet-wide trends.

6.3.1 Models incorporating other features

The role of communities other than the plankton should not be ignored in generalizations about lake production. Indirectly, they have been considered in equations attempting to predict fish biomass or fish yield (the harvest obtainable) from features of lakes. The morphoedaphic index [843] relates fish yield to $TDS^{0.5}/Z$, where TDS is the content of total dissolved solids, effectively an approximation to nutrient availability in pristine exorheic lakes. The inverse proportionality of mean depth (Z) may be a reflection also of nutrient availability and the recycling of nutrients regenerated by sedimenting material. It may also reflect the availability of shallow water and the associated aquatic-plant communities, which may provide feeding and spawning sites for fish. The index sometimes, though not always, predicts fish yield reasonably in limited areas, but the constant of proportionality varies greatly from region to region.

6.4 Eutrophication and acidification – human-induced changes in the production of lakes

There is a traditional belief that lakes become naturally more productive as they accumulate nutrients with time. This process is called natural eutrophication and it probably occurs very rarely, if at all (see Chapter 10). The processes by which phosphorus and nitrogen are either washed downstream, locked into the sediment or denitrified, ensure that the productivity of many lakes will largely reflect the contemporary nutrient supply and will increase or decrease in response to changes in this.

Mechanisms by which phosphorus can accumulate (see above) might seem to deny this, but nitrogen is so readily denitrified that it does not accumulate and becomes limiting if the phosphorus supply is abundant. Ultimately, lake production still depends on continued supply of one or other externally derived nutrient. In the last few decades, two human-induced processes have tended to cause major changes in lake production: artificial eutrophication and acidification. The former has increased and the latter has decreased the productivity.

6.4.1 Eutrophication

Artificial eutrophication (abbreviated here to eutrophication) is often a problem [682]. It includes any increase in nutrient loading, by either phosphorus or nitrogen or, more usually, both. The term is sometimes used to indicate only the problems arising from the increase in loading, with the ugly neologism 'nutrification' created to describe the increase in loading. The original definition is used here.

Increased loading of phosphorus comes from waste-water-treatment works effluent (Fig. 6.30), discharge of raw sewage (including that from stock- and fish-farm animals) and, increasingly, from arable land. Soils have traditionally been thought to retain phosphorus quite efficiently, but recent evidence [247] suggests that this capacity is now being exceeded, following decades of fertilization in some areas.

Nitrogen is also derived from excretal sources, but greater amounts come from cultivated land, where soil disturbance and fertilization mobilize the very soluble nitrate ions to the watercourses (Fig. 6.30). Nitrate may also enter from the atmosphere, following oxidation of nitrogen oxides produced by vehicle engines, and, in the tropics, from burning of vegetation to clear land for agriculture or to promote new growth for grazing.

Eutrophication in rivers was dealt with in Chapter 5, where problems of increasing aquatic-plant and filamentous algal growth were discussed. In

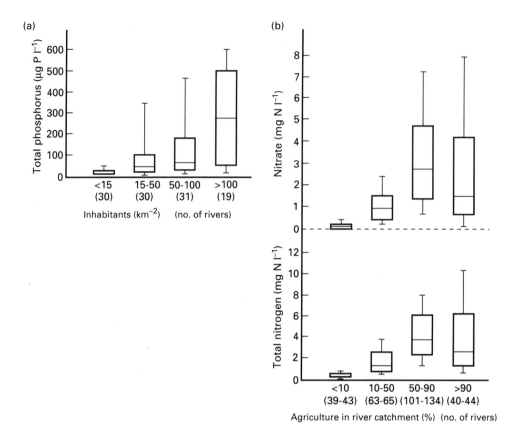

Fig. 6.30 Relationship between human population density and total phosphorus (P) concentration (a) and relationship between percentage of the catchment in agriculture and total nitrogen (N) and nitrate-N concentrations (b) in a large sample of European rivers (from Kristensen and Ole Hansen [536].)

lakes, the problem is reflected sometimes in increased plant growth but usually in increased phytoplankton growth. In shallow lakes, there may be eventual complete loss of all submerged aquatic plants, and sometimes also the regression of marginal reed swamps [74, 75, 678, 693]. Increased phytoplankton crops may make water turbid and unattractive, but they are also commercially important. They increase the costs of filtration for domestic supply and may cause tastes and odours, through secretion of organic compounds. They may produce substances toxic to mammals [152, 153, 398] and sometimes fish [419]. Fine organic detritus from plankton crops passing through the waterworks' filters may support clogging communities of nematode worms, sponges, hydrozoans and insects in water-distribution pipes. Dissolved organic matter secreted into the water by the algae may make chlorination more costly or produce astringent-tasting chlorinated phenolic substances in the domestic water-supply.

Increased phytoplankton crops increase the rate of hypolimnetic deoxygenation. Anaerobic, sulphide-rich hypolimnion water is unsuitable for water-supply, putting much of a lake or reservoir out of use for a long period in summer, when, in many areas, water is short. A deoxygenated hypolimnion also excludes some fish groups, particularly the coregonids (white fish) and salmonids. They may then be unable to live in the lake at all or to pass through it to up-river spawning grounds, because they are also intolerant of the high temperatures in the epilimnion. Such fish depend on cool, well-oxygenated hypolimnia for their summer survival, and frequently support valuable fisheries.

Fish tolerant of lower oxygen concentrations may increase in production as the coregonids and salmonids disappear (Fig. 6.31), but in extreme eutrophication, when dense algal growth outcompetes the marginal aquatic plant beds, the major loss of structure in the ecosystem may have severe effects. There is a loss of spawning and living habitat for the fish, some of which attach their eggs to aquatic plants or their detritus. Large plant-living invertebrates, such as snails and insect nymphs, are much reduced in numbers and this may lead to declines in the growth and survival of larger fish, which depend on them for food.

Fish-eating (e.g. bittern, heron) and plant-eating (swan, coot) bird populations may consequently decline, so that the entire conservation value of a habitat deteriorates. The loss of plants may also make bank edges more vulnerable to erosion by waves or boats, so that banks may have to be expensively and unaesthetically protected by metal or wood piling.

The degree to which eutrophication is considered a problem depends on place and people. A small lake heavily fertilized by the village sewage in south-east Asia may be a valuable source of protein, in the form of deoxygenation-tolerant fish (see Chapter 9) feeding on thick algal soups. A very modest degree of eutrophication by, perhaps, the effluent from a

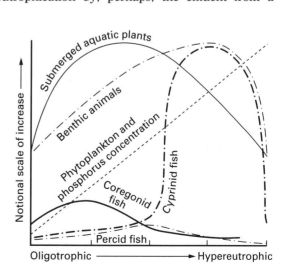

Fig. 6.31 General changes in north temperate lakes as they become eutrophicated. Changes in aquatic plants and benthic animals are considered in Chapter 8. In general, although yields of fish may increase, cyprinids are less desirable as a commercial catch than coregonids. (Based on Hartmann [387].)

couple of holiday cottages in Canada or Sweden may be a serious problem, because it has decreased the transparency of the lake from 5 m to 3 m and has been associated with a reduction in lake-trout catches.

In Great Britain, the water-supply companies have seen eutrophication as a problem only when its symptoms in a drinking-water reservoir cannot be treated in some chemical or mechanical way. More fundamental solutions have not been thought worthwhile, unless they significantly reduce filtration costs [157]. The threshold loading for this is far above that which has sent governments in North America and mainland Europe hurrying to pass antieutrophication legislation. Sensitivity to the problem has increased in the UK following the deaths of dogs and sheep from drinking water containing large concentrations of toxic blue-green algae in Rutland Water in the summer of 1988 [711], and pressure to alleviate the problem has also come from European legislation, promoted by countries with large numbers of lakes that have suffered eutrophication.

6.4.2 Eutrophication in the tropics – Lake Victoria

Eutrophication of temperate lakes is very familiar. Virtually all of the lakes of the agricultural and populated lowlands have been eutrophicated to some extent. In a survey of English lakes, even in protected areas – nature reserves of various kinds – Carvalho and Moss [134] found that three-quarters were damaged. It has been thought, however, that lakes in the tropics were not seriously affected, because of less intensive agriculture and a more diffuse distribution of excretal wastes. The case of the East African Lake Victoria [583] suggests that this is no longer so (Fig. 6.32).

Lake Victoria is one of the world's largest lakes (69 000 km^2; maximum depth 79 m). It has contained one of the unique species flocks of haplochromine fish, discussed in Chapter 2, and supports, through its fisheries, the protein needs of many people. Talling studied it in the late 1950s and early 1960s and published data [940, 941] on its annual cycles of thermal stratification, oxygenation and phytoplankton seasonal succession. The stratification pattern is shown in Fig. 6.11. In the 1960s, the hypolimnion was deoxygenated only in limited areas and then only just above the sediment. Diatoms dominated the phytoplankton, with some blue-green algae; nitrogen supply limited the phytoplankton growth.

By the early 1990s, the phytoplankton crop had increased 10-fold and the primary productivity fivefold [697]. There was a switch to dominance by cyanophytes, many of them nitrogen-fixing, with diatoms much reduced, a depletion of silicate from the epilimnion, widespread deoxygenation of the hypolimnion [402, 564] and frequent fish kills [729], associated with the deoxygenation. The phytoplankton, in bioassays, still shows evidence of nitrogen limitation; phosphorus concentrations in the water have changed little from the 1960s, but sediment-core analyses [402] (see Chapter 10)

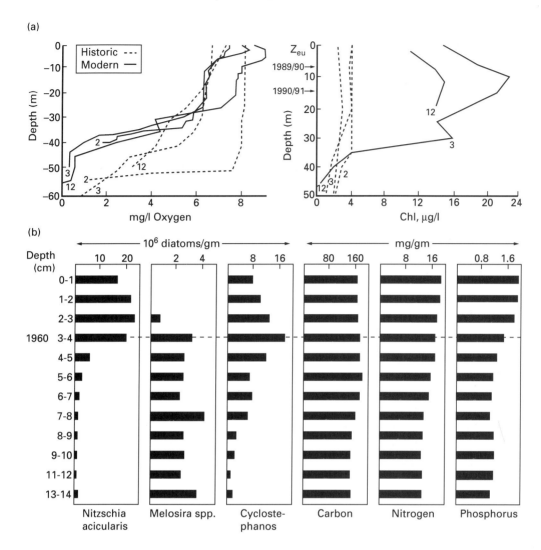

Fig. 6.32 Changes in Lake Victoria shown by (a) analyses made in 1960–61 and 1989–90 of oxygen and chlorophyll *a* (Chl) concentrations with depth in comparable months and (b) changes in diatoms and chemistry in a sediment core (see Chapter 10 for methods). Major changes in planktonic diatoms (*Nitzschia, Melosira, Cyclostephanos*) occurred after about 1960 and are matched by increases in carbon, nitrogen and phosphorus contents. The bottom of the core is dated at about AD 1800. Z_{eu}, euphotic depth. (Based on Hecky [402].)

suggest an increase in phosphorus loading to the lake during the last century.

Water enters the lake from direct rain (83%) on the surface and inflow from the catchment (17%). The concentration of combined nitrogen in the rain has not increased much, but the clearance of woodland, from about 1920 onwards, and increasing local populations, which have risen by about

3–4% year^{-1}, have resulted in enough run-off of nitrogen to have caused the changes. Leaching of nitrogen from the catchment may have initially stimulated diatom production, which led to loss, by sedimentation, of the soluble silica, with only slow current replacement from the catchment. The diatom growth maintained nitrogen-limited conditions, and increased run-off of phosphate decreased the nitrogen-to-phosphorus ratio in such a way as to favour nitrogen-fixing cyanophytes. The increased phytoplankton production led to deoxygenation of the hypolimnion.

This explanation is not entirely satisfactory without evidence of some increased supply of nitrogen from the catchment. Rival explanations involve changes in the food web of the lake [344, 350] following introduction of an alien fish, the Nile perch (see Chapter 9), and climate change. The symptoms could be explained without any increase in nutrient loading if the water column has ceased to mix as freely as it did [564], as a result, perhaps, of a warming-climate trend. There is some evidence for this, but it involves comparison of conditions over only 1 year in the 1960s and 1 year in the 1990s. Mixing in such a large and relatively shallow lake may vary from year to year, as a result or random vagaries of weather, without any underlying trend. Greater stability of the water column would lead to loss of diatoms, failure of silica to be returned to the surface layers, greater denitrification and hence decreased N-to-P ratios, which favour cyanophytes. Mixing disfavours cyanophytes (see Chapter 7) and stabler conditions might lead to their greater success and increased nitrogen loading to the lake through nitrogen fixation. Only data over several years on the temperature stratification will determine if this explanation has value.

6.4.3 Solving the eutrophication problem

There are two approaches: treating the symptoms and removing the causes. The former includes raking out plants, poisoning algae or altering the mixing and circulation to favour readily filterable algae or those which do not produce tastes and odours (see Chapter 7). Symptom treatment is attractive, because it is inexpensive in the short term, but it does nothing to tackle problems other than those of producing drinking-water or maintaining amenity. In the long-term, it is usually cheaper to remove the causes of environmental problems. In the case of eutrophication, this means limiting the nutrient supplies.

But which nutrient or nutrients should be removed? Increased algal production usually needs both nitrogen and phosphorus increases, because, even if phosphorus is currently limiting the potential crop, the nitrogen supply is generally not greatly in excess of algal need. To reduce the algal crop of a lake, however, should require reduction in only one nutrient. An analogy might be drawn with motor cars, which require lubricating oil, fuel and coolant to keep them moving, but which stop if they run short of any one of these.

Phosphorus can be most readily controlled, nitrogen less easily. Its compounds are very soluble, they enter waterways from many diffuse sources (every field seep) and there is a potential but uncontrollable supply from the atmosphere through nitrogen-fixers. Phosphorus is readily precipitated, enters mostly from a relatively few 'point' sources – large stock units and waste-water-treatment works – and has no atmospheric reserve. It is thus usual to attempt to control phosphorus.

This may seem pointless if the limiting nutrient is nitrogen. However, in many cases nitrogen is limiting because phosphorus has been artificially increased to a greater extent, and phosphorus control will ultimately restore the former state. In other cases, where phosphorus has naturally accumulated and where nitrogen increase has driven eutrophication, it may be necessary to control nitrogen.

This will be difficult and, although there can be legislative provision for removal of nitrogen from sewage effluents, most control will have to come from changes in agricultural practice. Leaving wide bands of natural vegetation (buffer zones) alongside the channels of inflowing streams can lead to some nitrate removal by plant uptake and denitrification. Constructed wetlands (see Chapter 5) can also be effective. But, ultimately, changes from the present intensive-farming practices will be necessary and these will need to be far-reaching to be effective.

Where phosphorus control is the sensible option, the questions to be asked are: what are the present supplies of phosphorus and how do these contribute to the total in the water; how is the total P concentration related to the algal crop; what size of algal crop is tolerable in the restored lake; is it possible to achieve a low enough phosphorus concentration to achieve that crop; how shall it be done; and what will the complications be?

6.4.4 What are the present supplies of phosphorus and their relative contributions?

Most sources of phosphorus are easy to identify, although some may be cryptic. Survey of the catchment will reveal obvious ones, but it may be expensive to measure them directly. An approximate budget may, however, be drawn up, as a desk study, using export coefficients (see Chapter 3). The calculated loadings from various sources can then be inserted in the Vollenweider equation (see Section 6.2.6) to relate loading to mean concentration and therefore to calculate what the reduction in loading should be to attain a desired concentration in the lake. The problems of measuring true loading, let alone those of measuring the retention coefficient, have meant that this approach has not been widely used (but see Dillon and

	Prediversion 1990–91	Postdiversion 1991–92
Inputs		
Main inflow	2250	504
Other small streams	55	42
Groundwater	7	5
Direct rainfall	21	19
Bird excreta	23	20
Net release from sediments		1060
Net change in the water mass	−100	−345
Total input	2456	1995
Outputs		
Overflow	1830	1990
Net loss to sediments	630	
Total output	2460	1990

Table 6.2 Phosphorus budget for Rostherne Mere, Cheshire, UK. Annual budgets are given for a period prior to the diversion of sewage effluent from the main inflow stream and for a period immediately after diversion. All quantities are in kilograms per year. (Based on Carvalho *et al.* [135].)

Rigler [211]). A typical budget for phosphorus loading to a north temperate lake is shown in Table 6.2.

6.4.5 Relationship of the phosphorus concentration to the algal crop

The concentration of phosphorus in the lake can be related readily to the expected mean or maximum chlorophyll *a* concentrations from relationships obtained from a large number of lakes (see Fig. 6.16). These relationships are used as a basis for planning a reduction in loading that will give the desired chlorophyll *a* concentration. What this target should be, however, is a problem not easily solved, for it concerns human perception.

Some limnologists divide temperate lakes into trophic categories – ultraoligotrophic, oligotrophic, mesotrophic, eutrophic, hypertrophic – which form a series of increasing fertility. Ultraoligotrophic lakes have very clear water, little phytoplankton, hypolimnia saturated with oxygen and coregonid and salmonid fish; hypereutrophic lakes have turbid water, dense algal growths, anaerobic hypolimnia (if they are deep enough) and cyprinid fish. Defining the intermediate categories of such a multivariate continuum, however, is much more difficult and not really sensible. But lake restoration has political as well as scientific aspects, and politicians and administrators need definitions for the prescribing of law. A working party of the Organization for Economic and Cultural Development therefore gathered views on the definitions of the above trophic categories in terms of phosphorus and chlorophyll *a* concentrations. It was then possible to decide modal chlorophyll *a* and total phosphorus concentrations for each trophic

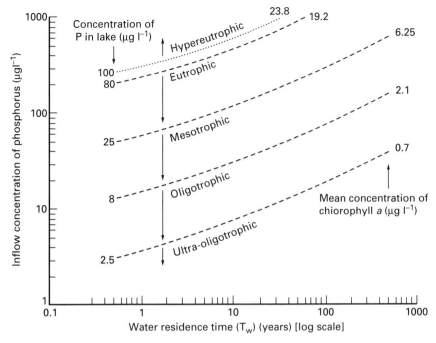

Fig. 6.33 Relationship between inflow phosphorus (P) concentration, water residence time and the consequent most likely phosphorus and chlorophyll *a* concentrations in a given lake. The longer the residence time, the more likely it is that phosphorus will be deposited in the sediment, so that lower in-lake phosphorus and chlorophyll *a* concentrations will be obtained for a given inflow concentration. The terms 'hypereutrophic', etc., are placed arbitrarily, on the basis of a consensus determined by the Organization for Economic Cooperation and Development (OECD). (Based on Vollenweider and Kerekes [988].)

category and then to relate these (Fig. 6.33) to combinations of inflow phosphorus concentration and T_w, the water replacement time. The degree to which the inflow concentration needs to be reduced to achieve a particular target for chlorophyll concentration can then be estimated.

6.4.6 Methods available for reducing total phosphorus loads

The best methods are those which act at source. These include diversion of sewage effluent to the sea, where it is diluted, and precipitation (stripping) of phosphorus from the effluent before it is discharged. Less fundamental or effective are the precipitation of phosphate by adding chemicals to the lake, aeration of the hypolimnion or removal of phosphorus-containing biomass.

Diversion of effluent is only possible where a lake lies near the sea, for pipelines are expensive. The effluent should not itself constitute a large part of the water-supply to the lake, or water residence times will be increased. Lake Washington (Fig. 6.34) is a well-known example of this method.

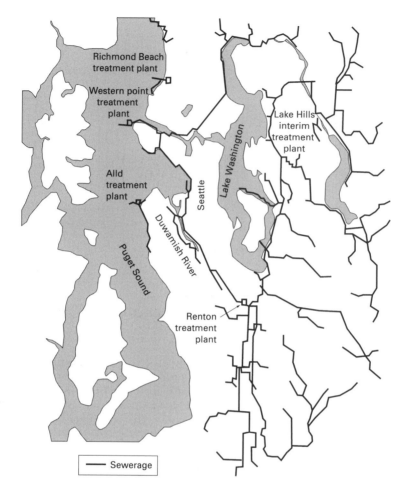

Fig. 6.34 Map of Lake Washington and the sewer system designed to remove effluent away from the lake and discharge it to Puget Sound.

Around the lake lie Seattle and its metropolitan suburbs. In 1955, a blue-green alga, *Oscillatoria rubescens* became prominent in the plankton, signalling a series of changes consequent on the progressive development of the area [235]. The lake was receiving sewage effluent (24 200 m^3 day^{-1}) from about 70 000 people and the effluent was providing about 56% of the total P load to the lake. By 1967, almost all of the effluent had been piped to the ocean in Puget Sound. The transparency of the lake quickly increased from about 1 m to 3 m and chlorophyll *a* concentrations decreased from 38 to about 5 µg l^{-1} [230]. Since the early 1980s (Fig. 6.35), the lake has improved even more [234]. These further improvements involve changes in the phytoplankton and its grazer communities and are discussed later.

Diversion is not always practicable, but all sewage-treatment works can employ phosphate stripping. The effluent (see Chapter 5) is run into a tank

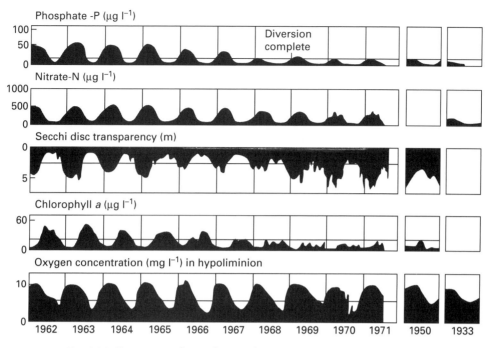

Fig. 6.35 Changes in Lake Washington between 1962 and 1981. Data for two earlier years are placed at the right side to suggest the possible end-point of the changes. The lines are smoothed to emphasize long-term trends. (After Edmondson [234].)

and dosed with precipitant. Aluminium salts work well but disposal of the precipitate in the sludge from the works as farm fertilizer may be precluded, because aluminium salts are poisonous. Iron salts are not, and ferrous ammonium sulphate is frequently the precipitant chosen. The costs of the process are largely in chemicals, rather than installations, and up to 95% of the phosphate can be easily removed, and more with greater difficulty. The US and Canadian legislatures require that effluents discharged to the St Lawrence Great Lakes should contain less than $1 \, mg \, l^{-1}$ of PO_4-P, compared with $10–20 \, mg \, l^{-1}$ previously, and this standard has also been adopted in much of Europe.

Results of phosphate stripping have not always been so dramatic as those for Lake Washington, but it is an effective approach. Despite either approach, the lake water may still contain a substantial background concentration from agricultural activities in the catchment and, in lakes with a long water residence time, it might take many years to obtain a marked improvement [210]. Many of the lakes for which stripping has been tested are shallow, extreme cases and, in these, return of phosphate to the water from the huge past stores in the sediment may frustrate any improvements (see below).

The costs of stripping can be lowered by reducing the amount of phosphorus reaching the sewage-treatment works, by concerted public campaigns. Figure 6.36 shows literature used in New South Wales, Australia, to encourage desirable practices. It centres on six main themes: wash vehicles on porous surfaces (e.g. lawns), away from drains and gutters; fertilize lawns and gardens sparingly; compost all garden and food waste; use zero- or low-phosphorus detergents; wash only full loads in washing machines; collect and bury pet faeces.

The promotion of non-phosphate detergents is a contentious issue, for a huge market is involved. Domestic detergents formerly contributed 40–50% of the total phosphorus in sewage effluent. Pressure and legislation have reduced this to about 15%, which is still regarded as unnecessarily high. Phosphate is, however, a particularly good detergent component. It is non-toxic, non-hygroscopic and non-corrosive [226], all features that favour its use in powders for household washing-machines. It is used as a 'builder', which binds calcium and magnesium ions in the water so that they do not react with the surfactants in the detergent and prevent them from removing fatty body wastes from the laundry. Most alternatives have disadvantages [252, 329] although zeolites (compounds of aluminium and silica, rather like artificial clays) work more or less as well in combination with polycarboxylic acids.

Fig. 6.36 Local communities in New South Wales, Australia, circulate newsletters and leaflets, all dedicated to reducing phosphorus inputs to waterways from homes and farms.

Even with phosphate-free detergents, the phosphate remaining in the effluent (from food excretion) would still need removal by precipitation. The costs of precipitation increase greatly as the concentration decreases, so that complete removal of detergent phosphate would not reduce the costs of stripping proportionately. Hence, why should we not continue to use *ad libitum* a substance whose properties are well known, for which the technology of removal is efficient and which is acknowledged to be an ideal product? The problem is that not all sewage is treated in waste-water-treatment works. Some is allowed to decompose in septic tanks, from which an effluent trickles. Often these are in remote areas, where watercourses are of high quality and thus very vulnerable. And phosphorus stripping is not yet a widely used process in the treatment works of some countries, including the UK. In some remote areas, removal of detergent phosphorus might obviate the need to carry out stripping for a considerable time; and, because reduction of phosphorus concentrations in lakes is proving more difficult than at first thought, any reduction must be deemed ultimately sensible.

6.4.7 In-lake methods

Once phosphorus is in the lake, three main methods have been used to remove it. The first is to treat the lake with a solution of aluminium or ferrous salt to precipitate the phosphate; the method may also coagulate particulate matter. Results may be immediately very good, with a clarification of the water [396], but ultimately negligible [288] if the treatment is not repeated periodically, so long as external inputs of phosphate continue.

A second method depends on the premise that entrainment of hypolimnion water contributes significant amounts of P to the epilimnion, and involves hypolimnetic aeration [55, 71]. Air is bubbled into the hypolimnion in such a way that the hypolimnion is oxygenated without mixing its water with the surface waters during summer (Fig. 6.37). This technique increases the volume of water available to fish, but its use in reducing the phosphorus content of the lake in its productive upper layers is equivocal, because little phosphorus may pass to the epilimnion in summer.

A third potential way of reducing the phosphorus content within the lake itself is to remove biomass that has accumulated in it [71, 444]. The problem is that, compared with the sediment, the biomass accumulates very little of the total load. Organisms like fish, at the top end of the food web, accumulate only small quantities and smaller organisms, such as invertebrates and algae, which accummulate more, are uneconomic to remove from the water. Removal of aquatic plants is easier, but they are bulky and accumulate phosphorus mostly from the sediments, whose reserves are very great. Substantial improvement in a lake must not be expected by biomass removal for several to many years, and then only if external supplies are also reduced [115].

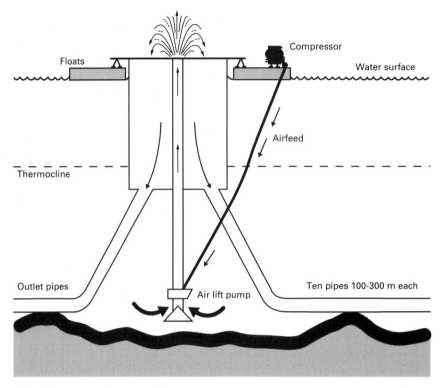

Fig. 6.37 Design of a hypolimnion aerator, which oxygenates the hypolimnial water, without mixing it with that of the epilimnion and thus bringing nutrients to the euphotic layers.

6.4.8 Complications for phosphorus control – sediment sources

There are few cases where phosphorus control alone has been dramatically successful. One reason is that inevitably it has been tried on the most severe examples of lake eutrophication, where sources of phosphorus in the sediment may be important. The Swedish Lake Trummen, for example, had become highly eutrophicated by 1981 after 30 years of discharge of sewage effluent into it [55, 70]. It was accumulating 8 mm year^{-1} of black, sulphurous mud and, during ice cover in the winter, its water column (2 m) became deoxygenated, with consequent fish kills. The sewage effluent and some industrial discharges were diverted in 1958, but by 1968 the lake was still suffering high algal crops, low transparency, deoxygenated and fish kills. Much of the algal crop was being supported by phosphate released from the sediment. There was an internal load of 177 kg P year^{-1}, compared with the external one, which had been reduced to 3 kg P year^{-1}.

In 1970 and 1972, surface sediment was sucked from the lake, settled in a lagoon and later disposed of as fertilizer, whilst the water, after treatment with aluminium salts, was run back to the lake. Following sediment removal, phosphorus concentrations decreased, the water cleared and winter deoxygenation no longer occurred, although in recent years there has been some regression towards the former state. This suggests that the sediment was a major supplier of phosphorus. Further evidence for a major role of sediment sources comes from lakes where the nutrient content of the water has been greatly reduced by sediment sealing.

A technique called the Riplox process has been tried at Lake Lillesjon, near Varnamo in Sweden [71]. Diversion of sewage failed to restore the lake, because of phosphate release from a sediment rich in P but not in iron. The Riplox process injected concentrated solutions of calcium nitrate, ferric chloride and lime from a specially designed harrow into the sediment. The nitrate acted as a substrate for denitrifying bacteria that oxidize organic matter in the sediment, thus creating conditions for the oxidized microzone to strengthen or re-form. The lime adjusted the pH to the optimum (7–7.5) for denitrification and the iron chloride precipitated phosphate. The transparency of the lake subsequently increased from 2.3 to 4.2 m; total phosphorus concentrations fell significantly. The treatment appears to have been successful, but, perhaps for reasons of cost, has not been widely used elsewhere. It will only work in the longer term, of course, where external nutrient supplies are also reduced.

There are snags to any presumption that the removal or sealing of sediment will inevitably restore a lake from eutrophication where external phosphorus control has failed. First, the regression of Lake Trummen suggests that the reasons for its improvement following dredging may have been complex and possibly little to do with sediment removal (see later); secondly, sediment sealing has not yet been shown to give a permanent cure; and, thirdly, in many cases, removal of large quantities of sediment has had negligible effects on the concentration of phosphorus in the water.

Sediment is variable and complex. Generalizations about its behaviour are premature, but some attempt to reveal any pattern is desirable. Current understanding is: that most sediments will release phosphorus under both aerobic and anaerobic conditions to some extent [418, 561, 735], though to a much greater extent in the latter circumstances [516, 954]; that this release is unlikely to be important in a deep, well-flushed lake but is significant in a deep, poorly flushed lake; and that it is likely to be a normal feature of even pristine very shallow lakes. In neither of the latter two cases is sediment removal or sealing likely to be permanently useful. Each of the three cases will be taken in turn.

Sediments in deep lakes

In such lakes, much of the phosphorus released to the hypolimnion is returned to the sediment or lost to the overflow following overturn. Phosphorus supplies from the hypolimnion may mix sufficiently into the epilimnion in midsummer to be useful in maintenance of the summer algal crops [160], but Schindler [865] found that the relationship between external phosphorus load and total phosphorus concentration in the water was the same for both stratified lakes and unstratified ones. He argued that, in unstratified lakes, if internal loading from the sediment is significant, there should be a greater concentration of total P for a given external load. But there was not.

However, there is some release from the sediments underlying the epilimnion into epilimnial waters during summer [632]. In Lough Ennell in Ireland, Lennox [567] found that 17% of the total phosphorus load came from the aerobic marginal sediments. Drake and Heaney [222] found that, at high pH (> 10), significant amounts of phosphorus were released from the aerobic marginal sediments of Esthwaite. But the high pH was created only over short periods by photosynthesis of large phytoplankton crops. Hence sediment release might be consequential upon rather than causative of the problem. Removal of sediment is thus unlikely to be necessary following reduction of external loads in such lakes.

If the lake is poorly flushed, the mechanisms discussed earlier for the West Midland Meres may come into play. In these, phosphorus accumulates to high concentrations and the sediments may become saturated so that further phosphorus cannot be absorbed. Removal of sediment is likely to be ineffective, for the high phosphorus concentrations present in the water must be saturating even new sediment accumulating in the lake. Eutrophication of such lakes must have been nitrogen-driven and demands nitrogen control.

Shallow lakes

In shallow lakes (with maximum depths, say, 3 m or less and the potentiality for most of the lake to be colonized by submerged plants), there is the complication that alternative states of aquatic plant or algal dominance can occur (see Fig. 6.29). In pristine circumstances, however, aquatic-plant dominance is usual. Overall production will be high and much organic matter will slough off from the plants and their associated periphyton communities and the surrounding emergent wetland plants.

Wind circulation of water to the bottom of a shallow lake is likely to be complete, and oxygenated water must be expected always to be in contact with the sediment surface. Such circumstances should create an oxidized microzone, which prevents much phosphorus release. Yet it is clear [744, 747] that prolofic phosphate release does occur. The reason may be that the supply of organic matter to the sediment is both copious and labile, not having had time to decay in the short water column. It supports a degree of bacterial activity that cannot be sustained by diffusion of oxygen from the water to the sediment. Thus, despite overlying aerobic water, the sediment surface becomes anaerobic, which allows both Fe^{2+} and PO_4^{3-} ions to escape. In the overlying aerobic water, they should reprecipitate, so that little net supply of phosphate should occur.

However, in intensely reducing sediments, reduction of sulphate to sulphide precipitates iron as sulphide, leaving phosphate to accumulate unhindered in the water. The phosphate concentrations in summer will then be high and non-limiting to either plants or phytoplankton. Such systems are probably controlled by nitrogen, because nitrates will be readily taken up by the plants or denitrified, in sediment conditions that favour this process. Eutrophication of shallow lakes may thus be nitrogen- rather than phosphorus-driven, and again removal of sediment is likely to be ineffective. Where the lake has become dominated by phytoplankton and has lost its plants, a similar mechanism should prevail, driven by algal-derived organic matter.

This leads to a discussion of how plant-dominated shallow lakes switch to the alternative state of algal dominance and how the plant communities may be reinstated. It also allows the anomalies of the restoration of Lake Trummen and the continued improvement in Lake Washington, beyond nutrient control, to be accounted for. Changes in nutrients are involved but are only part of the story.

6.4.9 Shallow-lake restoration

Plant-dominated shallow lakes can often cope with eutrophication, provided the system is not placed under additional stresses. The species composition may change, but plant dominance and clear water will be preserved. Further stresses (forward switches (see Chapter 8)) may move the system to phytoplankton dominance [693]. The greater the nutrient loading, the more easily such a switch may occur. To restore clear water, the nutrient loading must first be reduced as much as possible. Phosphorus controls have been used, but nitrogen control would probably be more effective because of inevitably high internal phosphorus loads from the sediments.

Then it is necessary to operate a reverse switch (see Fig. 6.29), which restores the activity of the zooplankton grazers, usually by reducing the predation pressure of fish on the zooplankters. This is most reliably done by removing the entire fish community, but some successes has been achieved by addition of piscivorous fish. The biological details of these interactions are dealt with in Chapter 7.

The results of fish removal, usually called biomanipulation [59, 680, 887], have often been dramatic, with water clearing within a few weeks (Fig. 6.38). The situation, however, is not necessarily stable because the fish populations will recover from survivors of the previous removal. Nonetheless, the technique can be used as a symptom treatment for many waters if the zooplanktivorous fish community is kept low by repeated fishing.

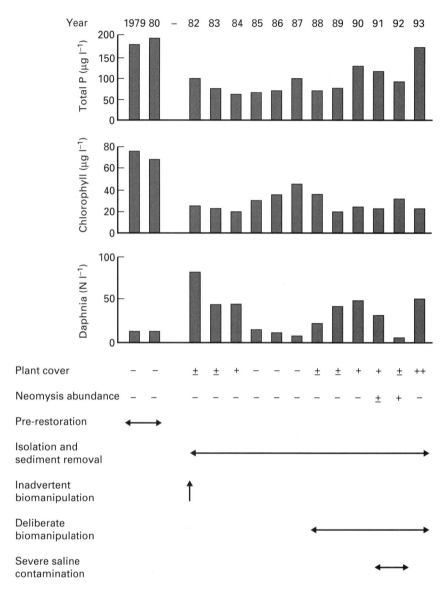

Fig. 6.38 Results of biomanipulation at Cockshoot Broad. The lake was formerly very turbid but cleared after 1982, when it had been isolated from the river and sediment had been removed. These were not the only reasons for its improvement, however. The engineering operations had scared most of the fish out to the river. Those remaining were too few to reduce the *Daphnia* populations, which increased and cleared the phytoplankton. As the fish community recovered, from 1985 onwards, the *Daphnia* declined and the phytoplankton chlorophyll *a* concentration rose. Fish were removed from 1988 onwards and the *Daphnia* population has again recovered, except for a period in 1991–92, when tidal surges up river brought saline water and *Neomysis integer*, an estuarine mysid predator on the *Daphnia*, into the lake. (Based on Moss *et al.* [694].)

The aim of restoration, however, should not be to maintain an inherently unstable situation but a stable one where the fish community is restored following clearing of the water and the re-establishment of aquatic plants. If the latter is achieved, the buffers stabilizing plant communities come into play, including the provision of refuges for the zooplankters against fish predation. Plants may spontaneously re-establish from remnant fragments, once the water has cleared, but it may be necessary to plant an inoculum and protect it from grazing birds, such as coot and swans. In later restoring the fish community, it is important to exclude, in areas where it is not native (all but eastern Asia), the common carp, *Cyprinus carpio*, which is particularly destructive of aquatic plants and disturbs bottom sediments to the extent that it constitutes a forward switch in itself. Figure 6.39 gives a strategy [693] for restoration of shallow lakes. It is still experimental, though increasingly used.

6.4.10 Twists in the tails

Although permanent reduction of eutrophication symptoms will always involve nutrient control, this is rarely the complete solution. Lake Washington continued to increase in clarity (see Fig. 6.35) after all the effluent had been diverted. The numbers of *Daphnia*, a zooplankton grazer, in the water increased after the diversion was complete, while those of a mysid shrimp, *Neomysis mercedis*, a predator on the *Daphnia*, declined. In turn, a predator on the shrimp, the long-fin smelt, increased. There appeared to have been a trophic cascade effect [127] in which increases in an organism at the top of a food chain result in alternating declines and increases of organisms in successive links. In this case, the ultimate effect was a further decline in algae and an increase in transparency as the *Daphnia* grazed. The ultimate increase in fish appears to have been due to an increase in spawning habitat, provided by engineering works designed to prevent slumping of roads along rivers flowing into Lake Washington, where the smelt spawn [234].

Such cascade effects are discussed more fully in Chapter 7. They may be crucial to the success of restoration measures against eutrophication. During the winter in which sediment was removed from Lake Trummen, there was a major fish kill under prolonged ice cover. *Daphnia* populations were subsequently high. It may have been this that triggered recovery of the lake, rather than sediment removal. For Lake Victoria, there is a school of thought that attributes the symptoms of eutrophication to changes in the fish community. A predatory fish, the Nile perch, was introduced to the lake for fisheries purposes in the early 1960s (see Chapter 9). There was little effect at first, but, in the 1980s, its numbers increased greatly and it now dominates the fauna, having reduced many native species to extinction. Loss of herbivorous fish may have resulted in build-up of algae, whose detritus

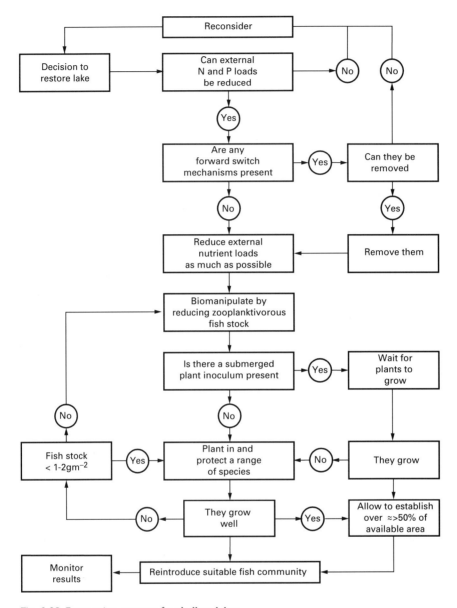

Fig. 6.39 Restoration strategy for shallow lakes.

now deoxygenates the hypolimnion. Alternatively, nutrient-induced deoxy-genation may have forced deep-water fish to the margins, leaving a large invertebrate community, formerly consumed by these fish, to support the more tolerant young of the Nile perch and to lead to marked population increase of this fish. In the former hypothesis, eutrophication is a conse-quence of changes in the fish community; in the latter, given the prior introduction of the Nile perch, the reverse is the case. The truth will probably have elements of both.

6.5 Acidification

In the uplands and remoter unpopulated regions, with their poorly weathered rocks and thin soils, acidification is generally more important than eutrophication [293]. It leads to a decline in productivity – sometimes called 'oligotrophication' – of already unproductive lakes. Although this may aid the filtration of water for the domestic supply, it may increase the corrosion of pipework, often of lead, in the cities to which the water is supplied. Acidification is also a problem, because of its effect on fish. The most obvious symptoms are the same as those affecting acidified rivers (see Chapter 4) [414]: aluminium and perhaps heavy-metal toxicity of adult fish and interference with the hatching of fish eggs.

Acidified lakes may become very clear, partly because mobilization of aluminium ions in the catchment soils leads to precipitation of phosphate, which then stays in the soils [923]. Aluminium ions also readily flocculate particles, including phytoplankton, in the lake water [710]. In the very clear water, aquatic plant growth may spread further across the lake bed than previously, but is of acid-tolerant mosses, *Sphagnum* spp., and filamentous algae, such as *Mougeotia*. Aquatic vascular plants usually decline. The biological diversity tends to decrease [16, 923], while the organic content of the sediment may increase, because litter washed in is not decomposed.

Lakes affected by acidification are described from Ontario [389], New England [571], the Galloway region of Scotland [278] and the Brecon Beacons in Wales. In Norway, an area of $13\,000\,km^2$ is devoid of fish, with lesser changes over a further $20\,000\,km^2$ [881]. About $18\,000$ lakes in Sweden (a fifth of the total numbers of lakes > 1 ha) have pH < 5.5 and fish stocks have been affected in 9000 of them [488].

6.5.1 Remediation

Ultimately (see Chapters 3 and 4), the atmospheric sources of hydrogen ions (H^+) must be reduced to solve the problem. Nonetheless, many catchments have been acidified for so long that high aluminium and H^+ run-off may occur for many years. These catchments may be limed – an expensive business because of their size and remoteness – or lime may be added to the lakes, and many of the changes that occur in acidified lakes can be reversed [248].

At Loch Fleet in Scotland, segments of the catchment of a lake moderately acidified (pH about 4.7) since the 1970s, with destruction of a trout fishery, were variously limed (Fig. 6.40) and the consequences followed [441]. About 40% of the catchment was limed in 1986–87. This raised the pH and calcium concentrations and decreased the aluminium concentrations sufficiently for trout to be re-established within 2 years. Conditions then declined and, by 1994, were returning to the former

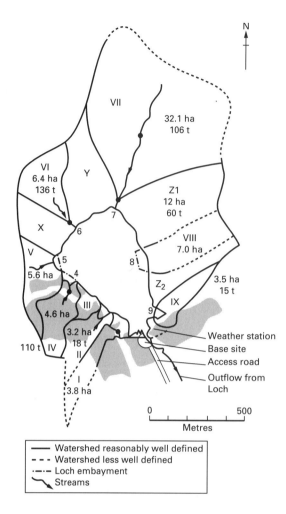

Fig. 6.40 Map of Loch Fleet and treatments applied to alleviate acidification. The area of each subcatchment is shown, with the tonnage of lime spread over it in 1986. Effects were subsequently monitored in the streams and the loch. (From Howells and Dalziel [441].)

acidified conditions (Fig. 6.41). Effects of liming on the catchment vegetation were severe only in one wetland area, where *Sphagnum* was lost. The results suggested that, for maritime upland areas with high flow-through, liming at a rate of about 10 ha^{-1} at 10-year intervals was appropriate, given current acidification levels.

Problems of the lime being readily washed-out or, in lakes with high concentrations of humic compounds, readily precipitated and made inactive might be mitigated by supplying carbonate to the sediments. In Lilla Galtsjön in Sweden, an injection of 40 t of 10% sodium carbonate solution into the sediments supplied a source of carbonate that slowly diffused out to neutralize the hydrogen ions over quite a long period [71]. Liming, however, is only a remedial measure. It does not cure the problem and it requires the indefinite expense of continued treatment. It also promotes the attitude that abuse of the environment is acceptable, because corrective treatments can be locally applied. The better approach (see Chapter 4) is to treat the

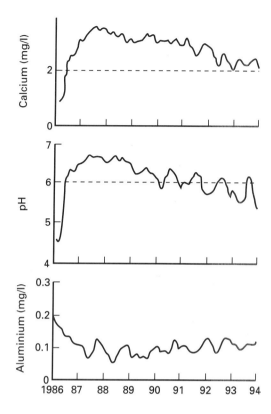

Fig. 6.41 Effects of liming Loch Fleet in 1986. Values are shown for pH and the concentrations of calcium and aluminium in the outflow. Horizontal lines show that target concentrations for re-establishment of brown trout. (Based on Howells and Dalziel [441].)

problem by removing acid emissions from power stations and industrial sources and finding ways of managing society that do not require such high levels of vehicle use as at present.

One approach has been to install flue gas desulphurization to power stations. The sulphur dioxide (SO_2) is neutralized by calcium carbonate, which is converted to calcium sulphate, a bulky product that must be dumped somewhere. Targets for removal of SO_2 have often been set in the past as percentage reductions on some base level – for example, the emission level in 1980. This at least begins to define the need to set targets but is unsatisfactory in that what is left after a certain percentage reduction may still cause severe ecological damage.

6.5.2 Targets

In a survey of British sites important for conservation in acid-sensitive areas, 141 out of 196 were damaged (phytoplankton, invertebrate and macrophyte community changes; declines in fish, amphibian, bird or mammal populations) [821]. Existing commitments (on a percentage basis) to reducing acidifying gases proposed under the European Community (EC) Large Combustion Plant Directive were shown to be insufficient to prevent further damage. A better approach to control is to define critical loads – amounts of acidity (as hydrogen ions) that can be permitted to rain down on specific areas without causing damage or without removing sensitive target species.

There are two approaches to defining critical loads, one using changes in diatom communities and the other using chemical calculations [45]. The former takes a series of lakes for which sediment cores have been examined (see Chapter 10). The onset of acidification is generally seen as a major change in diatom community in the cores. For these lakes, and for related lakes that have not shown such changes, the calcium concentrations in the water have been plotted against the sulphate concentrations (expressed as an equivalent amount of hydrogen ion in sulphuric acid). Lakes with a ratio of calcium (as milliequivalents per litre) to sulphate (as kequiv of H^+ ha^{-1} $year^{-1}$ delivered in rain) of greater than $94:1$ appear not to suffer ecological changes as expressed in the diatom communities. Lakes with ratios lower than 94 show such changes. The critical load can then be calculated from the calcium concentration in the water as the H^+ load that would keep the ratio > 94. Acidification may itself have altered the calcium concentration from former pristine values and allowance must be made for this.

Any current load that is greater than the critical load for a particular area is called the critical-load exceedance and it is this absolute amount that must be removed, rather than some percentage of current emissions. Critical-load exceedances must now also take into account the increasing amounts of hydrogen ion derived from nitric acid. The critical load is the same, no matter what the origin of the hydrogen ions. Figure 6.42 shows a current map of critical loads and critical-load exceedances for Great Britain. Considerable areas of the uplands are affected and also some lowland sites in southern England, where soils are sandy and base-poor.

The alternative, Henriksen method of calculating critical load is chemically based and widely used in Scandinavia [406]. It depends on estimation of the acid-neutralizing capacity (ANC) of the pristine soil water of the catchment and thus depends on the alkalinity (due largely to bicarbonate and dissolved organic salts) of the water. The ANC can be negative in already naturally very acid waters but is usually a small positive value. The ANC is then related to the occurrence or absence of key organisms (brown trout, a mayfly, *Baetis rhodani*) (Table 6.3) to define a threshold ANC for each target organism. The amount of hydrogen ion falling on the catchment that could be neutralized while leaving sufficient neutralizing capacity to meet the threshold for the target is then the critical load. Any excess is again the critical-load exceedance. The Henriksen method gives slightly higher critical loads (and therefore lower exceedances) than the diatom method. This may be because the diatom method uses a community approach, in which some very sensitive organisms are included, while the Henriksen method uses specific targets, which are common and well

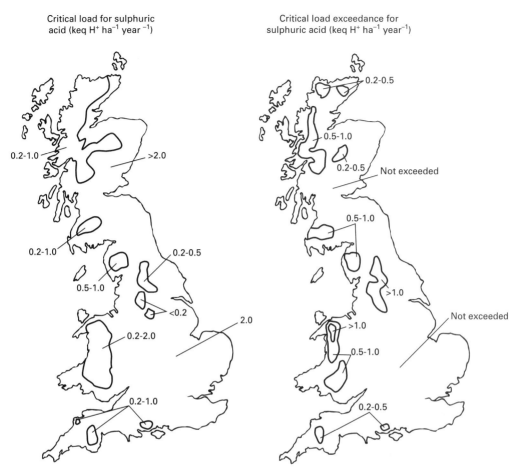

Fig. 6.42 Critical loads and critical-load exceedances for sulphur-based acids in Great Britain. The critical loads are generally high and not exceeded for much of lowland Britain, but low and often exceeded in Wales, Scotland, northern England and patches such as Dartmoor and the acid sandy soils of the New Forest. H^+, hydrogen ions. (Based on Battarbee [45].)

Table 6.3 Number of taxa of various invertebrate groups and population density of brown trout in relation to acid-neutralizing capacity (ANC) at sites in upland Wales. Values are means. (Based on Ormerod [737].)

ANC (μequiv l^{-1})	N	Mayflies	Stoneflies	Caddis	N	Trout density (n (100 m^2)$^{-1}$)
−200 to −50	6	0	6.2	3.3	4	0.73
−50 to 0	26	0.5	7.3	4.1	20	11.0
0 to +50	29	1.3	6.9	4.3	29	24.3
+50 to +450	25	3.4	7.7	7.1	25	39.2

N, number of sites examined.

understood but possibly not as sensitive to acidification as many other species.

6.6 Variations on the theme – other standing waters

This chapter has discussed the production and problems particularly of exorheic lakes – those regarded as 'normal' by limnologists in Europe and North America. It ends by being less parochial. Temporary rainwater pools, meromictic lakes and endorheic salt lakes are three groups that set the exhorheic lakes in a wider context and illustrate the continuum of variation in standing waters. In addition, for many parts of the world, endorheic lakes have a greater claim to be 'normal' than exhorheic ones.

6.6.1 Rainwater pools

Temporary puddles of rain form almost everywhere, although usually the rain soaks quickly into the ground and no particular ecosystem forms. Where the rain collects in depressions on rock, the pools may persist for weeks or months and develop a particular specialized community. Such pools have been studied in Malawi in Central Africa, where they form in depressions on the tops of rocky outcrops during the rainy season between November and March [616].

The depressions are generally bare at the start, though perhaps with some dried detritus from a previous wet phase. On filling, the pools, perhaps only a metre squared or less in area and a few centimetres deep, may be supplied with some nutrients leached from the previous sediment to supplement the meagre rainwater supply.

Phytoplankton may quickly develop [745] from dry cysts or spores, dispersed by wind. Most common are motile algae, such as *Euglena* and *Chlamydomonas*, which use flagella to remain suspended in the water and Chlorophyta, which grow on the bottom or attached to the surface-tension film (called neuston). If the rainfall is heavy and storms follow each other quickly, the pools may be flushed too rapidly for algae to develop at first, but, when they do, they are prolific, with biomasses, measured as chlorophyll *a*, of several hundred $\mu g\,l^{-1}$. Droppings from crows and other birds are washed into the pools, and wild cats, the civet and genet, favour the depressions for their nightly defecations [613, 614]. Phosphate and ammonium are leached from the dung to fertilize the water.

Fly larvae dominate the invertebrate community, feeding on the sediment and particularly its contained bacteria. There are three prominent species: *Chironomus imicola* and *Polypedilum vanderplanki*, which are chironomids, and *Dasyhelea thompsoni*, which is a ceratopogonid. All must complete their life histories to a stage where they are not killed if the pool dries out. This requires sufficient time and sufficient food.

Chironomus imicola can complete its life cycle from egg to adult in 12 days, but it cannot tolerate drying out. It favours the larger pools for egg laying and disperses efficiently. The larger pools last longer, allowing time for development of larger adults, which produce more eggs and can fly greater distances to deposit them in new pools as the old one dries out [615].

Polypedilum vanderplanki is not so effective at dispersal but extremely tolerant of desiccation (to only 3% of body weight). They can survive baking to high temperature (61 °C for a day) by the sun. On rewetting, the larvae resume feeding and growth in about an hour. *Polypedilum vanderplanki* can thus survive interruptions as its pool dries between rainstorms and refills. Selection for a short life history has not been so crucial. It can complete it in 35–43 days of uninterrupted growth, although with interruptions may take much longer, and it can survive in much smaller pools than *C. imicola*.

Dasyhelea thompsoni is intermediate [121]. It does not lay eggs in large pools with a stable water level, where it might be forced into competition with *C. imicola*, but on drying edges, where its larvae soon construct a cocoon against the bottom to protect them against drying out. It is not so tolerant of extreme desiccation as *P. vanderplanki*, but within the pools favours the sediment richer in dung than does *Polypedilum*.

To the rain-pool ecosystem can now be added a further component: the tadpoles of the savannah ridge frog, *Ptychadena anchiaetae* [614]. The tadpoles hatch from eggs laid in the pools and may number up to 1000 m^{-2}; they are not drought-resistant but rely on completing their life history before the pool dries out. If they do not complete it, their dried carcasses may be later scavenged by *Dasyhelea* [614] and thus contribute to the pool's nutrient supply.

The tadpoles feed on the sediments and mobilize nutrients by excretion more rapidly than would otherwise be the case [745]. In turn, these nutrients support greater suspended algal populations, which the tadpoles also eat and which speed the growth of *C. imicola*, increasing its chances of maturing before the pool dries out. Algae are a richer food source than sediment. However, tadpoles may also delay the final emergence of the adult flies by disruption of the habitat by their movements at a time when the insect must emerge through the water film.

This simple system encapsulates the main features of larger lakes. These include the importance of nutrient supply from the catchment (bird droppings and cat dung), the role of flushing in determining, with nutrients, the size of the suspended algal populations, the interaction between water and sediments and the impact that a species at the upper end of a food web can have on the stages below it.

6.6.2 Meromictic lakes

Some lakes have almost permanently unmixed layers, which usually become deoxygenated. Such meromictic lakes [451, 993] (as opposed to holomictic lakes, which mix to the bottom at some time during the year) generally have a bottom layer of high density. This may be due to ingress of sea water or mineral springs, to evaporative concentration in arid areas, to freezing and thawing of surrounding soils, or, in very deep lakes, to accumulation of salts following decomposition of sedimenting detritus.

The Hemmelsdorfersee near Lubeck in Germany has a deep, 45 m hole, which receives flood water from the sea every century of so (the last time in 1872). Subsequently, fresh water forms a layer on top. Floodplain lakes close to the sea may also acquire such salt injections [178, 977]. Some Norwegian lakes (e.g. Lake Tokke) have ancient salt layers, which were trapped at the end of the last glaciation in basins that were inlets of the sea. The basins were subsequently lifted above sea level as the pressure of the ice decreased on the land surface. Concentrations of brine in Arctic lakes during the formation of surrounding permafrost give similar layers [752].

The Dead Sea had a deep layer formed by ancient evaporative concentration, on to which less dense spring water had layered over several hundred years until the whole lake mixed in 1979 [921, 922]. The surface, spring-fed layers had relatively high radium-226 concentrations and, from the residual concentration in the monimolimnion (the non-mixing layer), it was calculated that the last mixing was 300 years ago.

In the tropics, Lake Kivu, formed by the damming of the Rutshuru River in the late Pleistocene by lava flows from the Virunga volcanoes, has a saline layer from about 70 m to its greatest depth (about 450 m). The salts are believed to have been leached from the lavas and the layer is anaerobic and rich in H_2S and methane. Fresh water has layered on the top of it. The deep lakes Malawi and Tanganyika [47, 204] also have anaerobic monimolimnia, although these are barely denser than the surface water, which, in major ion proportions, they resemble. Only the top 200–300 m regularly mix through wind action, which currently seems unable to disturb the lower layers (685 and 1470 m, respectively). The similarity of major ion composition with depth, however, suggests that complete mixing has taken place relatively recently.

In many oligomictic tropical lakes, the original hypolimnia may acquire denser and denser water as the products of decomposition accumulate. These lakes may then become biogenically meromictic (as opposed to ectogenically when the salt source is external, e.g. the sea or springs). Such lakes may then mix even less frequently – perhaps not for centuries. Examples may be some of the crater lakes in Indonesia and the Cameroons. Barombi Mbo is 110 m deep, with no oxygen below about 20 m. That this stratification is very old is suggested by the presence in this lake of a fish that seems adapted to extensive movements into the anaerobic water to catch its prey, a midge larva of the genus *Chaoborus*. The fish, *Konia dikume* [971], oozes blood when it is caught, for it is heavily supplied with blood-vessels. The blood has a high haemoglobin content – at 16.5 ± 0.9 g $(100 \text{ ml})^{-1}$, twice or three times as much as that of the other fish in the lake. It also contains more red blood cells of greater surface area per cell than the other fish [359]. Stratification in the lake is maintained by a temperature difference of only $1.3°C$ (29.3 to $28°C$) between the top and bottom, and is probably stabilized by meromixis.

When such deep tropical lakes mix, transfer of the deeper, usually deoxygenated, layers to the surface may cause mass deaths of fish [360] and even of local human populations. This happened around Lake Nios in north-west Cameroon in August 1986 [605, 659]. An odourless gas was released following heavy rainfall and blanketed a local village, causing 1500 deaths. The most likely contender is CO_2, which is heavier than air. Carbon dioxide concentrations built up under pressure in the monimolimnion from underlying volcanic activity, and bubbled out with release of pressure at the surface.

6.6.3 Endorheic lakes

Endorheic lakes are very widespread [1016], although for much of the time many may be dry salt-pans, dominated by chlorides, sulphates or carbonates, derived from catchment rocks of marine or volcanic origin. The diversity of their communities is low. Some are far more concentrated than sea water and may support single-species communities, perhaps of the pink, carotenoid-rich flagellate *Dunaliella salina*, or species of Halobacteria, fed upon by brine shrimps, such as *Artemia salina*. In the less saline lakes, some aquatic plants, particularly of the genus *Ruppia* [93], may grow.

The shallower endorheic lakes may be highly productive [947, 973]. The most widely quoted case is Lake Nakuru (Figs 6.3 & 6.43) in Kenya, where a dense population, with as much as $16\,000\ \mu\text{g}\,\text{l}^{-1}$ of chlorophyll *a*, of a blue-green alga (*Spirulina platensis*) may supply food for up to a million and a half lesser flamingo (*Phoeniconaias minor*), which filter the algae from the water; other birds filter crustaceans or feed on detritus and algae in the bottom mud. The spectacle of a blue-green lake trimmed at the edges with a dense pink fringe is one of the vivid cameos of African limnology.

Many endorheic lakes have high concentrations of soluble reactive and total phosphorus and perhaps also of nitrogen compounds [450, 649, 946]. This may simply be the consequence of evaporative concentration, but the inherent insolubility of phosphorus compounds and the fact that precipitants like iron should be equally concentrated make it likely that the explanation is more complex.

In some of these lakes, birds like flamingos, *Spirulina* and sediment may be parts of a closed cycle, in which the continual feeding and supply of excreta to the sediments prevents the formation of an oxidized microzone. Phosphorus may then continually diffuse from the sediment to support the

Fig. 6.43 Lesser flamingo on Lake Nakuru.

algal growth. The source of nitrogen is a problem, because sediments maintained in such an anaerobic state all of the time will contain large numbers of denitrifying bacteria. However, such conditions also favour nitrogen-fixing bacteria and blue-green algae. The nitrogen excreted by the birds will be converted from uric acid to ammonia, rather than nitrate, and will not be so vulnerable to denitrification. The high solubility of phosphorus compounds at the very high pH values (about 10) that are found in carbonate-dominated endorheic lakes may also be important. Such lakes may provide yet further exceptions to the principle that the fertility of pristine lakes usually depends on the nature of the catchment area at the time. They may have closed cycles, independent of the catchment, once some threshold of phosphorus accumulation has been reached, and perhaps represent the extreme cases of the warm lakes discussed earlier in this chapter.

The Aral Sea

Water balance is clearly crucial for endorheic lakes. Small reductions in their inflow in hotter, drier periods will be reinforced by increased evaporation to send water levels down and take the lake to complete dryness. Conversely, wetter periods will be followed by increased levels. Past water levels in endorheic lakes, if they can be deduced (see Chapter 10), are sensitive indicators of changing climate (see Chapter 11).

The levels of some of the world's larger endorheic lakes have been rising.

The Caspian Sea and the Great Salt Lake in the USA are examples, and this must reflect small climatic changes in these areas. The recent fate of most large salt lakes, however, has been reduced depth [1017] and area and increased salinity. This is not a consequence of climate change but of removal of water from their inflows for irrigation projects. The Aral Sea [760, 1018] is a sorry example.

The Aral Sea was the fourth largest lake on earth until the 1960s. It was moderately saline (about $10\,\mathrm{g\,l^{-1}}$, a third as salty as ocean water), huge in area ($68\,000\,\mathrm{km^2}$), draining a catchment of $2\,000\,000\,\mathrm{km^2}$, containing 32 million people of the southern Russian states, Afghanistan and Iran. It was not highly diverse but contained about twenty species of fish, twelve of macrophytes and 200 of invertebrates. It provided a fishery of $44\,000\,\mathrm{t}$ $\mathrm{year^{-1}}$, from a relatively shallow, clear-water, littoral-dominated ecosystem.

Its main inflow rivers were the Syr Daria and the Amu Daria, rising in the mountains of the Pamirs and the Tyan Shan to the south, and flowing across arid landscapes, eventually to evaporate in the shallow Aral basin. There were early interferences with the natural system when a number of fish, some fifteen eventually, were introduced to the lake. These included

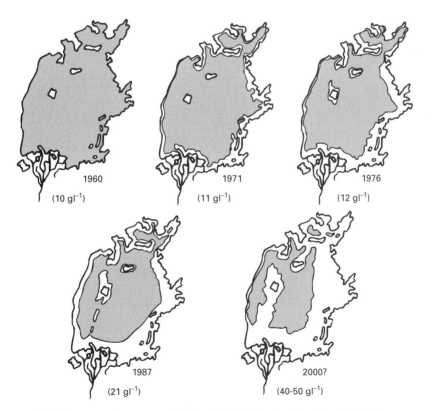

Fig. 6.44 Changes in the area and salinity ($\mathrm{g\,l^{-1}}$) of the Aral Sea since 1960 (from Williams and Aladin [1018].)

the Caspian sturgeon, the Caspian shad, the Baltic herring and the Chinese grass carp, which supplemented the native carps, percids and sturgeon. The consequences of these introductions are largely unrecorded, for a more serious impact was beginning. From the 1960s, the Russian government began to develop the lowlands of the catchment for cotton farming, using water for irrigation taken from the two main inflow rivers. There was little forethought or official concern for the effects this might have. They have been considerable.

The lake level has been reduced by 15 m or more, the lake area has been more than halved to 33.5 km^2 (Fig. 6.44), the complex shoreline and archipelagos have been smoothed and the salinity has increased threefold, to 30 g l^{-1}. Fish have disappeared and, with them, the fishery; a sequence of invertebrates of lesser diversity and increasing salinity tolerance has resulted in a current community of eight major taxa, of which an introduced bivalve mollusc, *Abra ovata*, is predominant. Of the former higher-plant macrophytes, only one, *Zostera nana*, remains, and former extensive reed swamps on the deltas of the rivers have disappeared.

Former lakeside villages are now 120 km from the water's edge, large boats rust on a bleak sand flat, the local economy has collapsed and storms of dust and salt plague those people forced to remain. Their health is poor, with lung diseases attributed to the salt dust, and problems linked with the drinking of salt water contaminated by organic chemicals used in the cotton schemes. Moreover, the irrigation water is poorly managed. Canals that should have taken it back to the rivers and thence the lake were never dug and it evaporates in depressions in the surrounding desert. *Sic transit gloria mundi.*

7: The Plankton and Fish Communities of the Open Water of Lakes

7.1 The structure of the plankton community

Lake water sparkling in sunlight hides a miniscule waterscape that is closer to a slum than a paradise. It contains millions of organisms, suspended passively or sometimes weakly swimming, in every litre. Some are photosynthetic, others feed on organic matter, live and dead, dissolved and particulate. The water contains their excretions and secretions, faeces and corpses, mixed with debris washed into suspension from the surrounding land. In this *mélange*, chemical and biological changes, both cyclic and irreversible, are taking place very rapidly.

Scaling one of these organisms to human size and considering the rest relative to it will help indicate the structure of this community. The rotifer *Keratella quadrata* (Fig. 7.1), a common small animal, has a body about 125 μm long (about half the size of a full stop on this page), with spines half as long held out behind. If the body of *Keratella* is scaled to the size of a tall man, the rest of the community ranges from lentils and peanuts to large houses, or, for fish, to the size of 'whales' 30 miles (48 km) long! The water in which the plankton lives is viscous relative to such small objects. In the scaled-up analogy, the viscosity of the fluid must also be increased, so the community must be imagined as suspended in light oil or glycerol.

Organisms that passively drift, maintained in suspension by water current; or float or swim weakly, comprise the plankton. They include heterotrophic bacterioplankton, the photosynthetic phytoplankton and the swimming animal zooplankton, some of which are grazers on smaller cells and detritus and others predators on smaller zooplankters. The larger, strongly swimming animals, such as fish, which are not at the mercy of movement by water currents, are called the nekton.

The plankton can also be categorized by size. Sometimes, for the smaller cells, it may be difficult to know whether a cell is photosynthetic or heterotrophic or uses both feeding modes, and, from the point of view of a grazing zooplankter, size is more relevant than function. The plankton has thus sometimes been divided into ultraplankton (< 5 μm), nanoplankton (5–20 μm), microplankton (20–60 μm) and net plankton (> 60 μm), although there is a continuum of size and these terms mostly reflect the pore sizes of filters available to separate the organisms.

Lately, more terms, femtoplankton (virus-like particles, 0.02–0.2 μm) and

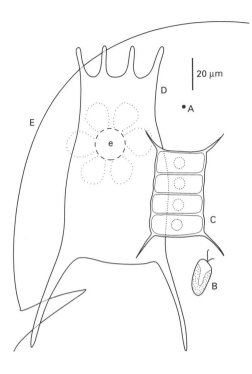

Fig. 7.1 Relative sizes of some major components of the plankton. (A) a bacterium; (B) *Cryptomonas*, a relatively small phytoplankter; (C) *Scenedesmus*, a moderately large phytoplankter; (D) *Keratella*, a small zooplankter; (E) outline of the head with eye (e), of *Daphnia*, a large zooplankter. The head constitutes about a quarter to a fifth of the total body size.

picoplankton, have been coined [893, 928, 929]. The latter describes hetero-trophic bacteria and blue-green and eukaryote algae of bacterial size (about 2 μm). Indeed, all of the size categories may contain heterotrophic bacteria or Protozoa) or photosynthetic organisms (prokaryotes or eukaryotes). Pico- and ultraplankters may carry out a substantial part of the photosynthesis of the plankton community in the oceans and in infertile lakes.

In very infertile waters where photosynthetic growth is limited by nutrient shortage, heterotrophic bacterioplankton, feeding on organic matter imported from the catchment, may be as important in providing food to the zooplankters and higher trophic levels as the photosynthetic organisms. Protozoa (often ciliates) may then be crucial in converting bacteria to particles sufficiently large for the zooplankters to feed. It is now usual to distinguish such detrital pathways, or microbial loops, from the more conventional grazing pathways based on *in situ* photosynthesis.

Carbon dioxide and photosynthetic algae form the bases of the latter, and the algae are mostly big enough to be eaten directly by the metazoon zooplankters. Where tiny photosynthetic algae are involved, there may be an intermediate protozoan stage and the detrital and photosynthetic pathways merge. Moreover, some phytoplankton species, particularly chrysophytes, cryptophytes and dinoflagellates, are phagotrophic, consuming bacteria, as well as photosynthesizing. The simple concept of a photosynthetic phyto-plankton base to a series of herbivorous animals and larger carnivores is still central but garlanded by numerous complications.

7.2 Phytoplankton

Phytoplankters are of many species, mostly oxygen-evolving prokaryotic blue-green algae (Cyanophycota) and Prochlorophycota and eukaryotic algae. The eukaryote groups of greatest importance are the Cryptophyceae (cryptophytes), Dinophyceae (dinoflagellates), Chlorophycota (green algae), Euglenophycota (euglenoids), Bacillariophyceae (diatoms) and Chrysophyceae and Haptophyceae (yellow-green or golden-yellow algae). Some examples are shown in Fig. 7.2.

In some transparent stratified lakes, light may penetrate to deep layers that have become low in oxygen or anaerobic. Communities of non-oxygenic photosynthetic bacteria (the purple and green sulphur bacteria) may be found there. They may form discrete layers, only a few centimetres thick, where their particular chemical needs are met (Fig. 7.3). They use sulphur compounds as a hydrogen donor for photosynthesis and hence do not evolve oxygen, but deposit granules of sulphur inside or outside the cells.

Such photosynthetic bacterial communities are relicts of a former anaerobic biosphere and are now quantitatively unimportant. Most attention has been paid to the ubiquitous and diverse oxygen-evolving phytoplankters of the surface aerobic waters (Fig. 7.2). In the scaled-up model, the sizes of unicellular phytoplankters range from those of lentils and peanuts (1–5 μm) to footballs and water melons (50 μm). Some colonies may be visible to the naked eye (several hundred micrometres) and would be scaled as heavy horses or elephants.

A common misconception is that phytoplankters float. This is generally not so, but certain blue-green algae, which are sometimes very abundant in highly fertile lakes, have organelles called gas vesicles. These comprise masses of protein-bound prisms with conical ends, contain air and give positive buoyancy. Under some circumstances, these blue-green algae may truly float at particular depths in the water column that favour their growth and, in other circumstances, may form a paint-like scum or water-bloom at the surface (see later). Similar gas vesicles are used by the anaerobic photosynthetic bacteria to maintain themselves in the most appropriate part of the redox gradient (Fig. 7.3).

Excepting *Botryococcus braunii*, a green alga, which may remain positively buoyant by storing large quantities of oil, all other phytoplankters are more dense than water. Diatoms, which have cell walls of silica, may be considerably more so. These phytoplankters are kept suspended in the water by wind-generated currents. Some species have flagella, and movement of these may help counteract the inevitable tendency to sink.

Non-flagellated species have evolved cell or colony shapes that decrease the rate of sinking. Flat plates, needle-shapes with curved ends, spines and projections all seem to be advantageous. However, too easy an acceptance of shape as adaptive should be avoided. Envelopes of mucilage, invisible

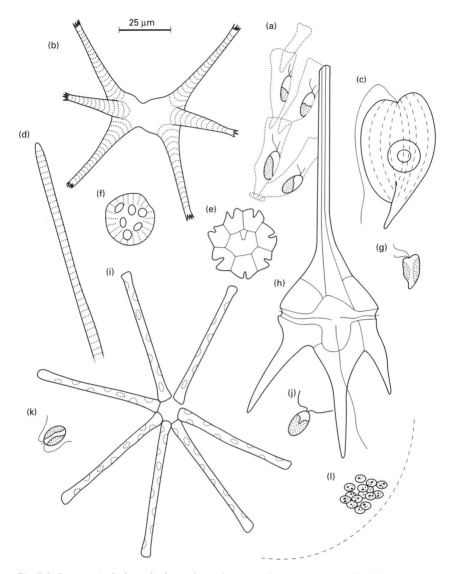

Fig. 7.2 Some typical phytoplankton algae, drawn to the same scale as Fig. 7.1.
Cyanophycota (blue-green algae): (d) *Oscillatoria*, (l) *Microcystis*; Chrysophyceae (yellow-green or golden algae): (a) *Dinobryon*; Chlorophycota (green algae): (e) *Pediastrum*,
(b) *Staurastrum* (a desmid), (j) *Chlamydomonas*; Bacillariophyceae (diatoms): (f) *Cyclotella*,
(i) *Asterionella*; Euglenophycota (euglenoids): (c) *Phacus*; Cryptophyceae (cryptomonads):
(g) *Rhodomonas*; Dinophyceae (dinoflagellates): (h) *Ceratium*; Haptophyceae:
(k) *Prymnesium*. *Microcystis* (l) is a very large alga, of which a diagram of the entire
colony could occupy as much as this page. Only a few cells are shown.

unless the cells are mounted in Indian ink (Fig. 7.4), may be thick enough
to give a spiny cell an effectively spherical shape and may even lubricate
passage through the water.

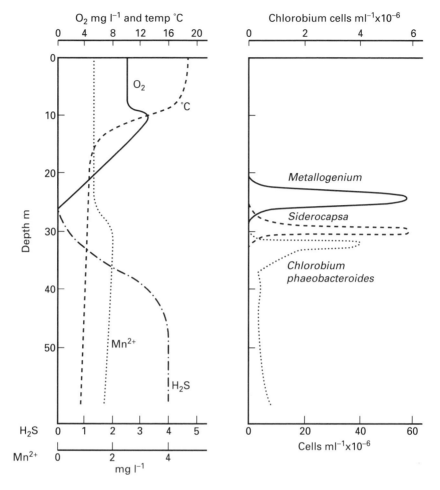

Fig. 7.3 Distribution of three bacterial species in the water column of Lake Gek-Gel in September 1970. The lake is meromictic, having a deep and unmixing saline layer from 30–40 m down, which is permanently deoxygenated. *Chlorobium phaeobacteroides* is a photosynthetic bacterium, absorbing light efficiently between 450 and 470 nm (green) and using hydrogen sulphide as a hydrogen donor. *Metallogenium* and *Siderocapsa* are chemosynthetic bacteria, probably using the oxidation of ferrous (Fe^{2+}) and manganese (Mn^{2+}) ions as sources of energy. (After Kuznetsov [537].)

Why have all phytoplankton cells not evolved positive buoyancy, when sinking, with its potential for loss of the organism from the euphotic zone, clearly has disadvantages? The answer is that it also has advantages. Phytoplankters need a supply of inorganic nutrients, which they absorb from the water layer, a few micrometres thick, immediately in contact with the cell. Molecular forces tend to preserve this layer intact and it soon becomes depleted of nutrients, which are not rapidly replaced by diffusion alone. Continuous movement of the cell through the water, as it sinks and is retrieved by upwardly directed turbulence, sloughs away the depleted

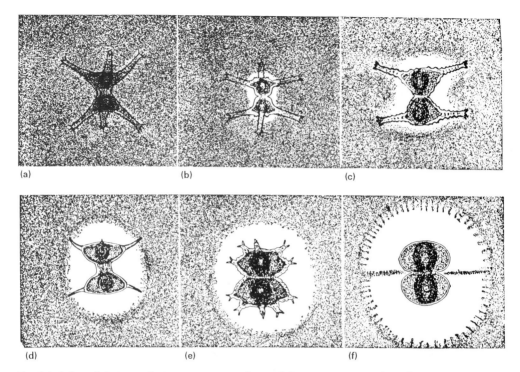

Fig. 7.4 Cells and their mucilaginous coverings of several *Staurastrum* species from the phytoplankton of the Lunzer Untersee. The cells have been mounted in Indian ink to demonstrate the mucilage. (a) *S. cingulum*, (b) *S. lutkemulleri*, (c) *S. manfeldtii* var. *planktonicum*, (d) *S. cuspidatum*, (e) *S. furcigerum*, (f) *S. brevispinum*. (From Ruttner [839].)

nutrient shell and replaces it by a supply of undepleted water [453, 703]. The cells must move relative to the water, if they are not to become nutrient-starved.

Phytoplankters have been assumed to satisfy their carbon and energy needs through photosynthesis alone. This is true for many but not all. Some require at least one preformed organic compound, usually a vitamin, such as cyanocobalamin (B_{12}), thiamine or biotin. In general, the blue-green algae, diatoms and desmids do not require additional organic compounds, but many algae of other groups are able to take up simple organic compounds heterotrophically. However, the concentrations of such compounds are normally low in natural waters and bacteria successfully compete for them. Some plankters of highly organic sewage-oxidation ponds (usually green algae and euglenoids) may depend as greatly on organic uptake for their energy needs as on photosynthesis.

The smaller phytoplankters (up to $10 \mu m$) may occur in very large numbers: up to 10^6 ml^{-1}, compared with about 10^2–10^5 for the larger phytoplankters. On the scaled-up model, the smaller species would appear as a population of objects the size of tennis-balls, mutually spaced at distances of about 2.5 m. The larger algae can be imagined as a similar constellation of water melons 27 m apart.

Several species may simultaneously be forming large populations in the water, but there is still a lot of space between the individuals. This might explain why, despite their ubiquity, parasitic fungi (frequently chytrids) and parasitic Protozoa (Fig. 7.5) [120] only cause epidemics when algal population densities are very large. Successful infestation of a host cell requires an encounter, which may be rare in an environment where hosts are well spaced and both host and parasite are continually moved by turbulence.

The very wide size range of phytoplankters (from lentils to elephants in the model) is remarkable, particularly because, in a nutrient-scarce medium (see Chapter 6), small bodies with high surface-to-volume ratios should be able to compete more effectively for nutrients. Large cells also sink faster and hence are more vulnerable to loss from the epilimnion in stratified lakes. Nonetheless, there are large phytoplankters. The advantage to large size is that big cells are less readily eaten by filter-feeding zooplankters, which are mechanically unable to manipulate large cells or colonies into their mouths.

Phytoplankters are also remarkable for their near-abandonment of sexual reproduction. They live vulnerable lives: most are readily grazed; all may be washed out of their habitat by incoming floods; there is a constant danger of sinking to deep, dark water, where photosynthesis is not possible and where they may be trapped in the sediments; and nutrients are scarce. Their habitat is an extreme one of rarefied resources and physical danger. Not surprisingly, there has been selection against sexual reproduction in such a habitat, for recombination might disturb the intricate adaptations already developed. Most phytoplankters reproduce only by asexual cell division, with generation times of hours or a few days.

A simple equation describes the conditions for increase in a population of a phytoplankter. It will increase its number (N) if its growth rate (b) is greater than the sum of its rates of loss by sinking and trapping in the sediment (v), grazing (g) and flushing out of the water body (w):

$$dN/dt = bN - (vN + gN + wN)$$

The existence of many thousand planktonic species – with several hundred in almost any lake – means that this equation has been successfully balanced in thousands of different ways. The next sections will look at the components of the equations.

7.2.1 Photosynthesis and growth of phytoplankton

Photosynthesis, the fixation of light energy into chemical bonds, and growth, the synthesis of new cell material, are different processes. Although the latter requires the former, photosynthesis is not necessarily followed by growth if the necessary materials are not available.

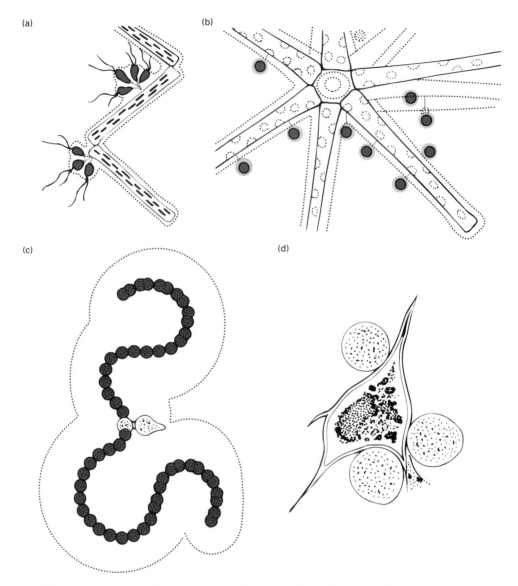

Fig. 7.5 Some protozoa and fungi associated with phytoplankton algae. Part (a) is the choanoflagellate *Codosiga* on *Tabellaria*. It feeds on bacteria and does not damage the alga. Fungi of the chytrid group, shown on *Asterionella* (b), however, will eventually kill the cells, as will the chytrids *Rhizosiphon* (c) and *Rhyzophydium* (d) on *Anabaena* and *Ceratium*, respectively. (Based on photographs in Canter-Lund and Lund [120].)

Measurement of photosynthesis was discussed in Chapter 5. If water from a well-mixed water column is placed in bottles and resuspended at a series of depths, and the photosynthetic rates measured by the oxygen-release method, a charateristic curve of gross photosynthesis with depth is obtained (Fig. 7.6).

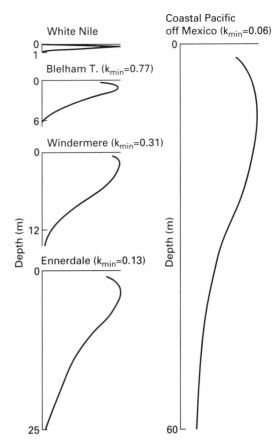

Fig. 7.6 Examples of gross-photosynthesis depth profiles in a series of waters of increasing turbidity, as indicated by the value of the minimum extinction coefficient (k_{min}) in each case. T, Tarn. (Based on Talling [943].)

Photosynthesis is often low at the surface, perhaps due to inhibition by ultraviolet light, with a peak at some depth below. This peak is set when the photosynthetic pigments are unable to abstract any more radiation (at light saturation, I_k). Its size and position depend on the light-saturated rate of photosynthesis per unit of biomass (P_{max}) and of the total biomass (n) of photosynthesizing cells present. Usually biomass is expressed as chlorophyll a.

Finally, the curve of photosynthesis with depth declines exponentially below the peak. The depth to which photosynthesis extends depends on the surface intensity (I_o) and the attenuation coefficients for light absorption in the water (k), usually expressed as the minimal of these (k_{min}) (see Chapter 6). As the biomass increases, it becomes a major contributor to the attenuation and the curve is displaced upwards. Such a curve can be described by an equation for the total photosynthesis per unit area of lake, (Σa), which Talling [943] has determined to be approximately:

$$\Sigma a = nP_{max} \times 1/1.33\, k_{min} \times (I_o/I_k)$$

The value of 1.33 was determined empirically and the term I_o/I_k approximately describes the position of the peak in the water column – the lower the intensity that photosynthesis becomes saturated, the deeper the peak. The value of I_k varies with different algae – it is often low for blue-green algae but higher for green algae – and the value of P_{max} tends to increase with temperature, with an approximate doubling for every 10°C. A similar photosynthesis–depth curve can be obtained even in stratified water columns, where the algae and photosynthetic bacteria may be layered, if the values of photosynthesis are expressed not per unit volume of water but per unit biomass of algae.

The effects of the algae in absorbing light can be described by plotting, for a lake, the minimum attenuation coefficient on several occasions against the chlorophyll a concentration. The intercept on the k axis (biomass = 0) gives the attenuation due to the background properties of the water and the slope of the line the attenuation per unit chorophyll a. The value of this varies with the nature of the algae (lower for larger cells than for smaller ones) but is often from about 0.008 to 0.021 natural logarithm (ln) units $m^{-1} (\mu g\ l^{-1})^{-1}$. A crop of $100\ \mu g\ l^{-1}$ of chlorophyll a, in a lake of background attenuance $k_{min} = 0.3\ m^{-1}$, could then reduce the potential euphotic zone (see Chapter 6) from 12.3 m to between 3.4 and 0.17 m. At high concentrations (above about $300\ mg\ m^{-2}$), the algae can shade themselves.

Mean values of gross photosynthesis, in g $(O_2)\ m^{-2}\ day^{-1}$, include, for example, 21 for Loch Leven, 15.6 for Lake Neagh and 57 for some tropical African lakes. Converted to yearly values on the basis of regular measurements and to units of carbon, the rather fertile lakes Leven and Neagh had annual productivities of $785\ g\ C\ m^{-2}\ year^{-1}$ and $1500–1800\ g\ C\ m^{-2}\ year^{-1}$ [484].

The actual growth of algae, however, is likely to have been much less, for these are values for gross photosynthesis or production. Net production is lower by the amount of respiration of the algal cells and even then may be expressed as carbohydrate or fat storage in the cells, rather than by true growth, the production of new ones.

7.2.2 Net production and growth

Algal respiration rates are likely to be high (and net production therefore low) relative to gross photosynthesis in early spring in a temperate lake, when the water column is vigorously mixing, and for much longer in very deep, well-mixed lakes, such as Loch Ness [510], when the cells will also be circulating between the illuminated upper layers and the layers below the euphotic zone (Z_{eu}). If the mixing depth (Z_m) is large relative to Z_{eu} (Fig. 6.9), they may spend long periods unable to photosynthesize but inevitably respiring. Until, on average, the gross photosynthesis made by the cells while they are in the upper layers exceeds the respiration while they circulate throughout the whole water column, no net production will be possible. This seems to determine the start of growth in spring. For example, in the deeper northern basin of Lake Windermere, growth starts about 2–3 weeks later than in the shallower southern basin, where Z_m is lower. In lakes in arid zones, higher concentrations of suspended silt particles may effectively reduce Z_{eu} considerably while not affecting Z_m. As an algal crop grows and becomes self-shading, the photosynthetic rate per cell, determined by light, may be decreased but the respiration rate, determined by temperature, does not and net production may then be low.

Even if respiratory demands are met, there may still be no growth, despite some net production, if nitrogen or phosphorus is scarce. Cells may then be forced to respire synthesized carbon compounds, to store them or to secrete them to the water. In one study, only about one-fourteenth to one-thirtieth of the carbon fixed in photosynthesis was incorporated into new cell material [812]. In unpolluted freshwater lakes, N or P is likely to be scarce, and the degree of scarcity may determine the rate of growth as well as the potential maximum yield of algae (see Chapter 6).

7.2.3 Nutrient uptake and growth rates of phytoplankton

Phytoplankters need about 20 elements for growth, but only C, N and P are likely to limit growth rates on any general basis. All are present in the water at lower concentrations than are required in the cell, so that active, energy-requiring mechanisms, involving enzymes, are needed to concentrate them into the cells. The efficiency of these mechanisms differs between species and can be measured as the half-saturation constant (K_s) for uptake rate (μ) in an equation [660]:

$$\mu = \mu_{max}S/(K_s + S)$$

The half-saturation constant is the concentration (S) of nutrient at which half the maximum rate, μ_{max} (when the enzymes are saturated to full capacity), can be achieved. Values of K_s should be lower for species growing in infertile lakes than for those in fertile ones. Coexistence of two species competing for the same nutrients may also be possible if the first is more adept at taking up one nutrient and the other a second nutrient. For example, Titman [964] studied the relationship between two diatoms, *Asterionella formosa* and *Cyclotella meneghiniana*, both of which grew in Lake Michigan, with *Asterionella* more abundant in the open lake and *Cyclotella* near shore. Both species require both silicate and phosphate. Using a continuous-culture apparatus, in which a nutrient solution of constant composition can be supplied, Titman determined $K_s(P)$ for *Asterionella* to be 0.04 µmol P and for *Cyclotella* 0.25 µmol P.

This suggests that, if phosphate is scarce, *Asterionella* will tend to compete favourably with *Cyclotella*. On the other hand, if silicate is scarce, the reverse is true, for K_s (silicon (Si)) was 3.9 µmol Si for *Asterionella* and

1.4 µmol Si for *Cyclotella*. For each species in turn, growth rates will be similar when:

$$\mu_{max}S(Si)/S(Si) + K_s(Si) = \mu_{max}S(P)/S(P) + K_s(P)$$

Hence both nutrients are in balanced supply for growth when:

$$S(Si)/S(P) = K_s(Si)/K_s(P)$$

For *Asterionella*, this ratio is 3.9 : 0.04, or 97, so that at Si : P molar ratios greater than 97 in the water, *Asterionella* will be phosphorus-limited. For *Cyclotella*, the ratio is 5.6, and above and below this the diatom will be phosphorus- and silicate-limited, respectively. When both species are present, both will be phosphorus-limited when the ratio is greater than 97, and *Asterionella* will tend to survive rather than *Cyclotella*. When ratios are below 5.6, both are silicate-limited but, being more efficient at silicate uptake, *Cyclotella* will predominate. But between 5.6 and 97, each diatom's growth rate is limited by a different nutrient and they should be able to grow together. In Lake Michigan, the Si : P ratio is between 200 and 500 in the open lake but is between 1 and 10 near the shore. The general predominance of *Asterionella* offshore and *Cyclotella* inshore is consistent with the laboratory findings [960].

In 1934, Redfield [803] suggested that cells which had a balanced nutrient supply would have a ratio (by atoms) of carbon : nitrogen : phosphorus of 106 : 16 : 1 (or, by weight, 42 : 7 : 1) and that severe departures from this ratio indicate a (physiological) nutrient limitation. A demonstration of the Redfield ratio in a phytoplankton population might suggest that it is not being limited by nutrients (note – limitation of growth not yield (see Chapter 6)). In the open ocean, Goldman *et al.* [341] found that the phytoplankton had ratios of C : N : P close to the Redfield ratio and that their maximal growth rates were at this ratio.

In fresh waters, departures from the Redfield ratio are common [403], as are properties of the cells which suggest that scarcity of N or P is keeping growth rates below the maximum. A variety of indicators may be used to demonstrate this [272, 397, 1000]. For nitrogen, high rates of ammonium uptake in the dark and increases in the ratio of carotenoids to chlorophylls suggest nitrogen deficiency. (Chlorophylls are nitrogen-containing compounds, which the cell may be unable to synthesize if nitrogen is scarce.) For phosphorus, an increase in the activity of acid phosphatase enzymes (which can break down organic phosphorus compounds), a high rate of isotopic ^{32}P uptake and an absence of free phosphate in the cells suggest scarcity.

Departures from the Redfield ratio in lakes, compared with the ocean, are interesting, because it is suggested that a very efficient system of remineralization of N and P takes place through grazing and bacterial activity in the ocean. This is said to maintain the availability of N and P in the correct ratio and sufficient always to allow high growth rates. In fresh waters, a similar system would be indicated by values close to the Redfield ratio in pristine lakes whose nutrient loadings have been unaltered by human activities – the situation which pertains, because of its huge extent, in the midocean. A survey [403] of a wide range of fresh waters, many of them in remote areas, however, revealed high but variable $N:P$, $C:P$ and $C:N$ ratios.

In fresh waters, because of their small size, local variations in loading rates, especially those caused by human activity, may have disrupted achievement of a system like that in the ocean. Eutrophication may have favoured large algae with low surface-to-volume ratios, which are not readily grazed, and the imbalance of artificially altered P and N loadings may itself create scarcities. Heavy phosphorus loading may induce nitrogen scarcity, for example. The implication is that the behaviours of many freshwater phytoplankton communities are artefacts of human influence.

7.2.4 Distribution of freshwater phytoplankton

Land vegetation obviously differs from place to place, in both its appearance and its species composition. Many algae are cosmopolitan and, among the freshwater phytoplankton, such major latitudinal differences are not found [777]. There are substantial differences among different lakes, however, and sometimes within big lakes, although they are not necessarily visually obvious.

Figure 7.7 shows how the biomass and productivity of phytoplankton can vary horizontally within a large lake, Lake Erie. The distribution depends primarily on the greater availability of nutrients at the eastern end, close to sources of urban effluent. There are also differences in the community composition, the 'appearance', of the community.

Appearance includes size, organization into filaments or colonies and possession of flagella or buoyancy vesicles. The environmental features associated with differences among these include the availability of nutrients, the degree of stratification and the incidence of grazing, among others. The phytoplankton of an infertile lake is likely to be of small organisms with high surface-to-volume ratios. These are readily grazed, so scarce nutrients might be rapidly recycled by excretion from the grazers before the cells can be lost to the sediment and their nutrients removed from use. Unicells are more likely to dominate than colonies or filaments (Fig. 7.8). Fertile waters, on the other hand, may be able to sustain greater proportions of larger organisms (Fig. 7.8). Nutrient supply is easier and the organisms may invest more energy in avoiding grazing, any loss of nutrient through sinking to the sediments being met by renewed supplies from the catchment or even return from the sediment (see Chapter 6). Table 7.1 gives some general associations of algae in waters of increasing fertility but many variants

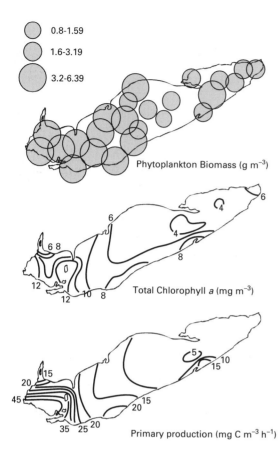

Fig. 7.7 Horizontal variation in phytoplankton biomass, phytoplankton chlorophyll *a* concentration and primary productivity in Lake Erie. (From Munawar and Munawar [702].)

occur. One feature might be picked out as an example for further discussion. This is the correspondence of a diverse desmid plankton with very infertile conditions.

7.2.5 The desmid plankton

Desmids (Fig. 7.9) are a group of the Chlorophycota (green algae), often said to be calciphobic. Grown in laboratory culture, however, they will often tolerate high concentrations of calcium and most of the other major ions that generally increase in concentration from the least to the most fertile waters. The exception appears to be bicarbonate, HCO_3^-, whose concentrations are linked with pH, carbonic acid (H_2CO_3) and carbon dioxide (CO_2) in natural waters.

Inorganic carbon associates with water (H_2O) in a series of equilibria:

$$CO_2 + H_2O \Leftrightarrow H_2CO_3 \Leftrightarrow HCO_3^- + H^+ \Leftrightarrow CO_3^{2-} + 2H^+$$

and the proportion of the total inorganic carbon concentration accounted for by each form depends on pH. Below pH 4.5, almost all is in the form of

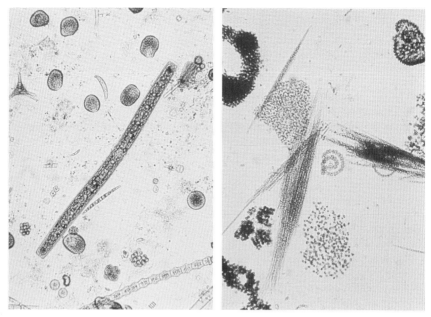

Fig. 7.8 The phytoplankton of infertile ('oligotrophic') lakes (left) is often characterized by chrysophytes, small diatoms, desmids (for example, *Pleurotaenium*, the largest cell, *Staurastrum*, the small triangulate cell, and *Spondylosium*, the filament) and dinoflagellates (*Gymnodinium*, the rounded cells). That of fertile ('eutrophic') waters may have other diatoms, green algae and blue-green algae. Shown right are three characteristic blue-green algal genera: *Aphanizomenon*, the bundles of straight filaments, *Microcystis*, the irregular colonies, and *Anabaena*, the coiled filaments. (Photos by E. Stoermer.)

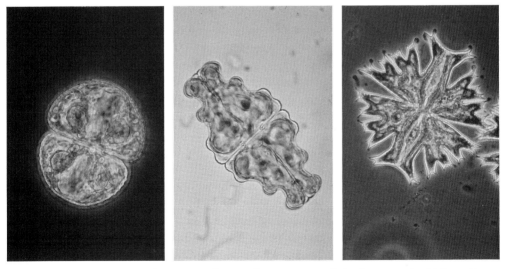

Fig. 7.9 Desmids, a group of green algae, are particularly characteristic of lakes with low concentrations of dissolved ions. Three characteristic genera are *Cosmarium* (left), *Euastrum* (centre) and *Micrasterias* (right).

Table 7.1 Some general associations of phytoplankton communities with waters of low and high fertility (based on Hutchinson [453]).

Infertile waters ('oligotrophic')

1 Dominated by a large variety of desmids (e.g. *Staurodesmus* and some *Staurastrum* species (small unicells), with some colonial green algae (*Sphaerocystis*) and thin-walled (*Rhizosolenia*) or small (*Tabellaria*) diatoms. Often in slightly acid waters, low in alkalinity and calcium

2 Dominated by small diatoms (*Cyclotella*, *Tabellaria*) and some filamentous diatoms (*Melosira* spp.) in neutral or slightly alkaline waters. Chrysophyceae may be major components, perhaps together with very small (< 3 μm) flagellate and coccoid blue-green algae

Waters of intermediate fertility ('mesotrophic'–'eutrophic')

Dominated by dinoflagellates (*Peridinium*, *Ceratium*) and diatoms (*Cyclotella*, *Stephanodiscus*, *Asterionella formosa*), with desmids but of the genera *Staurastrum*, *Closterium* and *Cosmarium*, other green algae (*Scenedesmus*, *Pediastrum*) and perhaps some filamentous blue-green algae, either fixing N_2 (*Anabaena*) or not (*Oscillatoria*)

Highly fertilized waters ('hypertrophic')

1 Dominated by *Oscillatoria* with diatoms in lakes

2 Dominated by small Chlorococcales (green algae) in small ponds

3 Dominated by *Spirulina* or *Dunaliella* in saline lakes

4 Dominated by Euglenophyta in small ponds heavily fertilized with organic matter

---

H_2CO_3 and CO_2; above about 8.4, increasing proportions of CO_3^{2-} can exist. Bicarbonate dominates in the middle range. Absorption of CO_2 or HCO_3^- for photosynthesis (CO_3^{2-} cannot be used) causes the equilibria to move to the left. The concentration of H^+ thus decreases and the pH rises. This resets the equilibrium proportions of CO_2, HCO_3^- and CO_3^{2-} to favour progressively HCO_3^- and CO_3^{2-}. At high pH, free CO_2 is very scarce. Talling [944] showed that phytoplankters could be ranked according to their ability to take CO_2 at lower and lower concentrations (and increasing pH).

Moss [674] found that several desmids from infertile waters ceased growth at pH values between 8 and 9, with a mean around 8.4. Other algae, including some desmids found in fertile lakes, would continue to grow at higher pH (some beyond 9.5) (Fig. 7.10). A pH of 8.4 was associated with an alkalinity (a measure of bicarbonate) of about 2.5 mequiv l^{-1}. Diverse desmid floras are confined to alkalinities below about 1.5 mequiv l^{-1} in nature. It seemed likely that these desmids of infertile waters were unable to use CO_2 at low concentrations or to use HCO_3^- as a carbon source for photosynthesis.

Desmids of infertile lakes could grow in fertile water if the pH was reduced to below 8.4 [675]. These findings suggest why many desmids do not compete readily with other algae in alkaline waters, but not why fertile lake algae do not displace the desmids from infertile lakes. Such species take up either CO_2 or HCO_3^- and hence should be at no disadvantage in an acid lake.

Indeed, many of the fertile lake species can be found (Fig. 7.10), although they are scarce, in low-alkalinity water, but do not compete effectively with the desmids. The reason may lie in the availability of phosphorus and nitrogen. If the rates of supply of these are low, low growth rates might allow a species to cope without becoming nutrient-limited. A genetically determined high growth rate, on the other hand, which might be sustainable in a fertile lake, might be a disadvantage in an infertile one. The maximum potential growth rates of the desmid species tested were lower than those of the species of fertile lakes.

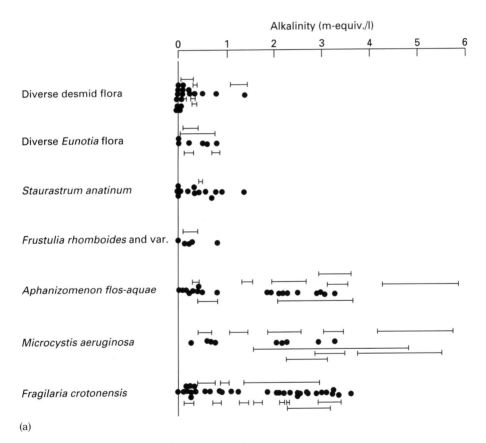

(a)

Fig. 7.10 (a) Many desmids are found only in low-pH, low-alkalinity waters. The same is true of certain diatom genera, such as *Eunotia* and *Frustulia.* Many blue-green algae (*Aphanizomenon, Microcystis*) and diatoms (*Fragilaria*) have much wider distributions. (b, *page 285*) Upper and lower pH limits for growth of a variety of fresh water algae. When grown in media of differing pH, species with distributions restricted to low alkalinities in nature (square symbols) cease growth at pH values not much greater than 8, while those of wider distribution continue growth at above pH 9 and possibly (arrows) even greater values. (Based on Moss [673, 674].)

7.2.6 Washout, mixing and stratification

Phytoplankton communities are influenced by washout, mixing and stratification, as well as by water chemistry. The effects of washout can be demonstrated in the Norfolk Broads, a series of shallow riverine lakes. These waters can be arranged in a gradient from rapidly flushed (say, once every two weeks or less) to poorly flushed (once or twice per year). In the most flushed, the phytoplankton is dominated by centric (radially symmetrical) diatoms (*Stephanodiscus*, *Cyclotella*, *Melosira* (*Aulacoseira*)), which have heavy cells, requiring much turbulence to maintain them in suspension. Where lowland rivers develop phytoplankton, it is also usually diatom-dominated. As the flushing rate decreases, small, bilaterally symmetrical, pennate diatoms, such as *Synedra* and *Diatoma*, join the centric diatoms, together with filamentous blue-green species (*Oscillatoria*).

Finally, in the least flushed areas, colonial blue-green algae (e.g. *Aphanothece*) dominate, with only pennate diatoms present. The blue-green algae are often large and do not, in this case, have gas vesicles. They should sink fairly rapidly, although the short water columns are well mixed at all times. Association of these species with low flushing might be due to their low growth rates, which cannot cope with high washout rates. It is impossible to be certain, however, for many environmental factors are correlated with low flushing. These include reduced nitrogen inputs, slightly higher pH, potentially greater numbers of grazers and greater deoxygenation rates at night. All of these factors have been shown to affect blue-green algal distribution to some extent. The links with flushing in the Broads, however, match similar associations in deeper lakes with mixing (high turbulence) or stratification (low turbulence). Diatoms are associated with mixing and blue-green algae, particularly gas-vesiculate forms and large motile flagellates, with stratification.

Gas-vesiculate and flagellate algae tend to be common in stratified lakes in midsummer, when nutrient supplies to the lake may be at their lowest for the year. Light is not scarce but accumulated algal crops are making large demands on any nutrients regenerated by grazing and decomposition within the eplimnion. Daily movements up and down by organisms such as large dinoflagellates may give access to greater nutrient supplies at the top of the hypolimnion [846, 907]. Vesiculate blue-green algae may also migrate to gain access to nutrients, and the paint-like scums or blooms that are sometimes seen at the surfaces of fertile waters may reflect breakdown of this mechanism.

7.2.7 Blue-green algal blooms

The word 'bloom' has been misused recently to mean any very large (or not even very large) growth of algae (or, indeed, of other organisms) and is

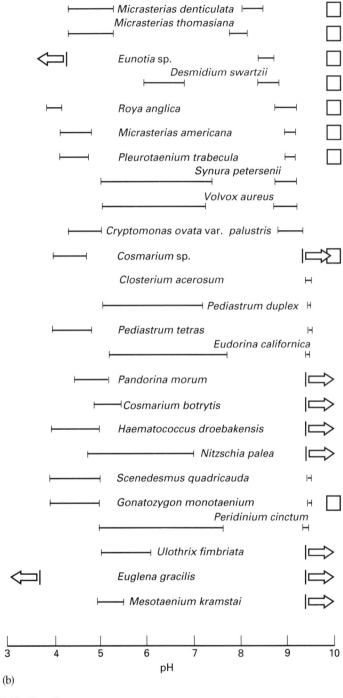

(b)

Fig. 7.10 (*Continued*)

rapidly becoming meaningless. It is used here with the precise meaning of a surface aggregation of blue-green algae (Fig. 7.11). The organisms concerned include, among others, species of *Aphanizomenon*, *Anabaena*, *Microcystis* and *Oscillatoria*, which can regulate their buoyancy with gas vesicles. The protein walls of the vesicles are permeable to atmospheric gases, but not to water, and they contain air. Some vesicles are weaker than others and can be collapsed by pressure (about 7 atm). If the vesicles are first removed from the cells, rather more pressure (about 11 atm) is needed to collapse them. This is because in the cell they are already subjected to considerable internal osmotic pressure.

Because they may occupy up to 30% of the cell volume, gas vesicles can make the cells buoyant so that they float upwards. This movement, like sinking, has advantages in promoting nutrient uptake. However, as the cells rise into regions of higher light intensity in the water column, their photosynthetic fixation of carbon increases and the concentrations of soluble organic substances, such as sugars, rise in the cell. This is accentuated if phosphorus and nitrogen supplies are low and new cell growth is prevented. As the concentration of soluble organic substances rises, so also does the osmotic (turgor) pressure in the cell. The weaker vesicles may collapse. The cell may then become denser than water and start to sink. This mechanism prevents the cells moving into the highest light intensities at the water surface, which, in summer, may be lethal for these organisms [810].

As the cells sink into the darker, deeper waters, the rate of gas-vesicle formation is greater than that of cell division. The concentration of gas vesicles in the cells increases and the cells become positively buoyant again and start to float upwards. The cycle apparently keeps them in water layers most suitable for their survival and gains them access to nutrients at the top of the hypolimnion.

Ganf and Oliver [310] showed that the development of vesiculate blue-green agal in Mt Bold reservoir, South Australia, at the time of thermal stratification, was linked to their access to nutrients in the hypolimnion. At

Fig. 7.11 Blue-green algal blooms – surface aggregations of buoyant algae – are now very common. This was photographed on a canal in Delft, the Netherlands.

the time, nutrient supplies in the epilimnion were insufficient to support growth of any of several algae tested. They placed water samples from the lake in dialysis bags (which allow free passage of small nutrient molecules) and showed that algae in the bags would not grow if the bags were suspended near the surface (0.2 m) or at 10 m, below the euphotic zone. However, if the bags were circulated daily between 0.2 m and 10 m, abundant growth was obtained, as the algae met their light requirements at one depth and their nutrient needs at the other. This demonstrates the feasibility of blue-green algal movements in gleaning nutrients, but does not absolutely prove it. Such an experiment has not yet been carried out, and is very difficult to conceive without labelling a hypolimnion with large quantities of isotopes. Some observations suggest that certain blue-green algae may obtain their nutrients from the sediments [472].

In extremely fertile water, the phytoplankton crop may, in summer, reach such a concentration that the euphotic zone is much shallower than the epilimnion. The blue-green algal cells may then spend much time at very low light intensities and the differential rates of gas-vesicle synthesis and cell production may lead to formation of cells so buoyant that they rise to the surface very rapidly on calm days. The mechanism by which increased turgor pressure bursts the weaker vesicles is unable to operate, for photosynthesis is inhibited at the high extreme surface-light intensity and the cells are trapped at the surface, forming the characteristic scum. This may cause odour problems as it decays or if it is later windrowed at the lake edge. On some lakes, the blooms are indeed dramatic, forming thick porridges and clogging the cooling-water intakes of boat motors [823].

At least, that is one explanation of how algal blooms form, and, in support of it, it is claimed that cells from the bloom are not viable if attempts are made to grow them in culture [810]. Others claim that this is not so and that surface-bloom formation is a device by which the algae cope with high pH and a shortage of CO_2 in very fertile waters packed with algae. Increasing the pH of suspensions of the algae caused movement to the surface and closest access to the supply of atmospheric CO_2, for example [77, 528, 750]. It may be that there are different sorts of blooms with different explanations.

7.2.8 Phytoplankton communities, toxic algae and drinking-water

In drinking-water reservoirs fed by fertile water, large growths of algae cause problems, because filtration to remove the algae, particularly blue-green algae, and produce clear, sparkling water is expensive. Increasing filtration problems, plus a realization that many blue-green algal populations are toxic, has led to considerable interest in their ecology.

Shapiro [885] synthesized much information pointing to carbon availability as the master variable controlling blue-green algal abundance. This followed his experiments in transparent polythene bags, filled with lake water and resuspended in the lake, in which adjustment of CO_2 concentrations downwards by increasing the pH led to predominance of blue-green algae, while the reverse led to green algal dominance. Shapiro linked other features of blue-green algal abundance (association with stratified conditions and deoxygenated hypolimnia; with low N-to-P ratios; with long residence time; with grazer removal of competing algae; and with relatively high temperature) to CO_2 availability, but experimental attempts to stimulate blue-green algal populations by adjusting the pH do not always work [50].

Perhaps because many species are concerned, no simple story has emerged. In general, blue-green algae are associated with high nutrient loading, although at the very highest loadings they may give way to green algae [481]. This may be because, in heavily eutrophicated lakes, there is frequently a large amount of organic matter produced and consequently a high availability of carbon dioxide following its decomposition.

The problems at the water works are best resolved by reducing nutrient loading (see Chapter 6). If this is not possible, the reservoir may be managed to favour species of algae that are more desirable (if still undesirable) than others. Blue-green algae not only may be toxic but may produce noxious ('pigsty') tastes if they decompose on the filters used for clearing the water. Diatoms may also produce 'fishy' odours, but are more readily filtered out. Algicidal chemicals (e.g. copper sulphate can be used but are costly and undesirable.

Reduction of blue-green algal growths is often possible through judicious mixing of the water. This may be done by bringing the inflow river water into a storage reservoir through jets placed near the bottom. The jets are angled so that the upward movement of the water causes the maximum turbulence [538, 919]. Bubbling with air may also be successful. Intermittent mixing and periods of stabilization may prevent build-up of very large crops of any alga by changing the environment frequently enough to prevent any particular group reaching its maximum potential [811].

Blue-green algal blooms are now frequent and even a small bloom can be concentrated by wind at the edge of a lake or reservoir to form a hazard, particularly to thirsty pets or farm stock. Humans are not at great risk, for such water does not look appetizing. This is fortunate, for the toxins produced by about 50–75% of blooms [554] are very potent [153]. They include nerve and liver toxins, capable of killing a large mammal in hours. The neurotoxins are often anatoxins (secondary amine alkaloids) or saxitoxins (cyclic polypeptides), also produced by marine dinoflagellates associated with red tides. The hepatotoxins come in over 45 variants, each a related cyclic polypeptide. Why they are produced and what their natural role might be are unanswered questions. There is increasing evidence that they may be passed into food webs and hence may have widespread effects.

7.3 Microconsumers of the phytoplankton – bacteria

Heterotrophy is the sole source of energy for most bacteria in lakes, and some evidence suggests that large proportions of the organic matter produced by the algae may be used by the bacterioplankton. The bacterioplankters are about the size of lentils in the model outlined earlier, though with shapes varying from rods and cocci to filaments and branched (prosthecate) forms. They are suspended freely in the water as cells or small colonies and commonly are studded on to a nucleus of detritus.

It is difficult to know their population densities, because methods of study are unsophisticated. The best methods of estimating their numbers or biomass use counting after fixation with glutaraldehyde in the field and then filtration on to very fine black filters, followed by staining with fluorescent substances specific to each main group and examination with a fluorescence microscope. Determination of adenosine triphosphate (ATP) [794], after separation of the bacterial size fraction by screening and filtering, is also useful. It is now possible to make antibodies to specific organisms by injecting their proteins into laboratory animals. The antibodies can be purified, joined with fluorescent substances and added to a water sample. They then attach to the appropriate organism if it is present and can be used to identify and count particular species. The technique (immunofluorescence) is expensive but could be used in studies of particular target organisms, either bacterial or algal, that are too small to be easily recognized visually [989].

Numbers or biomass are not good measures of activity, because the turnover of the populations may be rapid. Moreover, there is probably much specialization on particular organic substrates – cell or body detritus or secreted matter – among the strains present. Glycollate and peptides, for example, are secreted into the water by algae [279]. Secretion may amount to over half of total carbon uptake and may have some advantage to the organisms – for example, in chelating scarcely soluble nutrients, such as trace elements. However, it seems to happen to a greater extent in infertile than in fertile lakes and hence seems likely to be a function of the mismatching of energy and nutrient supplies.

The activity of the bacteria feeding on such material can be measured by separation of the bacterial cells from the water, using fine filters, and then measurement of the incorporation of tritiated (^{3}H) thymidine into the cells. Thymidine is a component of deoxyribonucleic acid (DNA) and the method must be calibrated against some measure of cell production. Using this method in Lake Michigan, Scavia and others [858] showed that secondary bacterial uptake of carbon accounted for more than the current fixation of carbon by photosynthesis. The bacteria were presumably drawing on previously produced reserves in the water.

Alternatively, the activity can be illustrated by the turnover rate of specific substances in the water [8]. The concentrations of such substances (e.g. glucose, acetate, glycollic acid) are often too low for convenient and precise chemical analysis. The life of a given molecule may be only seconds and the entire pool of some dissolved substances may be turned over every few minutes in summer.

7.4 Protozoa and fungi

Over thirty years ago, working with inshore sea water, Johannes [487] suggested that protozoans might be important. This was by grazing bacterial cells and detritus and regenerating their contained nutrients at much higher rates than bacteria or the larger zooplankters alone could do. More recently, also for the sea, Azam *et al.* [25] have suggested that bacteria and the smallest phytoplankton cells are fed upon largely by ciliate and flagellate protozoa. In doing so, they form larger aggregates, which may be more easily ingested by the zooplankton grazers than freely suspended cells only 1 µm or so in size.

This microbial complex is now thought to have a major role in regenerating nutrients for the marine phytoplankton. Freshwater protozoans may be similarly important, especially in infertile water, favouring growth of small algae and bacteria. Ciliates are certainly common in fresh waters [269, 555], as are flagellates and amoebae, and *inter alia* have been shown to eat small diatoms [119], blue-green algae [1035] and bacteria [783]. Even the distinction between autotrophs and phagotrophs is not clear in many algal groups. Some chrysophytes, for example, can themselves feed on bacteria, as well as photosynthesize [67, 68, 249].

That there is a considerable bacterioplankton, fed upon by ciliates and heterotrophic nanoflagellates, in fresh waters is now well established. Bacterioplankters number up to 10^{10} l^{-1} and heterotrophic nanoplankters (HNAN) up to 10^6 l^{-1}. Ciliates usually number a few hundreds per litre [556]. In general, all of these groups are most abundant in spring and summer, and there is often a link, sometimes weak, with photosynthetic algal production or biomass [66, 155]. Correlations of ciliates with algal chlorophyll *a* are often close [49], as are those of HNAN and bacteria. The overall implication is that it is algal excretion of organic substances that fuels the bacteria–protozoan pathway.

Not all results support this, however. In Loch Ness, a very infertile lake, populations of algae seem insufficient to feed even the small bacterioplankton population [509, 510]. Most of the organic matter comes from the catchment and numbers of bacterioplankton and Protozoa are linked with wet periods when more water enters from the streams. Elsewhere, in more fertile (though still infertile) lakes, correlations between algal and bacterial biomass or production may still be poor [749, 970] and the supporting organic matter may come from the lake catchment, surrounding wetlands or aquatic plants in the littoral zone. Coveney and Wetzel [166] found that bacterial and algal production were correlated in a small Midwestern lake in the USA, but that there was also a mutual correlation with temperature. When this was allowed for, the correlation disappeared, and it appeared that the bacterial production was greater than that of the photosynthetic algae by a factor of between 1.33 and 3.35, which suggests an external source of carbon to support the bacteria.

The open-water plankton community may thus be supported allochthonously in these rather infertile lakes, especially where the water is brown-coloured with humic compounds [408]. It seems likely that there will be a spectrum in which the detritus pathway, based on external sources of carbon, is of major to minor importance, based on the inherent fertility of the lake.

Phytoplankters are also attacked by internal parasites, including viruses, bacteria, chytrid fungi and Protozoa. Many of these are specific to single species of algae or groups of them, and there is usually a degree of infestation in any population. One recent study in the sea found that viruses caused similar mortality rates to bacteria as Protozoa [305]. Infected cells become part of the detrital aggregates of organic matter, bacteria and Protozoa, which, with the phytoplankton, form the food of the herbivorous zooplankters.

7.5 Zooplankton

Rotifers and crustaceans are the major groups of freshwater zooplankton (Fig. 7.12) other than Protozoa. Freshwater jellyfish, carnivorous on other zooplankters, some flatworms, gastrotrichs and mites may be common in particular lakes, especially in the tropics [265], but are scarce in temperate regions.

Rotifers are man-sized to horse-sized on our scaled-up model and are mostly suspension feeders. Their name comes from the rhythmically beating, apparently rotating 'wheel' of cilia close to the mouth. The cilia direct water, with its suspended fine particles, into the gut. Although some rotifers are predatory, grasping feeders, most take particles from about 1 to 20 µm, a range shared with the filter-feeding Cladocera (Crustacea), which can also take food a little larger, perhaps 50 µm.

The crustacean zooplankters include the Cladocera, most of which have a carapace covering the body. The group includes herbivores (water fleas, *Daphnia* and *Bosmina*) and carnivores on smaller zooplankton (*Leptodora* and *Polyphemus*) which are raptorial, (actively grasping their prey). The small-particle feeders have thoracic limbs bearing hairs (setae), on which are closely spaced (a few micrometres) setules, which retain small particles as the limbs beat and eventually convey food to the mouth.

There is controversy about whether the action is a simple filtering one or

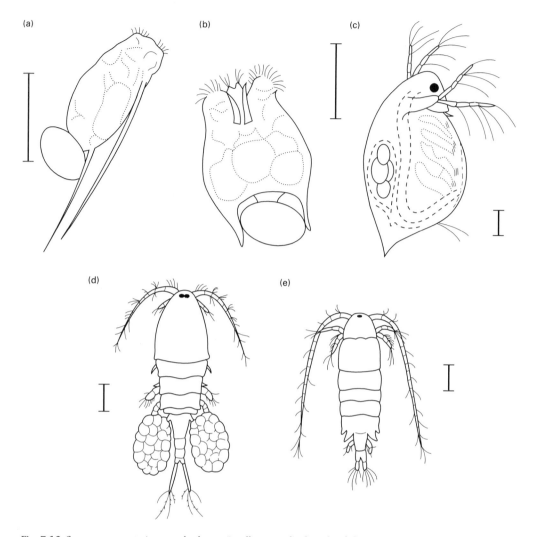

Fig. 7.12 Some representative zooplankters. In all cases, the length of the scale lines represents 100 µm. Rotifera: (a) *Filina*, (b) *Brachionus*. The corona of cilia can be seen in each case and, at the rear, a single egg. Crustacea (Cladocera): (c) *Daphnia*. The filtering limbs are enclosed by the carapace, which also contains the egg pouch, containing a few eggs. Crustacea (Copepoda): (d) *Cyclops*. The egg sacs are paired and the antennae, the lower pair of appendages on the head, are not branched; (e) *Diaptomus*, a calanoid copepod, in which the antennae are branched. When the animal is carrying eggs, these are contained in a single egg sac, in contrast to the paired sacs of the cyclopoid copepods.

one in which the creation of small currents pushes particles of food towards the mouth. A correlation between the sizes of particles taken and the spacing of the setules supports the former case [319, 349, 407], but theoretical considerations suggest that the water is too viscous to allow simple filtration [311]. Direct impaction of particles on the limbs may be important and the organisms can also reject unsuitable food – that which is toxic or too large to ingest. A claw on the lower abdomen can be used to prise out unsuitable food from the feeding groove between the limbs, along which food must pass to reach the mouth. Cladocera move through the water more actively than rotifers, rowing with their large, branched second antennae. On the scaled model, cladocerans, at most a few millimetres in length, would be as tall as church steeples in some cases.

The third important group of zooplankters, also crustacean, is the Copepoda, whose adult members are usually a little larger than Cladocera. They may be small-particle feeding (mostly the calanoid copepods, such as *Diaptomus*) or raptorial (the cyclopoid copepods, which include *Cyclops*). The prey of these may be smaller zooplankters, larger colonies or masses of phytoplankton. The copepods can tackle a wider range of bigger food particles (5–100 µm) than the non-raptorial Cladocera and rotifers. The calanoid copepods do not filter particles but actively select from those brought near the mouth by the movements of the limbs [751, 764].

The size of particles taken by zooplankters depends on the size of the organisms. This can be shown by feeding plastic beads of known size to them. But there is clearly some separation among the groups, in terms of the range eaten and the grazing rate. This is measured as the volume of water swept for particles per animal per day. It depends on many factors but is generally low for rotifers, 10 to 100 times higher in copepods and even higher for cladocerans.

Life histories vary among the zooplankton. Although the cytological mechanisms differ, both rotifers and Cladocera are parthenogenetic. Females asexually produce broods of eggs, which hatch into more females. This allows rapid replacement of populations that are especially vulnerable to predation (see later). The eggs are born in sacs by rotifers and in pouches within the carapace in Cladocera, and young are released that resemble the adults and soon grow large enough to reproduce. The rate of egg production is high, for a new generation of rotifers is produced in only a few days and each female produces up to twenty-five young in her lifetime of 1–3 weeks. Cladocerans take longer per generation (1–4 weeks), but each has a longer life expectancy (up to 12 weeks or so) and may produce up to 700 young per lifetime.

The genetic advantages of sexual reproduction are not lost in these animals, because most produce males during times of food shortage or other inclement conditions. Eggs are then fertilized, become thick-walled and do not hatch immediately. In some Cladocera, they are held in thickened egg pouches, called ephippia. Eventually, the eggs hatch to form a new season's parthenogenetic females.

The copepods are different. Each generation is sexual and, before the new mature adults are formed, there are eleven successive moults. The first six, after the egg hatches, are of juveniles, called nauplii, which look quite

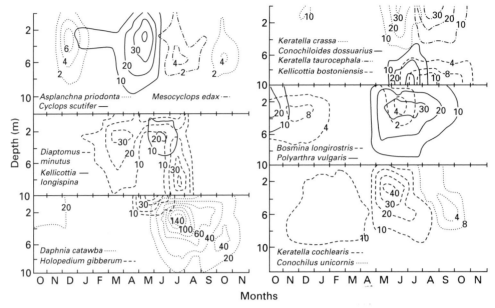

Fig. 7.13 Distribution of zooplankton species with depth and season in Mirror Lake, New Hampshire. Values are mean monthly production (μg l^{-1} month^{-1}). *Cyclops*, *Mesocyclops* and *Diaptomus* are copepods; *Daphnia*, *Holopedium* and *Bosmina* are cladocerans. Other species are rotifers. Further niche differentiation is given through food selectivity and vulnerability to predation. (Based on Makarewicz and Likens [627].)

different from the adults. The next five, the copepodites, do look like the twelfth stage, which is the reproductive adult. In terms of their longevity and fecundity, the copepods are similar to the cladocerans.

The zooplankton community thus includes a variety of forms and activities, many of which are ultimately related to the effects that predation, among themselves or by vertebrates, has upon their numbers (see later). They are also much more heterogeneously distributed (Fig. 7.13) than the phytoplankters. Zooplankters move actively, may shoal, both vertically and horizontally, and often go through diurnal vertical migrations, reaching the water surface by night and moving down by day.

Nets are used for sampling, because the concentration of zooplankters (at most 1–2 ml^{-1} for the larger ones) is small. Some of the larger, faster-moving copepods and mysids may be able to detect the shock wave that precedes a net as it is drawn through the water and avoid being caught. The mysids are a group of shrimp-like animals, including the genera *Mysis* and *Neomysis*, which are omnivores or predators on other zooplankters, relatively large (up to 2 cm), and which have sexual life histories.

7.5.1 Grazing

The herbivorous zooplankters feed on phytoplankton, bacteria and detritus

[764]. Comparison of the mean herbivorous zooplankton biomass or production with those of the phytoplankton often shows a general correlation. For example, McCauley and Kalff [602] determined that, in over 20 Canadian lakes:

$$\text{Log}_{10}\ \text{zooplankton biomass} = 0.5\ \text{log}_{10}\ \text{phytoplankton biomass} + 1.8$$

Not all the phytoplankton is readily available to the grazers and the proportion of those that are inedible, usually the larger forms and often the blue-green algae, tends to increase with increasing lake fertility. A better correlation is thus usually found between zooplankton biomass and biomass of the smaller (usually less than 30 μm) phytoplankters.

The implication of these correlations is that the potential maximum zooplankton crops are determined by phytoplankton production, much in the same way as the latter is set by the key nutrients. A comparison [904] of *Daphnia hyalina* in an infertile lake (Buttermere) and a fertile lake (Esthwaite) in the English Lake District, for example, showed that the maximum populations were six times as great in the latter, with higher birth rates and rates of increase. The animal was forced in winter to form resting eggs in Buttermere, but survived as adults in Esthwaite.

In Lake Constance also [543], *Daphnia* may at times be food-limited. Its rate of egg production depends on food concentration (measured as carbon in particles $< 50\ \mu m$ in size), with no eggs produced when food is $< 0.2\ mg$ $C\ l^{-1}$ and a rise in production as food increases to about $0.7\ mg\ C\ l^{-1}$, when egg production increases no further. Lake Constance has less than $0.7\ mg$ $C\ l^{-1}$ for much of the year. Other workers [661], noting an apparently full gut at all times in the animals, feel that temperature and other factors, rather than food availability, are more important.

There is a danger in generalizing from lake to lake and from species to species, for, particularly in the latter case, marked selectivity of food occurs. Some zooplankters are very fussy eaters. Tests with bacteria, a yeast and three small algae showed that, among three rotifer species, *Keratella cochlearis* would take all, but preferred *Chlamydomonas* (an alga), while two *Polyarthra* species would take only *Chlamydomonas* and *Euglena* (both algae) [76]. A cladoceran, *Bosmina longirostris*, also preferred *Chlamydomonas*, and other workers showed that it exerted up to a 13.7 times preference for *Chlamydomonas* over *Aerobacter* (a bacterium), while another cladoceran, *Daphnia rosea*, showed no preference at all [207].

Some animals prefer live food (e.g. *B. longirostris* and *Diaptomus spatulocrenatus*), others prefer dead detritus (*K. cochlearis*) and others (*Conochilus dossuarius*) have no preference [917]. Bacteria may be more readily taken, particularly by the larger animals, if they are attached to detritus. Among the algae, the assimilability of different species and hence their ability to support growth and egg production vary greatly and are not just a function of size. For *Daphnia* species in Lake Washington, Infante and Litt [467] found the greatest egg and biomass production with the flagellate *Cryptomonas erosa* and a small diatom, *Stephanodiscus hantzschii*, a middle range with larger diatoms, *A. formosa* and *Melosira italica*, and the lowest with the small-celled *Chlorella* and a thinner variant, *tenuissima*, of *M. italica*.

Many blue-green algae are poor food. Arnold [22] showed that maintenance of a *Daphnia* species is not possible on blue-green algae, but Schindler [869] found blue-green algae not the most readily assimilated, but not the least either. Part of the problem may be due to the size and difficult handleability of the larger forms [782], but sometimes they may be toxic, reducing feeding rates even if given as separated cells rather than as colonies [544].

Filaments of blue-green algae may be too large or awkward (too stiff) for ingestion. Some cladocerans may not have wide enough gaps between the edges of the carapace that cloaks their limbs to bring the filaments or colonies on to the limbs. If the filaments reach the feeding groove, rejecting them with the abdominal claw may pose heavy energy demands [781]. More palatable foods may be rejected at the same time, so that growth of the animal is further decreased [466]. This may give an advantage to small animals (which cannot take in the large blue-green algae at all, so do not have to reject them from the feeding apparatus) [1002]. It may be why small species are often predominant in midsummer in fertile lakes [333, 334], but there are other reasons, connected with fish predation (see later).

Aphanizomenon is a blue-green algal genus with stiff filaments, in which the individuals sometimes bundle together in parallel to form 'flakes' visible to the naked eye. Often there is an association between such flakes and an abundance of *Daphnia* [308, 424, 592]. *Daphnia* can feed on single-filament *Aphanizomenon* and even sometimes on small colonies. Larger colonies are not ingested and interfere slightly with feeding on small green algae, such as *Ankistrodesmus*, which *Daphnia* assimilates more efficiently (35%) than *Aphanizomenon* (11%). At high population densities (say, 10–15 l^{-1}), *Daphnia* may cause selection in favour of flake *Aphanizomenon*, by removing potential algal competitors with it for nutrients.

In Lake George, Uganda, the major zooplankter, *Thermocyclops hyalinus*, assimilates between 35 and 58% of ingested *Microcystis*, the major phytoplankter present, which is a blue-green algal species [665]. On the other hand, in experiments in enclosures in two contrasted lakes along the same river, *D. hyalina* alone survived, but not for long, in one, while it and two larger species persisted in large numbers in the other. The latter lake was dominated by diatoms, with only a few blue-green algae, while the former, with a longer retention time, had few diatoms but many blue-green algal filaments [688]. Reports of inabilities of temperate zooplankters to assimilate blue-green algae may reflect the present state of flux in many temperate lakes, due to eutrophication. This has stimulated growth of previously less common blue-green algae, for which the zooplankters may not yet have evolved suitable means of coping. Lake George is not a recently polluted lake, but has had a naturally large crop of blue-green algae, presumably for a very long time.

Studies in enclosures in lakes give clues as to how selective grazing may help to determine the composition of phytoplankton communities. Porter [779] enclosed 500-l samples of lake water in large polyethylene bags, which she sealed and resuspended for several days in the lake. From some bags, she had removed the larger zooplankters (crustaceans) by filtering the water through a 125 μm mesh net. In others, she increased the zooplankton population several-fold by adding animals.

The major effect of grazers, which included small-particle-feeding (*Daphnia galeata mendotae* and *Diaptomus minutus*) and raptorial (*Cyclops scutifer*) herbivores, was to suppress small flagellates, nanoplankters and large diatoms. Large colonial green algae were not affected or they even increased. They were ingested but not digested and may even have benefited from passage through guts. Phosphate released there by digestion of other species was taken up by the colonial green alga *Sphaerocystis schroeteri* as it passed through undigested and emerged growing healthily on copepod faecal pellets [780].

In a very fertile Scandinavian lake, Schoenberg and Carlson [875] asphyxiated existing fish and zooplankters in plastic enclosures by adding solid CO_2. Then they added *D. g. mendotae* to some enclosures and *B. longirostris* to others. The latter is small and unable to handle large filaments or colonies. In its enclosures, the agal crop increased, higher phosphorus concentrations were maintained in the water, the pH was forced upwards by CO_2 uptake and the proportion of blue-green algae in the phytoplankton community was increased.

The *Daphnia*, however, kept the water clear and the phosphorus concentrations and pH low and reduced the blue-green algal component of the algal biomass. Phosphorus was kept low, perhaps by sinking of phosphorus-rich faeces to the sediment. *Daphnia* may have directly grazed the blue-green algae (*Microcystis*) or, by grazing the other species, may have created conditions (lower pH, higher transparency, lower phosphorus availability) less favourable to blue-green algal growth. The potential impacts of grazing on the phytoplankton are thus varied and depend very much on the species present, of both animals and algae.

7.5.2 Feeding and grazing rates of zooplankton

Ingestion, or feeding rate, depends on the volume of water handled per unit time (the grazing rate) and on the food concentration. Burns and Rigler [114] fed a radioactively labelled (^{32}P) yeast to *D. rosea*. After a few minutes, the *Daphnia* were anaesthetized in saturated CO_2 solution, which prevents defaecation. The radioactivity incorporated into the animals was measured and grazing rate (G) calculated as:

$$G \ (\text{ml animal}^{-1} \ \text{h}^{-1}) = \frac{\text{radioactivity per animal}}{\text{radioactivity per ml of yeast suspension}}$$
$$\times \frac{60}{\text{time (min) of feeding}}$$

Ingestion rate could then be calculated as grazing rate times the concentration of yeast cells, as cells animal^{-1} h^{-1}. Grazing rate decreased almost exponentially as food concentration increased from about 1.5–2.0 ml h^{-1} at 25 000 cells ml^{-1} to about 0.2 ml h^{-1} at 500 000 cells ml^{-1}. Grazing rates increased with body length and temperature (up to 20°C), while ingestion rates increased with food concentration, up to about 100 000 cells ml^{-1}, where a plateau was reached. In general, grazing rate seems to decline above a threshold food concentration, so that there is an upper level of feeding rate, dependent on species and conditions.

These studies were carried out under ideal laboratory conditions, with an acceptable food source. In natural lake water from Heart Lake, Ontario, where *D. rosea* is common, filtering rates were lower, by a factor of 2–3, than those predicted from the laboratory studies. Reasons for this include a lack of particles of filterable size or interference with feeding by large particles. Removal of animals from a lake to measure their filtering rates may itself affect the rates measured.

Measures of the grazing rates of complete communities are scarce. Those for individual rotifers are up to 1 ml day^{-1}, while *Daphnia* species may filter 5–30 ml day^{-1} and *Diaptomus* about 35 ml day^{-1}. This means that the crustaceans, if abundant (say, 10–50 l^{-1}), may filter all of the water every day [545, 546], and this suggests a major potential impact of grazing.

An attempt has been made to link grazing and feeding rates to the factors that influence them, by multiple regression analysis [764]. Size of animal and food concentrations were the more important determinants. Grazing rates increase by some power of the body length. For example, for Cladocera [784]:

$$G = 0.54 \ \text{length}^{1.55}$$

where *G* is in ml animal^{-1} h^{-1} and length is in mm. The power to which length is raised may be greater for some species.

7.5.3 Competition among grazers

The open water of a lake is a relatively unstructured environment. Usually, however, several zooplankton species occur together and competition for what may become a limiting food resource, as grazing proceeds, might be expected [980]. To some extent, this may be offset by differences in food-particle size taken and seasonal and spatial differences in their distributions (Fig. 7.13).

Rotifer populations, however, are often sparse in the presence of large (> 1.2 mm) *Daphnia* species [207, 328], even at low densities. When *Daphnia* is removed by fish predation (see below), rotifers usually increase, and in laboratory cultures rotifers die in the presence of the bigger animals. This may follow from simple contest competition, the diets of both groups overlapping considerably but the larger *Daphnia* being able to filter more efficiently. There may also be interference competition, with small rotifers swept into the feeding chamber created by the limbs of *Daphnia*. Some may be ingested and others damaged before rejection. Eggs may be torn free to face predation as free-living bodies. Smaller cladocerans, such as *Bosmina*, may coexist with rotifers, because they also are not powerful feeders and are too small to sweep rotifers into their feeding baskets. Small cladocerans may also be more vulnerable to invertebrate predation, which prevents rises in populations to competitive levels.

There may also be competition among different species of the same genus. *Daphnia* species secrete substances into the water that inhibit feeding rates of other daphnids, and sometimes also of other individuals of the same species [641]. Such allelopathic effects have as yet been shown only in laboratory cultures.

7.5.4 Predation in the zooplankton

Herbivorous zooplankters inevitably consume protozoons [113, 475]. In truth, they are omnivores, and the strict separation of herbivory and carnivory is not possible at this level. However, these animals become prey to stricter carnivores, both invertebrate and vertebrate. The vertebrates, including amphibians, such as salamanders [425], wading birds in shallow ponds and especially fish, are bigger and tend to be more mobile than the invertebrates. They often have such overriding importance in determining the nature of zooplankton communities that the significance of invertebrate predation may be overlooked.

One way of establishing the extent of predation is through demographic analysis of the populations of prey. An example is the study of *Daphnia galeata mendotae*, carried out by Hall [373]. The method is applicable to any zooplankter, and depends ultimately on finding the birth and death rates of the animal as the year progresses.

The rate of change of numbers over a short time, t, is given by:

$$N_t = N_o e^{rt}$$

where N_o and N_t are the numbers at the beginning and end of the period, e is the base of natural logarithms and r is the intrinsic rate of natural increase. The term r equals $(b - d)$, the difference between instantaneous birth and death rates. Death may come from predation, parasitism and washout.

If, by regular sampling, the numbers of animals are established, estimates of r can be obtained for the period $t \to (t+1)$, between samplings, from integration of the above equation:

$$r = 1/t \times (\log_e N_{t+1} - \log_e N_t)$$

To understand the dynamics of the population, b and d must be estimated. The instantaneous birth rate, b, is defined by:

$$N'_{t+1} = N'_t e^{bt}$$

The term N' represents a potential population size in the absence of deaths and cannot be estimated. Hence b cannot be directly measured. However, a finite approximation to birth rate, B, can be defined as:

$$B = \frac{\text{Number of newborn (during interval } t \to (t+1))}{\text{Population size at } t}$$

$$B = (N'_{t+1} - N'_t)/N'_t$$

and because

$$b = 1/t \times (\log_e(N'_{t+1}/N'_t))$$

and for $t = 1$,

$$b = \log_e((N'_{t+1}/N'_t) - (N'_t/N'_t) + 1)$$

so

$$b = \log_e[1 + ((N'_{t+1} - N'_t)/N'_t)] = \log_e(1 + B)$$

The value of B can be independently estimated from the number of newborn per individual per day. Those about to be born are carried as eggs or embryos by the female, until they are released when the female moults.

$$B = \frac{\text{Number of reproductivity mature adults } (N_A)}{\text{Total population } (N_t)}$$

$$\times \frac{\text{Number of eggs and embryos carried per adult (brood size, } E)}{\text{Number of days for an egg to mature from production to release } (D)}$$

The values of N_A, E and N_t are readily estimated from sampling of the natural population. The value of D depends on temperature and is measured in laboratory cultures. In *D. g. mendotae*, it ranged from 2 days at 25°C to 20.2 days at 4°C. From the lake temperature at the time of sampling, an appropriate D value is selected.

The values of b and r can be used to calculate d, the instantaneous death rate:

$$d = b - r$$

Figure 7.14 gives these data for *D. g. mendotae* in Base Line Lake, Michigan, together with changes in population. From March to early June, the population increased and death rates were low. In midsummer, the popula-

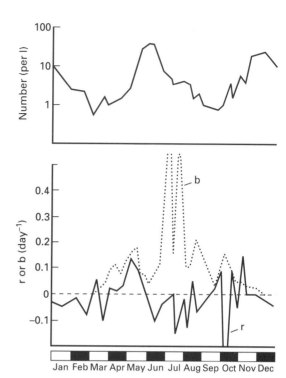

Fig. 7.14 Population changes of *Daphnia galeata mendotae* in Base Line Lake, Michigan. The upper graph shows changes in the total population, and the lower one the observed intrinsic rate of increase (*r*) and the calculated birth rate (*b*). The difference between the graphs gives *d*, the instantaneous death rate. (After Hall [373].)

tion remained steady but low, and yet birth rates (and therefore production) were high. Death rates were then also high. There was no evidence of large losses through parasitism or washout and the main mortality followed fish predation after young zooplanktivorous fish hatched in May and June. Although the population could double itself every four days, predation accounted for almost all of the production and kept the population low.

This approach assumes that all the mortality is of adults, but this is probably not so. Eggs carried by the adults when they were eaten are also eaten, so that the method overestimates birth rates and therefore death rates. Eggs carried may not always develop and newly hatched *Daphnia* might also be eaten by invertebrate predators. Threlkeld [959] has demonstrated potentially important effects of egg mortality, and sophisticated treatments of the method, which derives from work originally by Edmondson [229], are available [137, 231, 755, 789, 950].

Predation on zooplankters by zooplankters

Invertebrate predators of and among the zooplankton (Fig. 7.15) are rotifers (*Asplanchna*), Cladocera (*Bythotrephes*, *Leptodora* and *Polyphemus*), cyclopoid copepods (*Cyclops*, *Mesocyclops*), mysids (*Mysis*, *Neomysis*), larvae of the fly genus *Chaoborus*, coelenterates, including free-living *Hydra*, water-mites

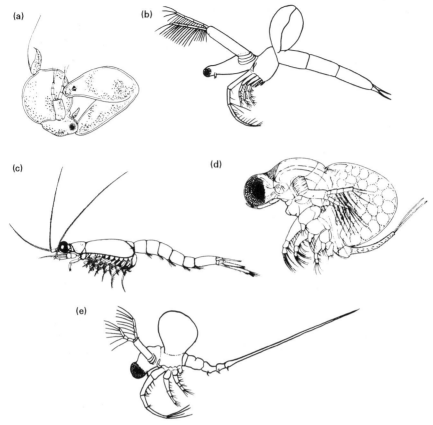

Fig. 7.15 Invertebrate predators on zooplankters include flatworms, such as *Mesostoma*, here (a) enveloping its prey, *Moina macrura*, in mucilage, and raptorial crustaceans, such as *Leptodora kindtii* (b), *Mysis relicta* (c), *Polyphemus pediculus* (d) and *Bythotrephes cederstroemi* (e) (based on Fryer [302] and Blaustein and Dumont [73]). *Mesostoma* is about 1.5 mm long, the others are between 3 and 7 mm.

(Hydracarina) and free-living flatworms, the rhabdocoeles, such as *Mesostoma*. The cladocerans and mysids appear to be confined to temperate lakes, the coelenterates and rhabdocoeles are more common in the tropics and cyclopoid copepods and chaoborids are worldwide.

Predator/prey relationships can be examined as a sequence of detection (encounter), attack, capture and ingestion or escape. Detection for invertebrate predators is usually by sensing small tremors in the water. Watermites, for example, will assume an attacking position in response to vibrating glass fibres but not to still ones [790]. They also rely on chemical signals, remaining motionless in water conditioned by the previous presence of prey but moving around in unconditioned water. The crustaceans and *Chaoborus* probably rely on random encounters, dependent on frequent movement. *Leptodora* has mechanoreceptors on its first thoracic limbs, which detect the prey, and the large eye of *Polyphemus* suggests that it may use visual cues. *Mesostoma lingua*, a tropical flatworm, uses several methods of encounter. It may trap prey in mucus, sit and wait for it to arrive, secrete neurotoxins into the water or actively search, especially in the surface-tension film.

Attack generally involves grasping and biting the prey by the predator and its effectiveness means the difference between ingestion or escape. The pattern of bite damage to the prey (all of which is usually not eaten) is often consistent [215] (Fig. 7.16), suggesting a particular way of attack. The production of spined or distorted (exuberant) forms may interfere with the handling mechanism of the predators [391, 392, 523, 728, 774] and allow prey to escape.

The exuberant forms may be genetically different or developmentally induced by substances (kairomones) emitted by the predator. The rotifers *Keratella tropica* and *Brachionus* spp. produce longer spines when their populations are declining and their predator, another rotifer, *Asplanchna*, is present [327, 357]. *Chaoborus* induces the production of a pedestal with extra spines (neck teeth) in *Daphnia pulex*, which increase the likelihood of escape but have a cost, in that less energy is available for reproduction. Small *Daphnia* species produced elongated heads (helmets) in response to predators in Lake Majick (Poland). Large species, too big to be handled by invertebrates, did not, *Chaoborus* also induced such helmet formation in *Daphnia* [377]. The substances concerned are as yet unidentified, although Parejko and Dodson [756] found that the neck-tooth-inducing substances from *Chaoborus americanus* were polar organic molecules, stable to heat and peptidase, destroyed by acid and base digestion and containing hydroxyl groups.

The impact of invertebrate predation on zooplankton communities is often difficult to establish, especially where vertebrate predation is also

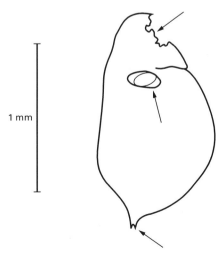

1 mm

Fig. 7.16 Typical damage to *Daphnia galeata mendotae* attacked by the copepod, *Diaptomus shoshone*. The antennae are usually broken and the tail spine and head bitten. (From Dodson [215].)

occurring. There are several approaches. The first is to examine gut contents and to look for inverse correlations in numbers of predator and prey in lakes. The second involves laboratory experiments in which single prey or a choice is fed to the predator, and the third is to carry out manipulative experiments in containers from which vertebrate predators have been removed. Advantage can also be taken of favourable circumstances, where fish predation is absent or invertebrate predators have newly invaded.

Gut analyses demonstrate the existence of a relationship but not its extent. Inverse correlations give that additional insight. For example, in Lake Michigan, a new predatory zooplankter, *Bythotrephes cederstroemi*, has recently achieved prominence [565]. It is a palaearctic species, whose introduction to this North American lake in the mid-1980s is obscure. Several *Daphnia* species were formerly common and the smaller herbivorous *Bosmina. longirostris* was scarce. The daphnids and *Bosmina* were preyed upon by *Leptodora kindtii*. *Bythotrephes* increase in the lake from 1987 onwards was associated (Fig. 7.17) with a decline in *Leptodora* and two of the *Daphnia* species and an increase in *D. galeata* and *Bosmina*. This suggests a release from *Leptodora* predation of *Bosmina* and possibly the daphnids and a possible predation of *Bythotrephes* on *Leptodora*. Vertical movements (see later) of the daphnids following *Bythotrephes* introduction changed so as to bring them into lesser contact with the *Bythotrephes*.

Inverse correlations between the flatworm *Mesostoma* and *Moina*, a cladoceran, ostracods and chydorid Cladocera have been shown in experimental enclosures [73]. However, enclosure experiments designed to test the predatory effects of *Neomysis integer* in a shallow lake in the UK showed negligible effects of this mysid on the zooplankton community [31], despite demonstration in the laboratory of a high ability to feed on both Cladocera and copepods [471]. This particular mysid is an omnivore and appears to have acquired its food by scraping periphyton from the walls of the enclosures! Other mysids, e.g. *Mysis relicta*, often introduced into North American lakes to boost salmonid production, have sometimes had the reverse effect by consuming most of the large daphnids [548].

A final example of the potential importance of invertebrate predation comes from the fishless Great Salt Lake in the USA (Fig. 7.18). A single zooplankton grazer, the brine shrimp, *Artemia franciscana*, feeds on a green alga, *Dunaliella viridis*, for part of the year and on a variety of other algae that develop associated with cast-off *Artemia* exoskeletons in the remainder [338]. The Great Salt Lake varies in salinity as rainfall changes and was only half as salty in 1985–86 as it had been before or was afterwards. The decreased salinity allowed a predatory bug, *Tricorixa verticalis*, to invade [1036]. The bug sucks fluids from its prey and attacked the *Artemia*, whose abundance was reduced, and other grazers (*Diaptomus*, *Cletocampus*, rotifers) appeared. The grazing efficiency nonetheless declined and algal concen-

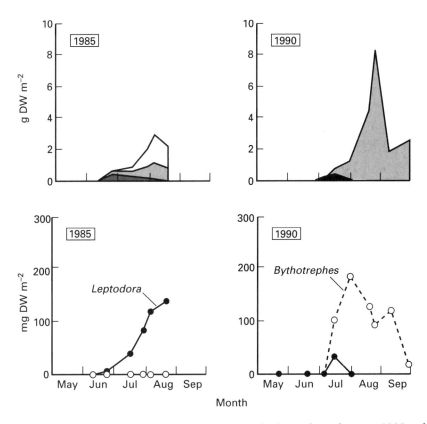

Fig. 7.17 Effects of *Bythotrephes* on the zooplankton of Lake Michigan between 1985 and 1990. Upper panels show biomasses of *Daphnia pulicaria* (white), *Daphnia galeata* (light grey), *Daphnia retrocurva* (dark grey) and *Bosmina longirostris* (black). Lower panels show biomasses of the predators, *Leptodora kindtii* (solid line) and *Bythotrephes* (hatched line). DW, dry weight. (Based on Lehman and Caceres [565].)

trations markedly increased. With the rise in agal population, there was an uptake of the remaining soluble nutrients in the lake and a decrease in transparency. This was a very clear demonstration of top-down effects operating through the food web, a theme that gains even greater predominance with a consideration of fish predation in the open water.

7.6 Fish in the open-water community

Fishes are very diverse. They include omnivores eating a range of foods, from invertebrates to plants, and specialists on diets such as the eggs of other fishes. Adult size ranges from the 12 mm of a Philippine goby to the (sometimes) 5 m of the European catfish, or wels (*Siluris glanis*). Big fish, however, grow from little fish. Diets and behaviour change as fish grow from tiny larvae to fingerlings to adults.

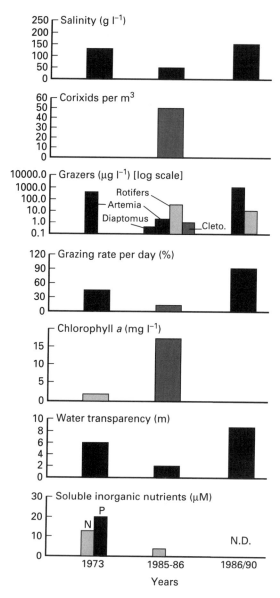

Fig. 7.18 Changes in the Great Salt Lake, USA, resulting from salinity fluctuations and the consequent increase in corixid predators on the zooplankton (based on Wurtsbaugh [1036]).

The greatest number of freshwater fish species lives in the tropics. This may reflect shorter generation times, which may permit faster evolution in warm waters, and the long history as permanent water bodies of many deep, tropical lakes. The fish fauna of islands such as Britain is depauperate. Glaciation eliminated the pre-existing community and there was little time for recolonization from mainland Europe before melt water raised sea levels and isolated the islands.

Temperate continental North America and Asia have fewer species than the tropics but many more than temperate islands. Table 7.2 lists some

Table 7.2 Some major orders of freshwater fish.

Order	Freshwater spp. (% of total)	British spp	Temperate US spp	Tropical African spp
Lepidosireniformes (air-breathing lung fish)	5 (100)	0	0	4
Polypteriformes (bichirs)	11 (100)	0	0	6
Acipenseriformes (sturgeons with primitive, partly cartilaginous skeleton)	15 (60)	1	14	0
Semionotiformes (gars)	7 (100)	0	7	0
Mormyriformes (elephant-snout fish; snout is often elongated and proboscis-like)	101 (100)	0	0	101
Clupeiformes (herrings, shads; often plankton feeders with long gill-rakers)	25 (9)	2	5+	2+
Salmoniformes (salmonids, pike, ciscoes, coregonids, grayling, smelt, believed ancestral to many other orders; often anadromous, breeding in fresh waters but spending most of life in the sea)	80 (16)	14	70	0
Cypriniformes (cyprinids (coarse or pan fish), carps)	3000 (100)	20	209	180
Siluriformes (catfish with sensory barbels on head and no scales)	1950 (98)	1	37	345
Atheriniformes (killifish, guppies)	500 (60)	0	186	0
Perciformes (percids, centrarchids, cichlids; extremely diverse, with spiny fins)	950 (14)	11	55	700+
Anguilliformes (eels, often catadromous, breeding in the sea but living in fresh waters)	15 (3)	1	?10	0
Total Total freshwater fish species is about 6850 or about a third of all fish	6559 (45)	50	583	1338

major freshwater fish orders, with some indication of their distribution. Few species breed, feed, grow and die entirely in the open water; most use the bottom and edge habitats (see Chapter 8) at some time and, except where there has been a long time for evolutionary specialization, many fish species can live in both rivers and lakes.

Newly hatched fish depend on the remains of the yolk-sac for food, but then feed on organisms of a size appropriate to their mouth gape. These may be algae or rotifers. Later, in their first year, the fish will move to larger prey: bigger zooplankters, such as Cladocera, or clumps of filamentous algae. Perhaps in their second year, they will need even larger items to meet their metabolic requirements. These items will be more varied (insect larvae, molluscs, other fish, filamentous algae, higher plants, for example) and reflect specialization in the species. As they grow, many fish move from the shelter they may find in the plant beds of the littoral zone to the edges of these beds on the fringe of the open water. As adults, they may favour particular parts of the lake, may move around a great deal in search of food or may be constrained to certain areas by the behaviour of their own predators, which include other fish, birds and reptiles.

7.6.1 Predation on the zooplankton and fish production

Almost all freshwater fish feed on zooplankters at some stage. This may be for a few weeks immediately after the fry have used up their yolk sacs, for a year or so, with switches to benthic or fish food as they grow, or, for many shoaling open-water species, for their lifetime. Fish production is thus unlikely to be precisely related to zooplankton production, but there should be some general link, as demonstrated by the small-mouth yellowfish, *Barbus aeneus*, in Lake Le Roux, South Africa.

The small-mouth yellowfish in Lake Le Roux

The small-mouth yellowfish is endemic to the Orange River system, which drains an arid land, subject, even in the absence of human activity, to much erosion in the wetter months. The river and the lakes formed by dams along it, including Lake Le Roux, are turbid with suspended clay and silt. Depending on rainfall, Lake Le Roux may have a Secchi-disc transparency varying (between 1977 and 1984) from only 80 cm to less than 30 cm. The silt absorbs light and the phytoplankton crops and their dependent zooplankton stocks are correspondingly low in turbid years (Fig. 7.19).

The yellowfish population (measured as catch per unit effort of fishing) is correspondingly low in these years, as also is its growth rate (Fig. 7.19), which depends on the availability of suitable zooplankton food but also on temperature. The turbid years are also ones in which heat radiation is intercepted higher in the water column, giving a lower water temperature

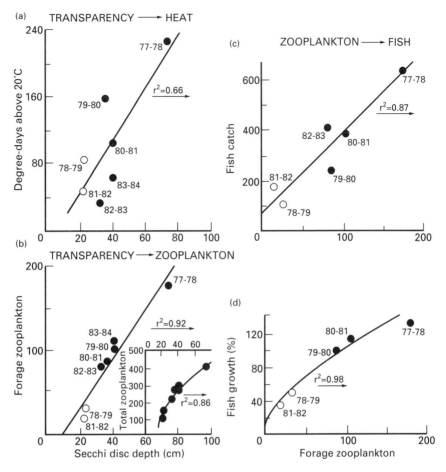

Fig. 7.19 Relationships between: summer heating and transparency (as Secci depth in cm, the greater the penetration of light and long-wave radiation, the warmer the water); zooplankton available to fish (forage zooplankton in relative weight units) and transparency (the greater the transparency, the greater the phytoplankton and consequently the zooplankton production); fish catch t (tonnes) and fish growth and available zooplankton, in Lake Le Roux between 1977 and 1984. 1978/79 and 1981/82 (open symbols) were relating wet years. (After Hart [384, 385].)

because more heat is radiated back to the atmosphere instead of being mixed down into the water column.

7.6.2 Predation by fish and the composition of zooplankton communities

The zooplankton community is often shaped by predation on it, and vertebrate effects seem to be more extensive than those of invertebrate predators. The first clues came from Hrbacek *et al.* [443] and Brooks and Dodson [103]. Effects of fish predation are clearly shown when fish have been introduced into lakes that previously lacked them. Figure 7.20 shows the marked shift in zooplankton species and sizes that followed introduction of the planktivorous fish *Alosa aestivalis* into Crystal Lake in Connecticut, which previously lacked such a planktivore. The zooplankton community was changed from one of *Epischura*, *Daphnia* and *Mesocyclops*, all usually more than 1 mm in size, to one of *Ceriodaphnia*, *Tropocyclops* and *Bosmina*, which are all smaller than 1 mm. *Cyclops*, at just under 1 mm, persisted in both situations.

This work led Brooks and Dodson [103] to state their size-efficiency hypothesis [374] to explain the size ranges of crustacean zooplankton communities. They believed that all zooplankters competed for particulate matter in the water but that the larger animals were most efficient. Small animals were thus excluded by starvation if large ones were present. Vertebrate predators (largely fish, but even wading birds in shallow pools [216, 449]), however, select the larger Crustacea (cladocerans, calanoid copepods) and, depending on the intensity of predation, allow smaller

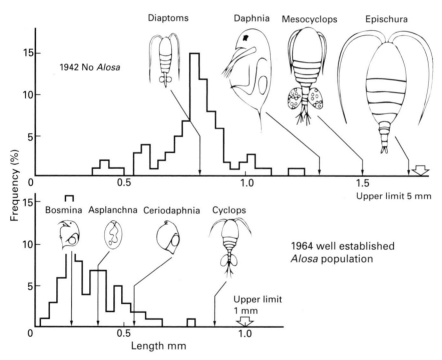

Fig. 7.20 Composition of the mainly crustacean zooplankton of Crystal Lake, Connecticut, before (1942) and after (1964) introduction of a planktivorous fish, *Alosa aestivalis*. Planktivorous fish had not previously been present. Specimens are drawn to scale and represent mean size for each species. The arrows indicate the size of the smallest mature instar of each species. The effect of the fish has been to replace a community of large species with one of much smaller organisms. (From Brooks and Dodson [103].)

zooplankters (smaller Crustaceans, rotifers) to coexist, up to the state of complete elimination of the large forms where predation is intense.

Fish do select the largest prey that they can catch and ingest. The larger the prey, the greater is the return of energy for that invested in catching it. They must see their prey to catch it, but objects smaller than about 1 mm are not readily seen. Rotifers, small Cladocera, such as *Bosmina*, nauplii and early copepodites will thus escape. The larger Cladocera are most vulnerable. They move slowly and probably do not have sensory mechanisms capable of detecting the shock wave of an approaching fish. In contrast, the copepods have sensory hairs on their antennae. With a flick of their abdomens, they can move away with great speed from the line of attack by the fish. An association of small Cladocera, rotifers and copepods with fish predation has been frequently shown (e.g. Lynch [591]). Selection of the larger animals can also be shown by comparison of fish-gut contents with the availability of prey in the lake [65] (Fig. 7.21).

7.6.3 Predator avoidance by the zooplankton

As with invertebrate predation, zooplankters have evolved ways to minimize the risk of being eaten by fish. Cladocera frequently produce individuals that are smaller, thinner and more translucent during the summer, when fish feeding is intense. These forms may be less visible [477], but have smaller brood chambers and produce fewer eggs. Within a lake where predation is concentrated – for example, at the margins, where fish may find refuges against their own predators – the larger, or sometimes just different, forms may persist offshore and the smaller ones survive more readily inshore. Green [358] has shown this for *Daphnia lumholtzi* in Lake Albert, while a similar contrast has been shown [1040] for *Ceriodaphnia cornuta* in Gatun Lake, Panama.

The smaller, thinner animals produced apparently in response to fish predation (although there is evidence that physical factors, such as temperature and turbulence, may also stimulate increase in their numbers [392, 401, 453]) often have protuberances, such as long spines, or, in Cladocera, extensions of the head (helmets). This is called cyclomorphosis, for the occurrence of these forms is often seasonal. The protuberances may, as suggested above, be devices largely to minimize predation by invertebrates, particularly for the rotifers and smaller Cladocera. Fish suck in their food and the shape of it seems not to matter very much, but many zooplankters may be coping with threats from both invertebrate and vertebrate predation.

There are also behaviours that minimize contact between the prey and the predator. Vertical migration has long been known in zooplankters, with the animals remaining at depth, in the dark, by day and moving to the surface at night to feed. Many explanations have been put forward for this.

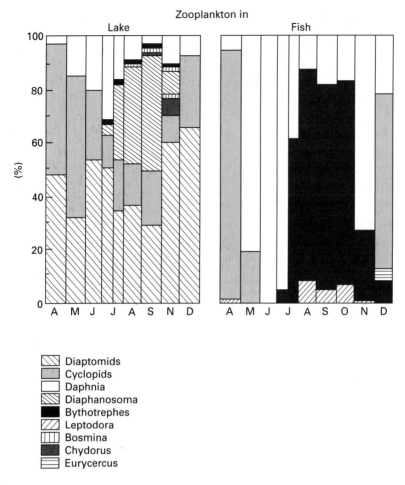

Fig. 7.21 Percentage composition of zooplankton in Lake Maggiore, Switzerland–Italy, and in the stomachs of zooplanktivorous whitefish (*Coregonus* sp.). Diaptomids and cyclopids are copepods, *Daphnia*, *Bythrotrephes*, *Leprodora* and *Eurycercus* are large Cladocera. The others are small Cladocera. (From de Bernardi and Giussani [65].)

It might be most economical, for example, to grow at low temperatures, which minimize respiration rates and to spend time in the warmer surface waters only when it is necessary to feed. However, the hypothesis that diurnal vertical migration is an antipredator strategy is now widely accepted [133, 1042].

Zooplankters are more likely to show vertical migration in the presence of fish predators than in their absence [335, 926]. By remaining in deeper darker waters by day, they avoid being seen by the fish but invoke certain costs, because the deeper water is often cooler and poorer in food, so that their growth and reproduction are inhibited. These costs are apparently balanced by the greater benefit of predator avoidance. The zooplankters become vulnerable only late in the day, when they must use light cues to

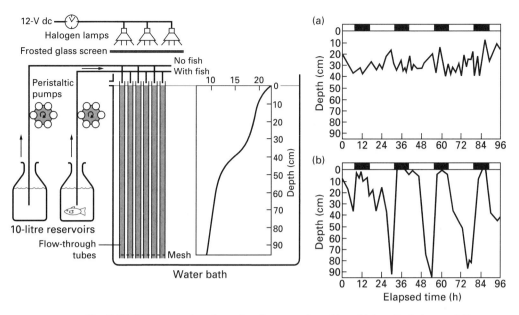

Fig. 7.22 Arrangement and results of an experiment in which a *Daphnia* population contained in stratified water columns was exposed to water from which fish had been excluded or in which they were living. With the influence of fish (b), the *Daphnia* migrated diurnally (and also grew less rapidly, through being exposed to relatively lower temperatures). In the absence of fish (a), they did not migrate (and grew more quickly). (Based on Dawidowicz and Loose [196].)

move upwards to graze during the ensuing night. They may then become outlined against the twilight sky to fish lurking below [1005].

An experimental investigation of vertical migration is shown in Fig. 7.22 [196]. *Daphnia* were placed in thermally stratified tubes, through which there was a constant drip of water. Some tubes received water that had had zooplanktivorous fish swimming in it; others had water that had not been exposed to fish. In the fish-conditioned water, the *Daphnia* migrated between the top and bottom of the tubes, while, in the fishless water, they remained permanently in the warm surface waters. Growth rates were twice as high in the non-migrating animals, suggesting a major cost to migration, but in turn at least as high a cost of being exposed to predation.

7.6.4 Piscivores and piscivory

A further complication in interpreting the processes going on in open water (and even more so in the littoral zone (see Chapter 8)) is that fish themselves have predators. Their activities cause changes in the community, numbers, size distribution, habitat occupied and behaviour of their prey. There are sometimes cascading consequences [621] for the zooplankton and thence the phytoplankton. Piscivores include many other fish, reptiles (turtles, snakes and crocodilians), birds, such as grebes, mergansers, herons and cormorants, and mammals, such as otters and, in Lake Baikal, freshwater seals.

Piscivorous fish may be so voracious that small-bodied fish cannot coexist with them, while in other cases the prey take refuge among structures, such as plant beds, where the predators may be less effective. In a Norwegian lake [85], when pike-perch were introduced, the previously large roach population of the open waters ($12\,000$–$15\,000$ fish ha^{-1}) was reduced to only 250 ha^{-1}, largely by movement into the littoral. Careful observations on a small Swedish lake [1019] suggested that birds, such as great crested grebes and mergansers, took 34% and piscivorous fish 66% of the fish eaten in summer, with the portion going to the birds becoming predominant in autum.

Discovering the effects of piscivores is difficult, because they are wide-ranging. Whole-lake experiments, where they are introduced or removed, give valuable but unreplicated information. In experimental enclosures, they may not necessarily behave as they would in the entire lake, but, if the enclosures are large enough, valuable insights may be obtained. Greenberg *et al.* [361] carried out such an experiment with combinations of two prey species, the active, schooling rudd and the sluggish, bottom-living crucian carp. They used two predators of different characteristics. Pike are solitary, clear-water, sit-and-wait (ambush) predators, which feed by day among vegetation; pike-perch are shoaling, 'wandering' piscivores, preferring turbid conditions and feeding by twilight in the open water. The predators and prey were introduced in various combinations in enclosures, with artifical vegetation (buoyant green polypropylene strips fastened to mats and occupying 17% of the enclosures) and lacking it, in experiments lasting 5–7 days.

In the absence of vegetation, predation rates (up to two prey per day) were higher for pike-perch; in the presence of it, pike were more effective. Both predators took crucian carp in preference to rudd when both prey were present, and pike in particular took the smaller fish. Crucian carp may have been more vulnerable, because of their sluggish behaviour and distinctive colour. Only one size of predator and two sizes of prey were used, in a restricted and simplified environment that lacked other prey species (and other predators). Whether the same results would be obtained on a larger scale cannot easily be established. We are very far from understanding the detailed of effects of piscivory.

7.6.5 A key role for *Daphnia*

Despite the complex relationship between zooplankters and their predators, one important generalization emerges. This is that the larger Cladocera, particularly *Daphnia* spp., are very vulnerable to fish predation, while also being very effective grazers on phytoplankton. A larger stock of zooplanktivorous fish should thus be associated with a dearth of grazing and development of dense algal populations. Husbandry of the zooplankton population by reduction of zooplanktivorous fish or increase in piscivores thus offers possibilities for reducing algal growth in eutrophicated lakes where only limited nutrient control is possible. The technique is called biomanipulation and is discussed later. For the moment, an interpretation of why domestic garden ponds with lots of goldfish are often bright green with algae may be obvious.

7.7 Functioning of the open-water community

The plankton community is very dynamic. Not only is the relative position of every particle, live or dead, changing from second to second, but dozens of chemical changes are simultaneously happening. We can divide the plankton into a series of 'compartments' to examine its activities. These are: phytoplankton, microorganisms (bacteria and fungi), detritus, Protozoa, zooplankton, dissolved substances and fish. A particular phosphorus or carbon atom may, in summer, find itself shuttled through several such compartments in a few minutes. Figure 7.23 shows the main pathways between compartments. Some of the processes – phytoplankton nutrient uptake, grazing, fish predation – have already been considered. Others are imperfectly understood, but the movements of phosphorus and nitrogen form a good means of illustrating some of them. The general picture is of frequent reuse of each element within the system, with inevitable losses to the sediment and overflow more or less matched by various inputs to the lake.

7.7.1 Cycling of phosphorus in the plankton

Phosphorus enters the cycle from the catchment or by release from the sediment (see Chapters 3 and 6). It leaves it when washed through the outflows or permanently incorporated into sediments as detritus or precipitates. During the plankton cycle between these events, there seem to be at least two main means of regeneration of inorganic phosphate: by zooplankton and by microorganisms.

The relative importance of these varies from lake to lake. In Lake George (Uganda), for instance, it is only by rapid recycling of nitrogen and phosphorus through the zooplankton that dense blooms of blue-green algae can be maintained [312]. When released by zooplankters, N and P are taken up so rapidly by the phytoplankton that their excretion can only be detected if the phytoplankton is experimentally removed.

Calculations based on the plankton of Heart Lake, Ontario by Peters and Rigler [766] suggest that about a quarter of the particulate matter (including its phosphorus) in the epilimnion is ingested by zooplankton each day in summer. Of this, about half is assimilated and, during summer, an

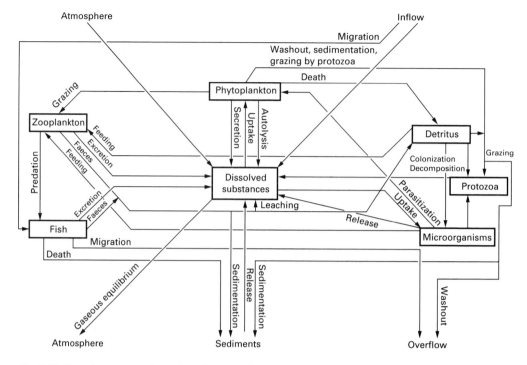

Fig. 7.23 Some important relationships between the major compartments in the plankton system.

equivalent amount is excreted daily. Thus, 0.5×25 or 12.5% of the phosphorus in the plankton is regenerated daily by excretion and a similar amount is daily turned into faeces, which enter the detritus compartment. Here, more of the phosphorus may be released by microbial activity.

The microbial complexes of small phytoplankton, protozoa, bacteria and detritus may be important, particularly in infertile lakes [948, 949]. Phosphate is taken up very rapidly by the smallest organisms, and the larger phytoplankters may reacquire it only slowly, following mineralization by the microbial complexes and release of inorganic or organic phosphates. Some algae have phosphatases on the outsides of their cells, capable of releasing phosphate from the organic phosphorus compounds.

Lean [558] showed some of the complexity of movements of phosphorus through microorganisms. He added radioactive inorganic phosphate to water from Heart Lake, Ontario. Within 1 min, half had been taken up by the smaller phytoplankton and microorganisms. Within 1 h, an equilibrium had been established in which only about 0.2% of the dissolved phosphorus was present as dissolved inorganic phosphate and 98.5% was in particulate form.

After filtering the water through molecular gel filters, two further sets of phosphorus compounds were detected. One, with a molecular weight of about 250, called XP, constituted 0.13% of the total and the other, with molecular weights greater than 5×10^6, constituted 1.16% of the total and was called 'colloidal phosphorus'. The XP was released from the particulate compartments, probably by bacterial rather than algal secretion, but certainly not through death and decay, as the process was too rapid (less than 3 min). The XP reacted with 'colloidal phosphorus', with release of inorganic phosphates (PO_4), but XP did not itself readily hydrolyse to inorganic PO_4. The time for passage of inorganic PO_4 through the particulate, XP and colloidal P compartments is likely to be only a few minutes in summer [819]. Some of the colloidal P ceases to react with XP after a few days and hence its contained phosphorus becomes unavailable to the plankton.

Phosphorus is thus rapidly cycled and reused by the smaller phytoplankton and bacteria. Organic phosphorus compounds may not always be available directly to the phytoplankton, but must be metabolized first by bacteria other than those which produce and secrete it. Zooplankters also cycle phosphorus, with a rather longer turnover time (days rather than hours), and may also be prime agents, through sedimentation of their faeces, of loss of phosphorus to the sediments. This occurs at the rate of a few per cent of the total phosphorus pool per day and implies again (see Chapter 6) that maintenance of plankton production relies on continually renewed supplies of phosphorus from the catchment, except in those, often shallow, lakes where there is substantial phosphorus release from the sediment.

At the scale which is important to these organisms (distances of micro- or millimetres), the environment is patchy [566, 804]. As a fish or zooplankter moves through the water excreting phosphorus, this substance becomes available to algae or bacteria in its plume, but, because they may take it up so rapidly, not perhaps to those only millimetres away.

7.7.2 The nitrogen cycle in the plankton

The behaviour of nitrogen in the plankton is as complex as that of phosphorus. There are four main dissolved inorganic pools – ammonium, nitrite, nitrate and molecular N_2. The first three are available for uptake by most phytoplankters and may be simultaneously absorbed. Some species, particularly the Euglenophyta [562], are able to absorb only ammonium, and other plankters may absorb ammonium preferentially, since it is energetically less costly to process in the cell. Ammonium, and particularly nitrite, rarely have large dissolved pools, however. Even nitrate may be barely detectable in summer, and, in lakes where the phosphorus loads are large, the rate of supply of inorganic nitrogen compounds may limit the growth of phytoplankton (see Chapter 6).

The turnover times of the inorganic nitrogen pools are largely unknown. Ammonia excreted from *Thermocyclaps hyalinus*, the dominant zooplankter of the equatorial Lake George, was measured by Ganf and Blazka [309]. The ammonia released was rapidly taken up by phytoplankton, with a turnover rate of about 1.5 day^{-1}. Parallel measurements of inorganic PO_4 turnover of about twice a day are likely to be underestimates [766]. The very small pools of inorganic nitrogen in the lake (no nitrate was detected at all) suggest that nitrogen should br cycled at least as rapidly as phosphorus and the turnover rate may be much greater than 1.5 day^{-1}.

Molecular nitrogen can be used only by certain blue-green algae and photosynthetic and heterotrophic bacteria. Blue-green algae possessing differentiated cells, called heterocysts, can fix nitrogen. Nitrogenase, the complex of enzymes responsible, is inhibited by oxygen, and heterocysts lack that part of the photosynthetic apparatus, photosystem II, which is responsible for oxygen release [260]. There is a high correlation between measured nitrogenase activity and heterocyst numbers [428]. Nitrogenase is also present in ordinary cells of some blue-green algae that do not have heterocysts [925], although it is active only under anaerobic conditions. A coccoid species, *Gloeocapsa*, has been shown to fix nitrogen in aerobic conditions [1037], despite a lack of heterocysts, and this may mean that nitrogen fixation is more widespread in lakes than currently believed.

In many lakes, nitrogen fixation seems, however, not to be an important source of nitrogen. Less than 1% of the total nitrogen income of Lake Windermere is provided in this way [427], but production is probably limited by phosphorus supply in this lake. In lakes with high phosphorus loadings, large crops of blue-green algae may account for about half of the total nitrogen income by fixation [428].

Blue-green and other algae and bacteria secrete nitrogenous organic compounds into the water, where they form part of a pool of many different dissolved organic nitrogen compounds, about which little is known [631]. The pool is also supplied by excretion of urea by fish, and perhaps with amino acids and urea by some zooplankters [479]. Walsby [994] showed at least twelve separate polypeptides to be produced by a single species, *Anabaena* cylindrica, and, in Lake Mendota, dilute pools (< 0.01 μmol) of free amino acids have been measured, with serine and alanine the most prevalent of ten acids examined. A ten-fold larger pool of combined amino acids was simultaneously measured and both pools increased during decomposition of large algal populations [313].

Part of the dissolved organic nitrogen pool, that of small molecules, such as amino acids, probably turns over rapidly, at rates similar to those of other simple organic compounds. Bacteria probably account for much of the uptake, because algae, although capable of using amino acids [3] and other organic nitrogen compounds [69] at high concentrations, do not compete effectively with bacteria at low concentrations [415].

The vitamins variously required by phytoplankton species are all dissolved organic nitrogen compounds and are rapidly used, once formed.

Pools of up to 8 ng l^{-1} of cyanocobalamin, 400 ng l^{-1} of thiamine and 40 ng l^{-1} of biotin were recorded in the Japanese Lake Sagami, by use of bioassay techniques, since chemical analyses are insufficiently sensitive [733]. The vitamins were secreted by bacteria and some blue-green algae, and were taken up rapidly by planktonic diatoms.

Dissolved organic nitrogen compounds, although not solely responsible, may also act as chelators. Chelators reversibly combine with metal ions in equilibria between free metal ions and soluble ion–chelate complexes. As ions such as Fe^{3+}, Mn^{3+} and Mo^{3+} are removed from the complexes by phytoplankton, the equilibrium moves to replace them. Such ions are readily precipitated inorganically as hydroxides or carbonates, and retention in soluble chelator complexes ensures a steady supply for the phytoplankton. The peptides secreted by blue-green algae can act as chelators [280].

7.8 Seasonal changes in the plankton

While the plankton hourly cycles substances within the community, the community itself changes on an annual or longer-term basis. These changes are imposed by climate, through its effects on temperature, rainfall and hence nutrient loading, by the earth's rotation, with its effects on day length and light intensity, and by internal changes, as the community reacts to the external factors.

Almost all species that increase their populations in the plankton at some time during the year are ever-present in the water as small residual populations. Some may form resting stages in the surface sediment and new ones may be brought in from time to time on water birds or by wind or floods.

There is, then, a great reserve of varied forms, each best fitted to exploit a particular set of conditions in the water, when its population will increase, and each less able to compete in other conditions, when its population will decline. The changing water mass throughout the year, in turn, selects the species better fitted for a particular time and, in turn, precipitates their decline. The result is a procession of overlapping, large populations against a background of small, declining populations.

If changes in weather are a major driving force in determining seasonal periodicity, the least marked periodicity must be expected in equatorial lakes. One such, Lake George (Fig. 7.24), has a very steady climate. Incident radiation, although irregularly intercepted by cloud, varies within a range of only ±13% of the mean, and the water temperature is always about 30°C. There are two dry seasons, but their potential effect in determining changes in nutrient loading is offset by the Ruwenzori Mountains in the catchment, whose high run-off permits a continuous inflow to the lake [986].

This constancy is reflected in the low diversity of the plankton. Over 99%

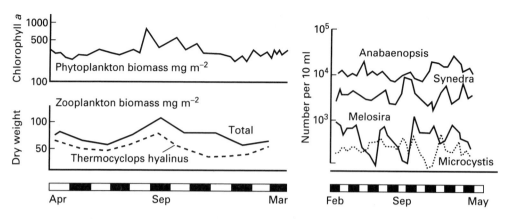

Fig. 7.24 Seasonal changes in the plankton of Lake George, Uganda. Details are given of the major zooplankter and four of the most abundant phytoplankters. *Melosira* and *Synedra* are diatoms, the rest are blue-green algae. (Based on Ganf and Viner [312].)

of the plankton biomass is phytoplankton and, of this, six species of blue-green algae comprise 80%. Only a dozen or so other species have been recorded, compared with hundreds in more seasonally variable lakes. There is a great seasonal stability in phytoplankton biomass and species composition, and also in zooplankton biomass, which is dominated by only two copepod species, with *T. hyalinus* comprising 80% by weight. Lake George is, however, likely to be unusual in the stability of its plankton communties. Away from the equator [645, 945], tropical lakes show as much seasonal variability as temperate ones.

A typical pattern of temperate phytoplankton periodicity in a moderately deep lake is shown in Fig. 7.25. A late winter/spring pulse of several diatom species is overlapped by one of Chrysophyta, largely *Dinobryon* species, as it declines. The onset of summer brings a wide variety of green algae, cryptomonads, dinoflagellates and, in mid- to late summer, blue-green algae. In autumn, diatoms may grow again.

This sequence of algae is superimposed on a set of environmental changes, which start, in winter, with short days and low light intensities and temperatures, but with relatively abundant nutrients. This is because run-off from the catchment is usually high, at least in late winter as the snows melt, and the incoming water replenishes the nutrient stock. As algae increase their growth in spring, with the lengthening warmer days, and the incoming nutrient supply is reduced, as run-off declines, nutrients may become very scarce.

The water column may also become more stable as stratification sets in. Zooplankters increase their populations later than the algae, because their growth is more sensitive to temperature, so that for a time grazing may be unimportant. Once it has begun, it may be effective in creating clear-water conditions for several weeks, but is then truncated as the newly hatched

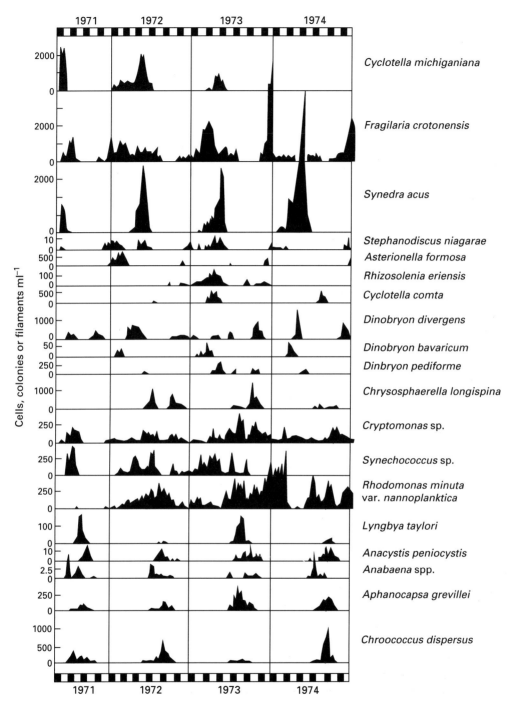

Fig. 7.25 Seasonal changes in the most abundant phytoplankton species of Gull Lake, Michigan, USA, over a period of several years. Numbers of organisms are the means of counts taken at several depths in the water column of this 30 m-deep lake. The top seven organisms are diatoms and the next four are Chrysophyceae. *Cryptomonas* and *Rhodomonas* are Cryptophyceae and the remainder are blue-green algae. (Redrawn from Moss *et al.* [684].)

young fish start to feed in June and July. In autumn, there is a general reversal of these trends, as inflows increase, temperature and light intensity fall, zooplankters enter resting stages and fish cease to feed.

7.8.1 Mechanisms underlying algal periodicity

Some algal growth goes on in winter, even under a thick ice cover. Ice is usually transparent, although a cover of snow is opaque. Winter algae from polar seas are adapted to low temperatures and low light intensities, although examples are not yet available from fresh waters. Such species will not grow at 10°C or above. In temperate lakes in winter, even so, growth is not great and the levels of available key nutrients are able to build up.

As temperatures, day lengths and light intensities increase in late winter, so also do cell-division rates of various diatoms, whose populations reach maxima in the spring. In Lake Windermere, UK, the diatom A. formosa (Fig. 7.26) is the first phytoplankter to form a prominent population after the turn of the year. Lund [584, 585] has shown that it is primarily low light intensities that minimize its growth in winter. Cells are always present in the water and will multiply in winter water if brought into more brightly lit conditions in the laboratory.

From around February, the *Asterionella* cells divide and exponential increase brings the population from about 1 cell ml^{-1} in January to perhaps 10 000 ml^{-1} by late spring [585, 587]. Concentrations of dissolved nitrate, phosphate and silicate, built up by the inflows in winter, all decrease during this growth, as the cells take them up, and in May or early June the *Asterionella* population suddenly declines. Some dissolved silicate (SiO_3) (about 0.4–0.6 mg SiO_3-Si l^{-1}), which is required in quantity for production of diatom cell walls, remains in the water, but the cells seem unable to take it up unless small amounts of phosphate are added [445].

On the other hand, addition of silicate without phosphate will also allow some further growth at the time when the population is declining in the lake. Clearly, there is a nutrient limitation, but the mechanism is complex. At this point, the cells cannot obtain enough silicate to complete the cell walls of their daughters. Weak walls are formed and there may be invasion by bacteria. Eventually, most of the population is lost to the sediments. Other diatoms, faced with low silicate concentrations, simply cease to grow, and they persist until silicate supplies are renewed [26, 908].

The major spring diatom growth of many temperate lakes does not always involve *Asterionella* and does not necessarily end due to silicate limitation, although nutrient shortage of one kind or another is often implicated. The genus *Melosira* (*Aulacoseira*) is common in many fertile lakes and seems to persist in the water only when mixing is vigorous. Its walls are thick and its cells heavy. The onset of stratification, whether inverse under ice during the winter or direct in late spring, leads to sedimentation of the

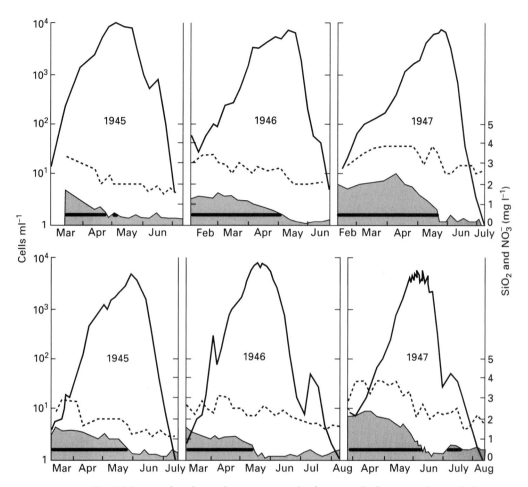

Fig. 7.26 Annual cycles in the spring growth of *Asterionella formosa* in the north (lower) and south (upper) basins of lake Windermere. Because of the lesser depth of the south basin and thus the greater exposure to light during the mixing season, growth begins earlier in the south. Interrupted lines show nitrate concentrations ($\times 10$) and the shaded areas show silica (SiO_2) concentrations, with the critical 0.5 mg l^{-1} value shown as a dark line. (From Lund [584, 585].)

cells. On the sediments, *M. italica* var. *subarctica* [586] survives in a quiescent state. Once mixing is vigorous enough to disturb the surface sediment, it is resuspended, expands its cell contents and divides, so long as light and nutrients are available.

Melosira species are also common in tropical lakes [945] and, again, their growth coincides with turbulent conditions. For example, *Melosira* was abundant in Lake Victoria when cooling by the seasonal trade winds mixed the water column [941, 942]. It may grow even in summer in temperate lakes, but usually in riverine ones, where water movements produce adequate turbulence.

Nutrient shortage and reduction in turbulence are not the only factors that may bring the spring diatom growth to an end. In one Connecticut lake, the growth could be reduced or prevented in different years, depending on the size of the blue-green algal population that had developed the previous year and overwintered [519]. Large populations of blue-green algae secreted substances, apparently highly specific to individual diatom species or even strains of a species, which suppressed their growth. In lakes with small species of spring diatoms, such as *S. hantzschii*, there may be a phase lasting several days or weeks of growth, followed by a rapid loss through grazing [547], as increasing populations of zooplankters respond to rising temperatures with increased birth rates. The diatom spring growth is often overlapped and succeeded by chrysophyte populations, often those of *Dinobryon*. These appear to cope with the nutrient-depleted water, following the diatom peak, by having very low half-saturation constants for P uptake (< 0.5 µmol P) [563].

Information on the interplay of populations of the many species that grow in summer is meagre. The water is more a self-contained system than in spring, when nutrients are still entering with the inflow water. Rapid cycling of phosphorus and nitrogen and the metabolically induced changes in pH and CO_2 concentrations, caused by high summer photosynthesis, make interpretation complex. Some green algae form summer populations, because they grow slowly. Although division begins in a sparse background population early in spring, its effects are not really shown until summer [588]. Other species – the cryptomonads, for instance – which require vitamin B_{12} and often thiamine also may benefit from increased bacterial activity at the higher water temperatures. The blue-green algae of late summer may exploit the microaerophilic zone of the metalimnion (see earlier) and, if dissolved inorganic nitrogen becomes scarce, nitrogen-fixing blue-green algae will be favoured in late summer.

Low nitrogen-to-phosphorus ratios may favour even some non-nitrogen-fixing blue-green algae [902], and others are adept at picking up CO_2 from very low concentrations. The increased gross photosynthesis and uptake of CO_2 in summer may force the pH value of the water to 10 or more in some lakes. Talling [944] showed that *Microcystis aeruginosa* and the dinoflagellate *Ceratium hirundinella* are both able to absorb CO_2 at such pH values more efficiently than *A. formosa* and *Fragilaria crotonensis*, which grow earlier in the year. Yet other late-summer species may be selected for their inedibility. As the young zooplanktivorous fish are reduced in numbers by piscivores, there may be some recovery of the cladoceran populations in autumn and, as the smaller algae are removed, large ones may be placed at greater advantage.

Superimposed on the phytoplankton periodicity is a sequence of zooplankton changes that is less easy to interpret. They may result from changes in availability of suitable food or from selective predation, as well as in response to a changing physical environment. Parasite-induced changes may also be important [112].

The periodicity of plankton usually shows a pattern, whose general features are constant from year to year, but whose details are subject to the vagaries of weather, expressed through inflow, nutrient loading and washout, temperature and the timings of stratification and mixing. Grazing adds a further dimension, for the zooplankters are temperature-dependent as are the success of fish recruitment and hence predation pressure on the zooplankton.

There may be several hundreds of species of algae waxing and waning throughout the cycle, and countless other microorganisms and protozoons. The plankton community is very diverse, especially considering the relative simplicity of its habitat. That it shows such diversity is a function of continuous change, Before processes like competition can lead to some sort of equilibrium, the environment has changed and the community starts to move to a new eqilibrium, which it never achieves. This responsiveness to change is one reason why it is relatively easy to manipulate the community as a management measure to reduce the amount of phytoplankton or change its nature. Such biomanipulation is the final topic of this chapter.

7.9 Practical applications of plankton biology: treatment of eutrophication by biomanipulation in deep lakes

Lakes are not just bowls of water variously fertilized with phosphorus and nitrogen and understandable purely in terms of inputs and outputs of these elements. They are ecosystems, which manipulate the N and P once they have entered. This opens up the possibility of treating at least the symptoms of eutrophication (see Chapter 6) by changing the structure of the open-water community. Nutrients may determine the potential production or crop of phytoplankters, but grazing may determine the extent to which this potential is realized. For a given P concentration, there can be a range of planktonic chlorophyll a concentrations in a particular lake (see Fig. 6.16). The aim of 'biomanipulation' [883, 884] is to achieve the lowest possible chlorophyll a concentrations for a given nutrient regime. It may also be possible to reduce the total phosphorus concentration by similar means.

That such changes are possible was suggested by whole-lake observations from Lake Washington (see Chapter 6) and can be shown in simple experiments, such as that done in tanks by Hurlbert et $al.$ [448]. Six replicate systems were set up in 30 cm deep × 2 m diameter tanks. To each was added a 3 cm layer of sand, tap water to 20 cm and 1 l of dried alfalfa pellets, as a source of organic matter and nutrients. A small sample of plankton from a nearby lake and an inoculum of $D.$ $pulex$ were added, and the pools left for a few weeks. Then 50 3–5 cm fish, $Gambusia$ $affinis$, were added to each of three tanks and observations were made on the water chemistry, plankton and benthos; some of the results are shown in Table 7.3.

The fish readily ate the larger zooplankters, *Daphnia pulex* and *Chydorus sphaericus*, which developed significant populations only in the fishless ponds. The less preferable, smaller rotifers increased to greater levels in the presence of fish, although they were also abundant in the absence of fish. In the tanks with fish, the reduction in zooplankton grazing allowed a dense population of a unicellular blue-green algae, *Coccochloris peniocystis*, to develop, which markedly reduced light transmission and the growth of *Spirogyra* sp. (a filamentous green alga) on the bottom. The fish also readily ate the benthic invertebrates, either in the sediments or as they emerged as adult insects, and probably took rotifers as other food supplies became short.

Table 7.3 Effects of *Gambusia affinis* on pond ecosystems. All differences were significant at the 0.1 level or better, except where indicated with an asterisk. (After Hurlbert *et al.* [448].)

	Without fish	With fish
Zooplankton (N per 10 l)		
Daphnia pulex (2 Dec.)	92	0.3
(3 Feb.)	1840	0
Total rotifers (2 Dec.)	1130	5930
(3 Feb.)	32*	4*
Phytoplankton (millions of cells per ml)		
Coccochloris peniocystis (10 Jan.)	0	117
(2 Feb.)	0	220
Colonial algae (2 Feb.)	79*	52*
Macroscopic bottom algae (g per tank)		
Spirogyra (6 Feb.)	312	29
Chara (6 Feb.)	24*	31*
Phosphorus concentration ($\mu g\ l^{-1}$) (3 Feb.)		
Inorganic phosphate-P	10	0.33
Organic phosphate-P	18	55
Particulate-P	12	271
Total P	41	326
Benthic invertebrates (no. cm^{-2} of tank bottom)		
Chironomid larvae (3 Feb.)	25	0
Oligochaete worms (7 Feb.)	14	0.7
Emerging insects (5 Nov. and 7 Feb.)	486	0
Light-extinction coefficients (m^{-1})		
Blue (425 nm)	5.1	64
Red (680 nm)	0.8	12

Significant too was the apparently greater mobilization of phosphorus in the ponds with fish, conceivably through feeding on the bottom fauna and release of phosphate in fish excreta. The presence or absence of an invertebrate-eating fish thus created two entirely different states of the ecosystem, despite similar initial physicochemical conditions.

This phenomenon can be demonstrated with communities less artificial than the ones discussed above. Lynch and Shapiro [593], for example, added increasing numbers of blue-gill sunfish, *Lepomis macrochirus* to enclosures in a lake and found an increase in chlorophyll concentration and a decrease in transparency with increasing numbers of fish (Fig. 7.27). The fish ate *Daphnia*, which otherwise grazed the algae. In this experiment, the total phosphorus concentration was unchanged, although more of it was in the available, soluble, reactive form when *Daphnia* grazing was most intense.

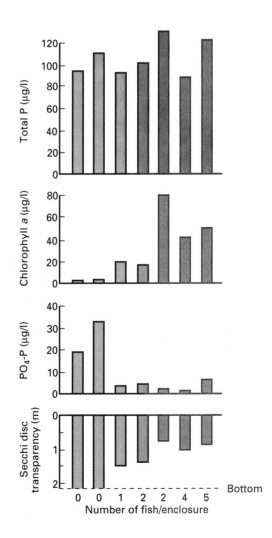

Fig. 7.27 Effects of adding bluegill sunfish to enclosures containing a plankton community. (Based on Lynch and Shapiro [593].)

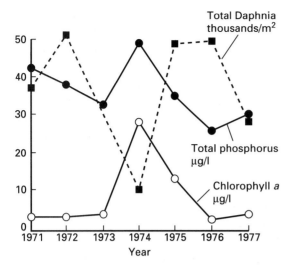

Total Daphnia thousands/m²

Total phosphorus µg/l

Chlorophyll *a* µg/l

Year

Fig. 7.28 Total phosphorus, chlorophyll *a* and *Daphnia* in Lake Harriet, Minnesota, in 1971–77. Mean surface values for the summer period are shown. (From Shapiro [884].)

Recruitment of fish is variable from year to year (see Chapter 10). It depends greatly on the water temperature just before spawning and hatching, and 'good years' are often associated with warm spring weather. In turn, the *Daphnia* populations can also vary greatly with inverse effects on the chlorophyll *a* and total phosphorus concentrations (Fig. 7.28). Packaging of phosphorus into readily sedimented *Daphnia* faeces is one mechanism by which the *Daphnia* is able to reduce the total P concentrations.

7.9.1 Experiments on whole lakes

The results of enclosure and tank experiments can be criticized, because they use only simplified communities. Often, only one species of fish is used, piscivores, and effects of changes in nutrient loading and washout are absent. The results, indicating strong top-down effects, may thus be artefacts of the experimental design. However, the existence of top-down phenomena is backed by whole-lake observations [557]. Lake Washington was quoted in Chapter 6 as an excellent example of a lake recovering after nutrient diversion. In the 1970s, however, when nutrient diversion had been completed for several years, the lake continued to increase in clarity and, indeed, became more transparent than it had ever been before. Populations of four *Daphnia* species had increased greatly, but those of a major predator on *Daphnia*, a mysid shrimp, *Neomysis mercedis* [707, 708], had apparently decreased. The nature of the food supply for the *Daphnia* had also changed, with a decline in numbers of blue-green algae. Both factors may have contributed to the *Daphnia* increase.

Neomysis is quite a large animal – about 1 cm long – and hence very vulnerable to fish predation. Its main predators, the sockeye salmon (*Oncorhynchus nerka*) and the long-fin smelt. (*Thaleichthys spirinchus*), which spawn in a major inflow, the Cedar River, increased in numbers in the mid-1960s [232, 234]. The river had been regularly dredged each summer before 1960, which must have disturbed the potential spawning beds. This was stopped in 1960 and revetments and piles of large boulders were placed along 30% of the banks to combat erosion and riverside flooding. Inadvertently, these may also have improved the spawning habitat.

Another example is Lake Michigan, where salmonid fish (the coho salmon, *Oncorhynchus kisutch*, the chinook salmon, *Oncorhynchus tshawytscha*, the steelhead trout, *Salmo gairdneri*, the brown trout, *Salmo trutta* and the lake trout, *Salvelinus namaycush*) have been introduced and regularly restocked to provide sport for anglers. Most of these fish are piscivores, and populations of the alewife, *Alosa pseudoharengus*, a major zooplanktivorous fish, have consequently been reduced. After 1983, there were marked increases in cladoceran zooplankton populations and clarity of the water [858]. However, these were short-lived and *Daphnia* declined from 1985 onwards, apparently through predation by a coregonid fish. Coupled with the complications brought about by the introduced *Bythotrephes* (see above), it has become difficult to disentangle the various effects of nutrient control and salmonid and other introductions in such a big lake [253, 857].

There are strong indications, therefore, that the phytoplankton crops can be adjusted by the interplay of *Daphnia* (and, of course, other grazers, although *Daphnia* appears to be particularly important) and fish predation. The case is strengthened by experiments on whole lakes, in which the fish community has been altered. The results can again be criticized, because the experiments cannot be replicated; the treatment has to be applied in one period, while the control against which the effects are judged has to be in a separate year. However, the combined results of all these approaches give confidence in the case.

Shapiro and Wright [886] followed changes in Round Lake, Minnesota, after it had been treated with a fish poison, rotenone, in autumn 1980 (Fig. 7.29). The former fish community was dominated by zooplanktivores, *Pomoxis nigromaculatus*, *Lepomis macrochirus* and *Lepomis cyanellus*, with relatively few piscivores. Rotenone is short-lived and the lake was restocked with a much greater ratio of piscivores (large-mouth bass, *Micropterus salmoides*, and walleye, *Stizostedion vitreum*) to zooplanktivores. Over the following two years, there were increases in transparency, in the body sizes of the zooplankton and in the proportions of *Daphnia* in the zooplankton community and decreases in chlorophyll *a* and total phosphorus concentrations.

Carpenter and his coworkers [127, 130, 243] have carried out experiments on three 5–10 m deep lakes (Peter, Paul and Tuesday) in Michigan, by transferring fish among them. Peter and Paul had large populations of the piscivorous largemouth bass. Tuesday had lots of planktivorous minnows. A reciprocal exchange of minnows and bass was made between Peter and Tuesday lakes, while Paul Lake was used as a reference. The

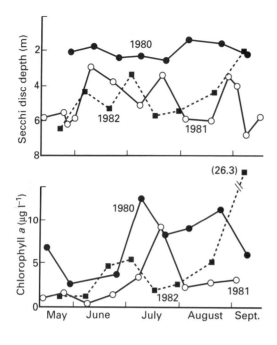

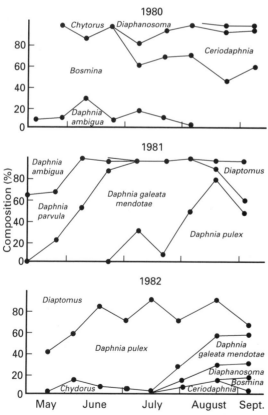

Fig. 7.29 Effects of biomanipulation by removal of fish from Round Lake, Minnesota, and restocking with a higher population density of piscivores. Round Lake is 12.6 ha in area and has maximum and mean depths of 10.5 and 2.9 m. The biomanipulation was carried out in autumn 1980. Secchi-disc transparencies were much greater in the two subsequent years and chlorophyll *a* concentrations declined as large-bodied Cladocera (*Daphnia*) replaced small ones (*Bosmina*). There was evidence of a reversion towards the end of 1982, as the fish community readjusted. (Based on Shapiro and Wright [886].)

result was that Tuesday Lake acquired large numbers of *Daphnia* and increased *Chaoborus* populations. In Peter Lake, the expected decline in *Daphnia* occurred, due not so much to the introduced minnows but to the small bass that had been left when the larger ones were transferred to Tuesday Lake. The minnows were forced out of the open water into the littoral zone by these bass.

In Germany, the Bautzen reservoir, created in 1974–75, began to suffer increasing algal problems, following removal of pike by anglers and an increase in zooplanktivorous perch. Zander (pike-perch) fingerlings have been added in most years since 1977 and the anglers' catch of these piscivores has been restricted. The zander and the larger perch have kept the zooplanktivorous fish populations low and increased threefold the number of *Daphnia* and also their average body size. Secchi-disc transparency has increased and there has been a shift to less edible blue-green algae (*Microcystis*) as a result of removal of competing smaller algae by grazers.

7.9.2 Scope of biomanipulation

This highlights a particular uncertainty about biomanipulation. Can enhanced grazing control blue-green algae? The answer is 'sometimes'. A well-established large blue-green algal population will probably not be removed [423, 688]; but, if the grazing is established early enough, grazing of the algal inoculum may be sufficient to prevent a large population developing. Biomanipulation sufficiently early in the year can prevent blue-green algal development [255, 644, 855], but the outcome is not entirely predictable.

Deliberate biomanipulation has rarely been used in deep lakes, although it is now common in the restoration of shallow lakes (see Chapter 8). There have arisen several rather polarized schools of thought about its efficacy. Perhaps the most outspoken came with DeMelo *et al.* [206], who surveyed a large number of papers purporting to show the effects of top-down control by fish and concluded that many did not show such effects: 'even the briefest perusal of the pertinent literature indicates that, far from being robust, the trophic cascade theory may be unsoundly based on many half-truths and much handwaving and overextrapolation of the data'.

DeMelo *et al.* [206] simply counted the number of papers showing significant and non-significant effects of fish manipulation. Such a technique perpetuates a statistical problem (type 2 error) in some of the original works, where an effect may have existed but was not detected readily because of low replication and the inherent variability of the systems used. A better approach is meta-analysis [88], in which the original data from all the studies are used, as opposed to the original authors' conclusions, and treated as the many replicates of a single grand experiment testing the hypotheses that fish removal enhances zooplankton abundance and that the latter causes a decline in algae. This approach, which avoids type 2 errors, gave strong support to the functioning of the trophic cascade. On average, fish addition led to a 29% decline in zooplankton and a 76% increase in phytoplankton. The effects were not necessarily strong in individual cases, but were very powerful in others. This is to be expected, considering the great variety of habitats and biological communities included in the analysis.

Some sort of consensus is now appearing. Top-down effects are likely to be more important in shallow lakes and will be dealt with in Chapter 8. In deep lakes, top-down effects occur but are strongest between piscivores and zooplanktivores, progressively weaken with steps down the food chain, while bottom-up effects are strongest between nutrients and phytoplankton and progressively weaken with steps up the food chain. In the middle, where the crucial relationships for algal control of zooplanktivores, zooplankton and phytoplankton occur, there are likely to be a variety of outcomes of fish manipulations, dependent on local circumstances. Indeed, Brett and Goldman [88] found that there was a 60% chance of not achieving reductions in phytoplankton in the set of 54 experiments they analysed. This uncertainty undermines confidence in the technique for lake management.

Some of the failures of biomanipulation in deep lakes, however, may be due to insufficient manipulation – too few fish removed, too few or inappropriate piscivores added. In some cases, the algal communities may have been dominated by inedible, particularly blue-green, algae. Other failures may be due to a belief that the biomanipulation is once and for all. In deep lakes especially, this cannot be true. The fish community will revert to a state that reflects the nutrient conditions, with a low ratio of piscivores to zooplanktivores at high nutrient loadings. This is probably due to the greater risk of low oxygen concentrations at high lake fertility. Piscivores are often fast-moving fish, with high oxygen demand. Any stable predator/prey relationship depends on the existence of refuges for the prey (either Daphnia or zooplanktivorous fish) from the predator (zooplanktivorous fish or piscivorous fish); otherwise both eventually become extinct. The simple structure of the open-water mass of deep lakes cannot provide the additional permanent refuge needed for maintenance of the altered ratios.

However, if biomanipulation is seen as a useful tool for temporary treatment (perhaps lasting for a few years) of problematic waters, it will gain acceptance. Its role may be where nutrient reduction is expensive or impossible – for example, where sources are diffuse or where nitrogen control is needed. It would be inappropriate on natural lakes containing unusual fish, but in man-made reservoirs, particularly those storing water pumped from distant, heavily enriched rivers in the lowlands, which are very artificial systems anyway, there can be little objection. The only problem comes where the reservoir owners are gaining income from recreational angling, in which case the relative benefits must be weighed.

In Britain, there would be resistance to removal of the initial fish stock, because of the scarcity of lakes for angling. Piscivore stocking would be more acceptable, but in Britain there is no native piscivore that specializes in feeding on small, open-water, zooplanktivorous fish. The closest possibility is the pike-perch, *Stizostedion lucioperca*, which is native to continental Europe and probably was to Britain prior to glaciation. However, where introductions of this fish have been made, local opinion has it that there has been a decline in other fish and that the angling has been spoiled. The declines may be much more likely to have been due to changes in the habitat through eutrophication and river management (see Chapter 5), but further introductions are, in any case, illegal. A future possibility, not involving fish removal or addition, is to apply substances to the centre of a lake that scare the zooplanktivorous fish to the edges [336]. Fish damaged by predators emit a variety of 'fright' substances. Dosing with freshly homogenized smelt tissue in a Polish lake led to such movement of smelt, but the technique has not yet been tried operationally.

8: The Littoral and Profundal Communities of Lakes

8.1 A variety of habitats

The edges and bottoms of lakes are more complex than the open water. There may be rocks, sediment and plants, colonized by sessile animals and algae, together with others that move freely but stay close to the interface. Activities in these benthic (Greek: bottom) zones may be affected by and, in turn, influence what is happening in the open water, for they are all part of a greater system. In very large lakes, their influence is probably small, but the smaller and shallower a lake becomes, the more the edges and bottom assume dominance. In ponds, the smallest of lakes, and in wetland lakes, where all of the lake bottom is capable of supporting photosynthetic activity, the benthic communities dwarf the open water in importance.

A classification of these habitats might recognize first the littoral and profundal (Fig. 8.1). Littoral (Greek: shore) in fresh waters is taken to include the part of the lake bottom and its overlying water between the highest water level and the euphotic depth of algae that can colonize the bottom sediments. The profundal (Greek: deep) includes the bottom below the euphotic zone, on which net photosynthesis is not possible. Profundal communities depend on energy supplied by a beneficent gravity as detritus, from the overlying water or washed down-slope from the littoral. The profundal zone is not usually taken to include the water that overlies it (the pelagial or open-water zone) but only the bottom sediments and rocks.

The profundal benthos (Fig. 8.2) is a relatively simple community of bacteria and protozoa and, if sufficient oxygen is present, invertebrates and fish. The habitat is usually a relatively uniform fine sediment of detritus, already processed by other, open-water communities before it reaches the bottom. Deoxygenation of the hypolimnion may confine animal growth to the cooler parts of the year, when growth is slow, and leave only anerobic bacteria and some Protozoa as active permanent residents.

The littorial supports a wider range of communities. Wave disturbance, in a large enough lake, first of all determines the nature of the bottom and divides the littoral into an upper disturbed part, with erosion of the finer particles, and a lower part, where particles may resettle. The parts grade into one another along a continuum of sediment particle size, from the upper coarse to the lower fine. In small lakes, wave action may be very small and fine particles may be deposited over all the littoral zone.

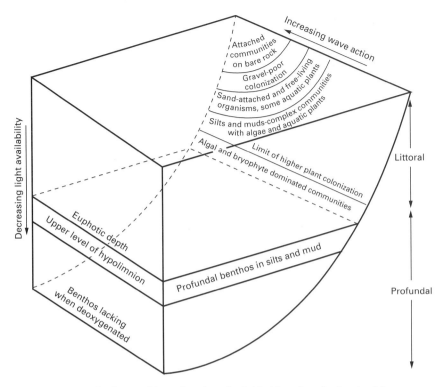

Fig. 8.1 A general scheme of littoral and profundal habitats in a freshwater lake.

In big lakes with much wave energy, the upper shore may be of rocks (Fig. 8.3), bearing attached algae, bacteria, Protozoa and sessile invertebrates, such as polyzoans and the larvae of some caddis-flies. Motile scrapers, such as limpets and mayfly larvae, will be present. In sheltered areas, the rocks become partly buried in gravel and sand, and there may be shores floored entirely by one or the other. Gravel is not readily colonizable. It is not as stable as rock, being moved by the waves, or as sand and finer sediments, which are only set down in less disturbed conditions. Moving

Fig. 8.2 Oligochaete worms are prominent members of the profundal benthos, indeed of sediments everywhere in fresh waters.

Fig. 8.3 The rocky littoral of lakes is dominated by invertebrates that scrape food from the biological films covering the rocks and those that can process the limited amount of litter debris lodging among the stones and gravels. In an upland lake, such as the Cow Green reservoir, with a catchment of poorly weathered rocks and sufficient wave exposure to prevent much sedimentation in the littoral zone, rocky habitats will predominate.

gravel may crush animals. In the quieter seasons, it may be important for egg-laying by fish, but its permanent community is small.

Sand, though less favourable than finer sediments, has a considerable community. The grains allow light penetration for as much as a centimetre or so into the bed and algae attached to them (called epipsammon) can photosynthesize. Other algae (epipelon) can move through the crevices and over the surface. There are also many attached and motile heterotrophic microorganisms (Fig. 8.4). The finer sands may be colonized by aquatic plants, including short-statured genera, such as *Isoetes* and *Littorella*. The community is completed by invertebrates and fish.

Finer sediments in the littoral have the most complex communities, with many bacteria, Protozoa and epipelic algae (Fig. 8.4). Where light and stability are suitable, emergent, submerged and floating plants add to the community, with their associated epiphytes (periphyton), invertebrates and fish. The greater productivity of this community (Fig. 8.5) may support herbivorous birds and mammals (e.g. coot, swan, muskrat, hippopotamus) and piscivorous birds, reptiles and mammals (e.g. bittern, heron, crocodile, otter). The upper part of a plant-dominated littoral zone may be dry in summer but flooded in winter and dominated by large emergent plants (helophytes, such as reeds and bulrushes; see Chapter 5). The lower part is permanently submerged, with emergent, floating-leaved and submerged plants, the former decreasing and the latter increasing in importance with depth. The emergent swamps may be extensive or may form only a narrow zone, dependent on the morphology of the basin. Their biology is similar to that discussed in Chapter 5 for floodplain swamps.

In subsequent sections, particular attention is given to the submerged

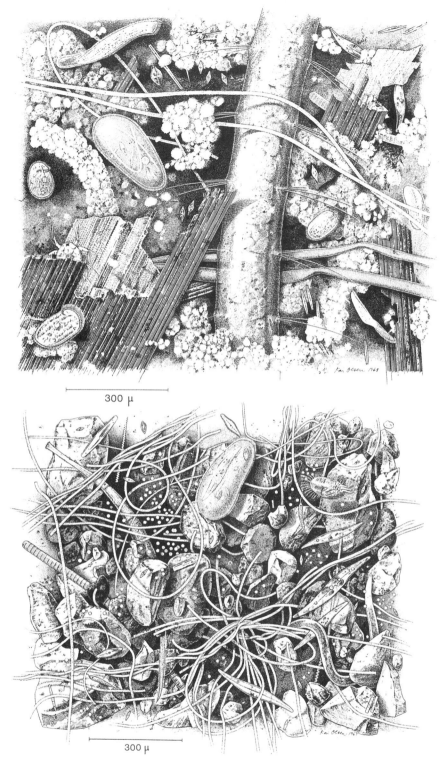

300 μ

300 μ

Fig. 8.4 Communities in sediments. The rich microbial communities include diatoms, filamentous algae and bacteria, coccoid bacteria and Protozoa. There are also small crustaceans, associated with plant debris, and relatively large, cylindrical oligochaetes. (From Fenchel [262].)

Fig. 8.5 The plant-dominated communities of the littoral zone support a diverse collection of invertebrates and vertebrates, and sometimes a veritable army of hippopotami (Flanders and Swann, [273]).

plant beds, communities of bare rock and the profundal benthos. The mutal effects of the littoral zone and the open water on each other and of the open-water communities on the profundal benthos will be discussed. The chapter ends by reviewing the particular problems of managing and conserving lakes dominated by littoral activity.

8.2 Submerged plant communities in lakes

There is often a more or less marked zonation of submerged plant species with depth. Vascular plants do not penetrate to beyond about 11 m for reasons of light penetration (see Chapter 6) [912] and possibly pressure. In Lake Tahoe (California–Nevada), prior to recent eutrophication problems, aquatic mosses and stoneworts (*Chara, Nitella*) were found down to 164 m and 64 m, respectively [289], while vascular plants reached only 6.5 m.

The overall depth of colonization of plants (both vascular and non-vascular) is often inversely related to the extinction coefficient of the most penetrative light in a simple way. Spence [911] has demonstrated this for a series of Scottish lochs (Fig. 8.6). Except where much inorganic turbidity (from glacial melt water or soil erosion, for instance) or organic colour (in peaty catchments) is present, the depth of colonization is also inversely related to the phytoplankton standing crop and hence to the nutrient

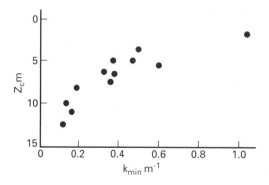

Fig. 8.6 Relationship between maximum depth of colonization, Z_c, of aquatic plants and the minimum extinction coefficient of light, k_{min}, in a series of lakes (based on data in Spence [911]).

Table 8.1 Depths of colonization of some *Potamogeton* species in Scottish lochs and the extents to which their photosynthetic rates were reduced on transfer from high light intensity (7.08 cal cm^{-2} h^{-1}) to low light intensity (3.29 cal cm^{-2} h^{-1}).

Species	Mean depth of colonization	Decrease in photosynthesis (%)
P. polygonifolius	9	42
P. filiformis	50	59
P. × zizii	120	24
P. obtusifolius	125	2
P. praelongus	190	10

loading. The most fertile lakes, with dense phytoplankton, are thus likely to have rather restricted submerged plant communities, although emergent vegetation may be prolific at the margins.

Zonation of submerged plants is common on shelving shorelines, controlled, at least partly, by the photosynthetic abilities of the species concerned [847]. Spence and Chrystal [913, 914] found a distinct range for each of several *Potamogeton* (pond-weed) species (Table 8.1) in a series of relatively transparent Scottish lochs. Leaves of each species were grown at high (7.1 cal cm^{-2} h^{-1}) light intensities under standard conditions and their rates of photosynthesis measured as oxygen evolution. The leaves were then placed in dimmer light (3.3 cal cm^{-2} h^{-1}) and their photosynthetic rates measured again. Those species which normally grew in shallow water photosynthesized in low light intensities at only a fraction (Table 8.1) of their rates at high intensity, while those normally found at depth maintained proportionately high photosynthetic rates, even at low light intensities.

The mechanism by which growth is maintained at low intensity was not examined, but in a macroscopic alga, *Hydrodictyon africanum*, which normally grows in low light intensities, the mechanism appears to be one of reducing respiration through the restriction of some energy-demanding syntheses [801]. One of these was the production of a photosynthetic enzyme, ribulose diphosphate carboxylase (RUBISCO), which forms much of the protein content of photosynthetic tissues. In restricting its production of this enzyme, *H. africanum* may reduce its own gross photosynthesis but evidently increases the difference between gross photosynthesis and respiration, allowing net growth at low light intensities.

Other mechanisms, including competition, the nature of the rooting substratum and the behaviour of seeds on dispersal, also affect the depth distribution of submerged plants. Seeds of two species of *Potamogeton* in African lakes float for different lengths of time after release. This seems to favour greater depth colonization of the one that floats longest before its lacuna-filled seed coat becomes rotten and waterlogged and the seed sinks to the sediments [915].

Unless the shoreline shelves steeply, the plant zonation is rarely regular; patchiness through local disturbance is one reason for this. Fishes may clear patches of sediment as nests. When the nests are abandoned, the patches may be recolonized by spore or seed-propagating species (e.g. *Isoetes* and *Elatine*), which cannot compete well with rhizomatous species. Continual disturbance of the nest sites allows these species to persist, but if the nest sites are permanently abandoned by the fish, the rhizomatous species eventually reinvade them [129].

Nutrient availability is less of a problem for rooted emergent and submerged plants than it is for phytoplankton or floating plants, such as the duckweeds. There is much phosphorus and ammonium in sediments, although injecting nitrogen compounds experimentally into sediments often greatly stimulates plant growth and a greater proportion of root biomass is produced in less fertile sediments [39]. The extent of aquatic-plant communities in lakes thus does not show any simple relationship with phosphorus loading or concentration, although there may be one with nitrogen loading.

Carbon availability is a significant problem for submerged plants in lakes because of the slow diffusion of carbon dioxide in water compared with air (by a factor of 10 000) [624]. A boundary layer of still water in which diffusion is very slow, easily forms in contact with the plant surface. This layer may be several millimetres thick in still waters. The flux of carbon dioxide (CO_2) is governed by Fick's law:

$$\text{Flux} = (C - C_c)D\ T^{-1}$$

where C is the concentration of CO_2 in the water, C_c that at the site of uptake within the plant, D the diffusion coefficient and T the thickness of the boundary layer. The value of T is greater for broad leaves, compared with thin or dissected ones, and increases from the leading edge of the leaf to the middle. Mechanisms have evolved that maximize the difference between C and C_c by using additional sources of CO_2 in the water, such as bicarbonate, by using aerial leaves with access to atmospheric carbon dioxide, as well as submerged leaves, and by using the high concentrations of CO_2 present in sediments and available through the roots (see Chapter 5).

A measure of bicarbonate use can be obtained by placing the plants in a sealed bottle and establishing the proportion of total carbon (CO_2 + bicarbonate) that has been withdrawn when photosynthesis stops for lack of available carbon. The ratio of final to initial concentration is small (0.3–0.5) in bicarbonate users. These include plants with a long history of evolution in water (many algae, and higher plants strongly modified for aquatic life, such as *Myriophyllum spicatum*). It is higher (0.6) in plants, such as *Elodea* and monocotyledons, that have more recently evolved an aquatic existence from land ancestors, and highest in partly aerial plants, such as the water lilies (0.91), floating duckweeds and the waxy, sediment CO_2 users of infertile waters, such as *Isoetes* (around 1.0).

Perhaps because of their relatively recent evolution (only 50–100 million years ago), vascular aquatic plants are less securely adapted to water than the more anciently derived algae. They are vulnerable to shortages of light and carbon dioxide and are less well protected against grazers than many land plants, which have energy to spare for producing protective devices, either chemical or structural. Nonetheless, they support within the littoral environment a complex of biological interactions, which help to maintain an environment in which they themselves can thrive. Some aspects of this environment will now be considered.

8.2.1 Microbial communities in plant beds

The periphyton community on submerged plants contains bacteria, fungi, algae and protozoa and other small animals. The number of bacteria can be determined by staining with phenolic aniline blue. Direct counting usually gives much higher estimates (sometimes 10–20 times) of the number of cells than traditional bacteriological methods based on growing the bacteria to form countable colonies, because not all cells will grow in laboratory media. On the other hand, some of the cells counted directly might be dead and such direct counts may sometimes give overestimates.

Many different species of bacteria are present, specializing on particular organic compounds. Hossell and Baker [430] identified *Pseudomonas*, *Flavobacterium*, *Acinetobacter*, *Moraxella*, *Xanthomonas*, *Aeromonas*, *Alcaligenes*, *Agrobacterium*, *Cytophaga*, enterobacteria and other Gram-negative and Gram-positive forms from the surfaces of *Ranunculus penicillatus*. Different plants almost certainly have different species, they certainly have different population sizes, and the populations may vary in different parts of the plant. In the submerged parts of watercress (*Nasturtium officinale*) and brooklime (*Veronica beccabunga*), the largest populations were on the undersides of the leaves.

The precise sources of organic matter on which they feed are not known either. The concentrations of labile organic matter in the water moving over a plant bed are generally low, but the high concentrations of bacteria on the leaf surfaces (perhaps 10^5–10^7 cm^{-2}, or 10^8–10^{10} ml^{-1}), assuming a layer 5 μm thick, to allow calculation on a volume basis for comparison with the populations (10^4–10^5 ml^{-1}) in the water, suggest a source from the plants themselves. Plants secrete organic matter under experimental conditions (see below) and ooze it if mechanically damaged. Colonizing bacteria may also help themselves by causing changes in the epidermal cells of the host – for example, internal swellings of the wall [827] and erosion of the cuticle [437].

8.2.2 Epiphytic algae

Dense communities of photosynthetic algae dominate the periphyton [6] (Fig. 8.7). Tangled, perhaps, in deposited calcium carbonate (marl), derived

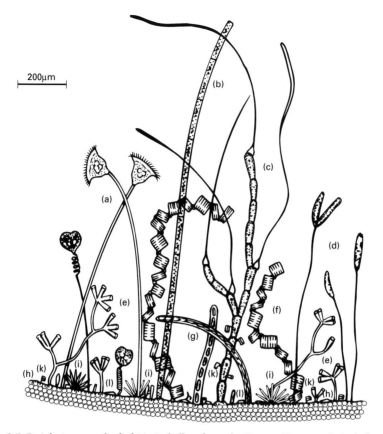

Fig. 8.7 Periphyton on a leaf of *Myriophyllum* from the Lunzer Untersee. It includes broadly attached diatoms (*Cocconeis*, l, and *Epithemia*, h), diatoms with mucilaginous stalks (*Synedra*, i, *Tabellaria*, f, and *Achnanthes*, k), sometimes branched (*Cymbella*, d, *Gomphonema*, e) and diatoms in tubes (*Encyonema*, g). There are long, filamentous, green algae (*Oedogonium*, b, and *Bulbochaete*, c) and Protozoa (*Vorticella*, a). Bacteria are not shown but are always present. (From Ruttner [839].)

from photosynthetic use of bicarbonate at the plant surface and mucous secretions of the bacteria, these epiphytic communities have a plant substratum that is not inert. Aquatic plants secrete organic compounds [761, 1013]. Such substances can be taken up by some algae, as well as by bacteria, in laboratory culture [9], but their importance in natural situations is hard to estimate. Allen [9] showed that $^{14}CO_2$, supplied in polyethylene bags sealed over the flowering stem, above the water surface, to naturally rooted *Scirpus subterminalis* in Lawrence Lake, Michigan, does appear in organic form in water close to plants whose epiphytes had been gently scraped off, or in the epiphytes themselves. In the former case, labelled organic matter could still be detected up to 3 m from the plants after 2 h.

Provision of such organic matter must help stimulate the growth of the epiphytic algal and bacterial community, which in turn absorbs much of the light that would otherwise be available to a submerged plant, thus retarding its growth [770, 827, 848, 850]. It may also compete for carbon dioxide, the diffusion problems of microalgae being smaller than for bulky plants. In *Lobelia dortmanna*, epiphytes cut off 67–82% of the light available and changed the spectral nature of the light penetrating to the leaf surface. Sand-Jensen and Borum [848] calculated that the plant, at present growing at 1 m depth or above, could have penetrated to 3.5 m without its epiphytes. It seems surprising, therefore, that plants should not, in the course of their evolution, have acquired means of restricting epiphyte colonization.

Some large algae, such as *Spirogyra* and *Zygnema*, produce mucilaginous outer cell walls, which almost entirely prevent epiphyte colonization, but these are exceptions. Vascular plants, mosses, liverworts and stoneworts are often thickly covered. Hutchinson [455] suggested that an epiphyte community may be advantageous, in that it diverts the activity of invertebrate grazers from the host plant itself. The plants do produce alkaloids and other substances, whose taste may discourage the grazers [748], but the view that aquatic plants are not much grazed has been challenged [478].

The epiphyte community includes most phyla of algae, although diatoms and green algae are often predominant. Some are attached by stalks or mucilage pads, others have basal systems of cells from which branched filaments emerge and yet others are free-living and loosely associated or embedded in marl or mucilage. Skeins of filamentous algae may stretch from leaf to leaf and shoals of flagellates may move through the quiet waters created in the interstices of this structure. Chudbya [151] found 220 algal species as epiphytes on blanket weed, *Cladophora glomerata*, in Poland.

This intricate community is easily disturbed and must be sampled delicately. The complex geometry of aquatic plant surfaces also makes it difficult to determine the surface area precisely for expression of the density of the periphyton community. These problems have stimulated the use of artificial substrata (glass or perspex slides, plastic netting or plastic aquarium plants) in experiments on epiphytes (Fig. 8.8). These substrata are much more convenient to use than living plants, but might not reproduce the metabolic effects of the real host. People using them have thus tried to show that such metabolic effects are negligible, and that live plants are simply platforms for growth, replaceable by the artificial substrata.

A study made largely on *Potamogeton richardsonii* in Lake Memphremagog, on the US–Canadian border, used plastic aquarium plants as replicas [138]. No significant differences from live plants were found in biomass of epiphytes per unit area (measured as the chlorophyll *a* concentration of cells that could be shaken from the surfaces) or the rate of photosynthetic uptake of ^{14}C-labelled CO_2 per unit area or per unit chlorophyll *a*. The latter is a measure of the physiological status of the epiphyte cells and might be expected to be affected by the plants' activity.

The activity of an enzyme, alkaline phosphatase, in the epiphyte commu-

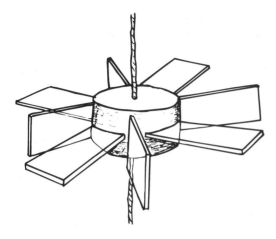

Fig. 8.8 Various artificial substrata have been used to sample periphyton, including plastic replicas of plants and, shown here, microscope slides inserted into a block of cork and suspended from a buoy.

nities was also measured. Alkaline phosphatase breaks down organically linked phosphate and is produced when easily available inorganic phosphate is scarce. The epiphytes on plastic plants in Lake Memphremagog showed higher activity of alkaline phosphatase than those on the live plants, suggesting that the latter were providing phosphate for their epiphytes. A later study [123] measured the release of phosphate from nine species of aquatic plants, grown in the presence of the isotope phosphorus-32 in aquaria and then replaced in the lake. The authors use the word 'only'– others might say 'as much as' – 3.4–9.0% of the P taken up by the epiphytes came from the host, and there was also some release of phosphorus to the water. Over 90% of this came from the epiphytes rather than the plant. In contrast, other studies [658] suggest that as much as 60% of the phosphorus taken up by epiphytic diatoms close to the plant surface can come from the host plant, although on average values were lower than 10%, falling to less than 5% at the outer fringe of the epiphyte layer.

The Lake Memphremagog studies did not seek information on community composition. Use of artificial surfaces implies that the colonizing community will be similar to the natural one in composition. Such expectations are never completely realized. Some studies show evidence of substantial determination of the epiphyte community by the plant itself [245, 741, 791]. Tippett [963] found fewer species of diatoms grown on glass slides than on adjacent plants, with maximum abundance at different times of the year on the different substrata. Fitzgerald [272] found that the nitrogen status of a host filamentous alga, *Cladophora*, directly affected the degree of epiphyte colonization. The more nitrogen it had, the more epiphytes.

Despite this evidence of an important effect of the plant on its epiphytes, many comparisons have shown that communities developing on submerged glass slides are often similar, at least in their more abundant species, to those on nearby plants [455]. Many of the studies made have been poorly controlled, in that the external microhabitat inevitably differs among the live plants and the artificial substrata, replication has often been low and access to grazers has not been standardized. In the most comprehensive and best-controlled study to date, Jones *et al.* [505] used two live species (*Elodea nutallii* and *Littorella uniflora*) and plastic replicas of them, planted in standard buckets in a controlled-environment room. They treated the substrata to three levels of nutrients and to the presence or absence of a snail, *Physa fontinalis*. The nature of the plant surfaces had little effect on the physiology or composition of the periphyton; the major effect was of the snail grazing.

This particular issue will not be resolved simply. The nature of the substrate may be important in some circumstances and irrelevant in others. A highly fertile water, for example, may supply so much nutrient to the epiphytes that any source from the plant becomes insignificant. In an infertile water, the nature of the plant host may be far more important.

8.2.3 Invertebrates

Invertebrates abound in submerged plant beds, but species lists do little to explain the nature of the community, because they do not tell of the distribution of the animals one to another and *vis-à-vis* the plant super-structure. In general, particular animals are not specifically associated with particular plants, although an apparent association, governed by differences in environment that mutually affect both groups, may be found.

There may not yet have been time for evolutionary development of such associations or the habitat may be too unpredictable, but the highly specific diets of many plant-associated terrestrial insects are not shared by aquatic invertebrates. The aquatic plant is grazed to some extent, but it is the associated periphyton and detritus that comprises most of the diet of plant-associated invertebrates. More species or higher biomasses may be associated with complex plant structures, especially if highly dissected [227, 524, 535], than with simple ones, but this is not always the case [187, 531].

Permanence of the plant and the amount of periphyton and detritus associated with it [937] will complicate the issue. Overwintering plants may have a very different fauna from those that die back in autumn, simply because they provide a permanent platform [532]. Two plants of similar structural complexity, *Chara tomentosa* (wintergreen) and *Nitellopsis obtusa* (autumn dieback) were compared [379]. The perennial had a diverse fauna, with many *Asellus* and *Gammarus*, together with chironomids, while the annual mostly had chironomids, which can disperse easily within a lake. Newly grown *Chara* was also chironomid-dominated but acquired the more diverse fauna as it matured.

Sampling and spatial distribution

Most sampling of plant beds for invertebrates is like the manoeuvring of some huge excavator over a city. Its jaws would enclose, for later sorting, the displaced office-workers, ice-cream vendors, shop assistants, teachers and students, hairdressers and many others among the rubble of high-rise office blocks, street furniture, department stores, schools and beauty parlours. Little can be deduced of the life of the city from this.

Commonly used samplers include nets, grabs and washings from artificial substrats. A coarse net swept through the beds misses many of the smaller (< 2 mm) crustaceans that move among the plants, while a fine net creates so much resistance that many animals move out of the way. Any net is selective; animals like leeches, which cling to stems and leaves, tend to be underrepresented, while those that normally move around are overemphasized. Dredges, such as the Birge–Ekman dredge, sample a standard area of plant bed by compressing the plants and then cutting them against the bottom between powerful spring-loaded jaws. They are likely to underestimate animals that can move away during descent of the grab. Macan [597] (Fig. 8.9) invented a rotary cutter, comprising two concentric cylinders, with teeth at their lower edges, which can move against one another when the cylinders are counterrotated by the observer. A cylinder-shaped section of vegetation is thus cut out and is retained in the cylinder by a plug of peat or sediment at the bottom and an airtight disc screwed to the top. Even with this sampler, however, there is considerable disturbance. The mere emptying of it destroys the spatial relationships of the plant and many animal species. It also mixes together species that are sediment-living and those that are plant-associated. There are distinct communities, with distinctive diets.

Heterogeneity in plant beds is a major problem for quantitative sampling

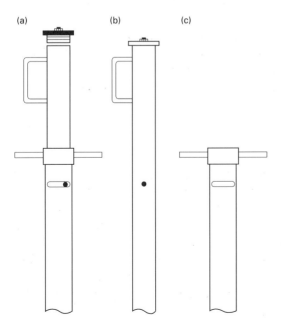

Fig. 8.9 Sampling of plant-associated invertebrates inevitably disturbs important spatial relationships. This sampler uses an outer tube (c) to delimit an area of plant bed. Tube (b) is then lowered into it and rotated to cut plant roots and rhizomes. A screw capped lid with a rubber washer is then fitted at the top (a) to create a seal and the core tube is pulled out and the contents sorted. (Based on Macan [597].)

and many replicates must be taken for a statistically acceptable estimate of invertebrate population size. To avoid this problem, Macan and Kitching [600] used 'artificial vegetation', made of polypropylene rope woven into a coarse mesh base, weighted with flat stones, to study the fauna of a small tarn in northern England. The polypropylene rope floats upright, giving, with 8 cm strands, a reasonable approximation to the rosette-like *Littorella uniflora* and, with 45 cm lengths, to *Carex*, which fringed much of the tarn. The mats of artificial vegetation were small enough to be retrieved with a pond net pushed under them. Although there was some bias in the composition of the communities contained in the artificial '*Carex*' compared with those in the natural vegetation, there was sufficient similarity to make this a useful technique, and it allowed demonstration of some valuable facts.

First, the density of artificial leaves determined the sizes of many animal populations, the total increasing greatly with thickness of vegetation. Secondly, the length of the 'leaves' was important. Nymphs of *Leptophlebia* and *Cloeon* (mayflies) and *Gammarus* (Crustacea) were much less abundant when the 'leaves' were shortened from 48 to 8 cm, thus indicating a preference by these animals for the upper parts of the long leaves and hinting at the distributional complexity which other samplers destroy.

The problem experienced with epiphytic algae in using artificial substrata of an inert versus a living platform may also apply to the colonization of the invertebrates. Again, some studies [909] have found few difficulties, while many ecologists will be doubly nervous of artificial substrata, especially where invertebrates grazing on the periphyton are concerned. One compromise for the sampling problem is to take many small samples of the invertebrates and plants by gently enclosing small plastic boxes around parts of the plant. A relationship between numbers of animals and amount of plant can be determined by regression analysis and used to predict the overall population from separate measurements of the gross plant biomass. Downing [219] found this approach to give higher estimates for more active animals (mites, amphipods, Cladocera, copepods, ostracods and caddis larvae) than grab or net sampling.

Diets

The diets of plant-bed animals range from periphyton to other invertebrates. Periphyton, because it comprises a mixture of epiphytic algae, bacteria, protozoa and detritus, provides a diet at least twice as rich in amino acids as underlying sediment. Chironomid larvae may move from their winter sedimentary habitat to feed on stems coated with periphyton in spring [533, 638]. Other periphyton feeders (Fig. 8.10) include mayfly (*Leptophlebia vespertina* and *Cloeon dipterum*) and stonefly nymphs, freshwater shrimps (*Gammarus pulex*), snails and caddis-fly larvae (Trichoptera), all of which normally cling to stems and leaves.

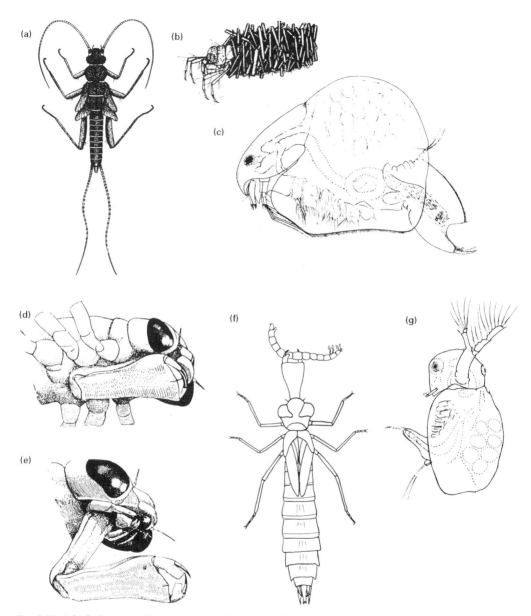

Fig. 8.10 A high diversity of organisms inhabits plant beds. Some are feeders on periphyton, such as the stonefly nymph, *Nemurella picteti* (a) and the caddis-fly larva, *Limnephilus rhombicus* (b). Others take resuspended algae and detritus, swimming freely (*Eurycercus lamellatus*, (c) or attaching to plants by a sucker at the back of the head (*Sida crystallina*, (g) to do so. Yet others are predators, such as the nymphs of the dragonfly, *Aeschna* (d–f), which lurk among the vegetation and then extend a toothed 'mask' to strike at passing prey, such as chironomid fly larvae. (Based on Fryer [302].)

In one Oxfordshire pond, Lodge [578] found allochthonous leaf litter to have few snails, but the aquatic plants to be associated with particular snail species. *Acroloxus lacustris*, a limpet, was most common on the smooth stems of emergent plants and the petioles of white water lilies (*Nymphaea alba*). *Planorbis vortex* was associated with aquatic grasses, such as *Glyceria*, and *Lymnaea peregra* with *Elodea canadensis*. For the latter two combinations, experiments showed a particular preference of the snails for the epiphytes specific to the plants: filamentous green algae on *Elodea* and diatoms and associated detritus on *Glyceria*. Grazing is very active, and the turnover rate of the periphyton food may be as much as 22–45% day^{-1} [513].

Grazing, in general, is an important determinant of the biomass of the periphyton growing on plants [504] and hence of importance in determining the production and survival of the host plant (see below). The grazing rate depends largely on the body size of the grazer and, when corrected for this, grazing by snails, caddis-flies, mayflies and chironomids is more or less equally effective [139]. As much or more of the periphyton may simply be dislodged as actually eaten.

Clinging, swimming and sometimes filtering fine material dislodged from the periphyton or swept in from elsewhere are small Crustacea (Fig. 8.10) (cladocerans, cyclopoid and harpacticoid copepods, ostracods). Some larger feeders on fine particles, such as sponges (Porifera), attach to the plants. These may be important filterers, helping in maintain an adequate light climate for the plants.

Carnivores in the plant beds include lurkers and hunters (Fig. 8.10), although the categories may overlap. The former remain stationary among the leaves, waiting for prey to pass. *Hydra*, a small (2–3 mm) coelenterate, extends tentacles armed with stinging cells, which can immobilize prey, such as small Crustacea, accidentally brushing against them. Leeches attach with a basal sucker, transfer to passing invertebrates, water birds or fish and extract blood or other fluids. Other lurkers include the larvae of the alder fly, *Sialis* (Megaloptera), and a stonefly (*Nemoura cinerea*) and the nymphs of damsel- and dragonflies (Odonata), which have a hinged lower mouthpart, or labrum, which is folded back under the head, but which can be snapped forward to seize the prey with two teeth borne at its tip. Yet others include caddis-fly larvae, such as *Phryganea*, whose heavy cases, made from pieces of reed leaves or roots, hamper movement and ensure that only slow-moving prey or eggs laid on the vegetation are taken. Members of another caddis family, the Polycentropidae, weave funnel-shaped nets among the leaves and stems and wait for the prey to become entangled.

Lurking allows a predator to remain concealed, not only from its prey, but also from its own predators, such as fish, but has the disadvantage that prey may not pass very often. Hunting reverses these features, with greater exposure to predators but readier availability of prey. The quieter water often to be found among vegetation allows hunting insects that live on the surface-tension film to operate. These include bugs (Hemiptera), such as the water-strider (*Gerris*) and the water-measurer (*Hydrometra*), which spear small crustaceans swimming just below the surface. The water-boatman, or backswimmer (*Notonecta*), feeds on aerial insects that become trapped in the surface-tension film. It hangs from the film by the tip of its abdomen, moves with paddle-like hind legs and seizes its prey with its front and middle legs before impaling it with its sharp, beak-like mouthparts.

Water-beetles (Coleoptera), both as adults and larvae, are voracious hunters; the adults are able to replenish, at the water surface, bubbles of air underneath their wing covers. With this rich supply of oxygen, their activity is increased. (Lurking carnivores, except water-spiders, which live in a 'diving-bell' of silk containing a bubble of air, are dependent on the more restricted supply of dissolved oxygen.)

There are larger predators, including many fish. Introduction of trout to a small tarn resulted in a reduction in numbers of water-bugs and a confinement of some species to the thicker parts of the vegetation [598]. Two further species, however, increased their numbers, perhaps because of reduced competition among the sparser community of water-bugs. Tench (*Tinca tinca*) and some sunfish feed particularly on snails, which are major grazers of periphyton.

8.3 Competition between submerged plants and phytoplankton

As overlying phytoplankton crops increase, through eutrophication, aquatic plants may become restricted in their distribution or disappear altogether. But it is not certain that shading is the factor responsible. The competition between plants and plankton is not a simple matter.

Infertile lakes often have short aquatic-plant species, including *Isoetes* and *Littorella*, where catchments are of hard rock and the sediments sandy and Charophytes and *Naias marina* in highly calcareous marl lakes, in which phosphorus is often scarce. On eutrophication, greater crops of aquatic plants tend to grow at first, but not of the short-statured forms. The larger pondweeds (*Potamogeton, Myriophyllum, Hippuris, Ceratophyllum*) tend to displace them [126].

Increased nutrient supply may stimulate growth of filamentous algae and epiphytes over the surfaces of the shorter plants [770, 892], thus cutting off light and decreasing their competitive abilities. The taller plants may need more nutrients [146] and may also have greater access to nutrients in the water than the shorter forms, because of their generally greater surface-area-to-volume ratio. The taller plants grow rapidly to the water surface, obviating the shading effects of moderate periphyton algal populations and, if heavy growths develop on their older leaves, the leaves are readily sloughed off and replaced. The replacement of the short plants is thus easily understood.

Despite increased nutrient loading, phytoplankton populations do not often grow to a large extent at first, especially in small lakes where aquatic plants cover large areas. The reason may be to do with competitive mechanisms between the plants and phytoplankton. Aquatic plants may secrete inhibitors that suppress phytoplankton growth [982, 1026]. Fertilization of enclosures or experimental ponds containing aquatic plants has resulted in little change with no increase in phytoplankton, though sometimes some increase of filamentous algae [33, 434]. The plants may take up the extra nutrient, especially nitrogen, as 'luxury consumption', or algal development may be prevented by cladoceran and other grazers, harboured in and around the beds [483, 549, 550, 876, 961].

The beds provide dark refuges, in which fish predation on the grazers is less efficient, allowing large stocks to build up, even if small fish are present, hiding from their own predators. At their densest, the beds may exclude fish, because oxygen concentrations within them may become quite low. The mass of organic matter and reduced mixing allow stratification to establish, often with high ammonia concentrations and near-deoxygenation at the bottom. Grazing of periphyton by invertebrates is also important [139, 505]. Snails keep plant surfaces clear for photosynthesis and, when snails are experimentally removed, plant growth is reduced [96, 97, 978].

Once established, an aquatic plant bed has considerable buffering powers, preserving it against changing nutrient loading, at least up to some critical point. Although the evidence is anecdotal, the change from plant dominance to phytoplankton dominance that sometimes occurs during eutrophication may take place relatively rapidly. It is as if a switch had been thrown, moving the ecosystem from one alternative state to another, with little or no change in external conditions.

Once this change has occurred and the aquatic plants have been replaced by phytoplankton, which is then able to take advantage of the abundant nutrients, a new set of stabilizing buffers comes into operation. Phytoplankton can grow earlier in the year, at lower temperatures, than the plants and, though circulated through the water column, on average, receives more light than aquatic plants developing on the bottom. The light compensation point (gross photosynthesis = respiration) for small cells is lower than for bulky organisms, as also is the CO_2 compensation point. Diffusion precludes rapid uptake into bulky tissues, so that, once established, the algae may compete very effectively for CO_2 [7, 594]. Uptake of CO_2, as it forces the pH to high values, may accentuate the problem [895].

Open water, particularly in shallow lakes, offers few refuges to large cladoceran grazers from fish predation. Deeper lakes may have refuges in the depths, but, where aquatic plants would otherwise be able to grow abundantly (less than a few metres), fish predation will be possible

throughout the water column. Once the switch to phytoplankton dominance has taken place, therefore, there is less possibility of grazer control of
the algal crops.

The problem then becomes one of what happens to operate the switch
from plants to plankton. Aquatic plants can withstand severe eutrophication
by using the buffers inherent in the community. It is not known whether
they can hold out indefinitely, but experimental attempts to cause the
switch to phytoplankton by fertilization alone have usually failed [33, 434].
Sometimes attempts to change phytoplankton dominance to plant dominance by reducing nutrient supplies, in Lund tubes (see Chapter 6) [683,
693] and entire lakes [679, 686], have also failed.

The buffers should operate if the plant community is intact, but not if it
is damaged, particularly at vulnerable times, such as spring, when growth
is starting, or autumn, when the overwintering turions and seeds have to be
produced. Mechanical clearance, boat damage, grazing by vertebrates,
particularly exotic ones with burgeoning populations (geese, coot, coypu
and common and grass carp), and deliberate use or accidental run-off of
herbicides can all precipitate loss of the plants.

Alternatively, the grazer community may be destroyed for long enough
for the phytoplankton to gain advantage. Eutrophication through intensification of agriculture is also linked with greater use of pesticides. The
Cladocera are particularly vulnerable to very low levels of some of these
[337, 884]. In the Norfolk Broads, a switch to phytoplankton took place
generally in the 1950s and 1960s, when organochlorine pesticides were
used with less control than at present, and there is evidence of a correlation
between pesticide residues and changes in the cladoceran community
recorded in the underlying sediments. In lakes close to the sea, such as
Hickling Broad, in Norfolk, UK, drainage of the surrounding land may result
in salt water being pumped into the lake. Daphnids are very sensitive to salt
and die at chlorinities greater than about $1000 \, mg \, l^{-1}$ (5% of sea water). In
Hickling Broad, a nutrient problem caused by large numbers of roosting
gulls has now solved itself by the gulls moving away, but the saline lake
water, despite conditions otherwise favourable to plants, remains phytoplankton-dominated for lack of daphnids [31].

8.3.1 Consequences of the loss of aquatic plants

Plant-dominated communities in lakes have important roles, which are lost
with a switch to plankton dominance. Recruitment of fishes, such as perch
(*Perca fluviatilis*) and tench (*Tinca tinca*), which lay their eggs on plants, may
be limited; populations of large invertebrates – snails and dragonfly nymphs,
for example – used by larger fish may dwindle, to the detriment of fish
growth, and birds may disappear. The herbivorous coot (*Fulica atra*) was
once so abundant on Hickling Broad in Norfolk that thousands were shot

annually [683]. The population fell to a few tens in the 1980s, after loss of the plants.

Loss of plants also removes amenity value; pea soups of algae are not so attractive as water lilies. And they are associated with other problems – blue-green algal toxicity, avian botulism and fish kills. Loss of submerged plants exposes the backing emergent plants to wave action, and this may lead to erosion of banks, because plants absorb the energy of wind and boats. In flat, coastal areas, such as the Netherlands, where lakes have been created as part of the process of reclaiming land from the sea, this may be serious, because protective engineering structures are expensive and unattractive.

Areas of reed (*Phragmites australis*) have disappeared (Fig. 8.11) from shallow lakes for a variety of reasons, most ultimately linked with eutrophication. If subjected to high nitrogen loading, floating mats produce greater amounts of shoot tissue and lesser amounts of the root and rhizome material that supports them. They may then become less stable and liable to break up under grazing pressure from geese or from other forms of disturbance. The sediments under rooted growths may become so severely

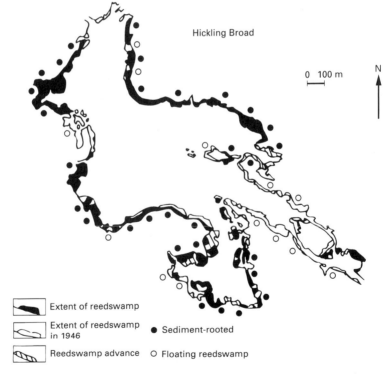

Fig. 8.11 Reed has been lost from many lakes for a variety of reasons. Losses from Hickling Broad have largely been from floating stands, which have become top-heavy, due to the vigorous growth resulting from nitrogen pollution of the lake. (Based on Boar *et al.* [75].)

inimical, through accumulation of sulphide, that the abilities of emergent plants to cope with sediment toxicity and deoxygenation (see Chapter 5) may be exceeded. Major losses of such swamps reduce the conservation value of a lake, because of the loss of habitat, particularly for birds. There are thus many reasons for wishing to restore plant dominance to a shallow lake and a strategy has been developed, based on understanding of the processes concerned, to facilitate this.

8.3.2 Restoration of shallow lakes back to plant dominance

The strategy for restoration of submerged and floating-leaved vegetation is based on the concept of alternative stable states, discussed in Chapter 6 [693]. It depends on a diagnosis of the forward switches that have been operating to destroy the plants and others that might be threatening, and an assessment of whether they can be removed. If they cannot, there is no point in proceeding. If they can, an assessment of the nutrient budget is sensible, followed by measures to reduce nutrient loading, for it is likely that these systems are more easy to switch back to plant dominance at low nutrient levels than at high ones [483]. Ideally, both nitrogen and phosphorus should be controlled (see Chapter 6). Biomanipulation is the next step and can mean addition of piscivores, although it generally involves removal of as many zooplanktivorous fish as possible for a few years. This may mean isolation of the lake from other sources of fish, such as interconnecting rivers. The populations of *Daphnia* usually respond rapidly and the water clears. Plants may redevelop from fragments or they may have to be reintroduced and perhaps protected, in the early, vulnerable stages, from damage by grazing birds, especially coot and swans. Finally, when healthy growth of plants has re-established, a new fish community can be introduced. This should not contain fish like the common carp, which are destructive of plants and stir up sediment.

The strategy has been constructed from attempts at restoration in Europe on both large and small lakes. None has followed it completely, for it has been developed with the benefit of hindsight; nonetheless, there have been some reasonable successes (Table 8.2 & Fig. 8.12) and some reversions. Mostly these latter have been because the background nutrient loading could not be sufficiently controlled or insufficient fish were removed at the biomanipulation stage.

Where emergent vegetation has been lost, some of the same principles apply. Forward switches must be removed or counteracted. This may mean lessening or diversion of boat traffic or mechanical protection of the emergent beds against wave erosion. Control of nitrogen loading is desirable. Biomanipulation will be needed only if very damaging fish, such as common or, particularly, grass carp, are present, but it may be necessary to stabilize the bank edges with matting ('geotextiles') for plants to establish.

Table 8.2 Summary of some restoration attempts on shallow lakes in Europe which have used biomanipulation. Based on Moss et al. [693]

Site	Country	Area (ha)	Cause of problem		Target	Approach	Current degree of success
			Forward switch	Nutrient sources			
Cockshoot Broad	UK	3.3	?Pesticides	Sewage effluent Agriculture (high)	Clear water Submerged plants	Isolation Sediment removal Biomanipulation (fish removal)	High
Lake Vaeng	Denmark	15	?	Sewage effluent (moderate)	Clear water Submerged plants	Biomanipulation (fish removal)	High
Zwemlust	Netherlands	1.5	Herbicides	Sewage effluent (very severe)	Clear water Submerged plants	Biomanipulation (fish removal and piscivore stocking)	High but deteriorating
Lake Wolderwijd	Netherlands	2700	?Salinity change	Sewage effluent Agriculture (moderate)	Clear water Submerged plants	Biomanipulation (fish removal and piscivore stocking)	Partial
Ormesby Broad	UK	54	Plants initially present	Agriculture (moderate)	Charophytes	Biomanipulation (fish removal)	Moderate
Lake Finjasjön	Sweden	1100	?	Sewage effluent (severe)	Clear water Submerged plants	Effluent stripping Sediment removal Biomanipulation (fish removal) Wetlands and buffer zones	High

Fig. 8.12 Cockshoot Broad from the air and the ground, after restoration of its aquatic-plant communities. The broad was isolated from the river by a dam and biomanipulated in the late 1980s.

Generally, the new plants will have to be reintroduced, usually as planted rhizome fragments, and choice of plant is important. Table 8.3 gives a comparative account of usable plants, based on Dutch experience [162, 163].

8.3.3 Ponds and pond loss

The ultimate problem in shallow-lake restoration is to counteract what is happening in many agricultural areas to small ponds. There is no formal definition of a pond, but they are usually small (< 2 ha), shallow and potentially completely littoral in nature. The word is also used for quite large lakes in the USA. European ponds were mostly dug to provide watering-places for farmstock or to extract calcareous drift material for liming the more acid fields, clay for bricks, sand for cement and mortar, or gravel for hard standings and tracks. Although each pond represents only a small part of the landscape, the tendency for them to occur in clusters, linked formerly by only lightly used land, created a distinctive 'patch' ecology, particularly valuable for amphibians. Many ponds, being isolated, are fishless and amphibian tadpoles do not easily coexist with predatory fish.

The number of ponds present in the nineteenth century has now been more than halved [79, 399] (Fig. 8.13) and, at present rates of loss, there will soon be few left. The reason is partly that they have been filled in, because farming practices have become more intensive and cattle grazing has been replaced by arable cultivation. Partly it is because stock are now watered through pipelines to drinking troughs and the ponds have not been maintained. With time, they inevitably silt up. The distribution of newts and frogs in the UK closely echoes the distribution of ponds and their loss is one of the reasons for declining amphibian populations.

Some new ponds and lakes are being created by gravel extraction for construction and road building and as storage for irrigation water, but at

Table 8.3 Characteristics of emergent plant species of relevance in choosing species for restoration projects. Species are arranged in order of their tolerance to increasing water depth. From Coops and Geilen [163].

Yellow iris (*Iris pseudacorus*)

Very tolerant of sediment type; prefers infrequent flooding, so plant at upper end of littoral zone. Seeds few but large with long buoyancy, so easily dispersed. Seeds germinate on wet but not totally flooded surfaces and the seedlings are intolerant of flooding. Spreads by large, erosion-resistant rhizomes in occasionally flooded soils. Best propagated in restoration projects by rhizomes rather than seeds

Reed canary grass (*Phalaris arundinacea*)

Tolerant of soil type but prefers sandy sediments; grows high on shore and prefers infrequent flooding. Produces many small seeds, which have brief buoyancy and germinate on wet but not flooded surfaces and are intolerant of flooding. Spreads rapidly by runners, which are easily eroded. Can be propagated by rooted cuttings or seeds sown on bare soil. Shade-tolerant and useful for exposed but infrequently flooded shores

Common reed (*Phragmites australia*)

Tolerant of a wide range of water depths and soils but prefers relatively fertile conditions. Produces many plumed seeds, which are very buoyant and germinate on wet surfaces, though usually very poorly. Spreads easily in dry ground or shallow water by rhizomes. Tolerant of wave exposure, especially in dense stands, and can form floating mats. Best propagated by planting of sections of rhizome or rooted cuttings at water depths of 0–50 cm

Reed sweetgrass (*Glyceria maxima*)

Prefers conditions around the water-line and shallow water, with silty and organic rich soils in fertile conditions. Produces a small number of small seeds with brief buoyancy, which germinate on wet surfaces and are reasonably flood-tolerant. Expands rapidly by rhizomes in shallow (< 5 cm) water and is tolerant of wave exposure, especially in dense stands. Best propagated by rooted cuttings at water depths of 0–30 cm on nutrient-rich soils. Tolerant of both shading and flooding

Bulrush, reedmace, cattail (*Typha latifolia*)

Prefers shallow waters with nutrient-rich, organic soils or unconsolidated silt. Produces many plumed seeds, easily distributed by wind and also very buoyant. Germination on wet and shallowly flooded surfaces. Spreads rapidly by rhizomes in shallow water (to 0.5 m). Rhizomes easily washed out of loose mud. Best planted as rhizomes in and just below the water-line. Susceptible to shading

Lesser bulrush, reedmace, cattail (*Typha angustifolia*)

Prefers permanent flooding on silty soils and peat. Produces many, small, plumed, wind-borne seeds with short buoyancy, which germinate on wet and shallowly flooded surfaces. Spreads by rhizomes in water up to 1 m deep, but is susceptible to wave damage. Rhizomes grow shallowly and are easily washed out. Best planted at 0–80 cm on silty soils and is susceptible to shading

Common club-rush, true bulrush (*Scirpus lacustris*)

Prefers permanent flooding, even tidal inundation in estuaries, on silty, organic soils, preferably with heavy new sedimentation. Produces few but large, poorly buoyant seeds, dispersed by water or animals. Seed germination on wet or shallowly flooded surfaces and spreads but slowly in water depths to 1 m. Tolerant of wave erosion, although rhizomes can be washed out in shallow water. Best planted by rhizomes or sods in 0–0.5 (1.0) m of water in fertile conditions.

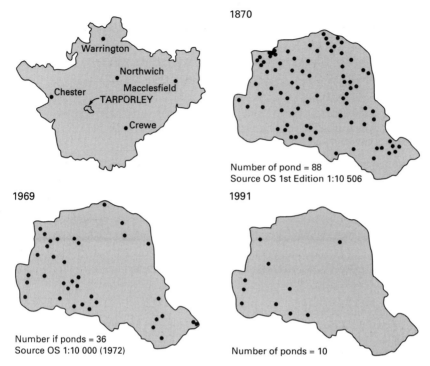

Fig. 8.13 Ponds are now disappearing from the landscape, as shown by changes in numbers in a single parish in Cheshire, UK. OS, Ordnance Survey. (From Boothby *et al.* [79].)

nothing like the rate of loss. Nonetheless, the new gravel workings can become extremely valuable wildlife habitats, particularly when the gravel is removed 'dry' from a pumped pit [330, 332]. If it is extracted from a flooded pit, the water tends to be turbid from disturbed clays and, when the pit is abandoned, often remains turbid, because aquatic plants cannot colonize, because of the poor light climate. Where it is kept dry but later flooded, aquatic plant colonization and clear water may be achieved.

8.4 Bare rocks and sandy littoral habitats

Wave action on a shore creates a habitat superficially similar to that of rocky streams (see Chapter 4), but the processing of allochthonous organic matter is not nearly so important as in streams. The food base is the community of algae, bacteria and protozoa that covers the stones. Scrapers and predators dominate the invertebrate fauna. This is to be expected, because wave action will move any loose organic matter down the slope of the lake shore to deeper sediments. Rocky lake-shore animals include nymphs of mayflies and stoneflies, triclads (flatworms), snails and limpets, *Asellus*, *Gammarus* and other Crustacea, caddis larvae and others. The shore is never uniform and protected pockets, where sediment may accumulate, leaves lodge or aquatic plants grow, will have a somewhat different fauna, characteristic of such conditions.

What determines the composition of the invertebrate community? Physicochemical factors, chance, predation and competition all contribute, as illustrated by the distribution of the triclads, or carnivorous flatworms.

8.4.1 Distribution of triclads in the British Isles

Flatworms occur in weed beds, as well as on rocky shores, but are certainly characteristic of the latter. Of the 11 British species, four – *Polycelis tenuis*, *Polycelis nigra*, *Dugesia polychroa* (then referred to as *Dugesia lugubris*) and *Dendrocoelum lacteum* (Fig. 8.14) – have been collectively studied [813, 814].

The first task was to describe the distribution of the four species in the British Isles. Flatworms were sought by hand on the rocky shores of 200 lakes. The searching time was standardized and the method is the best available, for flatworms are delicate, readily damaged and not easily dislodged by disturbance into nets. The total number of flatworms collected per unit time increased directly with calcium and total dissolved solids concentrations in the water. At high calcium ion concentrations, *D. polychroa* and *D. lacteum* were predominant, but, at $< 20 \, \text{mg Ca}^{2+} \, \text{l}^{-1}$, they declined and were usually absent from water with $< 5 \, \text{mg Ca}^{2+} \, \text{l}^{-1}$. At low

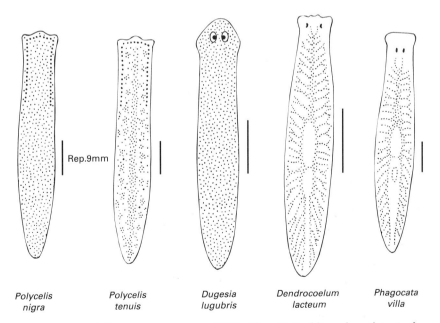

| *Polycelis* | *Polycelis* | *Dugesia* | *Dendrocoelum* | *Phagocata* |
| nigra | tenuis | lugubris | lacteum | villa |

Fig. 8.14 Five triclad species are native to British lakes. Vertical lines show the actual average sizes of adults. (From Reynoldson [813].)

calcium concentrations above 5 mg l^{-1}, *P. tenuis* was most abundant, but below 5 mg l^{-1}, if flatworms were present at all, *P. nigra* was either predominant or the sole flatworm species present.

This pattern reflects dispersal of the animals and recolonization of the lakes after the end of the last glaciation. When abundant food (slices of earthworm) was supplied, waters with the lowest concentrations of calcium and total dissolved solids supported breeding populations of all four species indefinitely. Lakes poorest in calcium tend to be at the furthest northern reaches of the British Isles and, in northern Scotland and the offshore islands, more lakes than expected had no triclads at all. Part of the triclad distribution pattern at low calcium levels may thus simply reflect a lack of time for colonization of the softer northern waters to have taken place.

Other features explain the change from *P. tenuis* predominance to that of *Dugesia* and *Dendrocoelum* in lakes where all four species potentially can occur. Predation on triclads by other organisms was unimportant [193]. Of seventy-five potential predators tested by serological methods, only a leech, *Erpobdella octoculata*, and a caddis larva, *Polycentropus flavomaculatus*, ate flatworms, but not enough to alter greatly the population size or balance of species. Serology involves testing the reaction of proteins present in the predator's gut against antibodies derived from the prey. These are produced in rabbit blood, following injection of protein from potential prey organisms. It allows the identification of prey that do not leave visually recognizable remains in the gut.

Increasing the food supply led to population increases of flatworms caged in the lake that were far greater than those of natural populations in the same waters. Competition for food might thus explain part of the distribution pattern. Concentrations of calcium and total dissolved solids, though not causing the distribution, are generally related to fertility in most lakes (see Chapter 3) and thence the production of both prey and flatworms.

Triclads feed carnivorously, by inserting their pharynx into soft parts of the prey, pumping in digestive fluids and sucking out the semiliquid consequences. All four flatworms will take a range of foods, but particular items are necessary for indefinite survival of each species. Serological work, examination of gut contents and choice experiments, where each species was presented with a range of foods, have shown that *Dendrocoelum* prefers the crustacean *Asellus*, which it actively hunts. Of the four species, it has the best-developed sensory systems and will attach to and coil around a moving nylon bristle the same diameter as an *Asellus* leg [53].

Dugesia polychroa eats snails, which are usually avoided by the others. *Polycelis tenuis* is the most active species, searching for damaged *Asellus* and attracted to it by oozing body fluids. The main food of *P. nigra* has not been determined. In experiments, the 'food refuge' of each species had to be provided for indefinite survival in either mixed or monospecific cultures, but *P. nigra* always declined, whatever food was given. In natural waters, introduction of *P. tenuis* to locations where *P. nigra* had previously been the sole flatworm species led to decline (though not disappearance) of *P. nigra* in favour of *P. tenuis*.

In lakes of increasing fertility, the proportion and abundance of *Asellus* and of snails tends to increase, and this seems to explain the coexistence of *Dugesia* and *Dendrocoelum* with *Polycelis* spp. at the higher calcium levels. This work also goes some way to explaining the changes in rocky-shore fauna noted by Macan [596] within Lake Windermere and in a series of lakes of the English Lake District (see Chapter 3).

8.4.2 Rocky-shore communities

Macan listed about thirty-six species of macroinvertebrates on the rocky shore of Lake Windermere, of which nineteen were common. The list included mayfly and stonefly nymphs, caddis larvae, beetles, Crustacea (*Gammarus pulex*, *Crangonyx pseudogracilis* and two *Asellus* species), two gastropod snails, *Ancylus fluviatilis* (freshwater limpet), six flatworm species, including all those discussed above, and nine leech species. Their distribution was not uniform among a large number of areas sampled around the lake shore. Two extreme sorts of community were found, linked by a continuum of intermediate ones.

At one extreme, there were few insects, but molluscs, crustaceans and flatworms were prominent. Stonefly (e.g. *Diura bicaudata*, *Nemoura avicularis*) and mayfly (*Ecdyonurus dispar*, *Centroptilum luteolum*, *Heptagenia lateralis*) nymphs were confined to the other extreme. All other species were found throughout the series of sites. The former community type was on shores fertilized with effluent from two sewage-treatment works, septic tanks and houseboats.

A similar pattern was recorded from a series of lakes in the area. Because of varying geology and the greater farming activity and human settlement in some catchments, these lakes form a series of changing fertility (Chapter 3) (Fig. 8.15). At the least fertile extreme, stonefly and mayfly nymphs were most abundant, while certain leeches, gastropods and flatworms reached their greatest abundance (based on a timed collection by hand and net) towards the fertile end. These distributions are linked with the interaction of fertility and predation [599]. As the fertility of a site increases, so does the productivity of the invertebrate community, including the predators. High fertility *per se* does not discriminate against nymphs of stoneflies and mayflies, but their eggs may survive less well under crowded conditions than those of animals that are common on such shores. The eggs, laid at the water surface, fall to the bottom and catch among the community of microorganisms on the stones. There they are vulnerable to the scrapers – molluscs and crustaceans – which continually work over the stones. In contrast, the eggs of molluscs are protected in lumps of jelly and

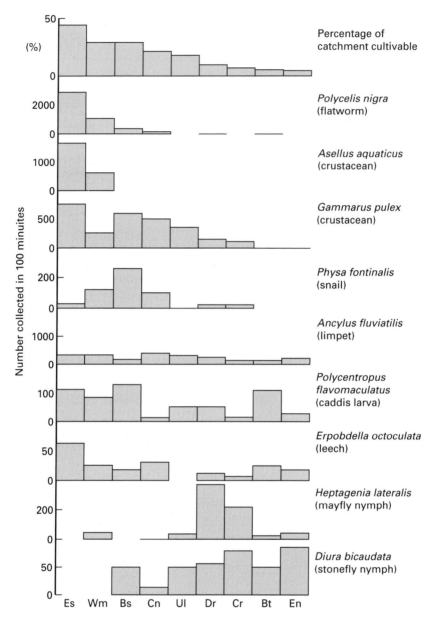

Fig. 8.15 Distribution of animals on comparable rocky shores of lakes of the English Lake District. Histograms show the number of animals of each of nine representative species collected by the same experienced observer in 100 min. A measure of the fertility of each lake is given by the percentage of its catchment area that is cultivable. Cultivability, in this case, largely reflects the availability of easily weathered rocks in the catchment. Es, Esthwaite; Wm, Windermere; Bs, Bassenthwaite; Cn, Coniston; Ul, Ullswater; Dr, Derwent Water; Cr, Crummock Water; Bt, Buttermere; En, Ennerdale Water. (Redrawn from data in Table 37 of Macan [596].)

those of leeches and flatworms in leathery cocoons; *Gammarus* and *Asellus* eggs are carried on the female's body until they hatch. Doubtless, this is only a first level of explanation of what is a complex pattern and yet it illustrates some of the factors that must be considered.

8.4.3 Specialization in the rocky littoral

Littoral habitats are far more complex than open-water ones, with an often fine division of niches. This can be illustrated by two examples from rocky shores. The chydorids (Fig. 8.16), a group of Cladocera [296, 297], includes animals so specialized that a species capable of living on rock or leaf surfaces is unable to manage in sand or sediment only centimetres away. They are animals associated with surfaces, although many can also swim. They move with their first pair of trunk limbs, collect food with their second, and manipulate it with the third and fourth and sometimes fifth pairs so that it passes forward in the ventral food groove between the legs to the mouth. The limbs are modified to handle particular foods: to scrape it, with the help of entangling excretions, in *Alonopsis elongata*, and to sweep loosely bound material, in *Disparalona* sp. The carapace is particularly important. In some species, it can be clamped down on a surface, like a

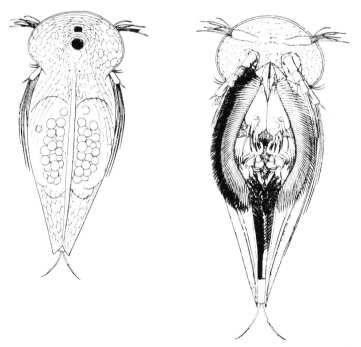

Fig. 8.16 The smaller invertebrates show intricate adaptations, no less than the larger ones. *Graptoleberis testudinaria* (dorsal side on the left, ventral on the right) is a chydorid cladoceran, with a carapace that allows it to clamp tightly to plants as it scrapes periphyton from them. (Based on Fryer [302].)

suction cup, allowing movement upside down on the underside of surfaces (e.g. *Graptoleberis testudinaria*), while others, with a flattened body, move more readily through soft deposits. One species is a parasite of *Hydra* and has armoured food grooves to obviate the effects of the latter's stinging cells; others feed as scavengers on dead Cladocera or in the surface-tension film. Some are very small (< 0.25 mm) and have access to material in the finer crevices of rocks, which bigger species (> 5 mm) do not. Fryer [301] argues cogently against the combining of such disparate animals into single categories for the purposes of creating models to explain ecosystem processes. The argument is unassailable, but generally ignored.

A second example concerns the fishes of Lake Malawi, which have evolved 'species flocks' – groups of species that are very closely related and yet distinct. Many are related to the cichlid genus *Haplochromis*, and similar flocks occur in Lakes Victoria and Tanganyika. The range of diet is not greater than elsewhere, but the specialities within the range are much narrower, presumably allowing a very efficient use of food. One flock feeds on the algae that form a felt on rock surfaces [304]. Local fishermen recognize, unwittingly, the close taxonomic relationships of these fish by calling all by one name, 'mbuna' (Fig. 8.17).

One species, *Pseudotropheus tropheops*, has small, close-set teeth, like a small file. It rasps the algal felt from vertical or steeply sloping rocks and bites off the rasped mass with large conical teeth at the edges of its mouth. In this way, it removes even firmly attached filamentous forms. *Pseudotropheus zebra*, on the other hand, is larger and more mobile than *P. tropheops*, but has longer and more spaced back teeth. It opens its mouth against the rock and combs the algae, so that only the looser algae are removed. A third species, *Pseudotropheus livingstonii*, has similar dentition and diet to *P. zebra*, but inhabits relatively deeper water.

A closely related genus contains *Petrotilapia tridentiger*, whose teeth are long, curved and flattened at the tips into spoons. The whole tooth is flexible. When the mouth is opened against the rock, the teeth comb the

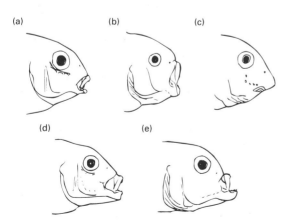

Fig. 8.17 Heads of some of the *mbuna* of Lake Malawi. (a) *Pseudotropheus tropheops*, (b) *Petrotilapia tridentiger*, (c) *Labeotropheus fuellebornii* (d) *Pseudotropheus zebra*, (e) *Genychromis mento*. (Based on Fryer and Iles [303, 304].)

looser algae, as do those of *P. zebra*, but *P. tridentiger* can feed on both vertical and horizontal rock surfaces, 'standing on its head' to do so. Lastly, *Labeotropheus fuelleborni* has rigid, strong jaws, strengthened in the plane of forward movement, and a mouth set on the lower side of the head. It hovers over horizontal rock surfaces and chisels the algae off with its teeth.

These are fine distinctions, based on attachment of the algae, angle of rock surface and water depth, but rock-scraping fish in Lake Malawi have also evolved into scrapers of other surfaces. *Genychromis mento* scrapes the body scales from sluggish bottom-living fish, but belies its evolutionary origin by occasionally rasping algae from rocks and points to future possibilities by biting pieces out of the fins of other fish. More specialized scale scrapers and fin biters are known from the lake, and scale scraping is sometimes accompanied by a mimicry that allows the scraper to move unrecognized among shoals of its prey.

8.4.4 Sandy shores

Like gravels, sands often have sparse fauna and flora; continuous agitation grinds organisms to death and washes away fine organic particles that might act as food. The least disturbed, yet still sandy, habitats do, however, have a characteristic set of communities, with microscopic algae and bacteria living freely among the sand grains and attached to them. There is also an intersitial fauna of Protozoa, nematodes, small Crustacea, oligochaetes, tardigrades (water bears) and mites.

Epipsammic algae are sometimes non-motile and sometimes of genera that can attach firmly to the grains but also move freely among them. Sandy beaches with large populations of such algae (largely diatoms) are intermittently rather than continuously disturbed by water movements. Diatoms attached to sand grains may then be buried to depths of several centimetres, where no light penetrates. The motile ones can then detach and move back to the surface, but the permanently affixed must either die or tolerate darkness for several days, perhaps weeks. There is little evidence of heterotrophic ability [704], but some that they can tolerate both darkness and deoxygenation for several days, while retaining their photosynthetic potential [676]. In contrast, epipelic algae (those that are always free-living and move in the surface layers of sand and other sediments), though tolerant of darkness if buried, are less resistant to anaerobiosis. They survive by rapid movement back to the surface. Deoxygenation is usual, even in sands, at depths of more than a few centimetres and in finer sediments below a few millimetres.

In muddier deposits under stiller conditions, epipelic algae become more important, as also does the supply of organic matter deposited from elsewhere. The community may be rich in species – several hundred epipelic algae and equally large numbers of small invertebrates, including the protozoa, rotifers, nematodes, oligochaetes and Crustacea. Strayer [932] found 322 species of animals in the sediments of one small North American lake and suspected that the total was 600. The microbial community is also complex, for a variety of redox conditions exist, not only with depth in the sediment, but also around decaying pieces of organic matter, such as leaves, animal bodies or faeces. Such habitats have within themselves all the complexity of the lake as a whole.

8.4.5 Zebra mussels – a problem

The zebra mussel, *Dreissena polymorpha* [610], is a small (2–2.5 cm), strikingly coloured animal that is native to the Black Sea. It lives on any hard substrate, often in colonies of several layers. Over the past 200 years, it has spread into northern Europe and is common in rivers like the Rhine. Around 1985 or 1986, however, it was inadvertently introduced into North America, probably in the ballast water of a cargo ship, and was found growing in the river that connects Lake St Clair with Lake Erie. By 1988, it had formed very large populations (up to 200 000 m^{-2}) in the western basin of Lake Erie and, in the early 1990s, it had similarly colonized parts of Lake Huron, upstream of Lake St Clair. By 1995, it had invaded eighteen American states and two Canadian provinces [492], largely via commercial shipping routes, and is likely to spread over most of the continent [933]. Populations in the Hudson River average 4000 m^{-2} in the freshwater tidal section, a total of 550 billion animals [934], and it is spreading down the Mississippi River. It has a planktonic larval stage (veliger), which facilitates its spreading, as well as a very wide tolerance of physical and chemical conditions [619], including those of brackish estuaries and other intertidal waters [648]. Small adult mussels can also detach and drift to new sites [635]. It can even invade soft sediments, once it has made an initial colonization of a small area of hard surface, by building out on its own shells [447].

All of this would be merely of interest were it not such a prolific animal that it now dominates the benthic communities of hard substrates wherever it has invaded, displacing native clams and other bivalves [371, 870] and making considerable inroads into the phytoplankton crops of the lakes, whose water it filters at rates far greater than those of the zooplankton [23]. Native unionid mussels may have dozens of zebra mussels smothering them. In Saginaw Bay in Lake Huron, it has reduced the phytoplankton crop by 40% and diminished the zooplankton population [256 –258, 489, 552]. It has caused major changes in water chemistry and increased the water transparency. In the western basin of Lake Erie, it has reduced the phytoplankton crop by 80% [720] and doubled the transparency [421]. These might seem desirable consequences, but, in switching the predominant energy flows in this part of the lake to the benthos instead of the plankton, it may disturb the recruitment of fish dependent on the plankton chains. Zebra mussels are also a threat to water-supply undertakings and aquaculture on the Great Lakes, for they clog the water-intake pipes. So far, it has not spread into many lakes off the main waterways, but there is a risk of its doing so, attached to recreational boats moved around on trailers [717]

Fortunately, it requires waters of reasonably high calcium content (> 15 mg l^{-1}) and pH values > 7.3. Chemicals are being sought to control the mussel, but it now covers such large areas that specific biological control may be the best approach. A further introduction from the Black and Caspian seas, a fish, the round goby (*Neogobius melanostomus*), eats the mussel [323] but may itself cause further biological problems if it is widely distributed.

8.5 Relationships between the littoral zone and the open water

The littoral zone links the catchment area of a lake and its open water. Potentially, it can influence the open water in three ways: as an interceptor or sink for materials that would otherwise move from the catchment to the open water; as a source of new materials to the latter that would not otherwise be supplied; and as a refuge or resource for animals that may move into it from the open water to escape predation or to feed. In any particular lake, these functions might have greater or lesser importance. For a huge, deep lake, there would probably be little consequence for the open water if the littoral was obliterated; but, for most of the world's lakes, which are small and relatively shallow, the influence of the littoral zone has probably been greatly underestimated [1010, 1012].

The littoral may intercept silt, remove nitrate by denitrification and act as a sink for major ions entering from the catchment. The high productivity of littoral swamps means that organic detritus is laid down as peat to a greater extent in lakes than in floodplains (see Chapter 5), where the flow may move much of the detritus down river. Peat contains minerals, and peat build-up must mean a sequestering of many elements.

For most of these, there is little functional importance, because concentrations of them are little changed and they are not limiting to the open-water community. Burial of phosphorus, however, could sometimes be significant [961] and the removal of nitrate by denitrification must usually be important, with concentrations of several milligrams per litre in inflow streams being reduced to perhaps less than 1 mg nitrogen (N) l^{-1} in the lake, even in winter [687]. Preservation of lakeside swamps is itself a useful device in combating eutrophication.

The littoral zone may also provide materials to the open water. Usually these are carbon compounds. Most of the dissolved carbon compounds, resulting from secretion or decay, that reach the open water from the swamps are highly refractive yellow substances of high molecular weight. Some can react with metal ions and phosphate to decrease their availability to organisms in the open water [508]. More labile compounds are probably taken up rapidly by the large bacterial populations within the littoral.

The yellow substances stain the water and influence light penetration (see Chapter 6), so that a small lake in acid, boggy terrain heavily endowed with swamp may have brown water, containing as much as 30–40 mg l^{-1} of such substances [205]. Such lakes have been described as dystrophic and there is some evidence that concentrations of humic substances have been increasing in some Scandinavian areas in recent years (Fig. 8.18) [283]. The reasons may involve changes in rainfall chemistry and their influence on the swamps or changes in forestry management in the catchments.

Small organic particles washed out of the swamp may be of greater importance in some places. Lake Chilwa in Malawi, for example, is a very shallow, saline lake, which dries out completely on occasions but has a clay-floored basin ringed with swamps of *Typha domingensis* [515]. The lake level rises and falls annually by a metre or so, with longer-term trends towards very high or very low mean water levels. At the higher levels, the water may be clear enough to allow a ring of submerged plants to front the *Typha* swamp. But, at the lower levels, wind disturbance brings clay into suspension and this severely limits the amount of light penetrating to support planktonic or benthic photosynthesis.

The swamp detritus may then support zooplankton and invertebrates living in the clay bed, together with their dependent fish. The zooplankton populations are aggregated inshore, close to the edge of the swamp, and fish catches are also greatest in this area. In the amorphous clay sediment, the chironomid, *Chironomus transvaalensis*, uses detritus from the swamp to construct its burrows [611], and it too has a distribution closely following the lake's perimeter (Fig. 8.19).

There has been much discussion about the possible role of the littoral zone as a source of phosphorus to the open water. It can be this if it is able

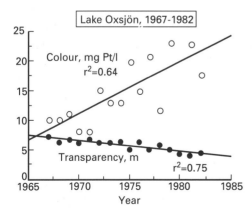

Fig. 8.18 Increase in colour and decrease in secchi disc transparency of water in Lake Oxsjön, Sweden, due probably to changes in forestry practice. Colour is based on comparisons with a standard scale of platinum salts. (Based on Forsberg and Petersen [283].)

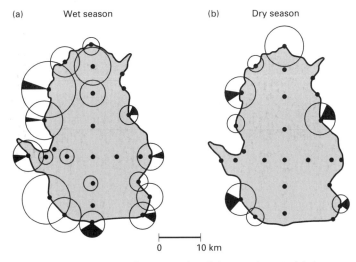

Fig. 8.19 Distribution of the mud fauna in Lake Chilwa, Malawi, in (a) the wet and (b) the dry seasons of 1969. Areas of circles are proportional to the biomass of the fauna, with the black sectors the proportion due to *Chironomus transvaalensis*. Solid circles indicate sampling sites. (Based on McLachlan [611].)

to mobilize phosphorus from particles entering from the catchment which would not otherwise be available. Otherwise, it must either be neutral or, if it is laying down peat, a sink for phosphorus. No studies have critically examined this balance, although some have shown an increase in total phosphorus in the open water when the littoral plant beds have been removed by herbicides and plant-eating fish [897], suggesting a role for the plants as a net sink rather than a source.

However, aquatic plants do mobilize phosphorus from sediments, either through uptake and subsequent decay in the water column or through destruction of the oxidized microzone (see Chapter 6) at the surface of the sediment within their beds. In temperate lakes, senescence and decay of the plants in autumn may release large quantities of phosphate to the water [339]. However, the released amounts will be diluted by incoming water and washed out or reabsorbed by the reoxidized sediment surfaces. Laboratory experiments are usually the means by which phosphate release rates are measured. Calculations of the potential contribution of this loading to the lake are largely misleading if season and hydrology are not taken into account. In tropical lakes, where growth occurs year-round, a steady leak of phosphate from the sediment through the plants might be significant in creating a consequent general nitrogen scarcity.

The littoral zone may be a refuge for otherwise open-water organisms or provide food at particular stages in their life histories. Plants and their detritus are used by many species of fish to lay their eggs. Species like pike (*Esox liucius*) lurk in the edges of plant beds for their prey. Older, larger fish take larger and larger invertebrates as food and these are to be found in the littoral zone, while smaller invertebrates, such as Cladocera, may find refuge there from open-water fish predation (see above and Chapter 6). The swamps of Lake Chilwa, lying closer to the inflows, act as refuges for fish when the central part of the lake dries out. A catfish, *Clarias mossambicus*, can bury itself in the damp mud of the swamp and, air-breathing, await the reflooding of the hard, cracked, clay basin which much of the lake becomes.

Intuition would suggest that, where they are extensive relative to the open water, littoral zones, with their considerable complexity and diversity, might have many overt and subtle influences on the open water, but good information is scarce and at present only their potential importance can be emphasized [128, 1010.]

8.6 The profundal benthos

The profundal benthos is much less structured than most of the littoral communities. To a casual observer, its existence is suggested only by the clouds of midges (chironomids) that emerge from the bottom through the water column to swarm and mate around the lake (Fig. 8.20) and which can sometimes be a problem for local amenity. A comparison between littoral and profundal benthic invertebrate communities recorded at the peaks of their development in the same lake illustrates just how much less diverse is that of the profundal [63, 498].

Fig. 8.20 Swarms of midges, when the adults emerge, can be very dense. *Tanytarsus gracilentus* emerging in May from Lake Myvatn, Iceland. (From Lindegaard and Jonasson [576].)

At 2 m water depth in Lake Esrom, in Denmark, the littoral plants and their underlying sediment provide habitat for at least forty macroinvertebrate species, including oligochaete worms, caddis-fly and dipteran larvae and gastropod and bivalve molluscs. A comparable sampling of sediment from under 20 m of water, in the hypolimnion, revealed only five species: an oligochaete worm, three dipteran larvae and a bivalve mollusc. Ciliate protozoa would be numerous in both habitats [266].

Oxygen concentration is important for benthic animals and there are strong correlations between the occurrence of particular species and the lowest oxygen concentration reached at the surface of the sediments in summer. Figure 8.21 contrasts the respiration rates [501] of four littoral animals from Lake Esrom and four profundal species. In the limpet *Theodoxus*, which grazes algae from the rocks of the wave-disturbed littoral, respiration rate is maximal only at full saturation and steadily falls with decreasing saturation. With littoral species from progressive depths, rates at saturation are lower and then fall with decreasing saturation to 40–60% before plunging to around 20 or 30%. In the profundal species, the rates are low, even at saturation, but are largely maintained as saturation decreases, until a final drop-off at saturations as low as 10%. Frequently, profundal animals are rich in haemoglobin, which allows continued respiration at very low external levels, by concentrating oxygen within the organisms, thus allowing the animal to continue feeding.

The low diversity of the profundal benthos, compared with the littoral, might thus be attributed simply to oxygen stress. It lives also at lower temperatures, for the deeper waters of many temperature lakes remain at 4–10°C year-round. Temperature is not crucial in reducing diversity, however. The hypolimnia of tropical lakes can be very warm (see Chapter 6) and yet are also species-poor. They are often severely deoxygenated, however.

Even so, there is a wide range of animals that can cope with deoxygenation, and other factors may be important. Reduced diversity in the profundal is a feature of very infertile lakes with well-oxygenated bottom waters, such as Thingvallavatn in Iceland [575] (Table 8.4). The clue may be in the relatively simple habitat structure of the profundal. It is a comparatively featureless mud-flat, on to which falls a rain of detritus, of varying food quality.

Many profundal communities are dominated by chironomid larvae (Diptera) and oligochaete worms, with sometimes some small bivalve mollusc species. Attempts have been made to classify lakes on the basis of their benthic invertebrates [91, 845] and, although these have been completely successful only for limited regions or purposes, a general trend in benthic fauna, related to lake fertility, can be seen. Well-oxygenated hypolimnia have a variety of larvae of *Chironomus*, *Tanytarsus* and other chironomid species, *Pisidium* spp. (bivalve molluscs), some Crustacea, such

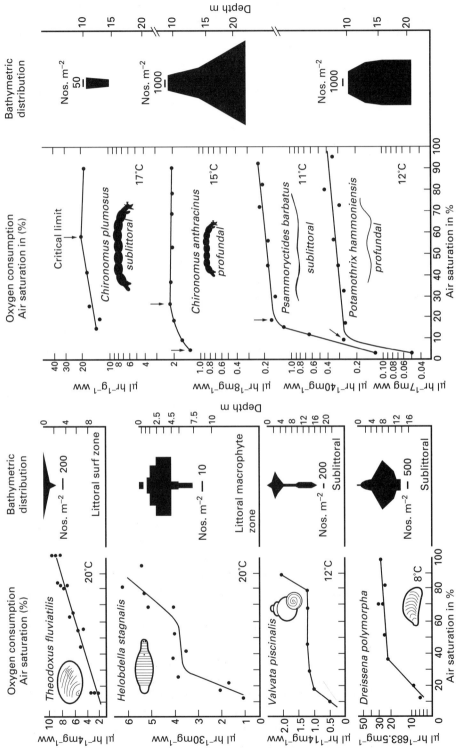

Fig. 8.21 Respiration rates of benthic animals that are common in Lake Esrom, Denmark. Values are given on a per animal (mean weight) basis and are collected from various sources. The major differences are in the shapes of the curves as respiration declines with decreasing saturation with oxygen. WW, wet weight. (From Jonasson [501].)

Table 8.4 Number of bottom-living animals, their biomass, annual production and turnover ratio (production : biomass) and species diversity in successive depth zones of Lake Thingvallavatn in Iceland. Biomass and production are given as ash-free dry weight m^{-2}. Turnover ratio is given in parentheses in the production column. Based on Lindegaard [575].

	Number m^{-2}	Biomass (mg m^{-2})	Production (mg m^{-2} year^{-1})	Species richness
Littoral surf zone, rocks and boulders (0–2 m)	21 000	3 900	14 000 (3.6)	36
Littoral rocks and boulders (2–6 m)	18 000	3 500	11 500 (3.3)	41
Littoral rocks and boulders (6–10 m)	10 400	2 600	8 000 (3.1)	39
Littoral rocks, sand and with aquatic plants (*Nitella* sp.) (10–20 m)	3 500	1 400	3 600 (2.6)	35
Profundal bare rock and sediments (20–114 m)	3 900	850	1 200 (1.4)	18

as the amphipod *Pontoporeia*, and sometimes insect larvae, such as *Sialis* (alder-fly), and oligochaetes.

At the other extreme of sediments covered by anaerobic water for part of the year, the chironomid fauna is reduced to one or two detritivorous species – for example, *Chironomus anthracinus* – and about the same number of species of larval predators. These often include *Chaoborus*, the translucent 'phantom larva', which migrates to the epilimnion to hunt zooplankton at some stages of its life history. There is usually also a much greater biomass of oligochaete worms, such as *Tubifex* and *llyodrilus*. The profundal benthos of such a lake, Lake Esrom, has been studied in some detail [498–500] and will illustrate the biology of these organisms.

8.6.1 Biology of selected benthic invertebrates

Chironomus anthracinus

Chironomus anthracinus in Lake Esrom (Fig. 8.22) starts life as an egg mass deposited between sunset and darkness in May on the lake surface. Mated female gnats dip their abdomens into the surface-water film. The egg masses are about 2×2.5 mm in size when dry but swell 100-fold when wetted, and are deposited in the lee of beech woodlands on the western shore. These create the calm conditions necessary for the flight of the gnat, but water currents subsequently distribute the eggs throughout the lake and they sink to the bottom.

By June, the eggs have hatched into instars less than 2 mm long, the first of four stages that precede pupation and emergence of the adult some 23 months later from populations in the profundal. Growth is rapid and the third instar is reached in July. The growth coincides with a ready supply of food falling from the epilimnion, where a major spring population of diatoms has formed (see Chapter 7).

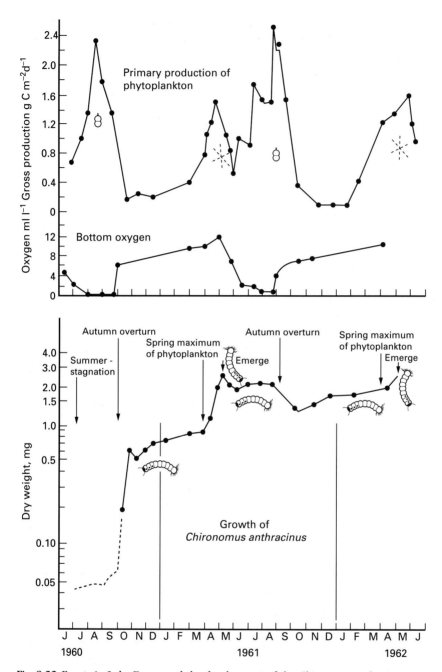

Fig. 8.22 Events in Lake Esrom and the development of the *Chironomus anthracinus* population. (Redrawn from Jonasson [500].)

The first instar is transparent and probably moves along beneath the sediment surface, swallowing sediment as it moves. The second has heavier musculature and some haemoglobin. It builds a tube, open at the surface and tapering at the base. In this it lives, pumping a current of water past its mouth by undulations of its body. The larvae may also emerge partly from the tube to gather sediment encircling it. The third-instar larvae build tubes, lined with salivary secretions, which project a little way above the sediment surface, and feed on the deposits around the tube. They spread a net of salivary threads over the mud, to which particles stick, and then drag the net down into the tube to eat it.

Lake Esrom is fertile and the dissolved oxygen (O_2) concentrations fall rapidly in its hypolimnion. By July, the lower few metres are almost deoxygenated and at around 1 mg O_2 l^{-1}, growth of the *Chironomus* larvae stops (Fig. 8.22). This may be because only limited respiration is then possible, or because the supply of sedimenting food has changed in quality or amount as the overlying plankton community processes it first. On the other hand, if growth of *Chironomus* stops, so does predation on it by bottom-feeding fish, which cannot tolerate the deoxygenation and move to shallower waters.

After the overturn of stratification in Lake Esrom in September and reoxygenation of the water, growth of *Chironomus* resumes and the third-instar larvae moult and change into fourth-instar larvae. These are bigger, over 1.5 cm, heavily muscled and bright red in colour. They produce tubes of sediment particles, glued with salivary secretion, which have 'chimneys' projecting 1–2 cm above the mud surface. The height of the chimney may be sufficient to reach water which, though only a few millimetres from the sediment surface, is much better oxygenated than that immediately in contact with the sediment.

Growth of the fourth-instar larvae depends on the supply of phytoplankton and detritus reaching the bottom in autumn, but is soon reduced as the shortening days diminish plankton production. Growth resumes the following spring, but most of the larvae are not mature enough to pupate and form adults. They must therefore remain as fourth instars to achieve further growth in the spring and autumn of the second year. Pupation follows in the succeeding spring when the pupae emerge from the tubes and float to the water surface. This happens in the evening; within about 35 s of reaching the surface, the pupal skin splits and the adult emerges, turning from the pupal red to black. The adults (imagos) rest on the surface until sunrise, when the rising warmth quickens their metabolism sufficiently for them to fly away and later to mate and lay eggs on the water. The adult life of the midge is very short, only a few days, compared with the 23 months spent as a larva in the profundal.

At lesser water depths, where there is no summer deoxygenation, *C. anthracinus* is able to grow sufficiently in one year to emerge the next. The shallow-water populations also have higher respiratory rates, which probably also accelerate development of the larvae. A few of the deep-water population do manage to emerge after only twelve months, in the spring following their birth, but they do not contribute any young to the population. Their eggs are mostly eaten as they reach the sediment by the large population of remaining fourth-instar larvae.

Ilyodrilus hammoniensis and Pisidium casertanum

Ilyodrilus is an oligochaete worm, a centimetre or so long when mature. Its development in the profundal of Lake Esrom is rapid (within about two months), so that several generations may be found in the sediment. It hatches from egg cocoons, each containing two to thirteen embryos, and populations of 25 000 worms m^{-2} may be present. The worms burrow, eating each day four to six times their own weight of sediment and continually expelling what they cannot digest.

They are thus important sediment processors, mixing the layer in which they live and maintaining the bacterial population in a fast-growing state by continual feeding. In the alternate years, when fourth-instar *C. anthracinus* are present, the oligochaete population is low, but it increases in the intervening years, when these smaller animals can compete more effectively with the younger chironomid larvae.

The tiny pea-shell cockle, *Pisidum casertanum*, lives in burrows running parallel to the mud surface. It draws a current of water to maintain its oxygen supply, except of course in summer, when, like the other profundal animals, it is inactive and does not grow. It sucks a current of watery sediment through a tube, or siphon, and passes it over the gills, where steadily beating flagella direct a stream of particles to the stomach. The life history probably lasts more than a year and the young are retained, until fully developed, within the shell of the hermaphrodite adults.

Carnivorous benthos – Chaoborus and Procladius

Carnivores in the profundal benthos of Lake Esrom are few (about 10%) compared with detritivores. There are two main species, both larvae of dipteran flies. *Chaoborus flavicans* (syn. *Chaoborus alpinus*) feeds mainly on the plankton and only partly on the sediment community, and little is known about the other, *Procladius pectinatus*. Both species ultimately emerge for brief adult lives, in which they lay eggs.

Chaoborus, the phantom midge, is so called because of its transparent body, punctuated only by its dark eyes and apparently black (an optical effect) air-sacs at the hind end. These possibly allow it to regulate its buoyancy when it moves between the epilimnion and the sediment. Eggs are laid in late summer by the adult midges and first-instar larvae appear in the plankton in September. Zooplankters, on which *Chaoborus* mainly feeds, after seizing them with its prehensile antennae, are reasonably abundant in autumn and the *Chaoborus* larvae quickly pass into their second and third instars. These spend some time in the plankton, particularly during overcast weather, when their visibility to prey is least, and some in the sediment. Fourth-instar larvae are generally produced the following spring, and these migrate nightly from the sediment to the surface water, where zooplankters are again abundant. In winter, *Chaoborus* spends most of the time in the sediment, possibly feeding on *Ilyodrilus*. Pupation and emergence are in July.

Procladius is a chironomid, known to be carnivorous and feeding probably on the smallest *Chironomus* larvae and *Ilyodrilus* juveniles. Its maturer larvae emerge in the spring, while less developed ones may migrate to the shallows, for none can be found in the profundal mud in summer.

8.6.2 What the sediment-living invertebrates really eat

In relatively shallow water, much of the organic matter arriving at the sediment surface is quite rich in energy and nutrients. In deeper water, processing by the planktonic community may have reduced this richness, leaving particles lower in nitrogen and energy content. Thus, the overall production of the benthos in Lake Thingvallavatn (Table 8.4) shows a lower biomass, a lower annual production and a lower turnover rate with depth. Partly this is due to lower average temperature with depth, but it is also due to the changing quality of food reaching the bottom.

Those organisms living closest to the sediment surface, such as *C. anthracinus* [496] and *Chironomus riparius* [799], have the advantage of better food quality, though the disadvantage of greater exposure to predators. This is also true of those filtering the particles out before they can be sedimented, such as *Chironomus plumosus* and *Glyptotendipes paripes*. The surface-deposit feeders often construct U-shaped tubes in the sediment, from which they rarely emerge, while the filter feeders construct J-shaped tubes, from which they protrude. Animals deeper in the sediment, such as the oligochaetes, have to process material that has already been ingested by the surface-livers and defecated, but are least vulnerable to foraging fish searching through the surface sediment.

The faeces of the surface feeders are still rich in carbon but not in other elements, and the remaining carbon compounds are likely to be refractory. It seems likely that processing by bacteria is important in converting this material into usable food. Studies by McLachlan *et al.* [617] have uncovered, in the invertebrate community of a bog lake in Northumberland, a story paralleling that of the feeding of shredders in streams (see Chapter 4).

Blaxter Lough is a shallow basin set in the peat of an extensive blanket bog. Erosion of the peat by waves at the windward edge provides a ready source of organic matter, but the major detritivore, *Chironomus lugubris*, is conspicuously distributed at the opposite site of the basin. The eroded peat is washed by water movement across the lake bottom to the leeward shore and, as it moves, it is broken down into smaller particles. However, although *C. lugubris* does have some restrictions on the size of particles it can eat, this is not why it does not colonize the area where the peat is freshly eroded. Suitably sized particles sieved from eroding or *in situ* peat would not support its growth.

On the other hand, if fresh peat is allowed to become colonized by microorganisms over a few days, it will support growth of *Chironomus*, whether in the laboratory or in chambers placed in any area of the lake. The natural distribution of the chironomids in the lake reflects the distance travelled by suitably fine peat particles in order for them to become colonized by palatable bacteria and fungi. The microorganisms absorb nitrogen compounds from the water and, by the time they become palatable, the peat particles have increased in calorific content per unit weight by only 23%, but have doubled in protein content.

Peat ingested by *C. lugubris* contains more bacteria than fungi, but the balance changes as the microorganisms pass through the gut. The faecal pellets are relatively large, coherent and dominated by fungi. They form the food source of a small cladoceran, *Chydorus sphaericus*, which can rasp material from them, presumably digesting most of the microorganisms, and producing fine faeces of its own. These are small enough to be nutritious again to *Chironomus*, once recolonized by bacteria. A reciprocal relationship exists between the two animals, resulting, with the help of microorganisms, in the ultimate breakdown of the peat. Initially, peat is poor food, much like the sediment of the profundal following a long passage through the plankton system or after it has been worked over by the surface feeders.

The bacteria produced on such material are, nonetheless, unlikely to be as rich a food supply as fresh algal material. Johnson *et al.* [497] studied the feeding of *C. plumosus* in a Swedish lake and found that, although the animal selectively fed upon bacteria and algae (*Microcystis* and *Melosira*) from the sediment, it assimilated five times as much carbon from the diatom, *Melosira*, as from the bacteria and did not readily digest the blue-green alga, *Microcystis*. A modelling study [339] suggests that only in infertile lakes, where algal material reaching the sediment is scarce, are bacteria likely to be a major source of carbon for the benthic animals. In fertile lakes, which are often also shallower, the abundance of algal material and its entry, barely degraded, to the sediment mean that bacteria account for only a few per cent of the invertebrates' food. This, then, places some importance on the nature of the sedimenting material in the relationship between the planktonic community and the underlying benthos.

8.7 Influence of the open-water community on the profundal benthos

In moderate- to large-sized lakes, the profundal benthos receives most of its food from the overlying plankton community. (The littoral may contribute the bulk in smaller lakes.) But the relationship is complex, as shown by the feeding studies discussed above. Work by Johnson and Brinkhurst [493–495] in the Bay of Quinte, on the northern shore of Lake Ontario shows in greater detail how the open water and profundal processes are linked.

The Bay of Quinte is long, narrow and winding. It is shallow (about 5 m deep) at its inner end, where several towns enrich it with sewage effluent, and opens out, over 100 km distant, into the 30 m-deep waters of inshore Lake Ontario, where depth and dilution have reduced the fertility of the water. At four stations (Big Bay, Glenora, Conway and Lake Ontario) along this gradient, the amount of sedimenting material and its fate was followed as it was processed by the bacteria, benthic invertebrates and fish. The results emphasize both quantity and quality of the sedimenting material as important in determining the productivity of the benthic animals.

In the inner bay, Big Bay station had a benthic community dominated by chironomids, while the third station, Conway, had a rich association of bivalve molluscs (*Sphaerium*), oligochaete worms, chironomids and crustaceans, with an intermediate community at the second station, Glenora. In Lake Ontario, the fourth station, there was a more diverse community of *Sphaerium* spp., Crustacea and many other species. In general, the diversity increased towards the main lake, along with the gradient of decreasing fertility and a gradient of summer bottom-water temperatures, which were above 22°C at Big Bay, but around 10°C in Lake Ontario.

The first task was to determine the rate of supply of materials to the bottom communities. The sedimenting seston was collected in 20-cm diameter funnels fitted into bottles to retain it and suspended about 1.5 m above the bottom. 'Seston' includes all the fine particulate matter suspended in the water. The traps were emptied weekly and the inorganic and organic parts of the sediment measured. Seston is colonized by bacteria while it is still suspended and these bacteria continually decompose it, even in the traps. The rate of this decompostion was found to be a few per cent per day and appropriate corrections were applied to give the true rate for the four stations.

Sedimentation rates decreased from the inner bay to the lake, though not steadily, for Big Bay had a much greater rate of sedimentation of organic matter than the others. The overall rate of processing of this material after it reached the bottom was measured as the community respiration, the sum of bacterial, invertebrate and fish respiration. The first two components were measured together as the rate of oxygen uptake from water overlying the sediment in small cores. The sediment and its overlying water were sealed from the air by a layer of oil and incubated in the laboratory at the appropriate natural temperature. The respiration rates of fish feeding on the benthic community could not be directly obtained. They were estimated from studies elsewhere as about 50% of the total invertebrate respiration (see below), and this was added to the measured

Table 8.5 Mean sedimentation rates and community respiration rates at four stations in the Bay of Quinte.

Station	Organic matter sedimented (g m^{-2} day^{-1})	Community respiration (g O$_2$ m^{-2} day^{-1})	Mean temperature (°C)
Big Bay	3.01	0.35	17.1
Conway	0.71	0.25	11.8
Glenora	0.29	0.22	10.8
Lake Ontario	0.28	0.15	9.1

sediment-core respiration to give total community respiration. Although community respiration decreased from Big Bay to Lake Ontario (Table 8.5), it did not do so to the same extent as sedimentation rate. This point will be returned to later.

It was then necessary to separate the activity of the benthic microorganisms from that of the macroinvertebrates (those larger than about 1 mm and therefore not the protozoa and organisms like nematodes, which had to be included with the microorganisms). This was done by measuring the respiration rates of invertebrates separately. Representative animals were placed in small jars with a substratum of sand, almost free of microorganisms, and their rates of oxygen uptake measured at a variety of temperatures. Numerous experiments produced equations relating respiration rate to size (as dry weight) of animals and temperature for all of the major species. This meant that, from routine samplings, in which temperature was measured and animals were counted and weighed, respiration rates for the whole macroinvertebrate community could be calculated.

It was also possible to find net production – the increase in amount of animal tissue per unit time – by keeping representative animals in small mud cores, freed of other animals by previous heating or freezing, and by measuring by how much their weight increased over several days or weeks. The method has the disadvantage that growth rates may be altered in the absence of competition with other species. Growth rates were expressed as percentage increases in weight of animal per unit time, and allowed extrapolation to the natural community, using the relationship:

Production = growth rate × biomass

Production of animals that formed distinct cohorts in their life histories was determined by a variant of the Allen-curve technique (see Chapter 4).

Assimilation rates, approximately the sums of respiration rates and net productivities, were then calculable and these are given in Table 8.6. Biomass of the macroinvertebrates increased towards the outer lake – this contrasts with sedimentation rate and community respiration – and assimilation, production and respiration all reached peaks at Glenora, the second station, and were generally similar or lower at Big Bay and the outer two

Table 8.6 Productivity of the benthic macroinvertebrate community in the Bay of Quinte. All rates in kcal m^{-2} $year^{-1}$ and biomass in kcal m^{-2}

Station	Biomass (B)	Assimilation	Production (P)	Respiration	P/B
Big Bay	5.45	108.7	74.3	34.3	13.6
Glenora	29.9	368	233	136	7.8
Conway	25.6	142	65.8	75.8	2.6
Lake Ontario	38.0	165	51	115	1.3

stations. The high sedimentation rate at Big Bay, therefore, did not support comparably high invertebrate production. The turnover rate of the community, measured by the yearly production-to-biomass ratio, did, however, follow inversely the gradient outwards, probably reflecting the decrease in mean temperature.

From all of these data, the diagrams in Fig. 8.23 could be constructed. They show the flow of energy through the sediment community (bacteria, invertebrates and fish) at the four stations, and all quantities have been converted to cal m^{-2} day^{-1} for rates and to kcal m^{-2} (in brackets) for the standing biomass of the various components. The incoming organic matter (IM) represents the amount of sedimentation.

Community respiration (R_{com}) degrades less than a quarter of incoming organic matter, and animal production uses less than a twentieth at Big Bay, so that much of it is not utilized (NU) and forms the permanent sediment. It is not used because it is refractory and difficult even for bacteria to degrade. Its low quality probably reflects the nature of the cell walls of blue-green algae, which are abundant in the phytoplankton there, but also the large import of fibrous organic matter left after terrestrial decomposition and washed into the Bay.

The amount of unutilized matter is small at the other three stations. This is reflected in the low organic content of the sediment at the Lake Ontario station, 3–4%, compared with 32% in the sedimentating material. The organic content of sediment has, in the past, been used as an indicator of potential benthic animal production in lakes. These studies indicate that it represents the net result of several processes of accumulation and degradation, and its use as such an indicator may be misleading.

These processes are amenable to a general treatment, which may have implications wider than those for the Bay of Quinte.

First, several measures of the efficiency of energy use may be deduced for the benthic community. The proportion of incoming energy used by the whole community (microorganisms, invertebrates and fish) is:

$$a = (R_{com} + E)/IM = U/IM$$

where E is the energy lost in the emergence of adult insects, which fly away, and U is the energy used, i.e. not stored in permanent sediment.

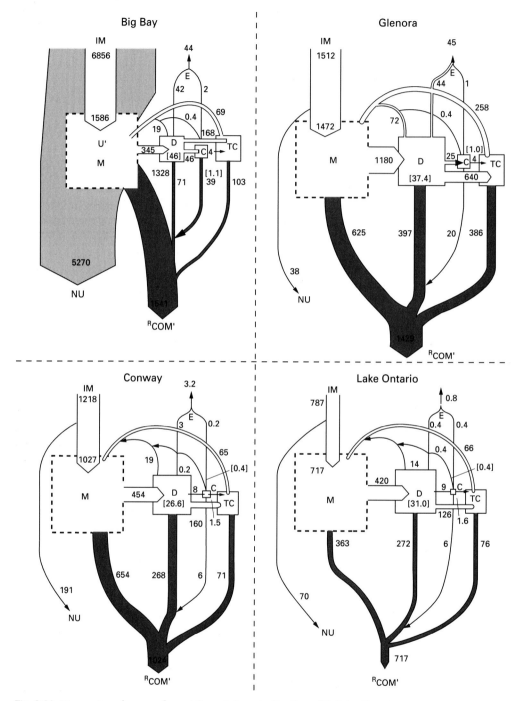

Fig. 8.23 Mean rates of energy flow at four stations in the Bay of Quinte. Boxes represent standing crops and stocks (kcal m^{-2}) and pipes the rates of flow, in cal m^{-2} day^{-1}. IM, incoming organic matter; M, microorganisms; U, amount utilized; NU, not utilized and stored in permanent sediment; E, emerging insects; D, detritivores; C, carnivores; TC, top carnivores (fish); R_{com}, community respiration. Heavily shaded pipes show respiratory losses. (Redrawn from Johnson and Brinkhurst [495].)

The proportion of usable energy channelled through the macroinvertebrates is:

$$b = (R_D + R_C + E)/R_{com} + E$$

where R is the respiration of detritivores, D, and carnivores, C.

Thirdly, E_{gc}, the net growth efficiency of the invertebrates, is defined as:

$$E_{gc} = [A_D - (R_D + R_C)]/A_D = P_{D+C}/A_D$$

where A is assimilation and P is production of both detritivores and carnivores.

Table 8.7 gives these quotients for the four Bay of Quinte stations. They show clearly the lower utilization of incoming matter at Big Bay ($a = 0.23$) and that the microorganisms at Big Bay take a greater proportion of the utilized material (($1 - b$) = 90%) than they do at the other stations (61–73%). This depends on the quality of the sedimenting seston. If it is difficult to degrade, a greater investment of its contained energy must be made by bacterial activity to convert it to a form (bacterial cells) usable by the invertebrates.

The product of a and b gives the proportion of incoming energy usable (U) by the invertebrates and is very low at Big Bay, 2.3%, and 23–25% at the other stations. A general equation for utilization by the animals is:

$$U = a \times b \times IM$$

The value of a probably decreases with increased allochthonous import of material (usually of a low-quality, refractory nature) from the catchment, while b reflects the cost of processing the material and is low when allochthonous matter or tough-walled algae, particularly blue-green algae, are present.

The term E_{gc} can now be incorporated into the model. Production of macroinvertebrates bears some relationship, c, to their utilization of organic matter:

$$P = c(U)$$

and

$$U = R_{D+C} + E$$

The insect emergence, E, is generally only a small proportion of the total

Table 8.7 Efficiency of energy use in the benthic communities of the Bay of Quinte.

Station	a	b	E_{gc}
Big Bay	0.23	0.1	0.68
Glenora	0.97	0.31	0.64
Conway	0.84	0.27	0.40
Lake Ontario	0.91	0.39	0.34

energy flow and may be neglected, so that:

$$P = c(R_{D+C})$$

Because

$$E_{gc} = P_{D+C}/A_D = P_{D+C}/(P_{D+C} + R_{D+C})$$
$$R_{D+C} = P_{D+C}[(IM - E_{gc})]/E_{gc}$$

Also, because

$$P = c(R_{D+C})$$
$$c = P/R_{D+C} = E_{gc}/(IM - E_{gc})$$

and because

$$U = a \times b \times IM$$

therefore

$$P = a \times b \times c \times IM$$

This is now a relationship that describes how production of invertebrates is related to the supply of incoming organic matter. The two quantities would be directly dependent only if a, b and c were constants, which they are not. They all decrease as IM increases; a and b decrease for the reasons stated above, c does so because, as the import of organic matter increases, the rate of deoxygenation in the surface sediment and the water just above it also increases. Animals must then use more energy in obtaining oxygen (in body movements to keep water circulating over the animal's surface or in production of haemoglobin). Or, like *C. anthracinus*, they may have periods when they lie quiescent, respiring, albeit at a low rate, but probably not feeding. This means that a smaller proportion of their energy supply is available for growth. The value of c should also be low at low levels of import, because of the extra activity then necessary in seeking food. The relationship between benthic invertebrate production and import of organic

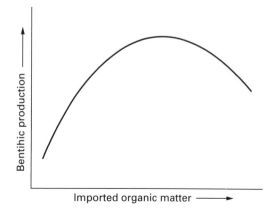

Fig. 8.24 Schematic diagram of a possible general relationship between production of benthic animal communities and the rate of supply of organic matter to them in lakes.

matter to the community may then have a maximum at intermediate import levels flanked by minima due to food scarcity, on one side, and to low food quality and to deoxygenation, on the other (Fig. 8.24).

The material not used by the invertebrates and microorganisms forms the permanent sediment. It contains a record, albeit a strongly distorted one, of activities in the lake and can be used to discern the lake's history, a topic covered in Chapter 10. Prior to that, however, there is more to be said about fish, which have figured in the planktonic, littoral and profundal systems, but which have traditionally occupied a branch of limnology of their own and which, through their movements and behaviour, also contribute to the stitching together of these systems.

9: Fish and Fisheries in Lakes

For many people, fish and fisheries are the reasons to study fresh waters. In the developed world, recreational fisheries are prized and, in the developing world, freshwater fish are the most important protein source for many people. Freshwater fish may be crucial to health, not only in diet, but also through the worm parasites (see Chapter 5) to which fish play intermediate host, particularly in the Far East.

Fish yields (the crops taken by man) are broadly related to fish production in a lake, which in turn is related to the overall productivity of the system (see Chapter 6). The yield is always lower than the total production, because some fish are unsuitable biologically for sustained fisheries and others may be difficult to catch or unacceptable as food for sociological reasons.

Maintenance of a fishery, without overfishing, when more fish are taken than are replaced by birth and subsequent new fish recruitment and growth, ideally requires three things. The biology of the fish should be understood; only those species whose populations will stand up to fishing should be taken; and there should be monitoring of the catch and imposition of regulations on the fishery to protect the stocks. The first part of this chapter will deal with these issues.

These requirements are not always achieved, particularly in the developing world. Expertise may not be available and simple collection of monitoring statistics from fishing villages widely spaced around a big lake may be impossible. Rough-and-ready methods of regulation may be more practicable than statistical models. Population and economic pressures now demand increased fish yields. Often these have involved fishing for inappropriate species or introduction of supposedly high-yielding exotic species. The problems that these have caused and the better solution of pond culture to the need for increase in fish yield are considered later. Finally, the impacts and issues of recreational angling are discussed.

9.1 Some basic fish biology

Fascinating differences are found among fish, with great implications for the suitability or otherwise of a species for a fishery. I have chosen the brown trout (*Salmo trutta*), the Nile perch (*Lates niloticus*), a 'tilapia' (*Sarotherodon niloticus*), the wall eye (*Stizostedion vitreum*) and the grass carp (*Ctenopharyn-*

goden idella) (Fig. 9.1) to illustrate the variety among fish, because of the ranges of their reproductive biology, diet and zoogeography.

Brown trout are carnivores, eating mostly benthic invertebrates; they are indigenous to Europe, North Africa and western Asia, but have been introduced to suitable waters elsewhere for their sporting qualities. Nile perch are voracious piscivores in the River Nile and its associated, or formerly associated, Great Lakes, Albert and Turkana. They are valued food fish in some places, being large (specimens weighing over 45 kg are common and lengths may be 2 m or more), and have been introduced to other African lakes (see later).

'Tilapia' is a name given to cichlid fish of the genera *Tilapia, Sarotherodon*

Fig. 9.1 Five fish species of contrasted biology (see text). (a) Brown trout, *Salmo trutta*; (b) Chinese grass carp, *Ctenopharyngodon idella*; (c) 'tilapia', *Sarotherodon niloticus*; (d) pike-perch, *Stizostedion lucioperca* (the walleye is a similar, though stockier, fish); (e) Nile perch, *Lates niloticus*.

and *Oreochromis. Sarotherodon niloticus* feeds on fine bottom detritus and phytoplankton, is also native to the Nile watershed and has been widely introduced to other African lakes and rivers. Walleye are North American piscivores, which feed on zooplankton when young. Some comparison will be made between walleye and their close European relatives, the pike-perch, or zander, *Sitzostedion lucioperca* [634]. Lastly, the grass carp is an avid feeder on submerged and even emergent aquatic plants. It is endemic to the River Amur and parts of eastern Asia but has been widely introduced to other areas of Eurasia for control of nuisance aquatic plants and pond culture.

9.1.1 Eggs

Fish eggs are mostly released into the water for their development. *Sarotherodon* species are exceptions, for the female may gather them, once fertilized, into her mouth for brooding. With such close protection, few eggs are needed. *Sarotherodon niloticus* protects the young fish in this way when predators approach, but may merely guard them, as eggs, in a shallow depression scraped on the sandy bottom, by sweeping movements of its tail. Trout also carefully excavate a nest, or redd, in gravel, much as described already for salmon (see Chapter 4) [292].

The Nile perch and grass carp take little or no care of their eggs and must produce large numbers to ensure survival of enough young to maintain the population. The eggs of both species are planktonic and kept floating in the water by incorporation of oil in the former and a water-filled cavity between the egg membranes in the latter. While *S. niloticus*, which guards its eggs, may produce only a few thousand, the Nile perch releases several millions to the open water.

Walleye also take no care of their eggs. They are scattered on to gravel in well-oxygenated water, having been fertilized during release from the female. The walleye's European relative, the pike-perch, is similar in morphology, but quite different in spawning behaviour. The male pike-perch excavates a nest on a muddy organic bottom, around the exposed roots or rhizomes of aquatic plants, to which the sticky eggs adhere. The female guards them and fans water over them with her tail; this may increase the rate of survival in a habitat generally unsaturated with oxygen. Many temperate fish lay eggs that adhere to stones or weeds in lowland rivers and shallow lakes, and often these eggs are unguarded.

Large numbers of small eggs tend to be produced by those species which do not hide or guard their eggs and smaller numbers of larger eggs by those that do. In the latter, the individual probability of survival is doubly enhanced, for a larger egg contains more yolk for sustenance of the fry. On the other hand, the spawning requirements of the 'protective' type may be more stringent and less easily available, while the larger egg, with its smaller surface area : volume ratio, may require greater external concentrations of oxygen to survive.

The general trend is not always followed. The eggs of the pike-perch, which are guarded, are smaller (0.8–1.5 mm diameter) than those of the walleye (1.4–2.1 mm), which are not. Furthermore, the walleye produces only 30 000–65 000 eggs kg^{-1} body weight, compared with 110 000–260 000 in the pike-perch. This reversal of the expected trend is probably related to the contrasted habitats in which these fish live. The walleye inhabits well-oxygenated waters and the pike-perch stagnant or slow-moving, productive ones, with low oxygen concentrations and high turbidity, where the nesting habitat is frequently associated with very organic sediments. The large numbers of eggs produced, even though guarded, may be necessary in a fertile habitat containing many cyprinid fish, which prey on eggs to a greater extent than the species coexisting with walleyes

9.1.2 Feeding

Fry hatch from the eggs after varying periods of development, dependent on temperature, among other factors. Trout spawn in autumn, walleye between March and June and the tropical Nile perch and tilapia probably throughout the year. The grass carp spawns in rising flood waters in spring, when increased turbidity may camouflage the eggs against predators. Hatching occurs at times – spring and summer in temperate regions – when food is available for the fry. The egg yolk is only sufficient to feed the fry for a short time and any increase in the amount of yolk produced per egg would mean a corresponding decrease in the number of eggs produced.

Small fish can eat only small portions; large fish can eat bigger items. Less energy is needed to find a given amount of food as large items than as small ones. The diets of fish thus often change markedly as the fish increase in size. Trout fry (alevins) dart from their shelter in the gravel to take small chironomid larvae and Crustacea in the weeks after their yolk is used up. When they have grown to about 4 cm, they station themselves in the water about 8 cm apart, defending their 'water-space' against neighbours by aggressive darts. They then feed on invertebrates moving past them in the drift (see Chapter 4).

Older trout actively hunt food along the bottom and, at lengths > 30 cm, may take fish fry and larger fish. Trout have wide mouths and many backwardly directed teeth, which efficiently hold prey once it is grasped. Other fish species feeding in midwater, but on smaller items, such as zooplankton, have a much narrower mouth, protrusible into a lengthened tube, with which they suck up prey, like a vacuum cleaner. The tropical elephant-snout fish (Mormyridae) are good examples.

The adult Nile perch is a voracious feeder, even on other piscivors, but smaller, fish, such as the tiger fish (*Hydrocynus vittatus*). When it is younger, however, it takes invertebrates. The fry in Lake Chad, 0.3–1.35 cm in length, feed on planktonic Cladocera and inhabit shallow, weedy areas [426]. At about 20 cm, they begin to take larger invertebrates, a bottom-living prawn (*Macrobrachium niloticum*), snails and some small fish. As they grow, they take larger fish. Nile perch have large heads, widely gaping mouths and serried ranks of backwardly directed teeth; they vigorously chase their prey. The dorsal and anal fins are situated well back (Fig. 9.1), giving the fish a powerful tail thrust, which, as in the trout, allows bursts of high speed.

Newly hatched walleye, only 6–9 mm long, feed in protective schools on plankton – large diatoms, rotifers, nauplii of copepods (see Chapter 7). Progressively, they eat larger zooplankters and *Chaoborus* larvae (see Chapter 8). Finally, they become piscivorous, seizing their prey and then manoeuvring it until it can be swallowed head first. Some favoured prey species, such as yellow perch (*Perca flavescens*), have spiny pectoral fins, which would lodge in the throat if the prey were swallowed tail first.

Walleye have elongate gill rakers (the strips of bone that protect the delicate gills from damage by large particles pulled in with the respiratory water current) and it is on these that spiny fins could catch. The pike-perch does not have such elongate rakers and is able to swallow at least some of its prey tail first. Spiny fins have some advantage for the prey, because, in the time taken for a fish, seized usually at the tail, to be manoeuvred into a head-first position for swallowing, there is a greater possibility of escape. Spines may also be used by a fish to avoid capture if they can be jammed into rock crevices.

The grass carp becomes predominantly vegetarian at lengths above about 30 mm. As fry, it eats rotifers and crustaceans, occasional chironomid larvae and perhaps some filamentous algae. Between 17 and 18 mm, it takes more chironomids and fewer of the smaller zooplankters. By 27 mm, higher-plant food becomes prominent in the diet. Although, thereafter, it unavoidably takes invertebrates associated with water plants, the grass carp is well adapted to a plant diet and this is unusual. Like other Cyprinidae, the grass carp is toothless, but has strong projections of the pharyngeal bones, which line the region between the mouth and the entrance to the oesophagus. These 'pharyngeal teeth' are serrated in younger fish but become flattened, with both cutting and rasping surfaces, with age. Young fish eat only the softer, submerged plants, whereas older ones can tackle more lignified emergent plants.

The pharyngeal teeth tear and rasp the plant food into particles 1–3 mm in diameter. Only the cells that are rasped and ruptured are digested and half of the food passes out, undigested, as faeces [411]. The pH of the gut secretion is quite high: 7.4–8.5 in the anterior part, around 6.7–6.8 in the rectum, and such values are not particularly noteworthy. They provide a comparison with *S. niloticus*, which, perhaps because of its extremely low gut pH – between 1.4 and 1.9 – is one of the few fish yet examined that can digest even the blue-green algae [664].

Fry of this tilapia may feed on insect larvae, but the fish soon turn to an algal diet. For example, in Lake Volta, Ghana, *S. niloticus* eats the weft of algae loosely hanging from submerged dead trees, and bottom detritus is also taken. In Lake George, Uganda, it takes blue-green algae, but whether from the water column or from dense aggregations on the bottom is not clear. Feeding makes little use of the teeth or jaws. Algae are sucked in with the respiratory current, entangled with mucus secreted by glands in the mouth and then passed back into the pharynx. The food is thus not filtered out, as it is in many other plankton feeders, by fine projections on the gill rakers.

The first food of the day is not well digested, but, as the stomach secretion falls below pH 2, digestion improves and 70–80% assimilation of the blue-green algae *Anabaena* and *Microcystis* has been noted in laboratory experiments [665]. Assimilation from natural phytoplankton populations in Lake George was lower, about 43%, but this is still considerable, in view of the long-held belief that blue-green algae are indigestible as live cells.

Sarotherodon niloticus shows some selectivity in its food. A comparison between the percentage representation of different foods in the gut (*r*) compared with that in the external environment (*p*) may be used to calculate an electivity index [474]:

Electivity = $(r – p)/(r + p)$

In Lake George, *S. niloticus* showed a positive selection for *Microcystis*, *Lyngbya* (blue-green alga) and *Melosira* (a diatom), but discriminated against the smaller species, *Anabaenopsis* (blue-green alga) and *Synedra* (diatom).

9.1.3 Breeding

Changes associated with spawning are major events. The gametes alone may constitute a quarter of the body weight, and the energy demands in producing them and in the act of spawning may be very great. A 'spent' fish (one that has just spawned) is weak and more vulnerable to predation, and the extensive migrations which some fish undertake before spawning may have exhausted them so much that bacterial and fungal infections are common. Nonetheless, most fish spawn in several successive years after reaching maturity.

Breeding occurs at different ages in different fish, presumably at a time which represents, for each, a compromise between several factors. If left too late, there is a high chance of failure to breed, through early death. If attempted too early, the fish may be too small and unable to cope with the energy demands. Trout first breed when they are three or four years old, walleye from two or three years and *S. niloticus* at only a few months. This reflects high growth rates at higher water temperatures and lack of seasonal food scarcity in tropical fresh waters.

In the main part of Lake Albert (Uganda, Zaïre), *S. niloticus* reaches a size of 50 cm and breeds when it has attained 28 cm. In the Bukuku lagoon, a part of the lake now completely isolated by a sand bar, it breeds at 10 cm and achieves only 17 cm at most. This appears to be a response to the extreme environment of the lagoon, which is very saline and must dry out almost completely from time to time. A fish able to mature earlier and therefore produce more frequent generations may have a greater chance of survival there than one whose breeding cycle is longer than the persistence of its environment.

Breeding rituals are common in fish. They preserve adaptive differences between closely related species and may need specific environmental conditions. Examples are provided by the European three-spined stickleback (*Gasterosteus aculeatus*) and the African *Haplochromis burtomi*, a cichlid fish. Sticklebacks (Fig. 9.2) are small and silvery-brown (5–8 cm). They eat small invertebrates. Sometimes they winter in estuaries. In spring, the males leave the mixed shoals and take on breeding colours: bright blue and red underparts and a translucent appearance to the scales on the back.

Each male chooses a small territory on a sandy or silted bottom and defends it against other males. He excavates a depression by sucking up sand or mud and expelling it some distance away. He then collects strands of aquatic plants or filamentous algae and, with the help of a secretion from the kidney, forms them into a tunnel-shaped nest about 5 cm long and broad, which lies in the depression. At this stage, he becomes responsive to swollen, gravid (bearing ripe eggs) females, though not to thin, spent ones. If a suitable female enters the territory, he first creeps through his nest and then performs a 'courtship' dance, in which he zigzags towards the female and then turns towards his nest.

This may not immediately attract the female, although she may swim nearby, adopting a characteristic posture with the head up. The male may then dart at her with his spines raised and usually she will then be led down to the nest. By inserting his snout the male points at the nest entrance and then backs off and swims on his side with his back towards the female. She appears to inspect the nest opening, and then may retreat to the water surface. The courtship process may be repeated several times before the female enters the nest, with something of a struggle, for the opening is narrow.

Eventually her head is well in and her tail sticks out of the entrance. The male then puts his snout against the base of her tail and quivers violently. The tail begins to rise as the male continues quivering and when it is raised high, the female releases a stream of 50–100 eggs into the nest. When the last egg is laid, the female, now much thinner, rushes out of the nest as the

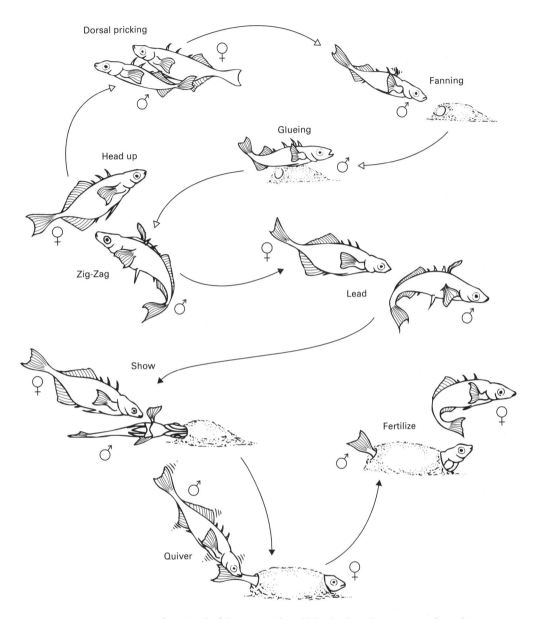

Fig. 9.2 Breeding ritual of three-spined sticklebacks (based on Wootton [1029]).

male bursts in and releases a cloud of spermatozoa quickly over the eggs. Thereafter, the female is chased away, the male prods the fertilized eggs deep into the nest, adds sand to reinforce and camouflage it and wafts water over the eggs with his fins. He remains guarding the fry until about 10 days after hatching, when his breeding colours have also faded.

Spawning behaviour in *H. burtomi* in Lake Tanganyika (Fig. 9.3). is equally complicated. Little is known of the preliminaries of courtship but, at

(a)

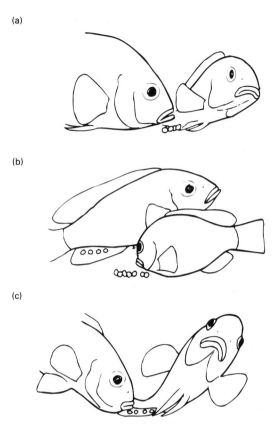

(b)

(c)

Fig. 9.3 Spawning in *Haplochromis burtomi*, a fish species of Lake Tanganyika. (a) The female has laid a batch of eggs; (b) she turns and takes them into her mouth before they are fertilized; (c) the male sweeps past the female, displaying the 'egg dummies' on his anal fin. In attempting to collect the dummies, the female takes in spermatozoa released by the male and fertilization of the eggs takes place in her mouth. (After Fryer and Iles [304], based on a film made by G.H, Wickler.)

its culmination, the female lays a very small batch of quite large eggs on the lake bed, while the male courts attendance. When she has laid the eggs, she quickly turns and scoops the unfertilized eggs into her mouth. Then the male sweeps past the female, displaying his anal fin as he does so. On it are light, circular markings, which, against the darker fin background, resemble eggs. The female is deceived by these and moves to pick up what she thinks are eggs. In doing so, she sucks in water from near the male's genital aperture, from which he has just released sperm and the eggs are fertilized in the female's mouth. The process is repeated several times, until her mouth is full of fertilized eggs, where they are protected until hatching.

These rituals appear often to have evolved to minimize the chances of hybridization in fish communities where specialization and speciation have resulted in minimization of competition for food or space. Unless each sex recognizes particular attributes in the other, mating does not occur. These colour and behavioured features are presumably genetically linked with gene combinations that code for the particular specializations that promote the chances of survival. In ancient lakes, such as those of East Africa, flocks of closely related species, particularly among the Cichlidae, have evolved (see Chapter 2) high degrees of specialization and endemicity (Fig. 9.4). Even

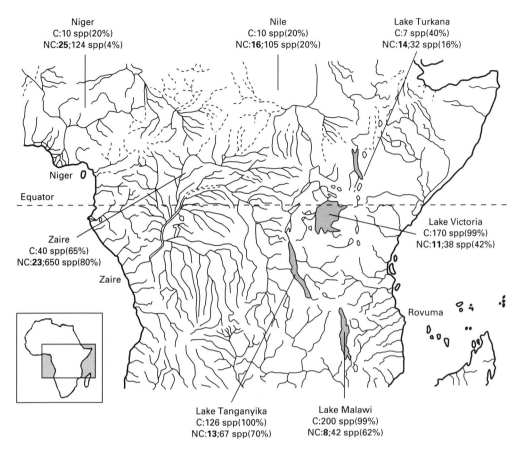

Niger
C:10 spp(20%)
NC:**25**;124 spp(4%)

Nile
C:10 spp(20%)
NC:**16**;105 spp(20%)

Lake Turkana
C:7 spp(40%)
NC:**14**;32 spp(16%)

Niger **0**

Equator

Zaire
C:40 spp(65%)
NC:**23**;650 spp(80%)

Zaire

Lake Victoria
C:170 spp(99%)
NC:**11**;38 spp(42%)

Rovuma

Lake Tanganyika
C:126 spp(100%)
NC:**13**;67 spp(70%)

Lake Malawi
C:200 spp(99%)
NC:**8**;42 spp(62%)

Fig. 9.4 Number of cichlid and non-cichlid species of fish in some of the major rivers and lakes of Africa. C is number of species of cichlid present, NC the number of other species, with the figure in bold type the number of families of these species. The percentage of species that are endemic is shown in parentheses. (From Greenwood [362].)

in temperate and near-polar lakes, something of the same process can be seen – for example, in brown trout.

Techniques that extract particular enzyme proteins and then characterize differences (polymorphisms) in the same enzyme have been used to distinguish differences between individuals or species in different individuals and populations by their electrical properties (electrophoresis) [264]. These studies have shown great variation in brown trout, with 54% of enzymes studied being polymorphic. This is consistent with the considerable colour and behavioural variation known from this widely distributed species. Indeed, since 1755, variants of the brown trout have been described as fifty different species. Variation is inevitable in a species spanning Iceland, arctic Norway, the Atlas Mountains of Morocco and the eastern region of Afghanistan, but it may be just as great within a single lake. Lough Melvin, in Ireland, for example, is only $22\,km^2$ but contains three brown-trout populations that are genetically distinct and do not interbreed. The gillaroo feeds on benthic organisms and spawns on the lake shores and outflow river. It has strongly marbled flanks. The sonaghen (or sonachen) is a midwater feeder and spawns in the smaller inflow streams. Its back is bluish and its marbling is less pronounced than that of the gillaroo. Thirdly, the ferox trout becomes piscivorous after its third year, lives longest and, in males, often retains, for all the year, the hooked jaw of its spawning period. It is redder in colour and spawns in the deeper waters of the main inflow to Lough Melvin.

These distinctive fish may have arisen from separate natural introductions, differentiation of some ancestral form by breeding isolation or artificial stocking. The latter is unlikely because the forms have been known for the two centuries that predate stocking operations. Evidence suggests a differentiation within the lough, based on the availability of different habitats and the custom of salmonids to return to spawn on the grounds of their own birth. Such local variation is a valuable resource for fishery management and breeding programmes, but has been greatly eroded in many parts of the trout's range by habitat deterioration and interbreeding with stockings of farm-raised fish. The status of trout, at least in the UK, appears to be declining [175, 242, 331, 652].

A second example of such rich variation is in the two genetic morphotypes, each with two subsidiary phenotypic morphs, of the Arctic charr, *Salvelinus alpinus*, in Lake Thingvallavatn, Iceland [853] (Fig. 9.5). One morphotype is a littoral, benthic-feeding fish, with small and large variants, all with overshot mouths and large pectoral fins, which aid in manoeuvring over the bottom. The larger form cruises over the bottom, while the smaller lives in interstitial spaces among the stones. The other morphotype is planktivorous in one morph and piscivorous in the other. All four types spawn in the littoral zone and feed on chironomid larvae at first. They avoid competition and remain distinct, however, because they spawn at different times (Table 9.1).

9.2 Choice of fish for a fishery

Knowledge of the natural history of fish is essential for wise choice of species for a sustainable fishery. In such a fishery, annual mortality due to natural causes and fishing combined should be no greater than annual growth of the stock plus recruitment of new fishes to it. A high-yielding fishery is thus one that uses fishes of prolific growth and recruitment matched by high natural mortality. Ideally, the natural mortality is replaced by removal in the fishery. A second requirement is that the fishery methods should not damage the habitat and hence interfere with spawning and recruitment.

Where only subsistence fisheries are concerned, with individuals or very

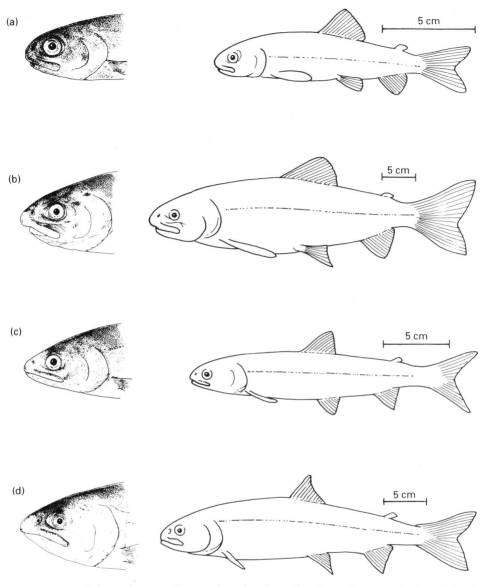

Fig. 9.5 Morphotypes of Arctic charr found in Lake Thingvallavatn, Iceland, with details of heads. (a) The small benthivorous form (SB), (b) the large benthivorous (LB), (c) the planktivorous (PL) and (d) the piscivorous (PI). (Based on Sandlund *et al.* [853].)

small groups using relatively inefficient methods, such as spears or hooks and lines, almost any fish species will sustain a limited amount of hunting. Where more people, using intensive methods, such as large nets, in a commercial fishery are involved, the most suitable fish are those of the open water (pelagic fish), followed by the bottom-living (demersal) fish of the profundal. Fish of the littoral zone are usually quite unsuitable.

Table 9.1 Some distinguishing morphological and ecological characteristics of four Arctic charr morphs in Lake Thingvallavatn, Iceland (from Sandlund *et al.* [853]).

	Colour, size range (cm)	Spawning period	Habitat and main food
Small benthivorous (SB)	Dark, 7–31	July–Nov.	Littoral; snails
Large benthivorous (LB)	Dark, 9–55	July–Aug.	Littoral; snails
Planktivorous (PL)	Silvery, 13–26	Sept.–Oct.	Whole lake; crustaceans, zooplankton, chironomid pupae
Piscivorous (PI)	Light silvery, 23–65	Sept.–?Oct.	Whole lake; three-spine stickleback, small charr

Pelagic fish live in an unstructured habitat, where, like the zooplankton (see Chapter 7), they are continually at risk from their own predators. Many drifting eggs are usually produced, and mortality at all stages is high. The fish tend to spawn early and grow fast. The 'structure' of the open water is not damaged by fishing methods. In contrast, the littoral habitat can be severely damaged by trawled nets, which, as well as disrupting nesting sites, may also wreck the habitat of fish feeding on specialist food sources (e.g. the rock-dwelling fish of Lake Malawi (see Chapter 8)). Egg guarding and hence low egg production and mortality mean low recruitment rates, and the cover provided by a well-structured habitat limits natural mortality rates. Demersal fish form an intermediate case, with a much less vulnerable habitat than the littoral, but one which, nonetheless, can be destroyed by too frequent disturbance.

Suitable fisheries fish should thus be adapted to a relatively unpredictable habitat, in which risks of natural death are high. Whole lakes, littoral zone included, may sometimes fall into this category, if they are subject to such disturbances as frequent drying out. Lake Chilwa (see Chapter 8), in Malawi, is one such. It has limited fish diversity (see Chapter 2), with species that are omnivorous and unfussy about their conditions for breeding. They are capable of living in a variety of habitats and of growing fast and reproducing quickly when they recolonise the lake from the rivers after a drying phase [662].

Fishes such as these have been naturally selected for *r* (high growth rate), as opposed to *K* (stable population), characteristics [303, 582]. These terms come from the logistic growth equation:

$$dN/dt = rN(K - N)/K$$

where N is number, t is time, r is the intrinsic rate of natural increase (see Chapter 7) and K is the 'carrying capacity' of the environment. Fish selected for r characteristics tend to invest much energy in growth and reproduction.

Examples are the five main fish compared in Section 9.1. Fish selected for
K characteristics make efficient use of the resources to which they are
closely adapted. Examples are the littoral fish of Lake Malawi and the other
African Great Lakes, discussed in Chapters 2 and 8.

9.3 Measurement of fish production

Production is estimated as the increase in weight of the population per unit
time and usually per unit area. The first problem is to estimate the absolute
population size per unit area of waterway. The most widely used method of
sampling fish is to net or trap a sample of the population. These fish are
then marked (by dyes or fin clipping) or tagged in some way to make them
identifiable and then released again. After a period to allow them to mix
randomly (it is assumed), the population is again sampled and the number
of marked and unmarked fish counted. The ratio of recaptured marked fish
(n_{rm}) to the total number recovered in the second sampling (N') is taken to
be equal to the ratio of the number of fish originally marked (n_m) to the total
population (N):

$$N = (n_m N')/n_{rm}$$

The value of N may be expressed per unit area of the lake. In a big lake,
however, the population may move over only part of it, which must be
determined by tagging fish with radio-trackers and following their movements. Practical problems of the mark–recapture method are in ensuring
that the samples are not selective of particular fish sizes and that the
marking technique does not alter fish behaviour or increase the chance of
death. These ideals are probably never attained.

Mark–recapture techniques are most suitable for small water bodies,
where there is a reasonable chance of recapturing sufficient of the marked
fish. In larger lakes, it is more usual to estimate the stock either by relative
methods (catch per unit effort of fishing (see below)) or by fishing of
representative areas to depletion. An area is isolated by nets and then
repeatedly fished in a standard way until no more fish are caught. The
cumulative catch is then plotted (x axis) against the sequential number of
fishings (y axis). The intercept of the asymptote of this curve on the x axis
gives an estimate of the total stock.

In capturing inshore fish for marking or depletion estimates, beach seine
nets may be useful. These involve the laying out of a small-meshed net in
a wide arc, with the start and end of the arc on the shoreline of a shelving
beach. The top of the net has floats to keep it at the surface and the bottom
is weighted to keep it down. After setting, the net is hauled in to
concentrate the fish (Fig. 9.6). Seines may also be set in the open water, but
must have a rope threaded into the bottom so that as the net is drawn in,
it is closed at the bottom into a 'purse'. This requires a mechanical winch

Fig. 9.6 Seine-netting a partly drained pond in France (left). Quite so much labour is not usually necessary. Gill nets (right) are efficient for food fisheries, but, because they kill the fish, must be used prudently in scientific work.

on the boat used. Fish escape around the seine nets, but this may be minimized by an elongated sock of netting bulging out at the centre and called a cod end. The fish tend to move into this as the net is hauled in. Trawling is also a relatively non-selective method for bottom-living fish. The trawl is a bag of netting, kept open either by its resistance to movement through the water or by a wooden beam across its entrance, and pulled by a sufficiently powerful boat.

Gill nets kill the fish and thus cannot be used for mark–recapture estimates but can be used for depletion fishing of limited areas. The net is floated passively in the water at a desired depth and fish encountering it may move through its meshes and pass out the other side, or they may be too big to penetrate it at all. Particular size groups, however, move part through it until the widest part of their body becomes jammed and their gill covers catch on the netting as they try to move back. Such nets, dependent on their mesh size and the slackness with which the net is set, are very selective for particular size ranges. A set (or fleet) of different mesh sizes may be used for experimental fishing to obtain an estimate of the balance of different-sized fish in the population and a relative estimate of the size of the fish stock.

Finally, but also destructive, depletion fishing has been used in delimited bays in some lakes by use of the fish poison, rotenone. This is extracted from particular leguminous plant species (*Derris*, *Lonchocarpus*, *Tephrosia*), whose dried roots may contain up to 5% of the poison. It acts by constricting the blood-vessels of the fish when applied at about 0.05–0.1 mg l^{-1}, and has a short half-life once in the water. However, not all fish killed by the poison come to the surface for weighing and rotenone may have long-term effects on other organisms, particularly invertebrates. It is a useful piscicide for certain management applications, but not ideal for biomass estimates [666].

9.3.1 Growth measurement

The individuals of most temperate fish species spawn over a limited period and the newly spawned generation or cohort can be treated as a unit, often recognizable in successive years because of the annual or other periodic rings laid down in the skin scales (Fig. 9.7) or in the otoliths (the ear bones). When a population is sampled, the captured fish are held in an aerated tank, while each individual, or a random sample if there are many, is weighed and its length measured. A length/weight graph may be used to avoid the need to weigh as well as measure in future samplings. The fish are usually lightly anaesthetized to calm them during these measurements.

Scales are removed from a part of the body, often high on the back just below the dorsal fin, where their rings are known to be clear. If this is done carefully, with blunt forceps, the fish should suffer little damage, although its chances of becoming infected by fungi, protozoa or bacteria subsequently may be increased. Removal of otoliths requires dissection and is fatal. The scales may be used simply to age the population and identify separately the different cohorts (often called year classes) or they may be used on an old fish to estimate the growth in previous years. The distance between the rings is approximately proportional to the growth in the year the ring was laid down. Rings, however, may be reabsorbed during periods of starvation.

For a given cohort, an Allen curve (see Chapter 4) may be plotted, relating the mean weight of fish (on the vertical axis) to the numbers remaining in the cohort for a sequence of sampling occasions (Fig. 9.8). At first there are many small fry, but, as individual weights increase, the numbers decline. The area under the curve as it approaches the abscissa, when the last survivor dies, gives the total production. Production in a given year can be found by determining the area under the curve between points representing the times in question. The curve is roughly hyperbolic, but has irregularities; numbers decline in winter in temperate lakes, although the

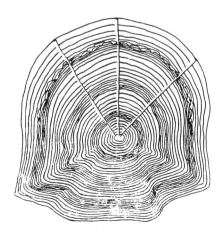

Fig. 9.7 Scale of roach showing three main annual rings. At the outer edge of the third, irregularities are associated with the first spawning. The annual rings are in fact tight groupings of subsidiary increments, laid down close together in winter and more widely in summer. The fish is in its fourth year (referred to as 3+) and was laying down the wide increments associated with summer feeding when the scale was removed.

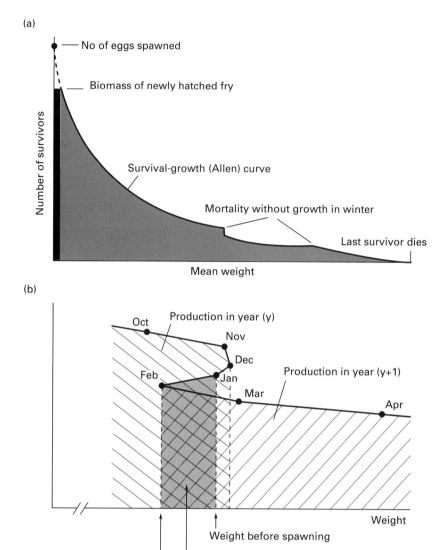

Fig. 9.8 Allen curve for fish. Number of animals in a given cohort is plotted against mean weight for a series of dates throughout the life of the cohort (a). The area under the curve gives the total net production. Irregularities in the curve arise from mortality without growth in winter and from spawning. Detail of this is shown (b). Spawning results in loss of weight that is part of the total production and must be allowed for in the calculations. (Based on Le Cren [559].)

increase in weight of the survivors may be negligible. Weight change may even be negative, as fat stores laid down the previous summer are used up. A similar loss of weight occurs during spawning, when the weight converted to gametes can be determined from the area under the curve as it reverses direction during the spawning period. By adding up the production of each of the several year classes present in a given year, and, by doing this for each species, a measure of the total fish production may be obtained.

9.3.2 Fish production in lakes

Sufficient estimates are now available to make some generalizations about fish production. In both rivers and lakes, it bears a general relationship to nutrient loading or concentration (Fig. 9.9) [220, 442, 765]. This suggests that the community as a whole is controlled by bottom-up mechanisms, reflected in the amount of primary productivity in lakes and perhaps the rate of processing of allochthonous detritus in rivers. Production (P), total

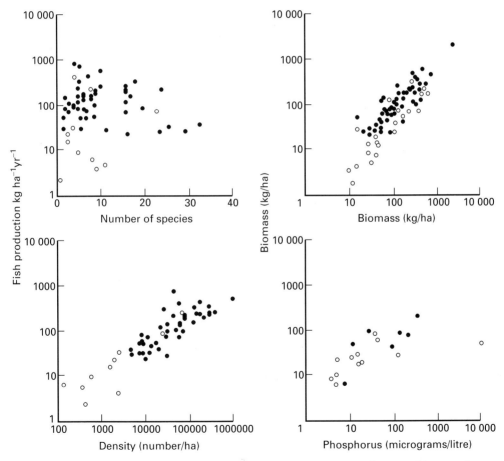

Fig. 9.9 Correlations between fish production and number of species, density and biomass in a series of lakes and rivers (closed circles – rivers; open circles – lakes). A correlation between biomass and total phosphorus concentration in the water is also shown. (From Randall *et al.* [796].)

Table 9.2 Comparison of fish production and related variables in rivers and lakes. Sample sizes were between 19 and 31 for variables in lakes and 42–58 in rivers. (From Randall *et al.* [796].)

	Mean		Ratio
	Lakes	Rivers	(rivers to lakes)
Production (P) (kg ha^{-1} year^{-1})	81.8	273*	3.3
No. of species	6.3	9.4	1.5
Density (no. ha^{-1})	5580	75 700*	13.6
Weight of fish (g)	28.9	4.1*	0.14
Total biomass (B) (kg ha^{-1})	83.8	146*	1.7
P : B ratio	0.78	1.63*	2.1

*Indicates a statistically significant difference at at least the 0.005 level (Mann-Whitney U test).

biomass (B), density (number of fish per unit area) and turnover (P : B ratio) seem to be significantly higher in rivers than in lakes [796, 842], but the average weight of the fish is greater in lakes. Number of species is also generally higher in rivers (Table 9.2). The range of annual production in a sample of fifty-five rivers was found [796] to be 26–2800 kg ha^{-1} year^{-1}, compared with 2–398 in a sample of twenty-two lakes.

Rivers may be more productive because they receive a greater load of energy and an immediate supply of nutrients from the catchment, concentrated into a much smaller area. Lakes receive their nutrient supplies after much processing and sequestering in the river systems, and, being usually deeper than rivers, their production is limited more by light availability. Rivers may provide a greater diversity of habitat structure, particularly where they retain natural quantities of woody debris, compared with the dominant unstructured open water of all but the smallest lakes. In turn, this may allow a greater diversity of species, although this is not certain. That lakes have larger fish could reflect a smaller gain of habitat in flowing waters, with a large variety of small species inhabiting crevices among rocks and debris or under banks, where flow is lessened. Equally, it might mean a greater degree of piscivory in lakes, where refuges are scarcer for prey. Predators, by removing small species or the smaller specimens of larger fish, tend to skew fish populations to larger sizes.

9.4 Commercial fisheries

Fisheries do not differ from any other commerical venture in that they must maximize the yield of their product, harvestable fish. The particular problem of fishing, however, is to sustain the yield from year to year. No more must be harvested than the equivalent of the annual increment of fishes becoming large enough to be worth catching, plus the annual growth of those already fishable. It is relatively easy for fishermen, when not subject to control, to remove more than this annual recruitment and growth, in which case the fishery will eventually fail.

Figure 9.10 shows the increase in biomass of a newly hatched cohort of fishes. The rate of increase is at first high, but declines as the maximum potential biomass, represented by the curve's asymptote, is reached. Simultaneously, some members of the cohort die and the rate of mortality is at first high, but declines as the fishes become older, bigger and less vulnerable to predators. The net effect of growth and death is to create the biomass curve; growth exceeds death in the early part of the cohort's existence and total biomass increases, but eventually mortality predominates and the biomass falls. There is a point at which the biomass of the cohort is maximal. The aim of a fishery is to catch that portion of the biomass which, after the maximum has been attained, would be lost to natural mortality. The more rapidly the fish can be removed after the maximum the better. The intention is to replace the natural survival curve *ab* with the fishing mortality curve *ac*, the area between the curves representing the yield. It is important that fishing does not remove fishes of an age at which growth exceeds natural mortality, for this will reduce the potential yield and may also remove fish that have not spawned.

Removal of pre-spawners may reduce the potential recruitment to the next season's fishery and is called overfishing. It is also important that fishing mortality should only ever replace natural mortality and never

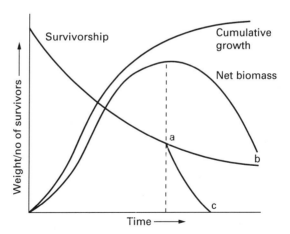

Fig. 9.10 Ideal exploitation of a fish stock. The net biomass represents the balance between cumulative growth and death, and at some time reaches a peak. Provided spawning has by then occurred, the peak represents the time from which most efficient removal of fish can take place. The natural survivorship curve, *ab*, is replaced by the steeper fishing-mortality curve, *ac*, as fishing removes biomass before it can be further diverted to disease organisms, parasites or other predators. (Based on Fryer and Iles [304].)

exceed it. Fish that can safely be removed are thus mature and should not be allowed to grow to their potential maximum individual size, by which time most of the biomass of the cohort will have been lost by natural mortality.

The year-to-year changes in fish population can be represented by the Russell equation [838]:

$$P_2 = P_1 + (R + G) - (F + M)$$

where P_1 and P_2 are the fish stocks (biomass) in two successive years, R is the annual recruitment of mature fishes to the fishery, G is the growth made by those already fishable but not yet removed, F is the annual mortility due to fishing and M the natural annual mortality. In a well-run fishery, P_1 is equal to P_2 and M is zero, but this is never achieved, for natural environmental fluctuations will alter R and G in ways that a fisheries manager cannot predict in time for him to regulate the amount of fishing (F). The value of P should, however, fluctuate around a mean, without showing a general tendency to increase (underfishing) or decrease (overfishing). The year-to-year changes in a fishery can be monitored and fishing methods regulated, through devices such as controls on net-mesh size, number of nets allowed and season for fishing, to keep F at the desired level. The term F is theoretically the maximum sustainable yield, if P is stable and M is minimized, but the concept is outmoded, not only for the reasons given above that P and M will be affected by natural environmental changes and are not constants, but also for economic reasons.

Figure 9.11 shows the relationship between yield of a fishery and fishing effort – the amount of fishing carried out. As the latter increases, the yield reaches a peak (the maximum yield), above which the yield is reduced, as overfishing interferes with recruitment or growth by removing spawning fish or fish not yet growing at their maximum rates. The yield curve is also a curve describing the value of the catch, assuming the sale price of the fish remains steady. As fishing effort increases, so proportionately does its cost (the straight-line graph in Fig. 9.11).

Overfishing eventually leads to cost being greater than value, and some fishermen go out of business. This is the usual state of an unregulated

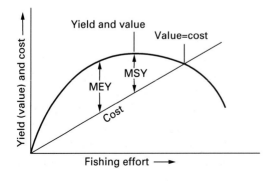

Fig. 9.11 Relationship between yield of a fishery (also its commercial value) and fishing effort (the curved graph) and between costs of a fishery and fishing effort (straight-line graph). MEY, maximum economic yield; MSY, maximum sustainable biological yield.

fishery. At an earlier stage, however, the greatest profit (value minus cost) is reached at a lower yield, the maximum economic yield (MEY), than the supposed maximum sustainable biological yield (MSY). A well-run fishery, in the interests of both fishermen and fish, will thus attempt to keep yields around this point.

Fisheries can be managed through sets of equations, or models. Much sophistication has been achieved in the development of such models for marine fisheries [184–186, 368, 776], where very good data can be obtained from fish landings. Most commerical freshwater fisheries, however, are subtropical or tropical and involve large numbers of fishermen operating individually or in small groups around the indented margins of large lakes. Catch statistics are then not easy to obtain. A former fishery for *Oreochromis* (*Tilapia*) *esculentus* around North Buvuma Island in Lake Victoria, however, was notable in that landings were made over a restricted area, where fishery scientists managed to account for changes in the population [315, 316]. The fishery has now been overexploited for political and commerical reasons, and thus provides a doubly interesting example.

9.5 The North Buvuma Island fishery

For good fishery management, the fished population should be recognizable and discrete. In a small lake, the entire population of a given species may provide recruits, but in a large lake there may be several different stocks or subpopulations, separated by geographical barriers, such as deep water, for inshore species, or stretches of swamp. Such subpopulations may be distinguished on the basis of slightly differing blood proteins, by minor morphological differences or, if neither of these exist, by careful observation of the movements of tagged fish.

The absolute size of such a 'unit stock' may be determined by the mark–release–recapture method, outlined above, but an absolute measure of population size, as opposed to a relative one, is not usually necessary for fisheries management. The catch obtained (numbers or weight landed) per unit fishing effort (for example, hours of fishing, numbers of nets set per night, total length of nets set or numbers of trawls made), called the stock density, is such a relative measure.

The aim of management is to predict future populations of the fish and the effects of different fishing intensities. This is best done separately for each age cohort and depends on recognition of separate cohorts. In temperate regions, annual scale or otolith rings, produced by the winter pause in growth, enable this, but for tropical fishes, such as the Buvuma *Oreochromis*, the scale rings may be irregular or laid down at 6-monthly spawning intervals. For the *Oreochromis* fishery, it was most convenient to consider the fish in successive length classes rather than age classes.

Consider a cohort of fish of a given length range, for example

23.0–23.9 cm, designated as b. N_b is the number of such fish and they die at the rate M_b from natural causes and F_b from the fishery. It takes a time t_b for a fish to grow through the length range b, and it is assumed that the mortality is exponential – it slows with decreasing numbers. The number of fish surviving into the next group $(b + 1)$ is then:

$$N_{b+1} = N_b e^{-(F_b + M_b) t_b}$$

where e is the base of natural logarithms. This represents an instantaneous abundance, for F and M are instantaneous mortality rates. A measure of the average abundance $\underline{N}_b$, is given by integration of the above equations for each length category in the range considered:

$$\underline{N}_b = N_b e (1 - e^{-(F_b + M_b) t_b})/(F_b + M_b)$$

The problem now becomes one of estimating t, M and F, so that future values of N_b can be predicted for successive length groups until the cohort is completely removed. The value of N_b is given by catch per unit effort, after a correction has been applied for any differences in selectivity of different sorts of fishing gear used.

9.5.1 Estimation of t_b, F_b and M_b for the Buvuma *Oreochromis* fishery

The value of t_b can be calculated from measurement of the rate of growth of the fish. Fish grow proportionately more slowly the bigger they become and this growth is often mathematically expressed by the von Bertalannfy equation:

$$L_{b+1} = L_\alpha - (L_\alpha - L_b) e^{-K t_b}$$

Where L is length and L_α the asymptotic maximum length attained by the fish as its growth rate drops in old age. The term K is a constant defining the rate of deceleration of growth. Fish in the length groups b, $(b + 1)$, etc. must be aged and L_α established from a plot of L against age. The value of K can be calculated from rearrangement of the equation. An estimate of t_b can then be made.

Fishing mortality is related to fishing effort and a graph can be plotted of total mortality for a length or age class against effort (Fig. 9.12) from several years' data. Total mortality is given by:

$$N_{b+1} = N_b e^{-(Z)}$$

where (Z) is the coefficient of total mortality (natural plus fishing) and N_{b+1} and N_b are the numbers (given by catch per unit effort) of the cohort at successive times. The intercept on the total mortality axis gives the mortality in the absence of fishing, M, and the curve gives the fishing mortality, F, for any given degree of effort. A conservative assumption is that M does not decline as F increases. Fishing mortality can also be inferred by tagging fish

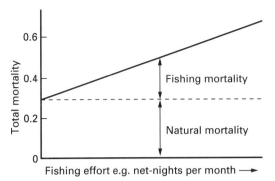

Fig. 9.12 The determination of natural mortality from the relationship between total mortality and fishing effort.

and finding the proportions of tags recovered by the fishery. The fishing mortality in the *O. esculentus* fishery could be regulated very closely by the use of different sizes of mesh in the gill nets that were used.

The original equation for N_b, above, can now be solved for all length groups (approximately age cohorts) in the fishery and the following variables calculated for future periods of the fishery: biomass (the product of abundance and average weight of the length group as a total for all length groups); numerical catch (the average abundance multiplied by the relevant fishing mortality in each length group); and catch in weight (the product of numerical catch and mean weight as a total for each length group).

Table 9.3 gives values for M_b and F_b for the lake Victoria *O. esculentus* fishery. It can be seen that 5-inch mesh (14.2 cm) does not catch fish below

Table 9.3 Mortality and net selectivity in the Buvuma Island *Oreochromis* fishery (from Garrod [315]).

Length (cm) (b)	Total mortality per year (Z_b)	Natural mortality per year (M_b)	Fishing mortality per year (F_b)	Percentage of catch of a particular length for nets of mesh size		
				4 in	4.5 in	5 in
23–23.9	0.28	0.01	0.27	9.9	–	–
24–24.9	0.40	0.01	0.39	38.6	–	–
25–25.9	0.33	0.01	0.32	82.5	4.4	–
26–26.9	0.37	0.01	0.36	98.9	19.1	–
27–27.9	0.40	0.02	0.38	65.5	51.6	1.0
28–28.9	0.46	0.02	0.44	24.0	89.5	11.0
29–29.9	0.52	0.12	0.40	4.9	98.0	32.0
30–30.9	0.63	0.27	0.36	0.5	82.6	73.5
31–31.9	0.76	0.54	0.22	–	46.9	99.0
32–32.9	0.93	0.79	0.14	–	25.5	91.5
33–33.9	1.32	1.02	0.30	–	10.2	63.5
34–34.9	2.0	1.12	0.88	–	4.1	34.0
35–35.9	5.0	1.14	3.86	–	–	17.5

26.9 cm length but is very effective in killing the 31.0–31.9 cm category, whereas 4-inch (11.4 cm) mesh kills fish optimally at length 26.0–26.9 cm and not at all above 31 cm.

Figure 9.13 shows the effects of various levels of fishing effort (which is proportional to fishing mortality) on the steady-state biomass of the stock, the weight of catch and the catch per unit effort. The effects are more extreme at the smaller mesh sizes, because they crop fish whose growth rate is still higher than the natural mortality. There would thus be advantages in using only 5-inch-mesh nets, for these keep the population biomass high, thus favouring the maintenance of spawning and recruitment and keeping the fish mortality high, between 0.5 and 1.0. Decreasing the mesh size further does not increase the catch, but decreases the stock slightly and the catch per unit effort markedly. This means an uneconomic use of labour and gear. For the fishery in question, a fishery mortality of 100% is equivalent to the setting of 46 000 net-nights per month or about 1500 nets per night. Each net is a standard length.

For various reasons, the fishery has not been rationally exploited, as it might have been with the help of the model outlined above. Mesh sizes much smaller than 5 inches have been used and events have taken an unfortunate course (Fig. 9.14).

Between 1953 and 1956, the fishery was fairly stable, with $F = 0.3$ and 153 000 net-nights fished per month. The model shows that a catch per net of 1.44 fish was obtained, which was marginally profitable for the fishermen.

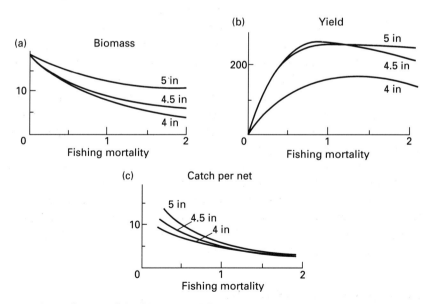

Fig. 9.13 Predictions of the fishery model for the Buvuma Island *Oreochromis* fishery for different levels of fishing mortality (related to fishing effort) and different mesh sizes of gill net (after Garrod [315]).

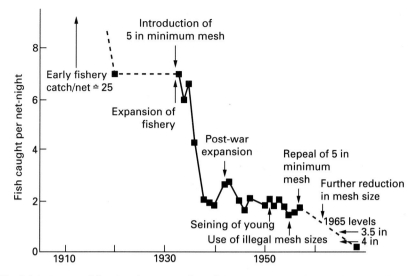

Fig. 9.14 History of the *Oreochromis esculenta* fishery in Lake Victoria (redrawn from Fryer [298]).

This was not the maximum yield that could be sustained biologically, but the amount of fishing needed to attain this is often uneconomic. In 1955 and 1956, for unknown reasons (perhaps poor recruitment), the catch per net fell. Legislation had been in force since the 1930s which forbade nets of less than 5-inch mesh. Illegal 4.5-inch nets were used and the yield temporarily went up because a section of the population previously unexploited was now being fished. The minimum 5-inch net legislation was repealed.

Under the new conditions, the model would predict a catch of 1.8 fish per net-night and catches of around 2.0 per net-night actually recorded provided part of the validation of the model. The increased catch provided incentive to the fishermen, which led to more nets and a renewed decrease in catch per net to 1.36. This was smaller than that previously, and meant a 25% reduction in biomass of the stock. It is believed that this reduced the rate of recruitment to marginal levels.

In recent years, both 4-inch and 3-inch nets have been used! These removed some fish before they had begun to breed and meant that the yield of the fishery could only be temporarily sustained at a very low level. The catch per net was 0.35 in 1968. More recent statistics are not available, but in any case the fishery has been disrupted by introduction of the Nile perch to the lake, a matter which is discussed later.

9.6 Approximate methods for yield assessment

The single-species approach to managing a fishery, described above, despite its limitations, offers the best method in principle. It has worked well in the regulation of some marine fish stocks by the developed nations, although it has been less used in fresh waters. It has not worked perfectly, however, for, despite its scientific sophistication, it does not account for sociological and political factors. Although an ideal fishing effort can be prescribed, it may be impossible to enforce it, especially on the open sea and under pressure from lobbies that wish to increase their yields, even though unsustainably. Most of the world's fisheries, in consequence, are overfished.

It is in the developing world, however, that inland fisheries are of the greatest importance, and here the diverse fish faunas and catches, lack of enough trained fishery biologists and more diffuse nature of the fisheries generally preclude sophisticated methods anyway. Events may also overtake the models, as has happened in the *Oreochromis* fishery described above. Frequently a *laissez-faire* attitude, with no regulation at all, leads to great variations in yield, with failures at some times, and the policies of governments may be towards maximum short-term exploitation, with little planning for the future. In these areas, some approximate method of predicting a safe yield of groups of fishes is better than no regulation at all [840].

Such methods predict the likely total stock of fish and then determine an annual yield from simple empirical equations. One way of predicting the yield directly is from the relationship between morphoedaphic index (see Chapter 6) and yield, determined from a series of lakes in the area and then applied to further lakes. A second way is to use the particle-size hypothesis of Sheldon *et al.* [890].

This hypothesis claims that approximately equal concentrations of biomass occur in a food chain for different particle sizes, segregated by order of magnitude in multiples of 10. In the equatorial Pacific Ocean, Sheldon *et al.* [890] found about equal concentrations of zooplankton (10^3 μm), small fish and squid (micronekton) (10^4 μm) and tuna fish (10^5 μm). Thus, if estimates are available of the biomass of one particle-size group – for example, the phytoplankton or zooplankton – that of the group of exploitable large fish can be inferred. Then this value (*B*) can be inserted in the empirically determined yield equation of Gulland [369]:

$$Y = kMB$$

where *Y* = yield, *k* is an empirically determined constant (about 0.4) and *M* is an estimate of the average natural mortality per year of the group of fish concerned. For Lake Nasser in Egypt, *B* was determined as 10^4 t [246] from zooplankton estimates. Data were available for *M* for a variety of fish (Table 9.4) from the ages of the fish caught.

To obtain the product *MB*, the individual mortality coefficients (*M'*) were multiplied by the proportion that each fish constituted of the total biomass (*B'*) and then $\Sigma M'B'$ was multiplied by 10^4, to give a final value for *Y* of about 1100 t [841]. This value was similar to that obtained by other

Table 9.4 Determination of long-term potential yield of Lake Nasser by use of the particle-size hypothesis (after Ryder and Henderson [841]).

Species	Mortality coefficient (M') $(year^{-1})$	Proportion of total biomass (B')	$M' \times B'$
Sarotherodon niloticus	0.25	0.56	0.14
Alestes sp.	0.4	0.18	0.072
Labeo sp.	0.25	0.09	0.023
Lates niloticus	0.1	0.07	0.007
Bagrus bagrus	0.15	0.04	0.006
Clariidae	0.3	0.01	0.003
Schilbeidae	0.6	0.01	0.006
Other	0.5	0.04	0.02
			$\Sigma\, M'B' = 0.28$

techniques and suggests that the particle-size hypothesis might be more widely applied. The power of this general approach seems either corroborated or rendered highly suspect by its ability to predict the stock of the Loch Ness monster [860, 889], an animal which almost certainly does not exist [669].

9.7 Changes in fisheries

Management of any fishery is vulnerable to the fact that fish growth and recruitment are subject to natural environmental change, particularly of climate. More often, however, problems are caused by changes wrought by human activities. These may be indirect, having an ultimate effect on the fishery, or they may be deliberate, but unsuitable, manipulations of the fishery itself. An example of interaction of both kinds is that of the North American Great Lakes, and of the latter kind some of the East African Great Lakes.

9.7.1 The North American Great Lakes

The North American Laurentian Great Lakes stretch almost halfway across the continent and are among the largest lakes in the world. They drain a huge area, originally of conifer forest to the north and deciduous forest to the south, and were once probably all well-oxygenated, clear-water lakes, with maximum total phosphorus concentrations probably less than $2\,\mu g$ $P\,l^{-1}$, very much towards the lower end of the fertility continuum (see Chapter 6).

The waterway they provided, from the Alantic Coast to the midwest and plains region in the USA and Canada, was a main route by which the continent was explored by Europeans in the seventeenth and eighteenth centuries. The discovery of minerals and cultivable land led to great increases in population. For example, the catchment of Lake Erie, which in 1750 was a largely unexploited wilderness supporting perhaps 100 000 people, now contains some 12 million people with their associated industry and agriculture. Lake Superior has changed the least in 200 years; its catchment still contains much intact forest. Changes in the lakes have been greatest around the shallow Lake Erie and intermediate in the other three Great Lakes, Michigan, Huron and Ontario. The most apparent change in the twentieth century has been eutrophication, although other changes have been equally significant and began in the eighteenth and nineteenth centuries. These changes are reflected in the commercial fish catches of the waterway [150, 900, 901].

Commercial fish yields over the past century have remained relatively constant in Lakes Superior, Michigan and Erie, but have declined in Lakes Huron and Ontario. The data mask the fact, however, that great declines in yields of prized salmonid, coregonid and other fish have been compensated by catches from a much less diverse fish community, supported by increased fertility and production in the water. The reasons for the changes, in the order in which they became significant, were: intensive selective fishing, modification of the tributary rivers, invasion by or introduction of marine species and, lastly, eutrophication.

The changes began in Lake Ontario, the lowest basin on the waterway, and have spread upstream so that the state of the Upper Great Lakes fisheries in 1970 was approximately that of Lake Ontario in 1900. Atlantic salmon, *Salmo salar*, were only ever present in Lake Ontario in the Great Lakes system, for upstream movement was blocked by the Niagara Falls. Fishing for salmon began in the 1700s, was reduced by 1880 and had ceased altogether by 1900. The fishing was intensive, but the species survived it for many decades and seems to have disappeared because of changes in its spawning sites in tributary streams. In the nineteenth century, forests were cleared over much of the catchment and dams were built on the streams to power sawmills. Waterlogged sawdust increasingly floored the streams.

Clear felling has two effects on drainage streams, other than chemical ones (see Chapter 3). It reduces flow in summer, because more water evaporates than previously, and it increases temperature, because the streams are no longer shaded. Repeated attempts to re-establish an Atlantic salmon fishery in Lake Ontario have failed, because the spawning stream waters are now too warm and the flows insufficient to maintain the cool, oxygenated water and gravel bottom which the salmon require.

In the upper lakes and in Lake Ontario, after the salmon declined, whitefish (*Coregonus clupeiformis*) and lake trout (*Salvelinus namaycush*) were both heavily overfished, although additional factors contributed to their decline, and fisheries for them had declined in Lake Erie by 1940 and Lake

Huron by the 1950s. A trout fishery is now maintained by annual stocking. Gradually, however, after these fisheries became less profitable, lake herring, or cisco (*Leucichthys artedi*), and deep-water ciscoes (*Leucichthys* spp.) were subsequently fished out. Currently, the lake fisheries depend on percids and other fish that have been favoured by different changes taking place in the lake.

The sturgeon (*Acipenser fulvescens*), although a valuable commerical fish, was deliberately removed, because of the damage it did to nets. By 1890–1910, it had almost disappeared, partly from overfishing, as its valuable by-products (gelatin, isinglass – a bladder extract used in clarifying beverages and sizing textiles – and caviare) were prized, and partly from ruination of its spawning habitat (see below). It is particularly vulnerable, with a low growth rate and late sexual maturity. Even following a ban on commercial fishing, it is now common only in parts of Lake Huron.

The changes in tributary streams, which affected salmon reproduction in Lake Ontario, became widespread elsewhere in the late nineteenth and early twentieth centuries. Most of the commerically exploited fish were those of shallow water, which entered streams to spawn. Congregation of sturgeon, coregonids and percids in the water below mill dams provided easy fishing with seines, dip-nets and even spears. Drainage of swamps and marshes, associated with the headwaters, removed a favoured breeding habitat for sturgeon.

Between 1860 and 1880, two marine species, the sea lamprey (*Petromyzon marinus*) and the alewife (*Alosa pseudoharengus*) entered Lake Ontario. They may have come up the St Lawrence River, which opens out at the northern edge of their ranges, or via the canal built in the early 1880s between the Hudson River and Lake Ontario. The Hudson River enters the sea at New York. The lamprey feeds by rasping fish flesh with its tongue after it has attached by a sucker, which forms its jawless mouth. Both species could have entered Lake Ontario at any time in the previous centuries, but, if they did, they were unable to establish significant populations.

Possibly the community changes caused by fishing and stream modification provided suitable niches for them. The Erie and Welland canals, both of which, for navigation purposes, bypass the Niagara Falls, removed an otherwise impassable barrier to migration of these species upstream to the upper lakes. The lamprey reached Lake Erie by 1921, Lake Huron and Lake Michigan in the early 1930s and Lake Superior in 1946. The alewife generally lagged behind, reaching Erie and Huron in 1931–33, Michigan in 1949 and Superior in 1953.

Both immigrants like deep water and are neither abundant nor problematic in the relatively shallow Lake Erie. In the other lakes, they have caused major changes. The pattern appears to have been one of parasitism by the lamprey, firstly of the larger deeper-water carnivores, such as lake trout, burbot (*Lota lota*) and deep-water ciscoes, and then of smaller species, until the lamprey population itself declined. Reduction of the large piscivores then apparently allowed increase of the alewife, which feeds aggressively on large zooplankters and benthic Crustacea, such as *Pontoporeia*. These are also the main food sources of the young piscivores, whose populations cannot then recover from the lamprey depredations. Alewives may even increase their competitive advantages by feeding on the young of the large piscivores, and have been among the commonest Great Lakes fish.

With all of these influences, it is difficult to separate the effects of progressive eutrophication this century. The mechanisms by which certain species disappear on eutrophication are not fully known. They may involve loss of gravel spawning habitat, as increased sedimentation covers it with organic deposits. The burbot lives and spawns in the deepest parts of well-oxygenated lakes, from where hypolimnial deoxygenation may force it at an early stage. Extensions of marginal plant beds, in providing cover, spawning habitat and abundant invertebrate food, may favour some species previously unable to compete successfully. Conditions leading to summer total phosphorus or chlorophyll *a* concentrations in the epilimnion of the order of 20–30 µg l^{-1} have led, in Lake Erie and in similar large lakes, such as the European Bodensee, to predominance of percid and cyprinid fish.

Attempts are being made to limit eutrophication by use of phosphate stripping (see Chapter 6) and restriction of detergent phosphate use. Some fish are being restocked and lampricide use has helped. Restoration of the former fish communities, however, is complicated by continuing changes. The introduction of Pacific salmonid game fish, including coho (*Oncorhynchus kisutch*) (see Chapter 7), is apparently reducing the alewife population and may still have unforeseen effects on other fish. These introductions have proved very popular with recreational anglers. The introduced salmonids are not as productive as formerly, however, because control of eutrophication, at least in the lower lakes, is reducing lake productivity overall, and this is causing concern among angling lobbies.

9.7.2 The East African Great Lakes

Tropical Africa has undergone many geological and climatic changes. Yet none has been so devastating and widespread for the freshwater fauna as the glaciations, which completely removed the freshwater habitat from much of the temperate land surface. African fish communities have thus had a long period of development, and speciation has occurred to a high degree (see Chapter 2). The East African lakes have been formed in a variety of ways (see Chapter 6) and their fisheries have developed from the technically crude but biologically sophisticated to the commercially advanced but ecologically unwise.

Initially the subsistence fisheries depended on one of five main methods [476]: addition of natural plant poisons to the water; spears and harpoons;

hooks and line; non-return basket traps; and baskets scooped through the shallows. Sometimes, these methods have been very cleverly used [1030]. On Lake Albert, the Banyoro tribe collected grass or brushwood and tied it into bundles, which they lowered to the lake bed in 6–10 m of water, attached to a line buoyed by pieces of ambatch (*Aeschynomene profundis*), a light corky wood. Overnight, the bundles became colonized by fish, mainly the cichlid *Haplochromis* species, taking cover from predators. The bundles were hauled up and the *Haplochromis* baited alive on small hooks on lines, with which the larger tiger fish (*Hydrocynus vittatus*) were caught. In turn, the tiger fish were used to bait large barbed hooks to catch Nile perch, the ultimate quarry.

Baskets, woven from papyrus, reeds or pliant tree branches, were widely used. Women of the Jaluo on north-east Lake Victoria moved into shallow water in groups of seven or eight, each with a small basket on her head and a much larger, wide-mouthed basket to hand. The women converged in a circle and simultaneously swept the large baskets through the encircled water, scooping out the fish and depositing them in the head baskets. Pelicans feeding in the same area employ similar methods, driving fish into a small area which they surround before scooping with their mouths.

Such methods are now rarely to be seen, for the demands of increasing population have brought about an expansion of artisanal fisheries, largely based in shallow water and catching the readily filleted and tasty tilapias. The keys to development of these fisheries have been the introduction of gill-nets and sailing-boats, by Arab traders in the nineteenth century. Nets were at first of twine and then flax or cotton, but rotted easily. Rayon, laboriously picked from the linings of old car tyres, provided a more durable material, eventually to be replaced by custom-made monofilament nylon nets. These gill-net fisheries have been extremely successful. They employ cheap gear and a large number of people, and exploit a group of fish that has many of the characteristics of the ideal fishery fish [303] in the offshore open water down to about 20 or 30 m. Artisanal fisheries are suited to the local conditions.

However, commerical ambitions, resulting initially from European colonists, have led to mechanized fishing. Greek settlers brought the skills of Mediterranean purse-seining to exploit the fish at the centres of the large lakes, where the bigger boats were less vulnerable to storms than canoes and sailing dhows. Species of sardine-like fish (Clupeidae), particularly *Limnothrissa miodon* and *Stolothrissa tanganyikae*, are successfully fished in this way in Lake Tanganyika. The sardines eat zooplankters, moving to the surface to do so. Predators (*Lates* spp.) of the sardines move with them and are also caught. Lights are used to attract the fish towards the boats, which use large nets closed and drawn in by power winches. Such a fishery is very efficient but also liable to overfishing; catches of four species of the predators were halved over seven years.

The main problem with mechanized fisheries, however, is that, because catches are large, big boats, large-scale docking, cold stores onshore, roads and fuel are all needed. This means investment of capital and also considerable unemployment of fishermen. A large boat run by a crew of six can catch as much fish as a fleet of over a hundred, three-man canoes. Because two different groups of fish are usually concerned, however – the tilapias inshore and the sardines offshore – it is possible for both fisheries to coexist.

Mechanized fishing by power trawls, however, is probably much more damaging than purse-seining [165]. It destroys the littoral zone and takes species that are unsuitable for a fishery as readily as those that are. It was introduced to Lake Malawi in 1968 and took as many as 160 species, 80% of them small haplochromines. Many of these are endemic and highly specialized [294] and have a very restricted distribution, perhaps over as little as 3000 m^2 of lake bottom in the cases of some species. Significantly, 20% of them disappeared from the catches in one three-year period and there is a danger that many may become extinct. If trawling is allowed to continue, extensive conservation areas will need to be set aside if this fish community is to retain its diversity. However, the sustainable yield cannot remain more than modest and a better policy would be to abandon trawling and to support the artisanal fisheries very strongly.

If mechanized fishing poses severe problems, a possibly worse aspect of the moves to intensify fishing in the African lakes is that of introduction of alien species. Fish introductions on a world scale have been extensive in the last forty years, with perhaps 160 species being transferred, *inter alia*, among 120 different countries [34]. Fishery officers have sometimes had the misapprehensions of those agriculturalists who believe that natural selection can be improved upon. The introduction of the Nile perch to Lake Victoria [12, 583] has, in some senses, been a disaster [36], which was foreseen at the time [295] by scientists, who were simply ignored.

The Nile perch was once a member of the Lake Victoria fish fauna, but this was in the geological past. Its bones are fossilized in sediments some hundreds of thousands of years old. It has been absent from the lake for a long interim, in which evolution has produced a remarkable flock of highly specialist small fishes of the genus *Haplochromis* (Fig. 9.15). The argument for introduction of the Nile perch was that it would feed on the small fish ('trash fish'), which were not marketable, and would package their flesh into its own very large and easily catchable body.

After an introduction, probably around 1954 (the details were not fully recorded), the Nile perch took a little time to expand its population and was not a problem even by 1971 [476]. Since then, however, it has become distributed throughout the lake, perhaps due to other changes, primarily eutrophication (see Chapter 6). It has demonstrated its omnivorousness and voraciousness by eating not only the small haplochromines but also other

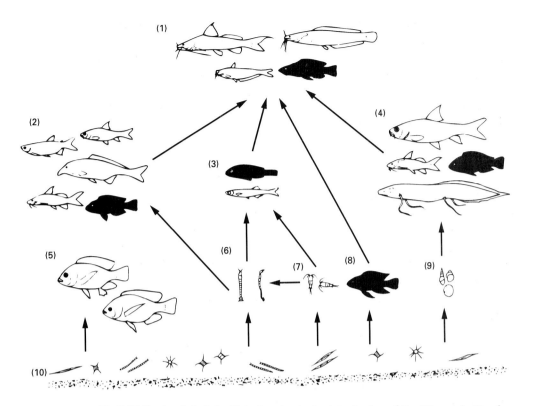

Fig. 9.15 Food web in Lake Victoria prior to the introduction of the Nile perch. Key for Figs 9.15 and 9.16 (haplochromid fish are shown shaded in black): 1, piscivores: *Bagrus docmak*, *Clarias gariepinus*, *Schilbe mystus* (all catfish) and haplochromine cichlid fish; 2, insectivores: *Alestes* spp., *Barbus* spp. (sardines), mormyrid fish, *Synodontis afrofischeri* (a catfish) and haplochromids; 3, zooplanktivores: *Rastrineobola argentea* (also 13) and haplochromines; 4, molluscivores: *Barbus altianalis*, *Synodontis victoriae*, haplochromines and *Protopterus aethiopicus* (lungfish); 5, algivores: *Oreochromis variabilis* and *Oreochromis esculentus*; 6 and 15, chironomids; 7 zooplankton; 8, detritivorous and phytoplanktivorous haplochromids; 9, molluscs; 10, phytoplankton and detritus; 11, adult Nile perch; 12, juvenile Nile perch; 14, introduced Nile tilapia, *Oreochromis niloticus*; 16, prawns (*Caradina nilotica*). (From Witte *et al.* [1023].)

commercially valuable fish, such as the tilapias, and has apparently reduced the fish stocks so much that it is now forced to take its own young and prawns (*Caridina*) as a major food (Fig. 9.16).

Catching Nile perch, which may typically weigh 35–50 kg and has reached 179 kg, requires large boats, so that some of the artisanal fishermen on tilapia have been put out of business. At first, it was disliked locally and less valued at the lakeside (1 shilling kg^{-1}) than the tilapia (30 shillings kg^{-1}). However, the mechanization it has brought has provided new jobs and the total yield of the fishery has increased five times between the 1970s and the 1990s. Frozen fillets find a good market in the large towns and substantial processing plants have been built. It is an oily fish,

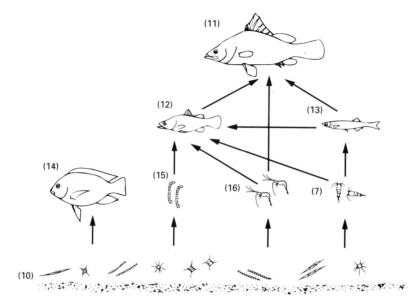

Fig. 9.16 Food web in Lake Victoria following the introduction and expansion of the population of Nile perch. For key see Fig. 9.15. (From Witte *et al.* [1023].)

which, for rural storage and distribution in an equatorial climate, needs to be smoked, which requires fuel (the less oily tilapias can simply be sun-dried). There is evidence of consequent deforestation, with its attendant erosion and other problems in the area.

The fish fauna of lake Victoria has been greatly changed (Figs 9.15 & 9.16 [1022]). Samplings in the Mwanza Gulf at the south end of the lake suggest that two-thirds of the small haplochromine species have been made extinct [342–344]. The endemic *Oreochromis* species, whose fishery was analysed earlier in this chapter, have also been replaced by introduced species, (*Oreochromis niloticus* and *Oreochromis leucostictus*). These changes are reflected in the fish landings around the lake (Fig. 9.17).

The present increased yield [732] is variously attributed to several factors, and it is difficult to separate them. There has been a change in fishing methods, coupled with an increased fishing effort; eutrophication of the lake (see Chapter 6) will have increased the fish production; and the main caught fish are now three *r*-selected species – Nile perch (60%), the introduced Nile tilapia, *O. niloticus* (14%), and the endemic open-water *Rastrineobola argentea* (see Fig. 9.16). Economically, the countries bordering the lake are probably benefiting from the changes at the present time. However, rather more processing capacity than is required by the present yields has been built and this may encourage overfishing. A fishery dependent on a species that is a top predator, whose current production depends on cannibalization of its own young and relatively few other prey,

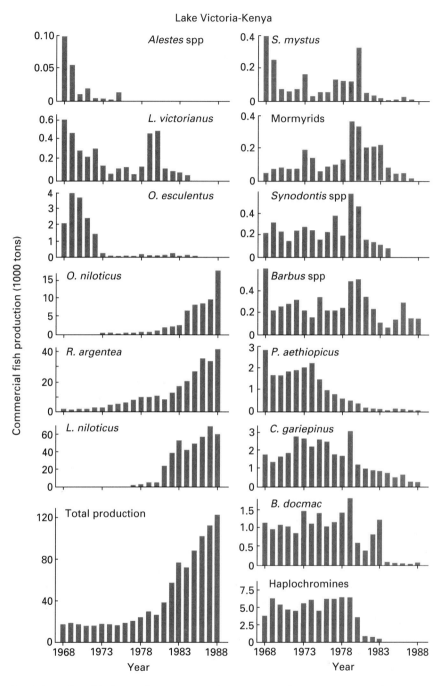

Fig. 9.17 Fish yields from Lake Victoria *L.*, *Lates*; *O.*, *Oreochromis*; *R.*, *Rastrineobolus*; *S.*, *Schilbe*; *P.*, *Protopterus*; *C.*, *Clarias*; *B.*, *Bagrus*. (From Oguta Ohwayo [732].)

is also inherently unstable, especially when imposed chemical changes to the lake are probably continuing. As a top predator, its yield must inevitably become lower than that which its prey previously provided. What the situation will be once some sort of equilibrium has been established is difficult to predict.

With this example and others, such as that of *Cichla ocellaris* in Lake Gatun (Panama Canal zone) [1041], it seems odd that introductions to the African lakes are still being mooted. *Cichla ocellaris* was introduced to ponds as a sport fish but escaped to the main lake in floods. One of its consequences, apart from a major simplification of the food web, has been an increase in malaria-carrying mosquitoes, whose larvae were eaten by one of the fish that *Cichla* has almost eliminated.

The problem is that some introductions have proved useful, although these have usually been to man-made lakes (see Chapter 10), where a previously riverine fish fauna did not have species capable of exploiting the open-water plankton. *Limnothrissa miodon*, for example, has been introduced to Lake Kariba (formed in the 1960s) without problem, and even to the older, but still comparatively recent, Lake Kivu, formed perhaps 20 000 years ago. Almost half the British fish fauna is also introduced and there have been few problems. The British fauna was severely depleted by glaciation and an early isolation of the islands by the English Channel prevented much recolonization from Eurasia.

It is a quite different matter to propose introducing the Lake Tanganyika clupeids to Lake Malawi, however, as has been suggested [974]. Lake Malawi has its own open-water sardine, *Engraulicypris sardella*, which some fishery biologists have claimed to be 'inefficient', because *Chaoborus* larvae, a potential prey of the fish, persist in the lake but are absent from Lake Tanganyika. The alleged inefficiency may be grossly misplaced [11, 228].

Introductions, once made, are usually irreversible. A programme of, it is hoped, destructive overfishing has been suggested for Nile perch in Lake Victoria, coupled with captive propagation of its threatened prey [815]. It is suggested that the lake could be restocked with the endemic *Haplochromis* species, once the Nile perch has been reduced to low numbers. However, quite apart from the theoretical problem of maintaining the endemic species in a habitat for which they have not been rigorously naturally selected, the Nile perch may rise again. The problem is now an indefinite one.

We are left, however, with a quandary. There *is* a problem of high population, short of protein, in Africa, as elsewhere in the developing world. In their attempts to meet this problem, fishery officers and government officials have doubtless felt that they have been doing the right thing in encouraging mechanized fishing or making introductions. Yet most of the wild fish populations of such ancient lakes are unsuitable for much increased exploitation. A better solution perhaps in an expansion of pond culture [410].

9.8 Fish culture

Pond culture of fish is an ancient art. The Egyptians were culturing tilapia in 200 BC and the Chinese have cultured fish from at least 500 BC [80]. Common carp were retained in ponds in medieval Europe to provide winter proteins. More recently, there has been interest in the culture of valuable marine and game fish in the developing world [935]. Trout and salmon farms have increased markedly in number in the UK in the last 30 years (see Chapter 4), to produce fish which can be sold for a high price (by Third World comparisons) to offset the high costs of production. Such farms depend on feeding high-protein food (generally of fish-meal made from small marine fish) and are net users of protein and energy. They provide no solutions to food problems, although they may be very important as sources of employment in areas such as western Scotland.

Greater fundamental significance attaches to fish farming in the developing world. Here there are two trends: one towards high-technology culture of tilapia [792], which could become a luxury industry serving rich subtropical countries, such as the Gulf States, and the other towards low-technology culture on a village basis [485]. The latter is far more important, for it not only helps preserve the wild fisheries, but provides protein for those who are really short of it.

Tilapia are ideal fish for both sorts of culture. They taste good, have no fine intramuscular bones, breed early and easily, thrive on cheap plant and algal food and hence produce high yields. They are tolerant of wide temperature and salinity ranges, are relatively free from parasites and diseases and hybridize readily. This latter helps breeding programmes.

Suitable species include *Oreochromis andersonii*, *Oreochromis macrochir*, *O. niloticus* and *Oreochromis aureus*. Because they breed frequently, large populations of small fish may result. This can be avoided by stocking predators, such as *Channa striata*, with them and by culturing first-generation hybrids, which are almost entirely male and do not breed further. Males also grow faster than females. Crosses between female *O. nilotica* and male *O. aureus* produce more than 85% males and, if *Oreochromis urolepis hornorum* is used as the male parent, 100% males can be almost guaranteed in the F_1. Treatment of the fry with androgenic steroids can also turn all the fish into males, but this increases the cost.

Cyprinid fish are used for culture in India and China, often in mixtures that allow a very full exploitation of the available food supply. For example, the major carps, catla (*Catla catla*), rohu (*Labeo rohita*) and mrigal (*Cirrhina mrigala*), will take zooplankton and other invertebrates, the white bighead or silver carp (*Hypophthalamichthys molitrix*) eats phytoplankton, the grass carp (*Ctenopharyngodon idella*) eats plants and the Chinese black roach (*Mylopharyngodon piceus*) and the common carp (*Cyprinus carpio*) will eat benthic invertebrates. Stocked in combination in ponds that can be cheaply fertilized with cow dung, grass cuttings, by-products of agriculture, such as rice bran, or even cheap artificial fertilizer, high yields can be obtained.

Even raw human sewage can be used, if diluted, and the ponds stocked with air-breathing fish, such as catfish and murrel, originating from deoxygenated swamp habitats (see Chapter 5). Fish from such ponds must be well cooked, because of the presence of various trematode parasites and tapeworms, for which the fish form an intermediate host. Pigs may be penned over ditches leading to such fish ponds, which they fertilize, and ducks may also be used to provide continuous fertilization [410]. Rice fields, after the rice has been harvested and the stubble left, can also be used for culture, although in recent years the use of pesticides to improve rice yields has prevented many fields from being used.

Low-technology culture of mixed species is very valuable and poses few problems, beyond those of parasite transmission. Some of the Indian major carps will not spawn easily but can be induced to do so by a simple injection of ground-up pituitary gland preserved in alcohol from a wild fish which was just about to spawn. There is a vast number of small ponds and ditches in the developing world that can be used – those created for irrigation, stock watering, water-chestnut cultivation and flood control, for example. Yields, particularly where the ponds are fertilized with village wastes, are much higher than those of wild fisheries. The latter might produce, in the tropics, up to a few hundred kilograms per hectare; intensive pond culture can realize as much per square metre, although this requires expertise and heavy feeding. On a village scale, yields 10–100 times those of wild fisheries would not be unreasonable, however. Such village polyculture represents a sensible and helpful approach to the problems of the poor, while complementing the wise management of lakes and their fisheries.

9.9 Still-water angling

Recreational fishing (angling) divides itself in the UK into game or fly-fishing and coarse fishing. Game fishing is for salmonids (salmon, brown trout, rainbow trout and sometimes grayling), uses active techniques, in which a replica fly is cast at the surface of the water to lure the fish, is associated more with flowing waters than lakes and is discussed in Chapter 4. Trout are frequently sought by such techniques from the edges of upland lakes and from artificial lakes stocked with rainbow trout. Coarse fishing, for pike, perch and cyprinid fish, such as roach, tench, bream and carp, is a more sedentary occupation, not confined to lakes, for it is practised on slow-flowing rivers and canals, but often a feature of them, in which baits of worms, maggots or nutritious pastes are dangled in the water for the fish to come to them.

The game-fish/coarse-fish distinction is peculiarly British, and carries certain social cachets, because game fishing is much more expensive than coarse fishing. Elsewhere, there is little distinction and, in the North American Great Lakes, anglers will cheerfully pursue salmonids from boats using baits rather than flies. In most western countries, recreational angling involves many people and substantial funds. This section is concerned with coarse, still-water fishing, which, in the UK, attracts upwards of 2–3 million people, rather more than attend football matches.

Such angling is only partly about catching fish. In the UK, they are thrown back anyway and not eaten. Social surveys have shown that some anglers enjoy the sense of freedom, the relaxation and the opportunity to reminisce, as much as the excitement of catching a fish. The surroundings are important; anglers regard themselves as natural historians, sensitive to the state of the water, and have traditionally been seen as allies of the conservation bodies, in lobbying for higher-quality waters and reporting pollution incidents that might otherwise have gone uninvestigated.

For such anglers, the management of the fishery is essentially that of managing access to the site, the number of anglers and their activities and gear, so as to minimize stress to the fish. Such management is regulated legally – for example, by the Salmon and Freshwater Fisheries Act (1975), by-laws set by the regulatory authority and the rules of the owner of the fishery or angling club [409]. The law bans certain gears, such as foul hooks, barbed gaffs and knotted keep nets, gives the opportunity for setting of a closed (close) season, when fishing is not allowed, and bans the transfer of fish, without licence, from one water body to another, so as to minimize the spread of diseases and parasites. The regulatory authority sets the close season (usually 15 March to 15 June for coarse fish), authorizes fish transfer and may prescribe that a fisherman use not more than one or two rods at a time. Local rules may determine the number of anglers, by confining fishing to a fixed number of places or pegs and by banning certain baits and practices.

Although angling has come under scrutiny as potentially a cruel sport, especially when the fish are put back, often damaged, after being kept for some hours in a submerged net, angling as described above is compatible with many other water uses and the maintenance of attractive habitats. Anglers, but no more than walkers and boaters, disturb birds [159, 180, 1038] and are as prone to leave litter as others [52, 237, 281]. Sometimes the litter can be damaging to wildlife – lead shot for weighting lines [696] and lengths of discarded monofilament nylon line, often with hooks still attached, as well as the drink cans and broken glass, sandwich wrappings and expanded polystyrene containers that mar many places. New laws have begun to resolve the shot problem by replacing lead with heavy steel, and educational campaigns about the effects of tangling line on birds are likely to have some effect.

There are developing problems, however. The increasing demand for angling, which can be inexpensive in a society where increasing numbers are unemployed or needing relief from a worsening work environment, leads to needs to accommodate more angling. This may mean habitat modifications – the provision of platforms and car parks, for example, which destroy natural habitat [626]. It may mean damage to bankside vegetation, as anglers cut reed that is in the way of their casting, trample the surroundings, dig turf for worms and even light fires to keep warm. These problems can be controlled by local rules and sensitive design. More insidious and very damaging, however, are changes in attitudes to angling, as, like many other sports in recent decades, it has become a competitive activity, with sponsorship money pandering to an apparently insatiable need to inflate male egos.

Competitive angling, in which he wins who catches the most or the biggest fish or the heaviest bag of them, now threatens the coexistence of angling with other water uses, not least wildlife conservation. It has brought with it increased demands to stock waters with more or different fish, practices such as live baiting, demands for removal of piscivores and competitors of target fish and a reduced sensitivity to the quality of the habitat and the needs of other users.

There has always been demand for stocking of waters, because freshwater-fish recruitment in the UK is annually very variable, being dependent on temperature in the early summer and other weather-related factors. Some cohorts of fish are thus naturally poor and others very strong. On average, the carrying capacity of a water will be achieved through a mixture of species of varying annual strengths. Stocking will simply lead to greater mortality of one component or another, unless it is to replace fish lost after a pollution incident. Regulatory authorities in the UK, which have carried out much stocking in the past in response to anglers' demands, have largely ceased this practice. Anglers themselves, however, are responsible for thousands of fish transfers each year, most of them illegal. Sometimes they involve fish brought in from mainland Europe, a practice that can introduce new diseases and parasites.

Most problematic has been the spread of common carp. This is not a native fish, but an Asian one, and it is extremely damaging. It is omnivorous, feeding as an adult particularly on bottom invertebrates, for which it forages in the sediment. In doing so, it disturbs the sediment, creating turbidity, and destroys rooted aquatic plants. Together with other bottom-feeding fish, such as bream, it can transfer considerable amounts of phosphorus to the overlying water [541]. It has been identified as a significant forward-switch agent (see Chapters 6 and 8) [134] in the conversion of plant-dominated shallow waters to algal-dominated ones. Carp are particularly prized by match anglers and those who seek fish of record size. They grow large and attempt to resist capture. In the UK, carp do not breed very successfully, because of the low temperatures. This has confined the problem to some extent. In Australia, however, where carp have also been introduced, they now predominate, to the detriment of native fish in many river systems, and are a major problem [116, 833, 888, 990].

Live baiting is used to catch some species and is inherently a cruel practice. It can also lead to the discarding of unused bait fish into a lake from which they were naturally absent. The most notorious example [626] is the introduction of ruffe into Loch Lomond in 1980, where it is now very common and predatory on the eggs of the powan, a rare native fish near-confined to the loch.

The pressures for predator control were discussed in Chapter 4 in relation to salmon fisheries, but they are no less for still waters. Birds, such as cormorants, herons, grebes, goosanders and mergansers, take 0.3–0.5 kg of fresh fish day^{-1} [651] and have been seen as competitors by the fishermen. Red kite and osprey populations were markedly reduced towards the end of the last century and the early decades of this, as were those of otters, by the demands of fishery owners, but there is little evidence that the fishing is improved by predator control. The piscivores take mostly small, diseased or damaged fish, or those reared in farms and stocked in reservoirs to which they are ill adapted. In one loch, where there was pressure for control of cormorants, which were believed to be feeding on trout, it was found that they were feeding on perch, for which there was also a control programme, because the perch were believed to be competing with the trout!

Predation can improve fisheries by skewing the size of the fish towards the larger end and, in any case, shooting some predators at a site rich in prey merely creates an opportunity for more of the same or different ones to move in. Fortunately, the high intrinsic value of such birds and mammals in the public perception has swung fishery-management opinion away from predator control and towards a broader view of conservation. The pressures generated by injections of large sums of money generally undermine any endeavour and should be resisted in angling, particularly because of the many other uses to which waters must be put. But this raises general issues of environmental management and use, to which I shall return in Chapter 11.

10: The Birth, Development and Extinction of Lakes

10.1 Introduction

Lakes have a distinct start; some of the ways in which their basins are formed were discussed in Chapter 6. They develop and change. Eventually, they may fill in and be lost to the landscape. This chapter is about the life histories of lakes; it seeks any common pattern that might accompany their development.

There are several problems in studying lake history. Most existing natural lakes were formed at least several thousand years ago and, although occasionally a new lake will be formed by a landslide (Fig. 10.1) or the retreat of a polar glacier, these events may not be readily amenable to study. The lifespans of lakes are much greater than those of freshwater ecologists. Direct observations of development are not possible, although sometimes the recent past can be reconstructed from oral history [824]. There is also the problem that any fundamental pattern in the development of lakes may be obscured by severe human interference.

On the other hand, in the last fifty years, we have created many new lakes by damming rivers for hydroelectric-power generation and the storage of water (Fig. 10.1). This continues a tradition beginning at least 2000 years ago with the ancient dams at Anuradhapura in Sri Lanka and Angkor Watt in Cambodia, although the number and size of such dams have recently increased remarkably. An amount of water equivalent to a third of that present in the Earth's atmosphere at any one time is now stored in reservoirs, some of which are so large as to be claimed the only man-made structures, other than the Great Wall of China, visible from space. The early stages of formation of these lakes have been studied and tell us about the infancy of lake development.

There is also an important human aspect to studies of man-made lakes. Their creation is usually heralded as an improvement in the lot of mankind [758]. For many people affected in one way or another by them, this may in retrospect be seen as the propaganda of a powerful dam-building industry. The lakes may create more problems – of social disruption and disease in the tropics – than they solve.

Long-term changes in lakes can be followed indirectly by studies of the sediment deposits. These are often laid down in an orderly sequence and contain much chemical and biological information, some of which allows us

Fig. 10.1 Observations on lakes that have been formed recently can give valuable insights into the early stages of lake development. Hebgen Lake in Montana (top) was formed when a landslide dammed a river during an earthquake. Most new lakes are created by damming for hydroelectric purposes (Lake Kariba (bottom)) or water storage.

to tease out the effects of natural and human-induced changes; and, when the lake ceases to exist as a body of water, its sediments remain to give details of its end. The remains of lakes are very widespread in the geological record, there having been many ancient lakes comparable in size to the present Caspian Sea [325]. They were particularly abundant in the Permo-Carboniferous Period (245–360 million years), the Lower Cretaceous Period (100–150 million years) in West Africa and Brazil, the Pliocene

Epoch (1.6–5.3 million years) in the Great Basin of North America and the Palaeocene and Eocene Epochs (58–66 million years) of China and North America. The Tertiary Period (from 66 until 1.6 million years ago) had vast lakes, which dried up in desert climates to give huge deposits of salt and other minerals, including lithium, boron and arsenic. The diatomites (diatom-rich sediments) often found in such ancient basins are sought after for industrial uses, including substitution for asbestos in heat insulation. Figure 10.2 gives some examples of the sorts of layering found in such ancient deposits by those concerned not only with exploration for minerals and oil, for ancient lakes and freshwater wetlands are also associated with hydrocarbon deposits, but with tracking global-wide climatic changes, which are reflected in the patterns of filling and drying out of these lakes. The resolution of the record for climate monitoring is much finer than that of marine sediments, because the huge size of the ocean mutes many temperature changes.

Studies of lake history over the past few hundreds or thousands of years from sediments also give an objective picture of the changes made to lakes by eutrophication and acidification. Used appropriately, limnogeological (from extinct lakes) and palaeolimnological (from still extant lakes) data [44, 64] can be used with studies on contemporary lake functioning (neolimnology) to influence decisions on how the environment should be managed. They can also be used to examine hypotheses, such as that frequently held that lakes become more eutrophic, in the sense of more fertile and productive, with time.

10.2 Man-made lakes

In the 1960s and 1970s, several very large lakes were created in Africa (Fig. 10.3), and much information resulted from the early stages of these [612]. Big dams are now being built or planned in China, Brazil and South-east Asia [140, 142]. The water rising behind a newly closed dam is often turbid, with suspended sediment in the river and from erosion of the newly flooded soils at the water's edge. The turbidity prevents much phytoplankton growth but, as the new lake becomes larger, the shoreline stabilizes and the river silt is deposited in deltas as the current flow is decreased by the water mass. In Lake Kariba (Zimbabwe–Zambia) there was no early turbid phase, for river silt is deposited in swamps just upriver from the lake; elsewhere it has been prominent.

Flooding kills the river-valley vegetation, which rots underwater, with two consequences. First, there may be deoxygenation of the bottom water, with production of hydrogen sulphide. In Lake Volta in Ghana (Fig. 10.3), the entire water mass was deoxygenated for a time, as flooded softwood forests decayed. Secondly, the decomposition and the waterlogging of the soils releases ions, giving a substantial internal loading of nitrogen and

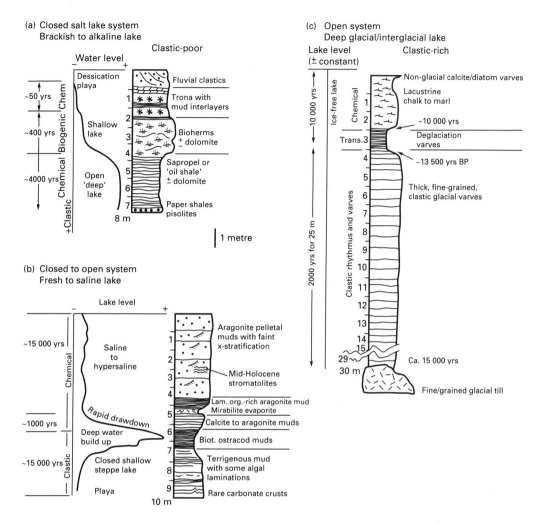

Fig. 10.2 Some typical sequences of sediments found in ancient lake basins, showing (a) drying out of an endorheic lake; (b) effects of a wetter period in coverting an endorheic lake to an exorheic one and then reversion to endorheicity in a drier period; (c) change from an ice-covered glacial lake through to a warmer open lake in a postglacial or interglacial period. For those not privy to the jargon of geologists, clastic means rock fragments (inorganic sediment), trona is hydrated sodium carbonate, a playa is a temporary shallow lake, often saline and found in desert conditions, a bioherm is a mass of material derived from living sources (e.g. coral), aragonite, calcite and dolomite are calcium carbonate minerals, pisolites are sediments with particles the size and shape of peas, stromatolites are layered mounds formed by blue green and other algae in the littoral zone through accumulation of sediment, and evaporites are deposits formed by evaporation. BP, before present. (Based on Gielowski-Kordesch and Kelts [325].)

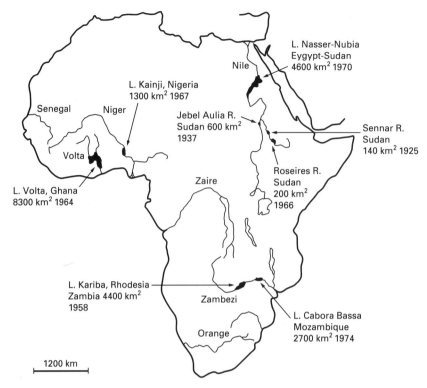

Fig. 10.3 Some major man-made lakes in Africa, with their areas and the dates at which the lakes began to form (based on Lowe-McConnell [580].)

phosphorus compounds. The rise in Lake Kariba was from 26 mg l^{-1} total dissolved solids in the river to 67 mg l^{-1} in the early lake. This, in turn, may stimulate increased growth of phytoplankton, aquatic plants and fish. Lake Kainji (Nigeria), which is frequently flushed out, consequently did not undergo these stages, nor did Lake Nasser-Nubia, where 80% of the basin was desert.

In Lake Kariba, however, there was a spectacular colonization and spread of a floating fern, *Salvinia molesta* (Fig. 10.4). It is a robust plant, up to 30 cm long, with overlapping leaves densely covered with hairs. These prevent wetting and waterlogging. *Salvinia* spp can normally reproduce sexually – sporocarps are produced on trailing underwater stems – but *S. molesta* produces stolons and quickly builds up a dense mat. This floating mat may support other species, such as *Scirpus cubensis*, rooted in it. *Salvinia molesta* sporocarps are mostly sterile. It seems to be a pentaploid hybrid of two related South American species, *Salvinia biloba* and *Salvinia auriculata*, perhaps of horticultural origin.

Salvinia molesta was a problem in the new Lake Kariba in the early 1960s, for it impeded navigation and the use of fishing nets. It also created anaerobic conditions in the shallow margins of the lake. *Salvinia* spp. do not

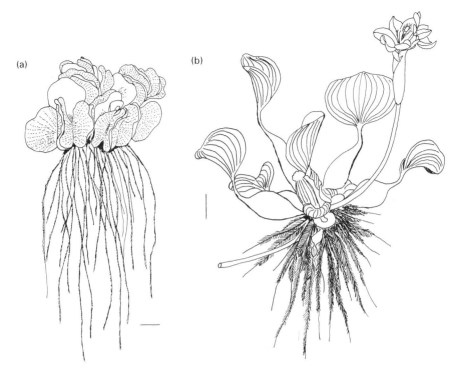

Fig. 10.4 Two floating aquatic plants that have caused severe problems in the early stages of new reservoirs. (a) *Salvinia molesta* (scale bar = 1 cm); (b) *Eichhornia crassipes* (water hyacinth) (scale bar = 3 cm).

form such monospecific stands in South America, where they are endemic, and it is a mystery how the hybrid has reached East and Central Africa and also Sri Lanka and Indonesia, where it is also a problem. Probably it was introduced by botanists, aquarists or water-gardeners, as were other noxious floating plants, such as the water hyacinth, *Eichhornia crassipes* (Fig. 10.4). Removed from contact with their endemic competitors and grazers, such plants have frequently become problems, demanding expensive control. *Salvinia* was known in the River Zambezi, but the river flow presumably prevented build-up of large populations until the lake started to form. By 1962, *Salvinia* covered a quarter of the (4400 km²) lake.

Around 1962, it seemed that *Salvinia* might permanently hinder navigation and fishing on Lake Kariba and might damage the turbines of the power station built underground next to the dam. At the Gebel Auliya dam on the River Nile, *Eichhornia crassipes* has continually to be controlled to prevent its spread into irrigation ditches below the dam. It is very bulky and would rapidly block the water flow in them. Masses of the South American *Eichhornia*, originally introduced as business gifts to Louisiana farmers (it has beautiful flowers), now clog canals in the southern states of America. Their control, to which an entire scientific journal is devoted, costs large sums.

Fortunately, *Salvinia* declined in Lake Kariba in the mid-1960s, as the initial burst of nutrients was flushed out. Its growth is now limited by the availability of nitrogen and it covers less than 10% of the lake, mainly in sheltered inlets. Wind disturbance may, in any case, have prevented its covering the entire lake and it is fortunate that plans to spray it with arsenate, and thus contaminate the lake, were abandoned. It now provides an extra habitat for wildlife, particularly wading birds, such as the jacana, which wander over it.

Serious aquatic-weed problems arise from availability of a fast-growing, usually exotic, species, lacking its natural competitors, and a copious nutrient supply [422]. Because aquatic plants frequently reproduce rapidly by vegetative means, problems caused by introductions may be very severe. Apart from the water hyacinth, the spiked water milfoil, *Myriophyllum spicatum*, a native of Europe, has caused clogging of ditches and littoral zones in the USA, while the pond weeds *Elodea* and *Egregia* have caused similar problems in the reverse direction. Currently, Great Britain is facing an increasing problem from the introduction, long ago, of the Australian swamp stonecrop, *Crassula helmsii* [754].

10.2.1 Fisheries in new tropical lakes

The filling of a man-made lake creates an extended period of rising river level for the original riverfish fauna (see Chapter 5) and, not surprisingly, most of the river species flourish initially. Production in the early, nutrient-rich phase, may be very high. As the lake level stabilizes, some of these fish become confined to the river mouths, if they need flowing water for spawning. In Lake Volta, the mormyrids (elephant-snout fish), which were important river species, are examples [769]. In contrast, species which feed on submerged plants, detritus and periphyton, such as *Sarotherodon galilaeus*, *Tilapia zillii* and *Oreochromis niloticus*, have much increased their proportion of the total fish population in the new Lake Volta.

Some niches in the lake may be unoccupied because there are no appropriate species to exploit new food sources. No zooplanktivore was present in Lake Kariba [378] until *Limnothrissa miodon* and *Stolothrissa tanganyicae* were introduced from Lake Tanganyika in 1965. These introductions might be seen as unwise (see Chapter 9), especially in view of the initial problems caused by *S. molesta*, but seem not to have been retrograde [633]. The fish are eminently suitable for a commercial fishery and were introduced to a relatively depauperate local fish fauna (twenty-eight species). The original riverine habitat was subject to severe draw-downs, which undoubtedly favoured rather generalist fish. As far as is known, none of these have been greatly affected by the sardine introductions and most grow better than they did in the river.

The early high fish production of new tropical lakes has been supported by high production of invertebrates in the littoral zone, where many of the original trees may remain, dead but standing, for several years. The submerged branches became covered with abundant periphyton, stimulated by the high nutrient levels, which, in turn, supported grazing invertebrates. Notable among these in Lake Kariba and Lake Volta was a mayfly, *Povilla adusta*. Its larvae burrowed into the bark and rotting wood of the softwood trees flooded in the southern part of Lake Volta and took advantage of holes bored by beetles (*Xyloborus torquatus*) in the hardwood trees around Lake Kariba.

About a fifth of the future basin of Lake Kariba was cleared mechanically of trees and bushes before inundation and the debris burned. This was to create areas free of snags for the nets of a future fishing industry. At Lake Volta, there was no clearance, because of cost and because it was thought that the softwood vegetation would soon rot down. Artisan fishermen in Lake Kariba have found it most profitable to fish where submerged trees still remain, because fish production is apparently higher there. The submerged trees are an ephemeral habitat, however, which must eventually disappear under the action of wood-borers, bacteria and water movement. The production they foster is a vestige of the initial productivity surge, which gives way, after a few years, to a less productive, relatively stable ecosystem.

The experiments unwittingly carried out in the creation of a new lake are of great significance. According to long-held ideas on lake fertility, the initially fertile water should have remained highly productive and gradually increased in fertility, as nutrients from the catchment area were presumed to accumulate in the basin. The fact that these lakes became less fertile as an initial supply of nutrients was washed out or fixed in sediments is further evidence that most lakes do not maintain their fertility unless an external loading of nutrients is continually applied (see Chapter 6).

10.2.2 Effects downstream of the new lake

Once a new lake has filled, the total river flow below the dam may be about the same as it was before. If water is removed from the lake for irrigation, the flow may be lower at some times but greater at others – in the dry season, when the irrigation network drains to the river. The seasonal pattern of flow is, in most cases, considerably changed, with consequences for the lower river ecosystem.

First, the silt carried previously by the river is deposited in the lake, and the turbidity of the outflowing water is reduced. In the River Nile this silt fertilized the delta lands and provided a detrital food source for a valuable inshore sardine fishery in the Mediterranean Sea off the Nile delta. This fishery has declined greatly since the closure of the Aswan High Dam, which impounds Lake Nasser-Nubia, and agriculture on the delta now requires increased use of artificial fertilizers [844]. The now-regulated River Niger, below the Kainji Dam, no longer drains some of the downriver swamps at times essential for rice cultivation in them [462].

10.2.3 New tropical lakes and human populations

The creation of new lakes in the tropics has led to some new human problems and intensified some pre-existing ones. The new ones are largely outside the scope of the book, for they are sociological and psychological. River valleys in the tropics are relatively densely populated and the new lakes have displaced as many as 50 000 people (Kariba), 42 000 (Kainji), 80 000 (Volta) and 120 000 (Nasser-Nubia). Although the moving of the villages was planned in advance, the modification of a culture closely geared to the seasonal flooding and shrinking of the river, and the fishing and farming opportunities presented by this, is no easy matter. During the period after such moves, there was increased mortality, not all of it attributable to infectious disease, but linked with the stress involved with a major change in lifestyle. It is, however, with water-borne diseases that limnological study can aid public-health measures. Water-borne diseases were, of course, features of the rivers before inundation. But the increased shoreline of the lakes and particularly the networks of irrigation channels which are associated with some – for example, Lake Nasser-Nubia – have exacerbated locally what is a widespread and serious problem.

10.2.4 Man-made tropical lakes – the balance of pros and cons

The larger projects have attracted much criticism and scepticism that their overall costs would exceed the benefits they would confer. Some of this, based on the early 'problem' period after formation of the lake, has not been borne out once the lake has reached some sort of equilibrium. Other criticism has been well founded, particularly that concerned with the problems of people displaced from their traditional homelands [345, 346]. There is often a general insensitivity on the part of governments to real human needs. Dams are usually justified on a balance of the costs of building against the commercial benefits. Properly constructed analyses should also include the social, environmental and health costs. Frequently, these are 'paid' by the local rural populations, while the benefits accrue to multinational companies, gaining cheap electricity in return for capital invested in the dam, and to urban élites in the country concerned [383, 727].

The Volta scheme could provide abundant cheap electrical power for Ghana. Its effects have been felt only in a minor way, however, for the loans of about $US 170 million necessary to build the dam are still being paid off and the non-Ghanaian Kaiser Aluminium, a company that financed the associated aluminium smelter and which uses much of the generated power, has enjoyed a pre-agreed low cost for its power for several decades. Nonetheless, Tema, where the smelter is situated, has established some new industry to supplement the national income and to reduce dependence on imported goods. It was expected that the annual fish harvest would be 20 000 t, but this was pessimistic for catches are currently running at 40 000 t year^{-1}, after the peak of 60 000 year^{-1} in the late 1960s. This has compensated for the reduced imports of beef protein from countries to the north affected by Sahelian drought [348].

In the long term, the costs of the Volta scheme should decrease, as the industrial benefits flourish. For the moment, the huge lake has disrupted land communications and water transport has yet to replace them. A planned irrigation scheme has not yet been constructed, but cultivation of the wet mud-flats left on draw-down of the water level allows cropping of maize, tomatoes, cow-peas and sweet potatoes at times when lack of rain prevents growth elsewhere in the country. Schistosomiasis has certainly increased, and it is as yet difficult to justify a case that the dam has been beneficial to the people of Ghana as a whole [356].

The Aswan High Dam in Egypt has provoked the greatest controversy, and its advantages and disadvantages are on a large scale [551, 844]. It has provided a constant supply of energy, which has allowed industrial development, and also supplies annually an extra 19×10^9 m^3 of irrigation water for crops in a country otherwise only 3% cultivable. The Aswan scheme has increased this by a factor of 1.6. Early fears that the lake would result in a net loss of water by seepage and evaporation have not been borne out.

The drawbacks of Aswan are more serious than those of the Volta, which may largely be those of delays in being able to take advantage of the opportunities offered. Egypt is arid and the evaporation of irrigation water is leading to increased soil salinity, which could threaten yields. The lack of freshwater flow to the Nile delta has led to encroachment inland of sea water and soil salination. Lake Nasser produces 10 000–13 000 t fish annually (see Chapter 9), but the detritus carried to the sea by the unimpeded Nile supported a sardine fishery of 15 000 t year^{-1} and probably other fisheries as well. The sardine fishing has declined completely, and the lake fish must be transported long distances to the centres of population near the coast. The fishery balance is certainly negative. The problems of erosion of the Nile delta and the need for artificial (and expensive) fertilizers to replace the once-free silt have already been mentioned.

Big dams are often heralded as triumphs of engineering. They may be anything but. A recent account from an engineering viewpoint [896] notes the progressive loss of capacity through sediment accumulation behind the dam, the onset of coastal erosion as sediment is denied to the river and the local coastline, erosion of the downstream channel and bank collapse, increased flood risk, due to alterations in the natural flooding regime, and risks of failure, due to ageing of the materials used to build the dam. As much land is taken out of production each year through salinization as is brought under irrigated cultivation; many dams are unsafe due to poor construction or human error in operating spillways; dams are associated with increasing local incidence of earthquakes. The power benefits and the food produced by irrigation usually go elsewhere than to the local population; the insensitive displacement and, it is alleged, destruction of rural people and cultures [143] cannot be regarded as a tolerable snag to the fuelling of international industrial markets, but only as an outrage. Climatic changes and almost universal underestimation even of identifiable costs have made most projects economic banes rather than boons.

Nonetheless, despite the clearly documented problems of the past, schemes for damming of even greater scope are being prepared. The Grande Carajos scheme is intended to industrialize one-sixth of Brazil, with parts of it flooding up to a third of a million hectares of Indian land, displacing 50 000 people [183]. And a start has been made on the Three Gorges Dam across the Yangtze River in China. It will be 100 m high and 2 km long, will consume the equivalent of forty-four Great Pyramids of material and will displace more than a million people [759].

10.3 The development of lake ecosystems

Studies of man-made lakes give insights into the early processes of lake development. But the lakes, with their early fertile phase followed by a decline in productivity, have been created on already mature landscapes. How relevant is the pattern described above to natural lakes, which developed on raw, often glaciated or volcanic, landscapes? The answers lie in their sediments.

Lakes inevitably fill with particles eroded from the land, as well as those produced in the water and sedimented to the bottom. The rate of filling depends on the locality, but is often a few millimetres per year. In water deeper than a few metres, the deposit is a grey-brown mud, rich in planktonic and littoral detritus, called *gyttja*. In very shallow water, it is often a more structured, fibrous peat. Peat-forming vegetation may encroach on the previously open water, with eventual succession (called the hydrosere) of the lake to a wetland or even dry-land vegetation.

This process is not inevitable. In a big lake, wave action may continually erode the peat deposit of a marginal swamp, so that no encroachment occurs. Photographs of the edges of Scottish lochs, taken over half a century [910], have shown no encroachment, except on river deltas, where allochthonous sediment deposition may have been greater than the rate of erosion by waves. Material eroded from edges, on the other hand, will be deposited in deeper water as lake mud and contributes to the filling process in this way.

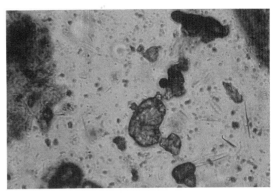

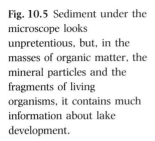

Fig. 10.5 Sediment under the microscope looks unpretentious, but, in the masses of organic matter, the mineral particles and the fragments of living organisms, it contains much information about lake development.

At first sight, lake mud is somewhat unprepossessing stuff: an apparently amorphous, dark-coloured, often evil-smelling, watery goo (Fig. 10.5). Nonetheless, it contains a record of changes, both natural and man-made, in the catchment area as well as in the lake. It may even harbour a record of global events, such as changes in climate and magnetic field.

Lake mud is laid down chronologically from the time of origin of the basin, but the sequence may sometimes be disturbed. In large lakes with steeply descending shores, there may be slumping and, in others, sediment may accumulate most abundantly in sheltered areas, which are subsequently eroded in storms. From most lakes, a continuous sequence of sediment can be obtained from the deepest part, using a corer. The determination of a lake's development then involves dating the sediment, analysis of suitable chemical and biological fossils and then interpretation of the data [64].

The simplest corer comprises a piece of plastic drainpipe. If the water and mud are both shallow, this can simply be pushed into the deposit until it penetrates the basin material and then pulled up. Success depends on the basin material and sediment sticking in the tube. For deeper water or deeper sediments, more sophisticated devices are needed. The Mackereth [607] corer (Fig. 10.6) is one of several such corers [1031]. Very recent, sloppy sediments can be best sampled by lowering a freeze corer [805, 882] into them. Freeze corers are boxes filled with solid carbon dioxide. The corer is lowered quickly into the surface sediment and the sediment freezes on to the outer faces without its vertical layering being disturbed. It can then be retrieved in an orderly manner while still frozen.

10.3.1 Dating the sediment

Sediment can be dated by several methods, mostly using radioactive isotopes. These usually involve measurement of concentrations of an isotope thought to be produced at a constant rate on Earth and of a derivative formed at a known rate (λ) from its decay. The time (t) during which an

(A)

(B)

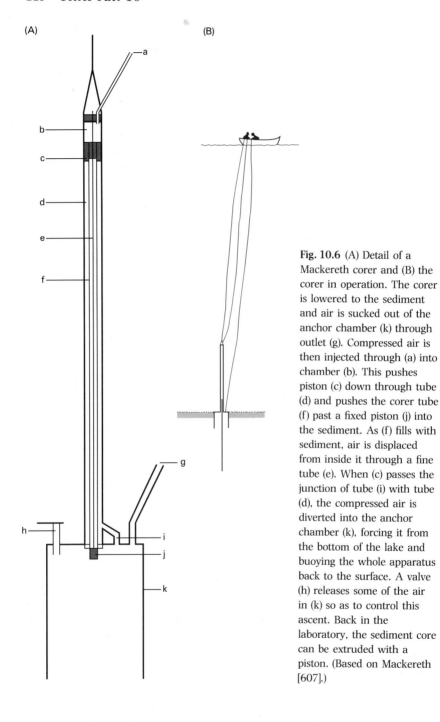

Fig. 10.6 (A) Detail of a Mackereth corer and (B) the corer in operation. The corer is lowered to the sediment and air is sucked out of the anchor chamber (k) through outlet (g). Compressed air is then injected through (a) into chamber (b). This pushes piston (c) down through tube (d) and pushes the corer tube (f) past a fixed piston (j) into the sediment. As (f) fills with sediment, air is displaced from inside it through a fine tube (e). When (c) passes the junction of tube (i) with tube (d), the compressed air is diverted into the anchor chamber (k), forcing it from the bottom of the lake and buoying the whole apparatus back to the surface. A valve (h) releases some of the air in (k) so as to control this ascent. Back in the laboratory, the sediment core can be extruded with a piston. (Based on Mackereth [607].)

amount A_0 has decayed to A is given by:

$$A = A_0 e^{-\lambda t}$$

and its derivative:

$$t = \lambda^{-1} \log_e(A_0/A)$$

Choice of isotope depends on the estimated age of the sediment. The older the sediment, the longer must the half-life (the time taken for half of a given initial amount to decay) be to ensure that it can still be detected. For sediments up to about 30 000 years old, the carbon-14 method can be used and for those up to about 150 years a method using lead-210 is useful.

10.3.2 Radiometric-decay techniques

The ^{14}C method depends on the steady formation of radioactive ^{14}C by bombardment of nitrogen (N) molecules by neutrons at the top of the atmosphere. The neutrons are part of the 'cosmic-ray' flux from the sun:

$$^{14}N_7 + {}^1n_n \rightarrow {}^{14}C_6 + {}^1p_1$$
$$\text{(neutron)} \qquad \text{(proton)}$$

The ^{14}C atoms are produced at the rate of about 10^2 cm^{-2} of the earth's surface min^{-1}, giving a ratio of ^{14}C to the stable ^{12}C of about 10^{-12}. It is assumed that carbon isotopes are taken up into living organisms in this ratio and that ^{14}C then decays, with a half-life of 5700 years. The $^{14}C : {}^{12}C$ ratio is measured in material to be dated and compared with the initial ratio, as follows:

$$({}^{14}C/{}^{12}C)_{current} = ({}^{14}C/{}^{12}C)_{initial} e^{-\lambda t}$$

For ^{14}C, it takes, on average, 8200 years for each atom to decay, so that:

$$t = 8200 \log_e(1/10^{12}[{}^{12}C/{}^{14}C])_{current}$$

where t is the age of the material being analysed.

The ^{14}C method is not usable for recent (< 200-year) sediments, since the burning of fossil fuels with very low $^{14}C/{}^{12}C$ ratios has upset the previously assumed constant ratio of 10^{-12} in the atmosphere. In fact, this ratio has never been entirely constant, and dates are corrected for this by using an absolute dating method, using tree-ring counts. By use of a series of wood samples, overlapping in age, so that the patterns in thickness of the growth rings can be matched, an absolute chronology has been constructed against which ^{14}C dates can be compared over several thousands of years.

The isotope ^{210}Pb is present in the atmosphere, ultimately as a result of naturally occurring uranium-238 in the Earth's crust. One of the products of uranium decay is radium-226, which decays slowly to the rare gas radon-222. This diffuses into the atmosphere, where it rapidly decays to

^{210}Pb. It is washed out of the atmosphere in rain, where its average concentration is around 2 pCi l^{-1}. Within a few weeks, it reaches the lake sediments and decays, with a half-life of 22.26 years, to bismuth-210. Measurement of the atmospherically derived ^{210}Pb can, then, give a useful method of dating recent sediments (Fig. 10.7). A correction has to be applied for the ^{210}Pb derived directly from ^{226}Ra decay in the sediment minerals, because the initial concentration of this varies from place to place.

10.3.3 Non-radiometric-decay methods

Radiometry is expensive. Other methods that are cheaper, if less versatile, are also available. The Earth's magnetic field has changed in both its horizontal and vertical components (declination and inclination) in past times, and a record kept in London since before AD 1600 shows a change in declination through some 30° since then.

Some minerals laid down in lake sediments become magnetized in the direction of the Earth's field at the time they are deposited, and some of this magnetization, the remanent, persists, even when the field direction of the Earth changes. The remanent direction of magnetization can be measured in core slices, kept orientated relative to a fixed line on the corer tube, with a

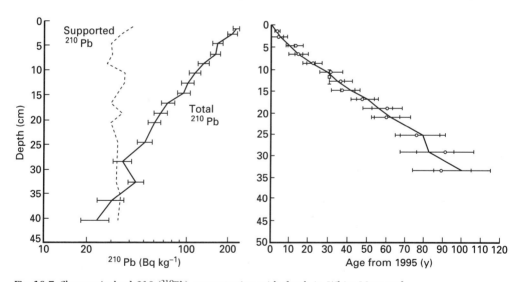

Fig. 10.7 Changes in lead-210 (^{210}Pb) concentration with depth in White Mere and dates calculated for levels in the core from these data. Lead-210 reaches lake sediments from the atmosphere, as described in the text, and from direct decay of locally washed-in mineral particles containing natural radium isotopes. The ^{210}Pb derived from the latter is measured as the equivalent radium isotope and referred to as the supported ^{210}Pb. The unsupported isotope is obtained from the difference between the total amount and the supported amount and used to calculate the dates. (From McGowan [604].)

sensitive magnetometer, and then plotted against depth in the sediment column. An oscillation in recent sediments, from east to west and back has been discovered and compares well with records kept at the London Observatory. For older sediments, comparison with ^{14}C dates shows an oscillation in declination with a period of about 2800 years, with the west and east peaks being contemporary for different lakes (Fig. 10.8). The pattern of remanent magnetization may therefore give a time-scale for other lakes where ^{14}C dating at frequent positions in the core proves too expensive.

Under anoxic hypolimnia, disturbance of the sediment by water currents and burrowing animals may be insufficient to obscure fine variations in deposition between winter/spring and summer each year. Light and dark layers, or paired varves (Fig. 10.9), may be detectable. The light member often contains carbonates deposited as a result of intense late-summer photosynthesis and marl formation (see Chapter 3). The pairs can be shown to be annual from spring and summer tree pollen in the appropriate members of each pair and cores can be dated by counting the pairs. Such visual varving is unusual, however.

Varving detectable by more sophisticated methods may be commoner.

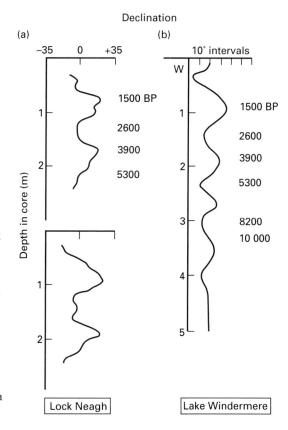

Fig. 10.8 Permanent magnetism, measured as the declination or horizontal component, can give a useful comparable dating technique between cores, once it has been calibrated against other methods. Two replicate cores are shown here from (a) Lough Neagh, Northern Ireland, and (b) one from Lake Windermere. Dates are based on carbon-14 determinations. BP, before present. (Based on Thompson [958].)

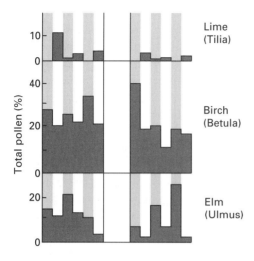

Fig. 10.9 Pollen analyses of two sets each of three paired bands of dark organic and light marl inorganic sediment from McKay Lake, Ontario. The bands are shown to be annual pairs (varves) from evidence, including the presence of summer-formed lime pollen in the marl bands and spring-formed birch and elm in the dark bands. (Based on Tippett [962].)

X-ray photography of thin longitudinal sections of cores from Lake Washington has shown prominent dense bands, which are not themselves annual, but their finer striations may be. Similar examination with a stereo-scan electron microscope has shown fine bands of diatom fossils, with pairs of layers characterized by spring and autumn species.

Ash falls from volcanic eruptions may result in distinctive sediment layers. If there is a historic record of the eruption, the layer can be dated (Fig. 10.10). A more recent but similar marker comes from the detection in the sediments of radioisotopes that do not occur naturally, but which have been made and discharged from human activities. The isotopes, caesium-137 and americium-241 were first introduced into the atmosphere in about 1954 from nuclear-weapons testing. Their amounts were greatest in 1962–63 and then declined. A new peak appeared in 1986, following an accident at Chernobyl, a Russian nuclear reactor, which released isotopes to the atmosphere. Detection of peaks of these isotopes in the sediments thus gives the dates of the 1954, 1963 and 1986 sediment layers (Fig. 10.11).

10.4 Sources of information in sediments

10.4.1 Chemistry

Both inorganic and organic particles accumulate in the sediments. Analysis of sediment for particular elements can thus give information on changes in and erosion of soils of the catchment, as well as production in the lake. For example, iodine is a scarce element in soils, but is present, from sea-water spray, in rainfall and is strongly adsorbed on to clay and humic colloids, which are washed into lakes. Its concentrations in sediments, therefore,

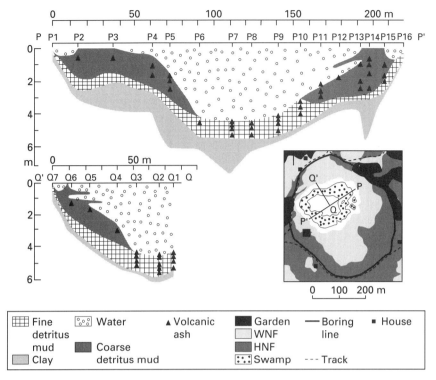

Fig. 10.10 Stratigraphy of a small crater lake at Birip in Papua New Guinea. Ash layers in the cores taken (numbered P1, Q1, etc.) are shown by triangular symbols. In the map, the crater rim is shown as a heavy line. WNF, woody non-forest vegetation; HNF, herbaceous non-forest vegetation. (Based on Walker and Flenley [992].)

reflect erosion of catchment soils, perhaps through climatic change or human activity, such as deforestation and agriculture [763].

Much information can also be determined from phosphorus in sediments. Because it is relatively insoluble, the phosphorus content is linked to the rate of loading on the lake, particularly when the rate of sedimentation is taken into account. It is sometimes possible to detect changes in sewage-treatment practice (diversion, expansion of the works) from the phosphorus content of sediments of lakes into which effluent is discharged [746, 930].

Sediments are often highly organic; gas chromatography, with mass spectrometry, is capable of revealing thousands of separate compounds. Many are derivatives of original compounds that have undergone change or diagenesis and cannot yet be associated with particular groups of organisms and these include hydrocarbons, fatty acids, amino acids, sugars, alcohols, ketones, steroids and plant pigments.

Sedimentary organic matter includes both refractory materials washed in from the catchment (leaf litter, 'humic acids') and autochthonously produced substances. There has been some question as to which source

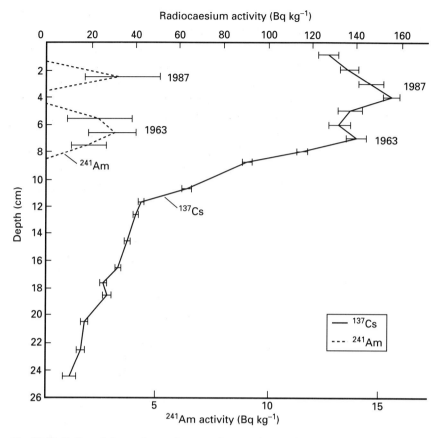

Fig. 10.11 Dating of the surface sediments of a core from Colemere, Shropshire, using the man-made isotopes, caesium-137 and americium-241. The lower peaks follow the peak of weapons testing in 1961–62 and the upper peaks the accident at the Chernobyl reactor in 1987. (Based on McGowan [604].)

contributes most. In the English Lake District, Mackereth [609] noted that the total sediment depth was broadly similar in both infertile and more fertile lakes. He concluded that the bulk of the sediment was catchment-derived and that organic matter produced in the lake was fully oxidized in the plankton or by the benthos and did not contribute much to the permanent sediment (see Chapter 8).

This hypothesis seems to hold for these lakes, except in the last few decades, when some lakes have been fertilized, particularly by sewage effluent. In the recent sediments of these, analysis for carboxylic (fatty) acids has shown a different array of n-alkanoic acids from that in older sediments and in the recent sediments of the unfertilized lakes [168]. Acids with 16, 22, 24 and 26 carbon atoms predominate in the former array, but the C_{16} fraction is absent in the latter. The C_{22}–C_{32} acids are abundant in soils but C_{16} acids are characteristic of algae. This suggests greater preservation of autochthonous material in very fertile lakes, probably associated with more severe deoxygenation of their sediment surfaces.

Analysis of the carbon skeletons of sedimented hydrocarbons also appears promising, for alkanes between C_{23} and C_{33} appear characteristic of higher plants and C_{31} is predominant in acid peat and C_{27} and C_{29} in base-rich forest soils [169]. Changes in the vegetation of the catchment may thus be detected in sediments and the information used to complement or further interpret pollen counts (see later). In Lakes Erie and Ontario, the $\delta^{13}C$ content (see Chapter 3) of organic matter in the sediment [861] correlates well with the non-mineral phosphorus concentrations and may give an index of changes in productivity that simple measures of organic-matter concentration cannot (Fig. 10.12).

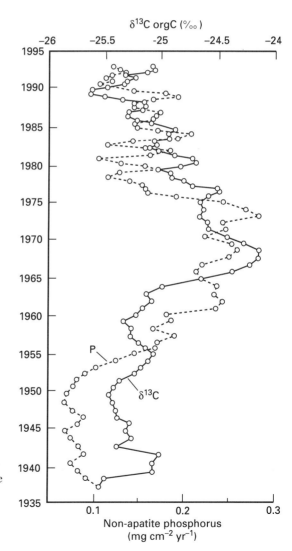

Fig. 10.12 Changes in non-apatite phosphorus concentrations and $\delta^{13}C$ (‰) in a sediment core from Lake Ontario. Values are three-point moving averages. (Based on Schelske and Hodell [861].)

The greater preservative properties of deoxygenated sediment surfaces are also reflected in the photosynthetic pigment derivatives found. Most pigments are quickly oxidized, so a greater pigment content and a greater diversity of pigments are preserved in lakes with anaerobic sediments [854]. Catchment vegetation contributes little pigment to the sediment, because the litter has been long exposed to the air before it reaches the lake bed. Extraction of chlorophyll derivatives and carotenoids with methanol or acetone and subsequent measurement by spectrophotometry thus helps confirm phases of changing fertility, but, if the pigments are separated chromatographically, even more valuable information may be obtained.

The carotenoids of different algal groups are often specific and, indeed, are used to help classify the algal families. Detection of particular pigments in sediments may, then, indicate changes in abundance of particular algal groups in the history of a lake. This technique has been applied to myxoxanthophyll and oscillaxanthin [107], which occur only in the blue-green algae. Sediment cores taken in 1967 from Lake Washington showed maxima of oscillaxanthin, derived from the large populations of *Oscillatoria rubescens* and *Oscillatoria agardhii*, which had appeared in the phytoplankton in the 1950s, as the lake became heavily fertilized with sewage effluent [366] (Fig. 10.13). The effluent was diverted from the lake by 1967 and progressively the *Oscillatoria* populations decreased. In new cores taken in 1972, the oscillaxanthin maximum was still present, but now buried under 5.45 cm of new sediments, which contained little oscillaxanthin [365]. The pigment thus preserved a record of the blue-green algal growth from the highly fertile phase which the lake had passed through. A deeper oscillaxanthin maximum recorded a phase when raw sewage was discharged to the lake for a few years before being diverted.

Some pigments are decomposed more rapidly than others, particularly fucoxanthin, diatoxanthin and diadinoxanthin, which characterize diatoms and dinoflagellates. Changes in abundance of these organisms are thus impossible to detect by these methods, Fortunately, these organisms leave morphological fossils. As well as blue-green algal pigments, those of green algae preserve well and β-carotene, which is common to many groups, gives a reasonable overall index of changing photosynthetic production.

Man-made chemicals may give evidence of changing pollutant loading. Heavy metals, such as mercury, and pesticide residues are easily measured in sediments. Oil-burning power stations produce smoke particles, often with characteristics that identify them to particular sources. These carbonaceous particles are preserved and carry with them a record of acidification of the lakes in many areas [364].

10.4.2 Fossils

Many algal and plant remains are preserved in sediments [64]. They include the heterocysts and sometimes whole filaments of blue-green algae, colonies of the green algae *Pediastrum* and *Botryococcus*, the silica walls of diatoms (Fig. 10.14) and the silicified scales and cysts of Chrysophyta, pollen from higher plants in the lake, its catchment area and perhaps further afield, and lignified cells, such as sclereids, fibres and xylem vessels, from aquatic plants. Animal remains include the tests of certain amoebae, sponge spicules, bryozoan statoblasts, carapaces of Cladocera (Fig. 10.14) and other small Crustacea, mollusc shells, head capsules of chironomid larvae, mite exuviae and even an occasional fish scale.

Because of the closer relationship that algae and plants, rather than animals, have with the physicochemical environment, it is generally easiest to interpret the meaning of fossils of algae, particularly diatoms, and of pollen. The former give information mostly about the lake itself and the latter about its catchment. The occurrence of particular animals reflects to a large extent their competitive and predatory relationships (see Chapters 7 and 8). Nonetheless, animal remains are useful and may eventually even throw light on the past operation of some of the predatory interrelationships.

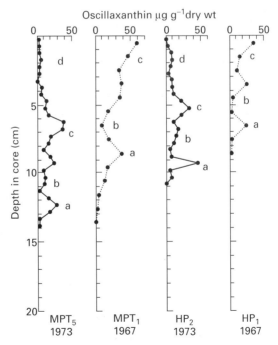

Fig. 10.13 Concentrations of oscillaxanthin in cores taken from two locations (MPT and HP) in Lake Washington in 1967 and 1973. The early cores show the build-up of blue-green algae as the lake became eutrophicated, with an early peak (a) due to raw sewage discharge, which was subsequently diverted. There was then a temporary recovery (b), followed by a later increase (c) as populations increased in the area and effluent volumes also increased. These effects are also shown in the later cores, with a subsequent decline (d) as effluent was diverted away from the lake and blue-green algal populations declined. (Based on Griffiths and Edmondson [365].)

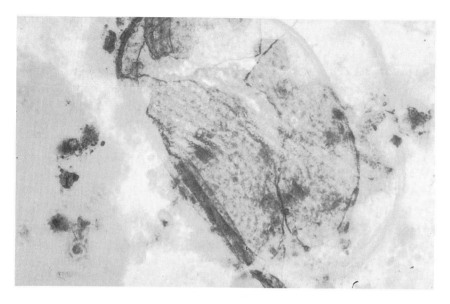

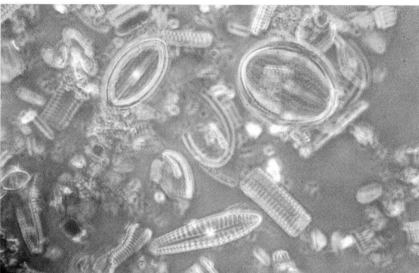

Fig. 10.14 Sediments are often rich in fossils. Upper photograph is part of a *Daphnia* exoskeleton, the carapace, overlying a fragment of moss leaf. Below is a range of diatoms and diatom fragments. All are from sediments laid down in Strumpshaw Broad, Norfolk.

10.4.3 Diatom remains

Diatom cell walls have patterns that characterize particular species. They have been studied for a long time and considerable reconstruction of past environments is possible from an analysis of the diatoms in sediments. Permanent microscope slides of diatom walls can easily be prepared, following oxidization of the sediment with hydrogen peroxide, chromic or nitric acid. The residual diatoms are resuspended in distilled water before being dried on to a slide and mounted in high-refractive-index mountant.

From the diatom species, much can be deduced about the balance of planktonic (see Chapter 7), epipelic (sediment-living) and attached communities (see Chapter 8) and hence about changes, for example, in water depth and the abundance of plant beds. Certain genera, such as *Eunotia* and *Frustulia*, are characteristic of infertile water and some species, such as *Aulacosira granulata*, of very fertile conditions. There are distinct marine and freshwater species and some indicators of brackish conditions. In African lakes, *Navicula elkab* and *Nitzschia frustulum* characterize saline inland waters and may be used in sediment analyses to recognize drops in lake level and periods of endorheicity [816].

Diatoms have been used to trace changes in the pH of waters believed to have been acidified and to help distinguish between rival theories of the cause of the acidification. Careful taxonomy is necessary, and study of contemporary lakes suggests that some species are associated with very high (alkaliphilic) or very low (acidophilic) pH. Others are associated with neutrality (circumneutral), while intermediates (acidobiontic and alkalibiontic) are also found. An index that compresses information on the characteristics of all the species from a lake can be calculated. For example, index B of Renberg and Hellberg [806] is based on counts of the species present and calculations of the percentages of each group of the total counted:

$$\text{Index B} = \frac{\%\ \text{circumneutral} + 5 \times \%\ \text{acidophilic} + 40 \times \%\ \text{acidobiontic}}{\%\ \text{circumneutral} + 3.5 \times \%\ \text{alkaliphilic} + 108 \times \%\ \text{alkalibiontic}}$$

The index can be related to the pH of lake water by regression of the measured pH against counts made from the diatoms in the contemporary sediment [276]. For example, in a series of 33 lakes in southern Scotland:

pH (log units) = 6.3 − 0.86 log (index B) (r^2 = 0.82, $P < 0.001$)

Changes in pH can then be calculated from measures of the index obtained from a sequence of sediments (Fig. 10.15).

Such indices have recently been refined to remove the partly subjective judgement involved in classifying a species as acidophilous or circumneutral, etc [2]. The new methods do not require such prior ecological classification and can be applied to any variable that influences the occurrence (not necessarily growth) of the species. The method, called weighted averaging, assumes that each species has an abundance that is related to the environmental variable in a Gaussian way. This means that it will be most abundant at some optimal value and decreasingly less abundant with increasing or decreasing values of the variable to each side of the optimum [924, 953]. A large set of lakes is sampled for the variable concerned and the relative abundance of each species is measured in living samples or contemporary sediments.

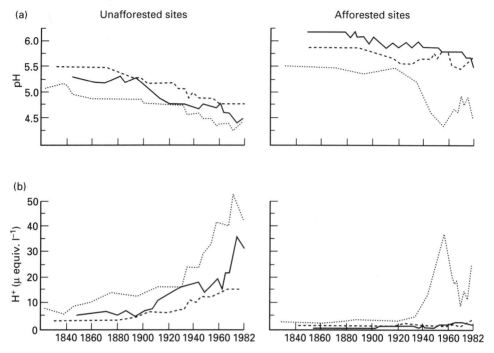

Fig. 10.15 The recent history of lake-water acidity changes, expressed as (a) pH and (b) hydrogen ion (H⁺) concentration, in three Scottish lochs where catchments have not been forested (left-hand side) and in three in which part or all of the catchment has been planted with conifers (right-hand side). The data (based on reconstructions of pH from the sediment diatom communities) suggest that factors other than afforestation have been responsible for the recent acidification, where it has occurred. (Based on flower *et al.* [278].)

The optimum for each species, U_k, can then be calculated from:

$$U_k = \Sigma(y_{ik} \times x_i)/\Sigma y_{ik}$$

where x_i is the value of the variable in lake i and y_{ik} is the abundance of species k in lake i. This establishes optima for a large number of species in what is called the training set. The diatom assemblage of a particular sediment layer in the same set or a lake broadly similar to those in the set (judged by its sharing a significant number of species) is then determined and the value of the variable x_i, associated with the conditions under which the assemblage was laid down is calculated from:

$$x_i = \Sigma(y_{ik} \times U_k)/\Sigma y_{ik}$$

The method is now being extended to other variables, such as total phosphorus [60–62]. Its sensitivity is likely to be greatest with variables, such as pH, which are well correlated with species distribution and which express a huge range of variation in the variable concerned. A change in

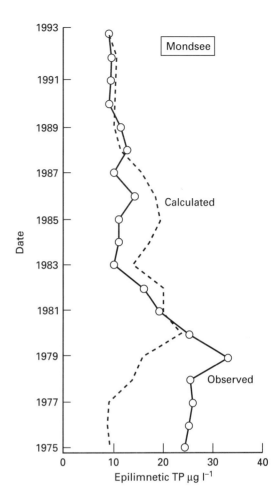

Fig. 10.16 Comparison between observed total phosphorus concentrations and those in the lake calculated from the diatom community in a sediment core from the Mondsee, Austria. Both sets of values are three-point moving averages. (Based on Bennion *et al.* [61].)

one pH unit represents a ten-fold change in hydrogen ion concentration. Where variables that have a more limited arithmetic, rather than geometric, range are concerned, the precision of the predictions possible from weighted averaging is likely to be lower, though still extremely valuable in showing relative changes (Fig. 10.16).

10.4.4 Pollen

The use of pollen preserved in peat and lake sediments has a distinguished history and has formed the basis for recognition that climate has changed greatly over periods of only thousands of years. The walls of many pollen grains are of resistant waxes ('sporopollenin'). Grains may be concentrated and separated from sediment samples by digestion with hydrofluoric acid. Genera, and often species, are easily recognizable, although grass pollens, for example, are not easily separable to species. However, major changes in catchment vegetation can be detected. Agricultural activity is recorded by the appearance of the pollen of characteristic annual herbs that accompany cultivation and evidence of deforestation may be found by a complementary decrease in tree pollen.

10.4.5 General problems of interpretation of evidence from sediment cores

Jack Vallentyne wrote [979] that:

no anatomist or physiologist in his right mind would ever base a study of the life history of an organism on the analysis of its accumulated faeces. This is, however, precisely the position of a palaeolimnologist with respect to the developmental history of a lake. Sediments are lacustrine faeces, the residue remaining after lake metabolism.

It is wise to examine some of the problems of handling sediment data before considering interpretations from particular lakes.

Sediment is not necessarily deposited uniformly nor are microfossils deposited evenly. They may also be redistributed by water movements. Davis [194] found that relatively heavy oak-pollen grains were evenly distributed over the sediment of a Michigan lake, but lighter ragweed pollen settled slowly and was carried to the windward edges before reaching the sediment. She also found that the surface few millimetres of sediment were resuspended during mixing periods and that material could be recycled upwards into the water perhaps four times before becoming part of the permanent sediment. This tended to even out irregularities in previous sedimentation, both in space and in season.

Benthic animals may also mix the surface sediments as they form, evening out annual variations into a sort of moving average and smoothing long-term trends. Davis [195] found that tubificid worms, feeding at 3–4 cm depth, moved sediments upwards as faeces and displaced larger particles, e.g. of pollen grains, less than smaller ones.

In general, diatoms are preserved well, but there is some selectivity. In deep lakes, only a fraction of the diatom-wall silica reaches the permanent sediment [757] and dissolution may occur in the surface sediments of shallow lakes. Dissolution is greatest in alkaline waters, and the bicarbonate-rich layers of meromictic lakes may allow little or no preservation. Breakage may occur of those that are not dissolved. Long, thin diatoms – for example, *Synedra* and *Asterionella* – break easily by abrasion with inorganic particles or in invertebrate guts and may then more easily be redissolved. Some thin-walled genera – for example, *Rhizoslenia* – are rarely preserved at all. Centric (radially symmetrical) planktonic and littoral diatoms seem to persist best. Comparisons between the diatom assemblages laid down in dated recent sediments and counts made on the living community for the same period show a very close correspondence [395], although the correlation might be expected to decrease as the sediments age.

In interpretation, much depends on knowledge of the current autecology of particular species, but this is usually imperfectly known. More confidence can be placed on the detection of arrays rather than individual species, a strength of the transfer-function approach, discussed above. A further problem of reliance on single 'indicator' species of diatoms is that the diagnostic patterning of the wall may change as dissolved silicate levels fall [51]. One 'species' may thus change into another. Sufficient laboratory work has not yet been done to determine how widespread this is.

Perhaps the greatest problem is in the way the results are expressed. Early workers often calculated the percentage representation of a particular species or group of fossils in their total count from a sediment sample of undetermined weight. This gave relative changes in frequency. Thus, a decrease in frequency of a species A would inevitably result in increases in frequency of other species. The reality, in absolute terms, could have been one of a variety of scenarios, in which A may have decreased or increased or stayed the same as others changed. An improvement has been the expression of absolute counts on the basis of per unit weight of sediment, because this gives some indication of absolute changes if the sedimentation rate has been constant.

However, where sedimentation rates have changed, an absolute increase in a species may have been accompanied by a decrease in its numbers per unit weight of sediment if it has been diluted by sediment from other sources. The only reliable way of expressing results is thus in terms of amount laid down per unit area of lake bed per year, which needs a comprehensive dating of the core. This is expensive and has not been done for most cores. All these limitations should be borne in mind in the following accounts of results from particular lakes, which will illustrate some of the features of lake development.

10.5 Examples of lake development

10.5.1 Blea Tarn, English Lake District

Hills of hard, Borrowdale volcanic rocks, up to 600 m high, surround Blea Tarn (Fig. 10.17) in the English Lake District. It is 3.4 ha in area and 8 m in maximum depth and is surrounded by tundra-like moorlands of grasses and sedges and by *Sphagnum* bogs. There are some plantations of conifers, but the only human disturbance of the catchment is a small sheep farm, whose effect on the lake is likely to be slight. Habitation of the area in the past was sporadic, so Blea Tarn is a lake largely under control of 'natural' events.

Blea Tarn was formed in an ice-carved depression in the debris eroded

Fig. 10.17 Blea Tarn, Langdale, Cumbrian Lake District.

from the hills by the glaciers and had water in it by about 14 000 BP (before present). A core of sediment from the lake [393] was about 350 cm long (Table 10.1). For the first 4000–5000 years, when the area still had much ice, biological production in the lake was low; relatively few diatom frustules and little organic matter are present in the clay and silty late-glacial sediments. Around 11 900 BP, there was a brief warming (the Allerød interphase, named after a site in Scandinavia in whose deposits it was first detected), in which the sedge, grass and dwarf-willow tundra was briefly diversified with birch trees and juniper bushes. The lake sediments were then slightly richer in organic matter and diatoms, which included few phytoplankters and were of a mixed collection, some characteristic of slightly alkaline water.

Final melting of the ice, a little more than 10 000 years ago, led to marked changes. Hazel (*Corylus avellana*) colonized the catchment, followed by birch (*Betula*), oak (*Quercus*), elm (*Ulmus*) and, in a wetter phase (the Atlantic period), alder (*Alnus glutinosa*). The lake sediments became more organic, brown, lake muds. The diatom fossils indicate high fertility, with a well-developed plankton, just after the ice finally melted. Species such as *Asterionella formosa* and *Rhopalodia gibba* and the genera *Fragilaria* and *Melosira* (*Aulacoseira*) were common and indicate a high nutrient loading.

This loading came from leaching, in a warming and wetter climate, of much-pulverized rock debris left by glacial action. The organic matter of the sediment was probably catchment-derived from the deepening soils of the woodlands near the lake. Between 10 500 BP and 7400 BP, the fertile-lake diatom flora was replaced by one of less fertile, more acid waters, characterized by *Eunotia* and *Pinnularia*, which has persisted to the present day, with little change in 7000 years.

The decrease in fertility of the lake is attributed to exhaustion of the readily available nutrients in the glacial debris and subsequent dependence on the weathering of fresh basal rock deep in the soil profiles. This is extremely slow; the composition of present-day waters of upland Lake District tarns differs little from that of rainfall [606].

Table 10.1 Changes in the history of Blea Tarn, Langdale, English Lake District, deduced from a core taken through its sediments (from Haworth [393].)

Depth (cm)	Date (BP)	Sediments	Pollen	Sedimentation rate (mm year^{-1})	Diatoms and interpretation
0–147	< 4900 (Subboreal)	Brown mud (*gyttja*)	Ash, elm declining	0.32	Very infertile conditions
147–227	< 7400 (Atlantic)		Oak, elm, birch alder		*Eunotia*, infertile, Acidophils increase
227–327	< 10 200 (Boreal)	Clay *gyttja*	Hazel peak	0.36	Fertile, alkaline indicators increase (*Asterionella*, *Fragilaria*, *Melosira*)
	POSTGLACIAL				
	LATE GLACIAL				
327–343		Pink clay	*Artemisia*, *Empetrum*, *Lycopodium*		Diatoms scarce, lake frozen over
343–349.5	< 10 700		*Artemisia*		
349.5–353.5	< 11 900 (Allerød)	Dark, organic silt	Juniper, birch		Alkaliphilic (*Campylodiscus noricus*)
353.5–357		Pink clay	Grasses, sedges, willow	0.07	Benthic diatoms (*Pinnularia, Cymbella, Fragilaria*) No plankters
Below 357	> 14 900	Clay	No pollen		No diatoms

10.5.2 Esthwaite

A glacier flowing down a small river valley, also in the English Lake District, carved out the basin of Esthwaite and dammed the lake with a moraine at the valley foot. It is larger (100 ha) than Blea Tarn and lies among more subdued scenery of softer Silurian rocks, more readily weathered than the Borrowdale volcanic rocks around Blea Tarn. For all but the last 3000 years, however, Esthwaite and Blea Tarn shared similar histories.

The late-glacial phases (Table 10.2) supported tundra around the lake, and clays were deposited in the cloudy water as the fine debris of the glacial till was eroded. The Allerød warming was marked by the appearance of Cladocera in the water, particularly species of Chydoridae, a group associated with the littoral zone (see Chapter 8), but diatoms were not well preserved, perhaps because of abrasion by the silt and clay. The concentrations of sodium, potassium, calcium and magnesium in the sediments

Table 10.2 Changes in the history of Esthwaite, English Lake District, deduced from cores taken through its sediments (data from Cranwell [167], Goulden [353], Mackereth [608] and Round [831].)

Depth (cm)	Sedimentation rate (mm year⁻¹)	Date (BP)	Pollen	Diatoms	Animals	Organic compounds	Calcium (mg g⁻¹)
0–150	5.6–11		Plantain	*Asterionella, Melosira,* fertile phase	Chironomids increase, also *Bosmina longirostris*	82% *n*-alkanoic, 18% branched cyclic acids	4–5
150–300	0.55	<5100	Oak, birch, alder	Infertile, acidic indicators (*Eunotia, Anomoeoneis*)	*Bosmina coregoni*		
300–410	0.46 Brown mud		Alder increase	Fertile (*Melosira, Stephanodiscus*)		Decrease in branched cyclic acids (13%)	
410–437	0.19 Banded clay/mud	<8900		Fertile (*Melosira arenaria, Epithemia, Fragilaria*)			
437–460	0.16 Grey clay		Pine, hazel				
460–478		<10 300		Poor flora			22
478–498					Littoral arctic forms		5–10
498–515	Grey silt/clay		Sedges, willow				
>515	Laminated clay						

were around 10, 32, 5–10 and less than 10 mg g^{-1} dry wt, respectively.

With the start of the postglacial period, the surge in production noted for Blea Tarn was also recorded. The leaching of broken rock fragments led to increases of calcium (22 mg g^{-1} dry wt) and phosphorus in the sediments, the latter increasing from about 3.5 to 5 mg g^{-1} dry wt. Diatoms (*Melosira arenaria*, *Epithemia*, *Fragilaria*) characteristic of a fertile phase and planktonic Cladocera numbers increased. Analysis of the organic component of the immediately postglacial sediments has shown a ratio of *n*-alkanoic to branched and cyclic monocarboxylic acids similar to that found currently in the surface sediments of fertile lakes.

About 5000 years ago, the regression to less fertile, more acidic conditions took place, as it did in Blea Tarn. Calcium concentrations fell to about 5 mg g^{-1} dry wt and those of phosphorus to about 2 mg g^{-1} dry wt. The diatoms (*Cyclotella*, *Tabellaria*, *Eunotia*, *Anomoeoneis*, *Gomphonema* and *Cymbella*) collectively indicate a decline in fertility. But, in Esthwaite, in contrast to Blea Tarn, the unproductive phase has not continued to the present day. In the last 3000 years, there has again been an increase in fertility. At first, this was moderate but, in the nineteenth century, it was marked. *Asterionella* and *Melosira* have reappeared in the diatom flora, and chironomids became abundant about 1600 years ago. At about the same time, the cladoceran *Bosmina longirostris* appeared and partly replaced the previously recorded *Bosmina coregoni*. Sedimentation rates increased in the nineteenth century to 5.6–11 mm year^{-1}, from a mean postglacial rate of only 0.48 mm year^{-1}.

The reason for these changes is to be found in human activity [596]. Pollen of a common agricultural weed, the plantain, *Plantago lanceolata*, is found in sediments younger than about 3000 years, and, around 3700 BP, Neolithic peoples are known, from archaeological evidence, to have moved into the Esthwaite catchment area, though not to any extent into that of Blea Tarn. Forest clearance for fuel and agriculture leads to release of nutrients previously stored in the woodland biomass (see Chapter 3), to greater soil erosion, and perhaps to greater leachability of nutrients from the excreta of domestic stock. The clearances were intensified by a new wave of colonization around 900 BP by Norsemen and Norse Irish, perhaps retreating from political disturbances in the Isle of Man. Many place names in the English Lake District are of Norse derivation.

During the nineteenth century, the English Lake District became popularized, particularly by the poetry of Southey and Wordsworth, who both lived there, and, since the building of a railway, completed in 1847, the populations of both residents and tourists have increased. In turn, this has led to improved mains sanitation, and it is the sewage effluent from the popular tourist village of Hawkshead, on the inflow river to Esthwaite, which has led to the latest phase of eutrophication of the lake in the last forty years or so.

10.5.3 Pickerel Lake

The prairies of north-east Dakota, USA, contrast with the mountains of the English Lake District, but their lakes were also formed by ice around 11 000 years ago. A large ice block, calved from the continental glacier as it melted back, was buried amid the rubble of limestone and shale washed out from under the glacier. When the block melted, it left a basin up to 25 m deep, which now holds Pickerel Lake.

The lake sediments are up to 8 m deep and record development of a lake [394] in an area now drier than north-western Europe. Four phases (Table 10.3) in the surrounding vegetation have been deduced from pollen analysis. After the retreat of ice, spruce (*Picea*) and tamarack (*Larix*) forest grew by 10 500 BP, and by 9400 BP had diversified to a forest of oak, elm, birch, alder, fir (*Abies*), sycamore (*Acer*) and ash (*Fraxinus*), probably with some grassy openings. The climate then became drier, for the forest was replaced by prairie, with grasses, including *Andropogon* (blue-stem), Compositae (*Ambrosia* (ragweed) and *Artemisia*) and annual herbs of the family Chenopodiaceae. This vegetation has persisted since 9400 BP, although since 4200 BP there has been some patchy deciduous woodland.

The drying of the climate is reflected in Pickeral Lake. In the early forest phases, the raw conifer-forest humus neutralized the alkaline groundwater of the calcareous glacial deposits, for the diatom flora was then mixed. There were indicators of neutral to acid waters (*Eunotia* and *Tabellaria*), which do not appear in the younger sediments, while genera of fertile, alkaline waters (*Fragilaria*, *Epithemia*, *Mastigloia*) were also present. Diatoms of strongly alkaline habitats were absent and the mainly benthic flora, coupled with the remains of aquatic plants (*Potamogeton pectinatus*, *Najas flexilis*, *Najas marina*), suggests poor plankton development.

As the deciduous woodland replaced the conifers and then itself was succeeded by grassland, the diatom flora changed and many molluscs appeared in the lake. Diatom indicators of neutral to acid water disappeared and the flora was generally sparser. Then as the prairie developed, there were more diatoms, particularly phytoplankters of fertile lakes – *Melosira ambigua*, *Melosira granulata*, *Fragilaria crotonensis* and *Stephanodiscus niagarae* – and some species characteristic of brackish waters. Prairie climates have summer drought, giving increased mineralization of nutrients and great fluctuations in water level and ionic content of the water.

The changes in water levels are indicated by layers of sand, eroded from the marginal high beaches, in the sediment profile and by seeds of annual higher plants characteristic of exposed wet mud. Concentration of ions by evaporation in summer explains the brackish diatom taxa and the increased nutrient loading the characteristic plankton species. This burst of production was not sustained, for, after 4200 BP, some woodland development may have stabilized the soil and decreased the erosion rate. It was

Table 10.3 Changes in the history of Pickerel Lake, Dakota, deduced from a core taken through its sediments (from Haworth [394].)

Depth (cm)	Date (BP)	Sedimentation rate (mm year^{-1})	Pollen	Diatoms	Interpretation
0			Ambrosia	Fewer, varied, alkalibiontic; increase in halophiles; some Melosira granulata, Fragilaria crotonensis	Some eutrophication from agriculture
100		1.05	Prairie grasses, some increase in woodland		
300	<2700			Many, but no halophobes; plankton decreases	
400	<4200	0.91	Blue-stem grass prairie, Ambrosia, Artemisia composites, chenopods; some wood; seeds of wet-mud plants	Many, alkaliphilous, no halophobes, more plankton, Melosira granulata, Stephanodiscus; some brackish taxa	Eutrophic; transition to open prairie with summer drought; fluctuation in lake level; increased erosion (sandy layers); evaporative concentration
600	<9400		Mixed deciduous woodland, with Quercus, Ulmus, Betula, Alnus, Abies, Acer, Fraxinus	Fewer, Fragilaria; benthic molluscs; acidophilous species decrease; many benthic, alkaliphilous, some halophilic	Change to more alkaline conditions with deciduous forest
700		3.82			
800	10 500		Forest: Picea, Larix	Acid to neutral species, some acidophilous; all halophobe, Eunotia	Spruce forest, acid soils, low nutrient loading

associated with a somewhat damper climate and linked with a mixed diatom flora, less dominated by indicators of high fertility.

Towards the top of the sediment profile is evidence of a further enrichment, though not as great as that between 9400 and 4200 BP. *Melosira granulata* and *F. crotonensis* have again increased and, coupled with an increase in ragweed pollen, suggest that this is associated with agricultural activity and fertilization of the land during the past two centuries.

10.5.4 White Mere

White Mere (Fig. 10.18), in the UK, was formed in the same way as Pickerel Lake, at the end of the last glaciation. It is one of a group of lakes that historical tradition [691] suggests have had blue-green algal blooms for at least several centuries and its palaeolimnology confirms this [604]. Diatoms are not well preserved in its deeper sediments, not for reasons of low lake productivity but more probably because their frustules dissolved in what is likely to have been somewhat alkaline water. They occur in the upper 2 m of sediment, corresponding to at least the last 1000 years. In the deeper sediments, as well as in the recent ones, there is an abundance of carotenoid pigments, which suggest (Fig. 10.19) rich blue-green algal growths, at least as early as 6700 years ago and probably from even earlier. The diatom changes at the top of the core suggest some recent eutrophication, for they contain small centric diatoms, associated with increased nutrient loading. But it is likely that the lake has had blue-green blooms for all of its history and that it has always been rich in nutrients, particularly phosphorus. It may have accumulated phosphorus, which, because the lake is groundwater-drained, was not easily flushed out (see Chapter 6). Nitrogen loading appears more important, and it is increase in this, from agricultural activities, which is behind the recent eutrophication.

Fig. 10.18 Preparing to core White Mere with a Mackereth corer.

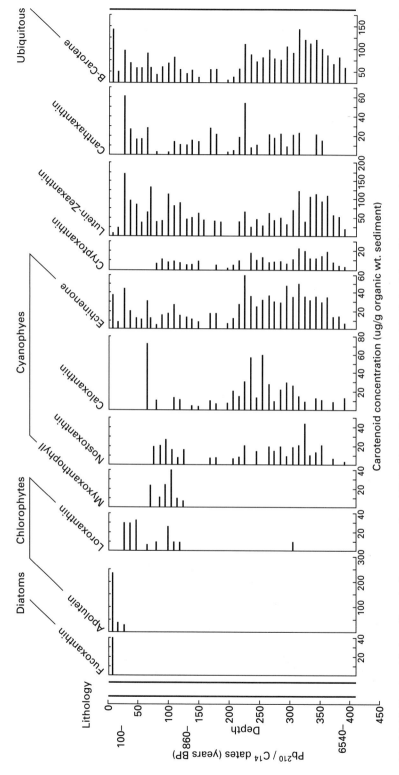

Fig. 10.19 Carotenoids in sediments from a core from White Mere, Shropshire. Carotenoids derived from blue-green algae are abundant in sediments laid down as early as 6000 years ago. (Based on McGowan [604].)

10.5.5 Lago di Monterosi

Not all lakes are as young as those formed by ice action. The Lago di Monterosi (Fig. 10.20) in the Campagna, 40 km from Rome, in Italy, is 26 000 years old. It lies in a volcanic crater and at present is moderately fertile, with extensive beds of aquatic plants (*Nymphaea*, *Myriophyllum*, *Ceratophyllum*) and a plankton typical of soft lowland waters, but it has a rich history [457].

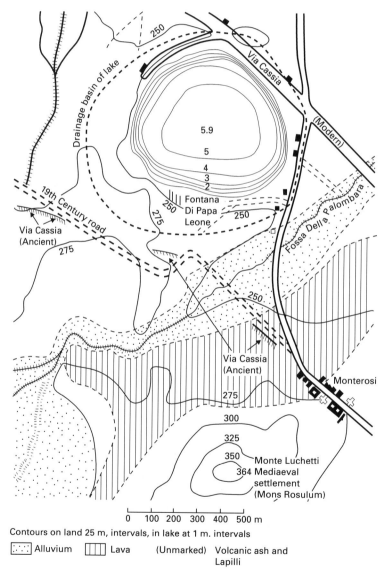

Fig. 10.20 Map of the Lago di Monterosi and its region (from Hutchinson *et al.* [457].)

When the basin was formed, northern Europe was still covered in ice and the Roman Campagna was a cold, dry steppe, with large mammals, such as the mammoth, grazing over it. Sediments from the lake in its first 3000 years (period A, Table 10.4) indicate an alkaline, rather fertile lake, with remains of *Gloeotrichia* (blue-green alga), *Chaoborus* and *Glyptotendipes* (Chironomidae) and aquatic plants. Sponge spicules, also of fertile habitats, remain from the invertebrate fauna of the plant beds. The crater rim and surrounding permeable catchment of fresh ash, lapilli and scoria provided fresh rock surfaces for leaching and provision of abundant nutrients.

Table 10.4 Changes in the history of the Lago di Monterosi, deduced from a core taken through its sediments (from Hutchinson *et al.* [457].)

Period	Date (BP)	Features	Interpretation
C2	0	Some blue-green algae, chironomids	Intensive farming and soil erosion Productivity less than in C1 but more than in B
	1200	Eutrophic diatoms, decrease in blue-green algae and chironomids	Decline in productivity
C1	2220 (AD 10–340 BC)	Increase in Ca, rate of sedimentation, chironomids, chydorids and blue-green algae	Dramatic increase in fertility due to road-building disturbance
	2800	More wooded catchment, with *Abies* and *Quercus*; pollen of Chenopodiaceae	Human activity increasing; probably less acid
B	5000	Increase in acidophil diatoms and *Sphagnum*; decrease in water plants; cysts; Chrysophyceae; few chironomids or chydorids	Lake shallow, low sedimentation rate; infertile (oligotrophic)
	13 000	Grass pollen, water plants, low sedimentation rate	Lake shallowing; well-developed plant cover
	23 000	Moderate sedimentation rate; few chironomids; diatoms of wide tolerance; sponges, pollen of *Myriophyllum*, water lilies and *Potamogeton*	Lake water dilute, slightly alkaline; catchment steppe-like, with *Aremisia*
A	26 000		Origin of lake by volcanic explosion

This early productive phase, seen also in other lakes discussed above, did not last beyond 23 000 BP, as the fresh rock deposits became exhausted and the run-off waters less rich in dissolved ions. There followed a long period (B) of reduced production and low sedimentation rates (0.044–0.075 mm year^{-1}), which lasted for 20 000 years with little change. *Sphagnum*, a moss genus of acid to neutral infertile waters, colonized. The sponges of alkaline waters declined and chironomids became scarce. Woodland of oak, fir and hazel developed and nutrient retention by the forest soils probably also reduced loading on the lake.

Then, dramatically, the lake again became very fertile, with maxima of blue-green algae, other algae (*Pediastrum*, *Melosira granulata*), characteristic of very-fertile-lake plankton and a rich cladoceran and chironomid fauna. Calcium was heavily deposited in the sediments and the sedimentation rate increased nearly sixfold to at least 0.38 mm year^{-1}. *Sphagnum* disappeared. The date for this transition, given by ^{14}C analysis, is 2220 ± 120 years BP and most probably falls between 340 and 110 BC.

No natural events could be found to explain the changes that occurred, but there was major human interference in the catchment at the time. The Roman consul, Lucius Cassius Longinius, built a road around 171 BC to give rapid passage between Rome, eastern Tuscany and the upper Arno valley, where there was unrest among the rural populations, which was to culminate in the second Punic war. This ancient Via Cassia ran around the southern edge of the catchment of the Lago di Monterosi, and the disturbance of road building on the edges of a catchment (Fig. 10.20) little larger than the lake itself is the only sensible explanation of changes in the lake.

The roadworks disturbed the vegetation and altered the drainage slightly, so that water previously draining away from the lake burst through as springs (the Fontana di Papa Leone) near its edge. This water percolates through calcareous deposits and explains the increased calcium deposition in the lake, while soil erosion, caused by removal of vegetation, explains the increase in sedimentation rate.

The effects of road building were short and the lake settled back into a less productive phase soon afterwards. It has never quite returned to its state in period B, because of the permanent alteration in the hydrology caused by the springs and because, in the post-imperial period, agriculture and deforestation caused further changes and disturbance. *Sphagnum* reappeared in the fifteenth century, at the time of a nadir in the fortunes of the Monterosi area, but the last 200 years of farming prosperity have seen a renewed small increase in the fertility of the lake.

10.5.6 Lake Valencia

Table 10.5 illustrates recent changes in an ancient Venezuelan lake, Lake Valencia, formed in a tectonic depression some time in the mid-Tertiary

Table 10.5 Changes in the history of Lake Valencia, Venezuela, deduced from a core taken through its sediments (from Bradbury *et al.* [87].)

Date (BP)	Features	Interpretation	Comment
0	High chlorophyll a	No overflow	Drier climate, evidence of eutrophication and diatoms not previously found
2000	Littoral and planktonic diatoms; Chenopodiaceae, grass and tree pollen; little aquatic-plant pollen	Overflow, some salinity increase	
3000/4000		Overflow, decreasing salinity	Maximum freshness
5000/6000	Stromatolites	No continuous overflow	Higher salinity
7000	Stromatolites (layered deposits formed by algae)	Overflow	Low salinity, wet period
8000/9000			Wetter climate

From this time on, freshwater lake with variable salinity (< 5‰); end of dry period

10 000	Diatoms and ostracods of saline water; evaporite mineral; little benthos; some tree pollen		

Saline lake for about 1000 years

11 000	Absence of diatoms, ostracods; pollen of emergent aquatic plant; low organic content in sediment		

Marsh at beginning of core and for next 1000 years

period, perhaps 50 million years ago. During the last 12 000 years, Lake Valencia has been, successively, a shallow marsh, a closed-basin saline lake and a freshwater lake with fluctuating salinity. Its sediments 12 000 years ago were of clay, containing few diatoms, but pollen of *Alternanthera*, *Ludwigia* and *Typha*, which are all emergent aquatic plants.

With the retreat of the polar glaciers, this tropical lake became deeper, around 10 500 BP, as the climate warmed and dampened, but it was very saline. Diatoms of salt water (e.g. *Cyclotella striata*) were present and also the minerals aragonite and dolomite, which are laid down at salinities greater than about 10 g l^{-1} of total salts if calcium and magnesium are abundant.

After about 8700 BP, the lake became even deeper and fresher, reaching its maximum freshness about 3000 BP. It had periods of low water, indicated by layered deposits (stromatolites), formed in the shallows about 5500 and 7100 BP, but was still a freshwater lake with an outflow when it was observed discharging to the River Orinoco in the early eighteenth century. At present, it is drying out again and has no outflow, but it is not known whether this is a reflection of natural climatic change or of changes in the local hydrology due to agricultural development of the catchment.

10.6 Filling in of shallow lakes

Sediment is deposited in all lakes, so theoretically all lakes should eventually be made so shallow that they are converted to areas of wetland or even terrestrial vegetation. In a very deep lake, however, this process may take a very long time. Lake Tanganyika, for example, has a maximum depth of 1500 m and a current sedimentation rate of around 0.5 mm year^{-1}. This should give it a future span of about 3 million years – enough for even geological events to disrupt the basin before such a state is reached. Lakes of only a few metres depth, however, may fill in not only by sedimentation, but also by peat accumulation at the margins and encroachment of swamp from the edges. Such processes may take only a few thousands or even hundreds of years. Malham Tarn Moss is an example

10.6.1 Tarn Moss, Malham

Malham Tarn, in Yorkshire, England, is a small lake situated in glacial drift and fed by calcium-rich spring water. In the past, it was larger, but encroachment of aquatic plants has now completely filled about half of its original area [772, 773]. About 8 m of peat have been laid down in parts where encroachment has been complete (Fig. 10.21).

The earliest deposits, from the late glacial phase, around 12 000–13 000 BP, are clay and gravels washed out from under the melting glacier. Pollen analyses show that tundra surrounded a lake turbid with suspended clay. Little light penetrated and aquatic plants were scarce. The water cleared as the clay was deposited and the sediments became silty and of marl (calcium and magnesium carbonates). The oospores of *Chara* became abundant in the deposit, together with shell fragments of at least six snail species. The marl deposited on the surface of the *Chara* (hence the common name, stonewort) probably formed most of the sediment.

In the remaining lake, similar conditions prevail. However, at the edge, emergent plants colonized the calcareous sediment and laid down peat. Grass and sedge remains have been found, but their peat is well

Tarn Moss, Malham

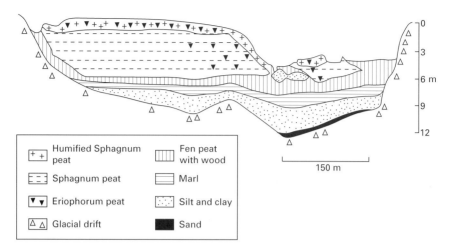

Fig. 10.21 Profile reconstruction from peat borings of part of Tarn Moss, which occupies part of the basin of Malham Tarn, a small lake in Yorkshire, England (based on Pigott and Pigott [772, 773].)

decomposed, amorphous and black. Such peats are characteristic of nutrient-rich conditions and are formed by a diverse flora called fen in the UK, alkaline bog in the USA and minerotrophic mire elsewhere. Fen conditions persisted for several thousand years at the edge of Malham Tarn, while the vegetation of the catchment changed form tundra to birch forest to hazel scrub.

The water-table in fen lies at the surface of the peat. Temporarily drier phases allowed some drying out of the peat surface and invasion by birch trees. The remains of these are to be found in woody (brushwood) peat within the fen peat. Fen vegetation, laying down fen peat, is still present at the tarn edges, where calcium-rich water from the lake still has access. However, some few thousands of years ago, in the middle of the fen mat, the peat built up above the level of the groundwater table. Rainwater then became the main source of water to the peat surface and leached the minerals in it downwards. The surface became acid and bog plants (which, by definition, occupy acid peat soils), such as *Sphagnum* species (Fig. 10.22), took over from the fen plants.

Many *Sphagnum* species are able to maintain the water around them acid, by ion exchange. They absorb cations, such as Mg^{2+} and Ca^{2+}, and release hydrogen ions (H^+). The *Sphagnum* peat, which is easily recognizable, sometimes even to species level, from the distinctive structure of *Sphagnum* leaves (Fig. 10.22), is about 3–4 m thick in the centre of the bog at Malham Tarn. As it grew thicker, it dried out and heather (*Calluna vulgaris*) colonized, leaving woody remains and leathery flowers in the peat.

(a)

(b)

s.t.d. (x7.5)

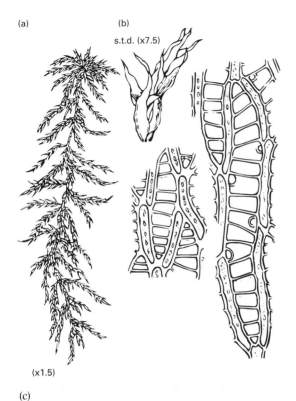

(x1.5)

(c)

Fig. 10.22 *Sphagnum* moss (a and c) has a characteristic leaf structure (b), with small green cells and large empty cells, on whose surfaces mineral ions may be exchanged for hydrogen ions. s.t.d.: shoot tip detail Sphagnum peat forms the uppermost deposits of many infilled lake basins that have succeeded to raised bog in wet climates.

In recent centuries, a general drying out has allowed cotton grass (*Eriphorum vaginatum*) to invade, and cotton-grass peat forms the most recent layer.

All small temperate lakes where vegetation is encroaching have not behaved exactly like Malham Tarn. If the local climate is dry, capillary action may provide sufficient salts to the fen-peat surface to prevent acidification and bog may never invade. Alder and willow woodland (carr) invade instead, as happens in the drier areas of eastern England and western Wisconsin. If the groundwater is acid, bog may invade immediately, without the intervention of a fen stage, as happens around many American and Scandinavian lakes set in sandy glacial drift. Either bog or fen stages may dry out sufficiently to allow colonization of hardwood forest, or, in very wet climates the bog may form the climax vegetation, as happens in the maritime seaboards in Maine, British Columbia and Sweden. Indeed, the bog may grow so well that it forms a floating mat or quaking bog.

10.7 Patterns in the development of lakes and the concept of natural eutrophication

For a long time there has been the idea that, as lakes fill in with sediment, they become more eutrophic, in the sense of productive or fertile. This is not an unreasonable proposition. Shallow lakes generally have higher total phosphorus concentrations in the water than deep ones. They also look more productive, because of their plant beds and the rich invertebrate fauna these often support. As lakes fill in, their hypolimnia become smaller and hence, inevitably, will become more deoxygenated in summer. Sediment coring sometimes reveals this in a change in the species of oxygen-sensitive chironomids preserved.

On the other hand, the higher productivity of shallow lakes is a feature of their catchments, which have fertile, easily eroded soils and often intense human activities, which, *inter alia*, have also favoured rapid filling in. Such cathments are generally of subdued relief and the lake basins formed in them may consequently never have been very deep. The apparent correlation between deepness and infertility and shallowness and fertility has thus led to an association of steady increasing fertility, or eutrophication, with the process of filling in. One most important result of palaeolimnological studies has been to counter this idea that lakes become naturally more fertile with time.

Sediments, themselves the agents of filling in, contain the evidence (Blea Tarn, Esthwaite, Lago di Monterosi) that, for long periods, lakes, if there is a common pattern at all, may become less fertile as they become more shallow. This 'oligotrophication' is a more reasonable general expectation than one of increasing fertility. The upheavals that form lake basins must initially expose readily weatherable minerals, but these are in finite supply and must eventually become leached out. Some parallel can be found in the conversion from fen to bog and, indeed, the words eutrophic and oligotrophic were originally coined to describe these stages. The early events in man-made lakes contain remarkable parallels.

On the other hand, just as 'natural' eutrophication is certainly not inevitable, neither is 'natural' oligotrophication. Pickerel Lake became naturally more fertile as climatic conditions changed to favour prairie rather than woodland in the surrounding catchment. The important point, however, is that this change in lake fertility was not an internal change in

an undisturbed catchment. It was a change consequent on major disturbance of the catchment by climate.

White Mere adds a further dimension. It had abundant blue-green algal
growths very early in its history and retains them. It has very high total
phosphorus concentrations, typical of groundwater-fed lakes in the area (see
Chapter 6) and its mechanisms of accumulating and retaining phosphorus,
coupled with the low $N : P$ ratios that seem to favour blue-green algae, have
prevented any oligotrophication and may have supported a natural increase
in fertility during the early few centuries or millennia. The key to this
history, compared with Blea Tarn and Esthwaite, is that it lies in deep fertile
deposits, which are not so easily depleted, perhaps, as the upland soils of
the English Lake District.

All of this underlies the general principle that what happens in a lake is
determined by what happens in its catchment. Often related to climate, the
direction of change may reverse, perhaps several times, as in Lake Valencia.
There is no single pattern in the development of lakes. What happens
depends on the nature of the original basin, local circumstances and
climatic change. The one feature common to most is the recent major
change caused by human activity, and this has resulted sometimes in
acidification, sometimes in eutrophication and sometimes in loss of the lake
(see Chapter 6, the Aral Sea). Thus the ground is laid for the final chapter
of this book.

11: Fresh Waters, the World and the Future

11.1 Introduction

In the analogy of the Earth as a cracked egg covered by the thinnest of films of moisture (see Chapter 1), the fresh waters, comprising only a tiny proportion of that film, must seem immensely vulnerable. Indeed, they are, as much of this book has shown. So what is their future and the future of the science that studies them? Is it to be a widespread recognition of the problems, followed by some wise solution, or is it to be progressive deterioration until catastrophe forces change, irrespective of human needs or ambitions? The portents of those who see these issues from different viewpoints are not good.

Richard Wagner's cycle of music dramas, *The Ring*, begins with the theft, by the deformed Alberich, of a store of gold from its guardian maidens in the River Rhine (Fig. 11.1). The gold corrupts all through whose hands it passes, even the gods themselves. It was fashioned into a ring, which gave great powers to control events. Erda, the primeval earth goddess, emerged from the rocks to counsel Wotan, the leader of the gods: *'Fly from the Ring's curse! Utter ruin past salvation, its gain will bring you.'* Wotan, ever-responsive to feminine charms, listened, but he did not learn. The final drama, *Götterdamerung*, leaves us with the immolation of the corrupt.

It is an analogy, perhaps, of the consequences when natural environments are raped in the interests of human greed, and it is not alone. From Jesus Christ to Mahatma Gandhi, the message emerges in different forms. Martin Holdgate, formerly head of the International Union for the Conservation of Nature (IUCN), used it [417] in an analogy of the horsemen of the Apocalypse, who were also bent on immolation, with fiery swords. He lists six currently unstoppable environmental horsemen: population increase, climate change, deforestation, desertification, pollution and loss of biological diversity. Paul Harrison [382] condenses the issues into a simple product. The multiplication of population number by resource consumption (and consequent waste production) by the powers of technology measures the impact of human activities on the biosphere (Fig. 11.2). And twined within each of these lies the availability and quality of fresh water.

Fresh waters are vulnerable for several reasons. First, water is subconsciously feared, perhaps through association with drowning or the hazards of inhospitable stretches of marsh. Wetlands have been seen as wastelands.

Fig. 11.1 Three Rhine maidens guarded the gold of the River Rhine, helped in their swimming, in an 1876 production of *Das Rheingold* at the Bayreuth Opera House, by quite a lot of their friends! (From Osborne [743].)

Secondly, lakes act as sinks for the consequences of human activity, while rivers are cheap drains for the removal of waste to the sea. Thirdly, water is nevertheless valuable and essential and often scarcest where it is required most. Featureless storage reservoirs have replaced floodplains and their previously more diverse ecosystems. Fourthly, the sediments underlying shallow wetlands often make fertile soils if drained; hence they are drained. And, lastly, as with other ecosystems, man has had the notion that by some form of management he can 'improve' them.

Natural ecosystems are the current products of continuous planetary change and adjustment. All chemical systems (including living ones) move towards a state of maximum homoeostasis when faced with changing conditions. This means that they develop mechanisms that minimize the effects of random changes (e.g. short-term weather) or anticipate or use predictable ones (e.g. tidal and photoperiodic changes). This is achieved through the operation of natural selection. It has resulted in arrays of species best fitted, in the prevailing environment, to maintain a system of maximal homoeostasis, while simultaneously retaining a capacity, through genetic mechanisms, to keep on adjusting to further change.

It is impossible to 'improve' an ecosystem formed and regulated in this way. We may not fully understand them, but there are always good reasons, in undisturbed ecosystems, why, for example, fish growth is moderate or aquatic plant beds are extensive. Attempts to increase fish growth or clear the beds will always result in a chain of repercussions, whose solution may need considerable expense. A river floodplain may only

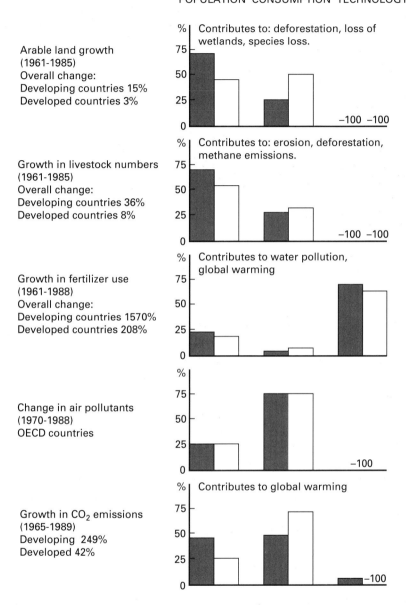

Fig. 11.2 Contributions of rising population, increasing consumption and increasing use of technology to several indices of environmental impact. Data are shown for overall absolute percentage change in developed and developing countries. Solid histograms are for developing countries, open ones for developed countries. In some cases, the use of technology has led to a reduction in impact. This is shown as a minus value. OECD, Organization for Economic Cooperation and Development. (Based on data in Harrison [382].)

be under water once every few years, but it is still part of the river bed, essential for the efficient and natural disposal of occasional high discharges. Conversion to farming, for example, must mean expensive engineering works to increase the height of the river banks to accommodate the flood water, with transfer of an increased flood risk downstream. Objective economic cost–benefit analyses have rarely been produced for such schemes. Where the general public good is concerned, the balance might be very unfavourable.

Changes in attitude may be helped by recent calculations of the value of the goods and services provided by fully functioning natural ecosystems on a world scale [164]. The current economic value of such benefits is estimated to be around $US33 \times 10^{12}$ year^{-1} (range 16–54, and probably underestimated). Global gross national product from human systems is only one-half to a third of this. Freshwater areas emerge as some of the most valuable habitats on an areal basis.

Inevitably, however, freshwater systems have been changed, either from a defensible need for water-borne disease control or for the establishment of farmland, from simple ignorance of the consequences of pollution and eutrophication or from the need to prevent flood damage to property built on natural floodplains through stupidity or greed. There is now a belated trend in public opinion that they should be 'conserved', but in the way of this is the inexorable product of population, resource use and technology of Harrison's [382] equation.

11.1.1 Population, food supply and water

The human population is currently around 5.8 billion people and will rise to at least seven, conceivably eleven billion by the middle of the next century [106]. That the causes of early death, which formerly reduced the rate of increase, have been extensively overcome by simple public-health measures is cause for celebration. But matching of the decreased death rates by lowered birth rates is something not yet achieved on a wide enough scale to prevent considerable impacts on the environment and, through poverty, on the people themselves. Control of population is not a simple issue of contraceptive technology; children are essential to survival in many parts of the world, for they provide security – hands to work, insurance for old age, prestige where there is little other dignity. Stabilization of the population will require attention to issues of land tenure, social justice, women's rights and education. In the meantime and in any case, a large population has to be fed. The portents for our ability to do this are not good [105].

From 1950 until 1990, world food production steadily increased, as more land was brought into production, usually through irrigation and fertilization. Grain production rose from 631 to 1780 million t year^{-1}. Since 1990, the increase has slowed and the annual production per person has fallen from 346 to 313 kg. Grain prices doubled in 1996, reflecting scarcity, and stocks fell from a comfortable 70 to 80 days' supply to below 50. The marine harvest reached a peak in 1989 and, with virtually all the sea-fish stocks overfished, will continue to decline.

There is a need, therefore, to increase agricultural production, but the trend is downwards. Much of the land brought into cultivation in the last few decades was barely suitable without copious irrigation water. The supply of such water is now lower than even the current demand. Underground aquifers, such as the Ogallala, which supplies much of the arable farming of the North American great plains, are running low and failing to recharge; the Colorado River, rising in the Rocky Mountains, rarely reaches the sea in California but peters out in the Arizona desert. The Yellow River in China has not entered the sea since 1972.

Much irrigation water is diverted to the cities, as they expand, and on a world basis and there has been a 7% drop in irrigated area per person since 1979. Large areas of former cropland have become severely eroded or salinized and have had to be taken out of production (Fig. 11.3). Desertification, a process by which formerly cultivable or grazable land is ruined by unsuitable use, now affects 20 million ha, or 15% of the land area. New land for agriculture is not available, for reserves like the wetland areas that could formerly be drained, have been used up. Spain, for example, has lost two-thirds of its wetlands since 1965. Terraceable hillsides are already terraced; shallow marine waters that are reclaimable have largely been reclaimed. There is pressure in many countries to grow water-demanding crops, such as soya, that have a high protein content for feeding to poultry and other domestic stock. The available food produced per unit of water thus declines.

Much irrigation water evaporates, and that which is returned to the rivers is heavily eutrophicated or loaded with agricultural chemicals or both. Expensive treatment is needed to purify it for further consumption. Water shortage is leading to resurgences of infectious water-borne diseases in dry countries. The threat of disease, especially those which lead to infant

Fig. 11.3 Overgrazing has led to severe erosion in this southern African landscape, with a future risk that loss of vegetation cover will result in changes in microclimate and desertification.

mortality, is a main driver of high birth rates. Many children still die from water-borne bacterial and viral infections, ultimately due to water pollution from poor sanitation. In many ways, rising populations thus threaten the very freshwater supplies on which they depend.

11.1.2 Resource use and water

In the developed world – Europe, North America and most of Australasia – and in the industrializing countries of the Pacific rim, it is consumption and waste production rather than population increase that threaten the freshwater resource (Fig. 11.4). Western society uses vast amounts of water for industry, domestic purposes, washing cars and irrigating sports fields and golf-courses. Many of the river catchments have been disrupted by development and an increasingly car-orientated society leads to greater areas of land being concreted for shopping areas, roads, car parks and additional housing and thus unavailable for soil percolation and purification of the water and its storage.

Hundreds of new chemicals enter the environment each year. There is no possibility that they can be adequately screened, under present arrangements, for the damage that they might do. Hydrological patterns are upset and river regimes, to which fish have adapted through a slow evolution, are disrupted. Much has been done to resolve problems of gross pollution, such as that by raw sewage, which bedevilled nineteenth-century Europe, but more subtle forms are now extant – trace organics with oestrogenic activity, for example. The strictures on availability of agricultural land and food security apply, but most of the developed world is not yet as short of water as the arid parts of the developing and third worlds. Nonetheless, water rationing and payment of its full costs must soon replace traditional treatment of it as a free good [1]. Conflict over water-supply is

Fig. 11.4 In developed countries, water is used profligately. In very poor, arid countries, collecting a minimal supply, the equivalent of one flush of a lavatory in the western world, may take a significant part of the day. In countries like India, supplies are intermediate, but multiple use at the local well may result in contamination problems.

already bringing nations close to war (Israel and its Arab neighbours over the Jordan, Ethiopia, Sudan and Egypt over the Nile, Turkey, Syria and Iraq over the headwaters of the Tigris and Euphrates).

11.1.3 Technology and water

It is in technology, as well as in consumption, that the developed world has its greatest impact on fresh waters. Engineering works have enormous scope to modify freshwater systems (see Chapters 5 and 10). Technological impact is further expressed through changes in atmospheric composition and in the radiation shield provided by atmospheric gases. The atmospheric concentration of carbon dioxide has increased by 25–30% since the last century to a current 360 ppm. The enhanced greenhouse effect of this and other gases, such as methane and nitrous oxide (N_2O), which have also increased, is associated with an unprecedented rise in temperature. This is occurring at roughly 10 times any previous known rates, with a rise of 0.9°C since 1866 and a further 1–3.5°C predicted by 2100 unless major reductions in the burning of fossil fuels, such as oil and coal, are made [89, 431]. The last time that such rises occurred, in a previous interglacial period, hippopotamuses frequented the UK. A similar precipitous rise about 11 600 years ago resulted in major alterations in ocean currents, and hence weather patterns, and the drying up of many endorheic African lakes.

Due to expansion of the mass of ocean water, sea levels have already risen 18 cm in the past century and are currently increasing by 0.1–0.3 cm year^{-1}. This affects the salinity regimes of the lower reaches of rivers, mandating greater flood defences and modification of river floodplains. For the 120 million people who live on river deltas, where population densities are very high and agricultural systems frequently depend on the seasonality of river flooding, sea levels rising the projected 50 cm or so during the next century will be disastrous. Together with the increasing frequency of extreme weather (storm surges, droughts, major floods) associated with rapid climate change, these consequences will affect one-third of all croplands and more than 1 billion people. The impacts of millions of environmental refugees, coupled with a near-doubling of population many of whom will be forced to migrate to find employment, shelter and food, are very great for the entire world.

Patterns of rainfall will change, again altering hydrological regimes in ways to which organisms may be unable to adjust quickly. Changing temperatures may alter competitiveness between species, leading to community changes, premature drying out of small water bodies, with loss of spawning habitat for amphibians [787], diversion of temperature-sensitive salmonid fish to sea areas away from the entrances to their spawning rivers [793], promotion of thermophilic species, such as many blue-green algae, and spread of exotic species originating from warmer countries. Since 1985, a 1°C increase in temperature in the European Alps, at altitudes of

2000–2900 m, has been associated with a shorter period of snow cover and greater leaching of glacial-rock flour, leading to increases in pH, conductivity, silicate and some major ions in a group of fifty-seven Alpine lakes. Nitrate concentrations have decreased, due perhaps to an increased ice-free growing season and increased phytoplankton production [906].

The increase in ozone-destroying substances, such as the chlorofluorocarbons, has resulted in increased amounts of damaging ultraviolet (UV) radiation reaching the ground and the fresh waters. Penetration of UV radiation into clear, high-altitude lakes may be to several metres, with inhibition of photosynthesis and other biological effects. In most lakes, UV radiation does not penetrate very deeply for it is quickly absorbed by the yellow and brown humic substances in the water. These are mostly brought in by inflow water from the surrounding wetland and catchment. In an interesting synergism between climate change and ozone destruction, lakes in Ontario have received about 25% less rainfall in the last 25 years. Their concentrations of humic substances have fallen and UV radiation now penetrates significantly more deeply than before [868].

11.2 Trends in freshwater science

Freshwater systems are already extremely vulnerable to all these trends. Many of the goods and services that freshwater habitats have been able to supply are being progressively lost. The writing of this book has, in one sense, been depressing, for, in the ten years since the last edition, there has clearly been widespread degradation, despite moves towards restoration in some places. How, then, is freshwater science responding to these quite staggering changes? What new tools and approaches are being brought to bear and will they be effective in reversing the changes?

11.2.1 New genetic technologies

The last two decades have seen major advances in genetic technologies. It is now possible to compare deoxyribonucleic acid (DNA) closely enough to determine differences among individuals and to recognize separate breeding populations within apparently uniform habitats. These approaches will be more widely applied, both to prokaryotes, many of which from fresh waters cannot yet be cultured and hence characterized in other ways [856], and to eukaryotes.

The certain conclusion will be that living communities are far more diverse than can at present be demonstrated. Already, it is recognized that communities of soil bacteria may have as many as 1000 different strains in a gram or so of soil [965, 966], and the same must apply to waters and sediments. Such work strengthens the contention that diversity must be important – otherwise, natural selection would have acted against maintenance of such high levels – and thus supports moves to conserve such diversity. It is unlikely, however, that the case will be strengthened to the extent that any more effective action will follow under present arrangements. Large sums of research money will undoubtedly be spent on such currently fashionable technologies, for research is much cheaper than action and 'the need for research' can be used as a mechanism by governments to delay action.

The new genetic technologies will also bring potential problems, in that there is now much interest in the engineering of new organisms to accomplish specific industrial tasks. There are immense profits to be made from this and equally large threats to the world environment [972]. Such organisms may include crop plants made resistant to particular diseases, fish and farm animals of very high productivity, if lesser resilience, and bacteria capable of degrading specific organic wastes. Such engineered organisms will inevitably escape, due to human error, into the greater environment. Many will perish very quickly and others will be big enough for quick discovery and removal.

The threats from engineered mammalian farm stock may be small, but the situation with plants, fish and particularly microorganisms is much less certain. There have already been many instances of the introduction of 'new' genotypes in the form of exotic species. Almost all have been disastrous. Some – for example, the zebra mussel (see Chapter 8) – were accidental and may have been unavoidable, but deliberate introductions (those of crayfish, for example [416, 726]), of whatever origin, should now be seen as extremely unwise. There is general agreement that, once released, genetically engineered microorganisms (GEMs) [663] are not recoverable and, if they do not perish, they may do considerable harm, either by disrupting natural processes or by passing genes to other strains, which may do likewise. The outcomes are not predictable and are unlikely to be, given the complexity of environmental processes. It is particularly alarming that such organisms are being designed to cope with particular industrial wastes, such as hydrocarbons spilled into the land. These must certainly enter the rest of the environment. The issue is doubly jeopardous in that it engenders an attitude that wastes can be dealt with simply by additional technology. More desirable is an attitude that there is no such thing as waste; everything can and should be recycled for use, rather than dumped in the environment.

11.2.2 Ecotoxicology

The same danger lies in a second major trend in freshwater research, the development of ecotoxicology. The need to recognize the presence of pollutants has long been a feature of pollution control. Early in this century, standards for sewage effluents were set and simple chemical measures – oxygen concentration, biological oxygen demand and ammonia concentration – used to monitor organic pollution. The limitations of chemical methods have been recognized and biological measures – for example, the biotic score (see Chapter 5) – developed. Gross organic pollution is well understood and increasingly under control, and industrial pollutants, particularly heavy metals, are now recognized and being controlled under licence-to-discharge systems.

There is, however, always the danger of accidental or criminal negligence resulting in polluting discharges; increasingly, these come as a result of fires in factories where regulations on the storage and containment of chemicals have not been honoured, and the discharge of fire-fighting water. Limiting the damage to river systems of discharges demands early warning. New approaches include the use of automated biosensors [367]. Water from the river is continually directed through a tank in which lives a fish. The activity (amount of movement, frequency of movement of the gill covers) of the fish is changed by the first traces of pollutant, and these movements can be sensed by infrared or ultrasound beams crossing the tank and converted into a signal to trigger appropriate action. Such systems can be very sensitive but are still prone to malfunction.

The sensitivity of organisms might also be exploited in the use of biomarkers [286, 601]. The key pollution problems in the developed world now come in the variety of new, often organic, chemicals. Their effects may be subtle but chronic, rather than gross and acute. Biomarkers are features of a test organism that show changes in response to a particular pollutant long before overt effects, such as death, are apparent. Such responses might include stress indicators, production of chemical detoxicants, inhibition of particular metabolic processes or enzymes, compromise of the immune system or changes in various tissues.

These systems take advantage of new understanding of the operations of cells and enzymes, but their application is not yet well developed. They may be specific to a particular organism, of no general ecological relevance, slow to develop or expensive to monitor. Examples are few, but herring gulls in the North American Great Lakes have been shown to produce 30–40 times as much porphyrin in their livers as herring gulls on the Canadian seaboard [287], and this is seen as a biomarker for some pollutant, although it is not certain which.

Again, there is an element here of the oversimplification shown by laboratory-orientated science to the complexity of ecological systems. Such approaches convey the message that, if the pollutant can be detected, the problem is soluble. This undermines the ultimate reality that the solution to pollution is not to discharge pollutants. Detection methods are always needed to produce evidence that illegal activities have taken place, but the unspoken view that pollution is inevitable should be curbed. Such new techniques, like those of molecular biology, will add great detail to knowledge of the problems we face, but they will only monitor the problems in a different way. They will not solve the problems. They may even make them worse, by getting in the way of proper solutions. Not surprisingly, environmentally insensitive governments tend to promote such approaches through their research policies.

11.2.3 Levels of approach

Problems at the ecosystem level cannot be understood by knowledge from the suborganismal or subcellular level. They must be studied at the whole-system level to reveal the properties of the interacting systems. Approaches here range from the use of mesocosms (Fig. 11.5) – experimental streams, large polyethylene bags or enclosures, limnocorrals, Lund tubes and netting cages – to replicated experimental pond systems, in which the size of each pond is perhaps up to 50 m square to lakes divided into parts, to whole lakes, to whole catchments. The more encompassing the system used, the more likely it is that results of real relevance will be obtained.

There is a problem in replicating the experiments so that statistically significant results can be shown. Nonetheless much of our deeper understanding of freshwater ecology has come from work on single systems, coupled with insight on the part of the investigator. A further problem is that the bigger the system, the greater the cost and the need for cooperation between large numbers of people. This does not lend itself to the system of organizing science at present, which rewards competition and individual prowess.

Two alternatives to whole-catchment experiments are the use of comparative methods and models. Comparative approaches use data gathered in a standard way from a large number of water bodies to reveal, through statistical techniques, common patterns that illuminate key processes (Fig. 11.6). The difficulties where a large number of people are involved are in ensuring that data are collected in a standard way. This is not easy.

Fig. 11.5 Replicated mesocosms can provide a compromise between well-controlled but often irrelevant laboratory experimentation and the desirability, but unreplicability, of whole-lake experiments.

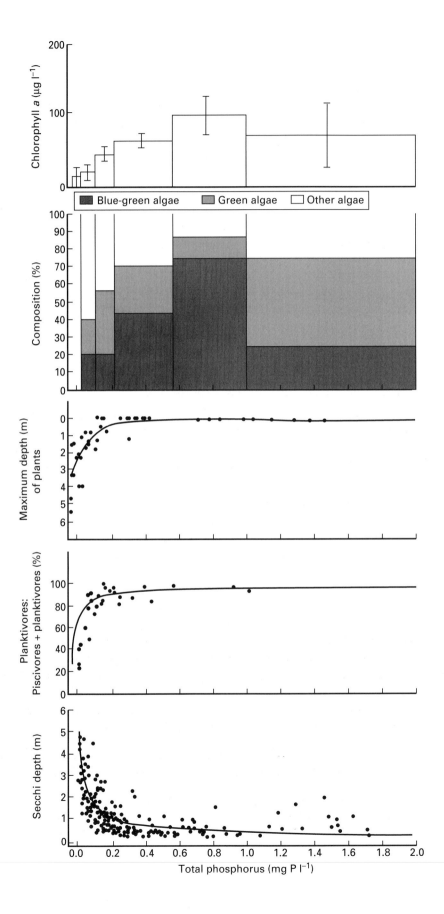

Human nature is wayward; individuals like to do things their way! We are all guilty. However, if pains are taken to involve everyone in understanding the project, or if the most sensitive sets of analyses are done in the same laboratory by people closely involved, much insight can be gained.

Modelling is cheaper than both the comparative method and empirical experimentation and is becoming increasingly easy with the development of user-friendly computing systems. Models take a limited set of data and use mathematical relationships between the variables to conduct notional experiments, by changing one variable after another. This has been useful in confirming insights and adding credibility. It may sometimes reveal interesting counterintuitive phenomena. Models, however, are only as good as the data put into them and, if they lack data on some variable that is important but unrealized, they will be misleading. Ultimately, empirical experimentation at the largest relevant scale is the most valuable approach to understanding further the operation of ecosystems, but comparative and modelling approaches both have valuable roles to play in supporting this.

11.3 Advances in monitoring techniques

Future sophistication will come in the recording of the state of freshwater habitats. Society needs to monitor the state of its environment to discover new problems and to ensure that the valuable natural functions ecosystems provide are not being compromised. Classification systems are needed to order these data. However, the simple characterization of flowing waters solely by their oxygen or biological oxygen demand (BOD) concentrations, or even by their present communities of bottom invertebrates, is no longer acceptable when gross organic pollution is being superseded by the influences of river engineering and nutrient, acid and other forms of pollution.

There are two important aspects to establishing the state of a fresh water. One is to ask if all the relevant features are being accounted for; the other is to ask what a classification means. New systems are being developed to assess characteristics that have not previously been monitored, such as habitat complexity, for conservation purposes. Examples are the river habitat survey (RHS) [714] and the system for evaluating rivers for conservation (SERCON) [78]. The riparian, channel and environmental (RCE) system (see Chapter 5) is similar.

Early interest in classifying lakes laid good foundations for the development of limnology [689] but was largely abandoned when it dawned on

Fig. 11.6 (*Opposite*) The comparative approach used with data from a large number of Danish lakes. Trends, with increasing total phosphorus concentrations, are clearly shown. Contrary to expectation, green algae, rather than blue-green algae, are predominant at the highest phosphorus concentrations. The ratio of planktivorous to piscivorous fish also increases greatly with nutrient concentration. (From Jeppesen *et al.* [482].)

scientists that there was continuous rather than discrete variation among water bodies, Recently, the deterioration in lakes has prompted the development of systems for assessing lake status, as well as that of rivers. There is thus a range of attempts to incorporate more comprehensive sets of features into systems for assessing and monitoring the state of fresh waters.

Built into many of these systems is an understanding that simple classification is of limited value. All fresh waters differ along continua of many variables. Each is unique, and some decision has to be made to emphasize particular characteristics over others. Even then, arbitrary divisions have to be made along the continuum to define classes – excellent, good, fair, bad. There is little merit in this, because the class boundaries have no absolute meaning. Many rivers could be promoted to excellent or demoted to fair by a simple change of the class boundary in each case.

What really matters is the extent to which the systems have changed in absolute terms from some pristine or baseline state. Pristine states are those in which human activity is absent and have probably not existed for the past 3000 years or more in Europe. Baseline states allow for human use of catchments and waters, but, for such a state to act as an absolute reference, such usage must be indefinitely sustainable. This must mean that land use reflects natural geological, climatic and topographic constraints and not the uses that can be temporarily achieved by unsustainable inputs of energy and chemicals. These have been widely used in the past few decades by an agriculture subsidized by governments to use them, and have caused severe problems [388]. Sustainability also means that substances introduced into the catchment that are alien, such as pesticides, herbicides, industrial products and waste products, must be removed at similar rates to those in which they are introduced, It also means that non-alien substances, such as phosphates, must also be in similar balance and that essential functional roles provided by freshwater systems, such as the role of floodplains in accommodating floods, must be restored.

Monitoring systems that compare the present state of a system with a baseline sustainable state have been developed and show that there is currently considerable deviation from sustainability [692, 695] (Fig. 11.7). They are called value-changed systems, as opposed to the spatial-state (arbitrary classification) systems traditionally used, and give a much more transparent picture of the state of the environment.

Value-changed systems can accommodate any number of variables, as the system depends not on the actual value of a single or closely related group of variables, as do spatial-state systems, but on a common unit of percentage change for every variable. An average percentage change can be calculated for an indefinite number of variables. Any variable whose value can be determined or set for the baseline state can be included. This is not always easy, but not impossible. Value-changed systems are not yet widely

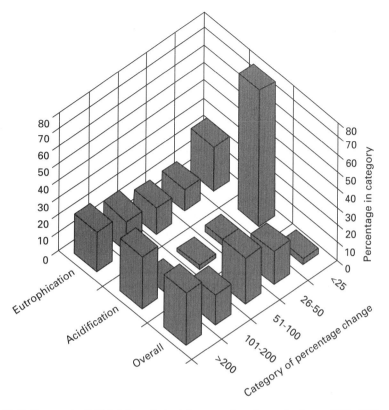

Fig. 11.7 Changes in British lakes, derived from a value-changed scheme [692, 695].
Ninety lakes were included, covering a range of geography and usage. The percentage
change from a sustainable baseline state is shown for variables linked with
eutrophication, for those linked with acidification and for all variables, including those
associated with marine seepage and with sedimentation through excessive erosion. More
than half of the lakes had deviated from the baseline by over 100%.

used, but will spread as the concept of a baseline sustainable state is better
understood. Empirical research at the catchment level is essential for this.

11.4 Solving the problems

Such approaches, again, will help to define environmental problems better,
but they will not solve them. Just as problems of freshwater systems can
only be revealed by work at the system level, so also can solutions be found
only at this and higher levels. The higher level concerns the ways society
views its environment and the regulations it imposes on itself to govern that
use. We already know what the environmental problems are; there is no
shortage of examples to demonstrate the degradation of this or that river,
lake or wetland. The issues of availability of water of sufficient quality and
quantity are widely acknowledged at levels as high as the United Nations

(UN) and among informed people there is widespread concern about the state of the environment. The problem is in putting the information we already have into operation to achieve solutions. Clearly we are failing badly to do this. What tools do we have to achieve it? How good are they? Are they the most appropriate ones?

11.4.1 Treaties

The priorities of a society are expressed in its legislation; the effectiveness of the legislation comes in how it is used. Legislation is mostly nationally based and varies widely. Some countries are environmentally extremely committed, with laws that are plainly expressed. The Danish Consolidated Environmental Protection Act (No. 590, 27 June 1994) is one example:

Part I Scope etc. 1.(1) The purpose of this Act is to contribute to safeguarding nature and environment in Denmark, thus ensuring a sustainable social development in respect of human conditions of life and for the protection of flora and fauna.
(2) The objectives of this Act are in particular:
(1) to prevent and combat pollution of air, water, soil and subsoil, and nuisances resulting from vibration and noise,
(2) to establish regulations based on hygienic considerations which are significant to Man and the environment,
(3) to reduce the use and wastage of raw materials and other resources,
(4) to promote the use of cleaner technology, and
(5) to promote recycling and reduce problems in connection with waste disposal.

Others couch their legislation to give maximal flexibility for governments to use it to a greater, or perhaps lesser, extent. The British Environmental Protection Act is a good example:

4.-(1) It shall be the principal aim of the [Environment] Agency (subject to and in accordance with the provisions of this Act or any other enactment and taking into account any likely costs) in discharging its functions so to protect or enhance the environment, taken as a whole, as to make the contribution towards attaining the objective of achieving sustainable development mentioned in subsection (3) below.
(2) The Ministers shall from time to time give guidance to the Agency with respect to objectives which they consider it appropriate for the Agency to pursue in the discharge of its functions.
(3) The guidance given under subsection (2) above must include guidance with respect to the contribution which, having regard to the Agency's responsibilities and resources, the Ministers consider it appropriate for the Agency to make, by

the discharge of its functions, towards attaining the objective of achieving sustainable development.

Environmental problems concerning fresh water are worldwide, however. Some 300 major river basins cross national boundaries and many waterfowl populations are internationally migratory. What runs down nationally owned rivers eventually reaches internationally owned seas. The increasing recognition of environmental problems as global has resulted in a number of conventions, treaties and protocols designed to foster international action and cooperation. One of the first was the Ramsar Convention on wetlands. A potentially more far-reaching one was the UN Convention on Environment and Development held in Rio de Janeiro, the 'Earth Summit'. A third is the Montreal Protocol for the control of chlorofluorocarbons (CFCs). How effective have these been?

The Ramsar Convention, fully called the 'Convention on wetlands of international importance especially as waterfowl habitat', arose from a meeting at Ramsar in Iran in 1971 and was the first modern instrument designed to conserve resources on a global scale [640]. Its provisions include a definition of wetland areas (marsh, fen, peatland, water that is permanent or temporary, fresh, brackish or salt, static or flowing, including seas less than 6 m in depth). It defines waterfowl as birds dependent on such habitats. It requires its signatories, which now include seven-five countries, or about half the UN, to designate at least one suitable wetland to a list maintained by the Bureau that manages the Convention. A suitable wetland is one of ecological, botanical, zoological, limnological or hydrological importance, especially to waterfowl.

By 1993, 590 had been designated, covering 36.7×10^6 ha. It took eight years to draft and agree on the text, a reflection of the difficult political situation at the time and also because the text puts obligations on its signatories. They must consider their international responsibilities for conservation and management, formulate planning to promote conservation, inform the Bureau of any threats to their wetlands, make and adequately warden nature reserves on wetlands, create compensatory wetlands if any are lost and encourage research and the exchange of data. The advantage of belonging is a mutual benefit because of the migratory nature of many waterfowl.

The operation of the convention has been driven by non-governmental organizations for the most part, particularly the World Conservation Union (IUCN) and, in the words of its Secretary General, it has 'a vibrant programme, a well-endowed budget and a sizeable permanent secretariat' (in Matthews [640]). It broke new ground by introducing the principle of 'wise use' of natural resources to a habitat type that has been much abused.

How effective has the Convention been? This is difficult to answer, for it is impossible to know whether actions attributed to it would have been carried out in any case. Probably they would not, but even the Bureau is hard put to be specific. Only Canada has a policy on wetland conservation, strongly based on Ramsar principles, which is countrywide, and even this applies only to nationally owned lands (which do comprise 40% of Canada, however). It commits the federal government to 'no net loss of wetland functions', mitigation of the impacts on wetlands of federal actions, cooperation with non-government organizations, native groups and the public and the development of a sound science base and research for wetland management [836]. The USA has a plethora of legislation to protect wetlands, following the Convention, and a 'no-net-loss policy', but the conclusions of a recent conference on wetlands in Ohio in 1992 [657] were that the current net loss rate of wetlands in the USA is 80 000–160 000 ha year^{-1}, with total losses in the lower forty-eight states now over 50% since presettlement times. Considerable progress in the classification and inventory of wetlands is acknowledged to have been made through the Ramsar Convention, but, 'in spite of the increased recognition of the importance of wetlands around the world, little progress has been made in halting their destruction' [657]. This conclusion is echoed in a number of recent books [223, 224, 603] on wetlands.

The most recent review [274] of the consequences of the Earth Summit at Rio de Janeiro in 1992 and its famous Agenda 21, which is a 40-chapter plan of action designed to achieve an environmentally sustainable global economy, concludes that very little has happened yet. Governments agreed to form national commissions for sustainable development to develop strategies to achieve this. Most reports of these commissions (117 so far) are 'broad, rhetorical and self-congratulatory' and describe existing progress and nothing new, except on problems identified long ago, such as local air and water pollution.

In the years since the Rio meeting, the population has increased by 450 million, a new peak of 360 ppm of atmospheric carbon dioxide has been measured, 1.3 billion people receive less than their basic needs for food and shelter and, despite a recognition that poverty and environmental degradation are closely linked, aid from the rich world to the poor has fallen to less than half its value in 1992. Few governments have even begun policy changes, only six have levied environmental taxes to conserve resources and many still subsidize clear-cutting of forests, inefficient energy and water use, mining and unsustainable agriculture.

In contrast, the Montreal Protocol, on substances that deplete the ozone layer, has, in a proximate sense, been a great success [290]. Chlorofluorocarbon compounds were invented in the 1920s and proved to be non-toxic, non-inflammable and non-corrosive ingredients for use in aerosol sprays, refrigerators and air-conditioning equipment. Their stability, however, means that they persist in the atmosphere long enough for them to reach the stratosphere, 10–20 km up, where a layer of ozone protects the earth by absorbing potentially damaging UV radiation.

In the early 1970s, it was realized that these substances could be decomposed in the upper atmosphere by solar radiation to produce chlorine, which, in a catalytic manner, reacts with ozone to produce oxygen. There was concern about this, which led to some restriction of CFC use in Canada, Norway, Sweden and the USA, but it was not until 1985, when measurements over Antarctica showed ozone depletion to be far greater (a reduction by 40%) than had previously been anticipated, that there was serious international concern.

A meeting in Montreal in 1987, attended by representatives from many countries, not only brought scientists and policy-makers together, but heralded the start of international environmental diplomacy. It also took on board the precautionary principle, by which action moves ahead of the existence of complete proof where it seems that inaction could be dangerous. The protocol was ratified by 130 countries, who agreed to a 50% cut in CFC production by 1988, with some derogations for developing countries, who were allowed a longer period.

However, by 1990, ozone loss was even greater, with a 95% reduction over Antarctica and it was clear that the provisions of the protocol would be insufficient to control the problem. It has since been strengthened, as depletion rates remained much higher than predicted. Following a meeting of the parties in Copenhagen in 1992, it was agreed to phase out CFC production by 1996 in industrial countries. There were again allowances for developing countries to be able to reap some rewards for investments they had previously made, and the setting up of a fund to aid such countries to convert appropriate industries.

At first, the chemical industry had denied the existence of any problem, but it progressively accepted its existence and then turned enthusiastically to the production of substitutes for CFCs. These included HCFCs (hydro-chlorofluorocarbons), which still had some ozone-destroying abilities, though only a few per cent of those of CFCs, and which also will be phased out by 2030, and hydrofluorocarbons (HFCs). The latter do not destroy ozone, but, like CFCs and HCFCs, are powerful greenhouse gases, with 1300 times the effect of equivalent amounts of carbon dioxide, such that their unconfined use would contribute, by 2050, at present rates of increase of use, as much heat-retaining capacity as the current fossil-fuel consumption of most of western Europe.

The provisions of the Montreal Protocol, with its later amendments, have, by and large, been followed. The former Soviet Union countries are having difficulties meeting their agreed obligations and there is an illegal black-market trade, with CFCs coming into the USA, probably from Russia, China and India, to service the repair of automobile air-conditioners. But, clearly, the international community has responded effectively and reasonably quickly to an environmental threat.

The success of the Protocol is attributed to several factors: a clear, well-defined problem; the use of the precautionary principle; involvement of both developing and industrial countries, represented by scientists not governmentally fettered, as well as by professional diplomats; and opportunity for industry to continue to profit through the production of substitutes. With the available substitutes, however, it is a case of one problem solved but another increased. It is notable, also, that nowhere near the progress has been made on a protocol to limit energy generation to mitigate climate effects. In this case, the coal and oil industries remain unyieldingly obstructionist.

Ultimately, the Montreal Protocol has worked because it allows continued economic growth, but with different substances. Solution to the world's environmental problems, however, demands sustainable economies, and these are incompatible with continued economic growth, no matter what weaselling of words is used to cover up the issue.

11.4.2 Consequences of evolution?

Here lies the sticking-point. Societies in the industrial and industrializing world depend on a particular economic system, which encourages competitiveness, acquisition of resources, increased consumption and reward for the aggressive individuals best placed to acquire more than a fair share of resources that are finite. It is arguable that this is what might be expected from the evolution of an animal, ourselves, by exactly the same mechanisms that produced all other species. Natural selection undeniably favours those individuals capable of acquiring resources most effectively. There are two possible fates for such species if ultimately they damage the environment on which they unwittingly depend. The first is extinction; the second is further change if they can change rapidly enough to cope.

It is possible that mechanisms additional to natural selection may act to stabilize the biosphere. The composition of the atmosphere and oceans is far from the equilibrium to be expected from chemical considerations alone. Otherwise, the atmosphere would be carbon-dioxide-rich and nearly anaerobic; the sea would be exceptionally salty and rich in nitrate; the temperature would be above $100°C$; and water would be prevented from boiling only by a crushingly large atmospheric pressure [579, 874]. That the Earth is an equable planet is due to modification of all of these conditions by living organisms in such a way that palpably does maintain stability.

How this is achieved by vast numbers of different kinds of organisms, each conditioned by natural selection, is not clear. Cooperation is implied, but has only been demonstrated between different individuals in the same species when they share a significant proportion of their genes, and between different species when each gains directly from the cooperation. Many genes for fundamental biochemical and cellular mechanisms must be shared among many species, and clearly maintenance of an equable biosphere represents a mutual pay-off for all. But how such cooperation may be achieved is still a mystery to biologists.

The lesson, however, of both the Montreal Protocol and the maintenance of the biosphere is that cooperation pays far more than destructive competition. Yet our environment and, ultimately, we ourselves suffer from a system that depends on competition between human individuals. Can we change it? Are there alternatives? Although our freshwater environment, like the rest, suffers from such a system, perhaps some of the lessons gained from a study of that environment can be used to diagnose the problem of why we do not change, despite a unique human ability to see ahead and direct the course of our activities. In particular, the model of alternative stable states [221], used to explain the functioning of shallow-lake systems and the changes in them in Chapters 6 and 8, might be used to explore the possibilities of alternative ways of organizing human societies [681].

11.5 Alternative states and human societies

Do ways of organizing human societies fall into alternative states, each stabilized by buffers? There are many ways of organizing society. There are the resource-demanding systems we presently have in the western technological world and which powered such city states as Rome, Sumer and Alexandria. In contrast, there are, or were a vast number of resource-efficient systems, frequently called traditional societies [32, 522, 643, 802]. These are the peoples whose lives Thomas Hobbes (1651, *Leviathan*, Part 1, Ch. 13) described as 'solitary, poor, nasty, brutish and short', and which John Dryden referred to as (1672, *Conquest of Granada*) 'noble savages'. They are neither of these. Rather, they are complex interacting societies, generally using scarce natural resources very efficiently and stabilizing their existence by intricate rules.

In American Pacific coastal tribes [376, 771, 802, 938], before European contact, salmon (Fig. 11.8) and other fish and timber were exploited and processed by small groups, which 'owned' particular areas. After the annual harvest, a group that was well endowed would give gifts (potlatches) of food or other natural resources to groups less fortunate. The size of a gift denoted prestige to the giver. This might seem a system designed to overexploit the resource, but there was a regulating mechanism. The gift had to be returned with interest at some time in the near future and failure to do so brought shame. Because it had thereafter to be returned, again with interest, by the first party, the size of the orginal gift and the degrees of interest had to be kept low enough for the environment to be able to support them. Inflation of the gifts to an extent that could not be met by the

Fig. 11.8 Potlatch society. The significance of resources to a society is often reflected in its art, such as this totemic carving of a male salmon, the crest of the Kut clan in the Tlingit village of Gunaho, Queen Charlotte Island, British Columbia. From a photograph in Bancroft-Hunt and Forman [35].)

available local resource was ultimately controlled by the risk of downgrading of an individual or group within the prestige structure of the tribe.

Rural Balinese societies provide a second example [54, 318, 802], in which a very large population density is maintained by wetland rice culture in a mountainous terrain, from which water runs rapidly to the sea. Systems of bunds and sluices retain the water sequentially in different land holdings from top to bottom of the mountain so that it is used several times over in the paddy fields before it finally drains away. These systems are controlled by a religious-festival calendar that promises divine retribution if the water is not passed on from one user to the next at the appropriate times. In this way, three rice harvests are obtained each year, where a selfish use of the water by only one user would produce only one. Similar arrangements govern a system of alternate fish farming and crop growing in shallow drainable basins in the Dombes region of southern France (Fig. 11.9). The basins are filled with water and a crop of carp is taken after a year or two. The water is then passed to a downstream set of basins for carp culture, while the first set is planted with maize. No external fertilizer is needed, because naturally derived silt and the ploughing action of the carp keep plentiful nutrients in circulation. Regulations dating back to the medieval period successfully maintain the system.

Many other such societies [552] clearly have stabilizing buffers that conserve resources and use them wisely. They contrast with the alternative of the western model. This depends on a considerable mobilization of natural resources to maintain systems from which some of its members benefit considerably, at the expense of others within the society and even more so those outside it.

Stabilizers of this system include the encouragement of competition, rather than cooperation, and acquisition of social prestige through the gaining and retention of material wealth, rather than from the sharing of the resource. Consumption of goods is encouraged, not only through the social

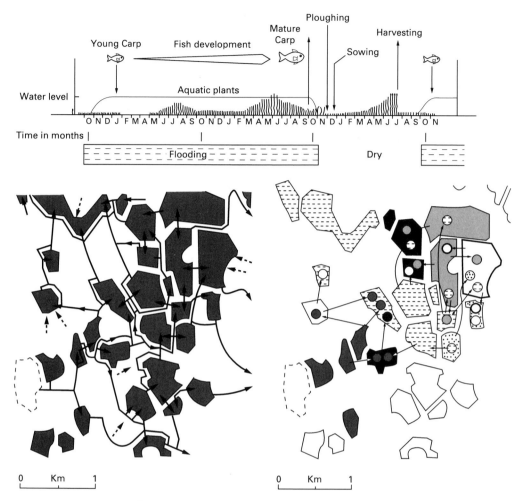

Fig. 11.9 The many ancient shallow basins of La Dombe in southern France have been traditionally managed to give alternate yields of fish (mostly carp) and cereal (formerly oat or wheat, now often maize) with no inputs of fertilizer (upper diagram). Transfer of the water is governed by traditional custom, enforced by honour, rather than a formal legal system. The lower diagram shows some of the ponds in the region of Le Bouchoux, the sluice-controlled system of possible water transfer (arrows) and something of the rather complex pattern of water transfers.

structure (keeping up with the Joneses!), but through powerful and insidious advertising. The rising problem of municipal-waste disposal is but one demonstration of a failure to conserve resources. This system is maintained by a concentration of decision-making among those who gain most from it. Kidron and Segal [525] note that, at the beginning of 1991, only seven (Iceland, Czechoslovakia, Austria, Romania, the Netherlands, Namibia and the Philippines) out of 165 governments were in power because an absolute majority of their citizens of voting age wanted them there.

We have allowed a domination by 'hawks', although this term is abusive to those birds whose existence depends on natural buffering mechanisms that limit their populations. This contrasts markedly with traditional alternatives, where decisions are spread much more widely and guided by an ethical code. There are parallels, also, in the structure and diversity of these human alternatives with those of the alternative states of plant-dominated and turbid-water-dominated shallow lakes.

Our western technological system is equivalent to the turbid-water system. Increasingly, it has lost diversity in its customs, its products and its ways of thinking. Consider the now dangerously restricted genetic base of our main cereals, vegetables and domestic livestock [469] and the threat that, through genetic engineering, it will become even more restricted [388]. Look at the uniformity imposed by huge companies, reflected not least in the depressing sameness of international hotels and the insidious march of fast-food chains and their like. Note the breakdown in the complex human relationships of society, reflected in high crime rates, cultural alienation of many of our young people and our dependence on preprepared entertainment. One recent British prime minister, Margaret Thatcher, a keen advocate of the western system, went even so far as to deny the existence of 'society' in favour of an aggressive self-interest. Note also the extreme loss in diversity in the habitats associated with heavily industrialized societies and the continuous battle needed to conserve even a few of them.

In contrast, the alternative, low-technology traditional societies preserve strong cultural identities, personal support systems, strong ethics and associations with still very diverse natural and seminatural habitats (even when human population densities are high). There is a greater tendency of complex plant-dominated systems to be switched to less diverse turbid systems when the nutrient loading is increased. The parallel switch from intricate human societies to our increasingly uniform one with the greater exploitation of natural resources is an interesting one (Fig. 11.10).

Where complex traditional systems have broken down, there has generally been strong western influence. The introduction of trade goods (beads, blankets and iron knives) undermined the potlatch system. Their availability was not linked with the sustainability of local natural resources and, when they came to dominate (rather than supplement) the nature of the potlatch gifts, the system was inflated to destruction for lack of any control. This constitutes a switch mechanism parallel to those that determine the alternative states in shallow lakes. By imposing features of our own society on these other societies, coupled with a parallel encouragement for increased consumption, we have as effectively destroyed the alternatives as we have replaced high-diversity plant-dominated lakes with the now much more common turbid ones.

Propaganda may also facilitate a switch and has been powerful and insidious. We undermine other societies by referring to them as primitive;

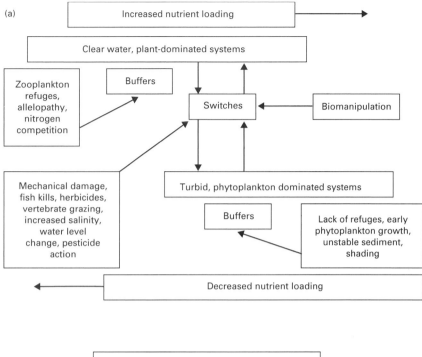

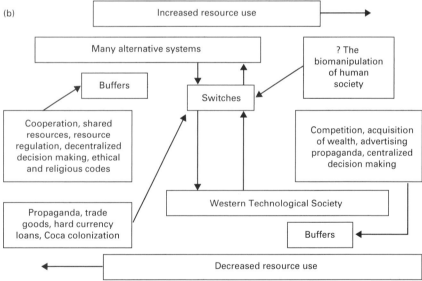

Fig. 11.10 Analogy between the alternative stable (persistent) state model of (a) shallow-lake systems and (b) a parallel with human societies (based on Moss [681]).

we emphasize tribal warfare, forgetting that we pursue it to a greater extent; we point out the incidence of disease, while failing to note the equal incidence of different sorts of diseases – those of affluence – among ourselves. Other switch mechanisms have been equally strong. We have undermined the natural support systems of alternative societies. The wanton destruction of bison on the North American great plains in the nineteenth century [284] is one example and the clearing of tropical forests [32] another. We have removed previous authority systems, imposing different systems of control and decision-making when previous ones were often more appropriate. The history of European colonialism provides many examples. Currently, individuals and governments who will pursue our agenda are supported in the effort of building 'prestige' projects – dams, airports and armies – with the promise of comfortable lifestyles for all. Meanwhile, their countries are drawn into immense debt to the western economies and any benefits are to the powerful few at the expense of the subjugated many. We compound the problem by introducing our trade goods in ways that undermine the local economies and diets – the process of Coca-Colonization.

11.5.1 The future

But the crux of the parallel of alternative states for those of us in western and westernizing societies is what the future of our own system – we of the 'turbid' state – will be. The buffers of any 'stable' alternative can be exceeded. Some authors prefer the term 'persistent' to 'stable'. There are switch mechanisms for movement between states in both directions.

Probably none of us with knowledge of the workings of the biosphere believe that our present rate of consumption of resources can continue. Many of us have subconsciously noted that the worst consequences of this will probably occur after we are dead and comfort ourselves with that, cowardly though it is. Some of us think that we can steadily move back to a sustainable resource use by a slow and evolutionary process. They see bottle banks, summits at Rio de Janeiro and elegantly produced coffee-table books lamenting the loss of wetlands and wildlife as welcome indicators of this. I believe that these rationalizations are wrong. I believe that we are failing to read the message of our own research.

If we are to move back to the truly stable, far more desirable, 'plant-dominated' analogue of diverse, sustainable human societies, a switch mechanism has to be designed and operated. We biomanipulate – something that is quite drastic – to make the switch in natural lakes. The effectiveness of this measure is increased if first we reduce the nutrient load – limit the resource use. The intensity of the biomanipulation needed is lowered the more we reduce the nutrient concentrations. The question is what 'biomanipulation' of our western society will be needed, either in the event of our curbing our resource use or in order to force that change as well. For, without a planned switch, nothing much can alter until some catastrophic switch forces a change in a much less palatable way.

Unplanned and chaotically operating switches might follow from the impacts of climate change, of millions of environmental refugees on the move in search of basic needs, of major crop failure consequent on lack of genetic variety, of war over the availability of water or some other resource. In 1991, about 40 wars were being fought. A planned switch to a sustainable system does not mean that return to living in mud huts, without sanitation, or to hunter-gathering. It simply means a wise use of resources and a fair allocation of them. It means much more restraint on the part of the western world, which now consumes the most. The key to a planned switch, I think, is in the different emphases apparent in traditional and western societies.

The former put survival of the group first; all their systems are geared to this. The latter emphasize the individual's right to self-gratification and reward the selfish. The traditional societies appear to have emancipated themselves from slavery to natural selection; we have reverted to it. The human societies of the future, if they are to survive with minimal uncertainty, stress and unhappiness, will not be competitive ones run by hawkish entrepreneurs self-interestedly manipulating huge resources from centralized power bases. They will be cooperative ones, organized by a greater cross-section of people on a regional or local basis and governed by strong ethical principles. Only then will our freshwater and other natural resources be wisely used.

Socrates is guilty of corrupting the minds of the young, and of believing in deities of his own invention instead of the Gods recognized by the State.
(Plato, c. 428–347 BC, Apologia 24b)

The State was insecure and forced Socrates to drink poison.

References

1 Abramovitz, J.N. (1997) Valuing nature's services. In: Brown, L.R. *et al. State of the World 1997*. Earthscan, London, pp. 95–114.

2 Agbeti, M.D. (1992) Relationship between diatom assemblages and trophic variables: a comparison of old and new approaches. *Can. J. Fish. Aquat. Sci.* **49**, 1171–5.

3 Algeus, S. (1950) The utilization of aspartic acid, succinamide, and asparagine by *Scenedesmus obliquous. Physiol. Plant.* **3**, 225–35.

4 Allan, J.D. (1978) Trout predation and the size composition of stream drift. *Limnol. Oceanogr.* **23**, 1231–7.

5 Allan, J.D. (1995) *Stream Ecology*. Chapman and Hall, London.

6 Allanson, B.R. (1973) The fine structure of the periphyton of *Chara* sp. and *Potamogeton natans* from Wytham Pond, Oxford and its significance to the macrophyte-periphyton metabolic model of R.G. Wetzel and H.L. Allen. *Freshwat. Biol.* **3**, 535–42.

7 Allen, E.D. & Spence, D.H.N. (1981) The differential ability of aquatic plants to utilize the inorganic carbon supply in fresh waters. *N. Phytol.* **87**, 269–83.

8 Allen, H.L. (1967) Acetate utilization by heterotrophic bacteria in a pond. *Hydrol. Kozlony* **1967**, 295–7.

9 Allen, H.L. (1971) Primary productivity, chemoorganotrophy, and nutritional interactions of epiphytic algae and bacteria on macrophytes in the littoral of a lake. *Ecol. Monogr.* **41**, 97–127.

10 Allen, K.R. (1951) The Horokiwi stream: a study of a trout population. *Fish. Bull. N. Zealand Mar. Depart.* **10**, 1–231.

11 Allison, E.H., Irvine, K., Thompson, A.B. & Ngatunga, B.P. (1996) Diets and food consumption rates of pelagic fish in Lake Malawi, Africa. *Freshwat. Biol.* **35**, 489–515.

12 Anderson, A.M. (1961) Further observations concerning the proposed introduction of Nile perch into Lake Victoria. *East Afr. Agric. Forest. J.* **26**, 195–201.

13 Anderson, N.H. & Sedell, J.R. (1979) Detritus processing by macroinvertebrates in stream ecosystems. *Ann. Rev. Entomol.* **24**, 351–77.

14 Anderson, N.H., Sedell, J.R., Roberts, L.M. & Triska, F.J. (1978) The role of aquatic invertebrates in processing of wood debris in coniferous forest streams. *Am. Midl. Nat.* **100**, 64–82.

15 Anderson, S. & Moss, B. (1993) How wetland habitats are perceived by children: consequences for children's education and wetland conservation. *Int. J. Sci. Educ.* **15**, 473–85.

16 Andersson, F. & Olsson, B. (1985) Lake Gordsjon: an acid forest lake and its catchment. *Ecol. Bull. (Stockholm)* **32**.

17 Andreae, M.O. & Barnard, W.R. (1984) The marine chemistry of dimethylsulfide. *Mar. Chem.* **14**, 267–79.

18 Armillas, P. (1971) Gardens on swamps. *Science* **174**, 653–61.

19 Armitage, P.D. (1978) The impact of Cow Green reservoir on invertebrate populations in the River Tees. *Ann. Rep. Freshwat. Biol. Assoc.* **46**, 47–56.

20 Armstrong, A.C., Caldow, R., Hodge, I.D. & Treweek, J. (1995) Re-creating wetlands in Britain: the hydrological, ecological and socio-economic dimensions. In: Hughes, J.M.R. & Heathwaite, A.L. (eds) *Hydrology and Hydrochemistry of British Wetlands*. Wiley, Chichester, pp. 445–66.

21 Armstrong, W., Armstrong, J. & Beckett, P.M. (1990) Measurement and modelling of oxygen release from roots of *Phragmites australis*. In: Cooper, P.F. & Findlater, B.C. (eds) *Constructed Wetlands in Water Pollution Control*. Pergamon, Oxford, pp. 41–52.

22 Arnold, D.E. (1971) Ingestion, assimilation, survival, and reproduction by *Daphnia pulex* fed seven species of bluegreen algae. *Limnol. Oceanogr.* **16**, 906–20.

23 Arnott, D.L. & Vanni, M.J. (1996) Nitrogen and phosphorus recycling by the zebra mussel (*Dreissena polymorpha*) in the western basin of Lake Erie. *Can. J. Fish. Aquat. Sci.* **53**, 646–59.

24 Asman, W.A.H. & Runge, E.H. (1991) Atmospheric deposition of nitrogen compounds in Denmark. In: Ministry of the Environment, Denmark (ed.) *Nitrogen and Phosphorus in Soil and Air: A. Abstracts*. Ministry of the Environment, Copenhagen, pp. 287–312.

25 Azam, F., Fenchel, T., Field, J.G., Gray, J.S., Meyerreil, L.A. & Thingstad, F. (1983) The ecological role of water-column microbes in the sea. *Mar. Ecol. Progr. Ser.* **10**, 257–63.

26 Bailey-Watts, A.E. (1976) Planktonic diatoms

and some diatom–silica relations in a shallow eutrophic Scottish loch. *Freshwat. Biol.* **6**, 69–80.

27 Baird, D.J., Barber, I., Bradley, M.C., Soares, A.M.V.M. & Calow, P. (1989) The *Daphnia* bioassay – a critique. *Hydrobiologia* **188/189**, 403–6.

28 Baird, D.J., Barber, I. & Calow, P. (1990) Clonal variation in general responses in *Daphnia magna* Straus to toxic stress I. Chronic life-history effects. *Funct. Ecol.* **4**, 399–407.

29 Baird, D.J., Barber, I., Bradley, M.C., Soares, A.M.V.M. & Calow, P. (1991) Interclonal variation in responses to acute stress in *Daphnia magna* Straus. *Ecotox. Environ. Safety* **21**, 257–65.

30 Baker, A.L., Baker, K.K. and Tyler, P.A. (1985) Fine layer depth relationships of lakewater chemistry, planktonic algae, and photosynthetic bacteria in meromictic Lake Fidler, Tasmania. *Freshwat. Biol.* **15**, 735–47.

31 Bales, M., Moss, B., Phillips, G., Irvine, K. & Snook, D. (1993) The changing ecosystem of a shallow, brackish lake, Hickling Broad, Norfolk, UK. II. Long-term trends in water chemistry and ecology and their implications for restoration of the lake. *Freshwat. Biol.* **29**, 141–66.

32 Balick, M.J. & Cox, P.A. (1996) *Plants, People and Culture: the Science of Ethnobotany*. Scientific American Library, New York.

33 Balls, H., Moss, B. & Irvine, K. (1989) The loss of submerged plants with eutrophication. I. Experimental design, water chemistry, aquatic plant and phytoplankton biomass in experiments carried out in ponds in the Norfolk Broadland. *Freshwat. Biol.* **22**, 71–87.

34 Balon, E.K. & Bruton, M.N. (1986) Introduction of alien species or why scientific advice is not heeded. *Environ. Biol. Fish.* **16**, 225–30.

35 Bancroft-Hunt, N. & Forman, W. (1988) *People of the Totem*. University of Oklahoma Press, Norman.

36 Barel, C.D.N., Dorit, R., Greenwood, P.H. *et al.* (1985) Destruction of fisheries in Africa's lakes. *Nature* **315**, 19–20.

37 Barko, J.W. & Smart, R.M. (1980) Mobilization of sediment phosphorus by submersed freshwater macrophytes. *Freshwat. Biol.* **10**, 229–38.

38 Barko, J.W., Murphy, P.G. & Wetzel, R.G. (1977) An investigation of primary production and ecosystem metabolism in a Lake Michigan dune pond. *Arch. Hydrobiol.* **81**, 155–87.

39 Barko, J.W., Gunnison, D. & Carpenter, S.R. (1991) Sediment interactions with submerged macrophyte growth and community dynamics. *Aquat. Bot.* **41**, 41–65.

40 Barlocher, F. & Kendrick, B. (1973a) Fungi and food preferences of *Gammarus pseudolimnaeus*. *Arch. Hydrobiol.* **72**, 501–16.

41 Barlocher, F. & Kendrick, B. (1973b) Fungi in the diet of *Gammarus pseudolimnaeus* (Amphipoda). *Oikos* **24**, 295–300.

42 Baross, J.A. & Deming, J.W. (1983) Growth of 'black smoker' bacteria at temperatures of at least 250°C. *Nature* **303**, 423–6.

43 Barrett, C.F., Fowler, D., Irving, J.G. *et al.* (1982) *Acidity of Rainfall in the United Kingdom: a Preliminary Report*. Warren Spring Laboratory, Stevenage.

44 Battarbee, R.W. (1991) Recent palaeolimnology and diatom-based environmental reconstruction. In: Shane, L.C.K. & Cushing, E.J. (eds) *Quaternary Landscapes*. University of Minnesota Press, Minneapolis, pp. 129–74.

45 Battarbee, R.W. (Chairman) (1995) *Critical Loads of Acid Deposition for United Kingdom Freshwaters*. Report of the Critical Loads Advisory Group, Department of the Environment, UK.

46 Bayless, J. & Smith, W.B. (1967) The effects of channelization upon the fish population of lotic waters in eastern North Carolina. *Proc. Ann. Conf Southeast Assoc. Game Fish Commissions* **18**, 230–8.

47 Beadle, L.C. (1981) *The Inland Waters of Tropical Africa*, 2nd edn. Longman, London.

48 Beauchamp, R.S.A. (1964) The rift valley lakes of Africa. *Verh. int. Verein. theor. angew. Limnol.* **15**, 91–9.

49 Beaver, J.R. & Crisman, T.L. (1988) Distribution of planktonic ciliates in highly coloured sub-tropical lakes: comparisons with clearwater ciliate communities and contribution of mixotrophic taxa to total autotrophic biomass. *Freshwat. Biol.* **20**, 51–60.

50 Beklioglu, M. & Moss, B. (1995) The impact of pH on interactions among phytoplankton algae, zooplankton and perch (*Perca fluviatilis*) in a shallow, fertile lake. *Freshwat. Biol.* **33**, 497–509.

51 Belcher, J.H., Swale, E.M.F. & Heron, J. (1966) Ecological and morphological observations on a population of *Cyclotella pseudostelligera* Hustedt. *J. Ecol.* **54**, 335–40.

52 Bell, D.V., Odin, N. & Tornes, E. (1985) Accumulation of angling litter at game and coarse fisheries in South Wales, UK. *Biol. Cons.* **34**, 369–79.

53 Bellamy, L.S. & Reynoldson, T.B. (1974) Behaviour in competition for food amongst lake-dwelling triclads. *Oikos* **25**, 356–64.

54 Belo, J. (1970) *Traditional Balinese Culture*. Columbia University Press, New York.

55 Bengtsson, L. & Gelin, C. (1975) Artificial aeration and suction dredging methods for controlling water quality. In: *The Effects of Storage on Water Quality*. Water Research Centre, Medmenham, pp. 313–42.

56 Bengtsson, L., Fleischer, S., Lindmark, G. & Ripl, W. (1975) Lake Trummen restoration project I. Water and sediment chemistry. *Ver. int. Verein. theor. angew. Limnol.* **19**, 1080–7.

57 Benke, A.C. (1993) Concepts and patterns of invertebrate production in running waters. *Verh. int. Verein. theor. angew. Limnol.* **25**, 15–38.

58 Benke, A.C., van Arsdell, T.C., Jr, Gillespie, D.M. & Parrish, F.K. (1984) Invertebrate productivity in a subtropical blackwater river in the importance of habitat and life history. *Ecol. Monogr.* **54**, 25–63.

59 Benndorf, J. (1992) The control of indirect effects of biomanipulation. In: Sutcliffe, D.W. & Jones, J.G. (eds) *Eutrophication: Research and Application to Water Supply*. Freshwater Biological Association, Ambleside, pp. 82–93.

60 Bennion, H. (1994) A diatom–phosphorus transfer functon for shallow, eutrophic ponds in southeast England. *Hydrobiologia* **275/276**, 391–410.

61 Bennion, H., Wunsam, S. & Schmidt, R. (1995) The validation of diatom–phosphorus transfer functions: an example from Mondsee, Austria. *Freshwat. Biol.* **34**, 271–83.

62 Bennion, H., Juggins, S. & Anderson, N.J. (1996) Predicting epilimnetic phosphorus concentrations using an improved diatom-based transfer function and its application to lake eutrophication management. *Environ. Sci. Technol.* **30**, 2004–7.

63 Berg, K. (1938) Studies on the bottom animals of Esrom Lake. *K. Danske vidensk Selsk. skr.* **7**, 1–255.

64 Berglund, B.E. (ed.) (1986) *Handbook of Holocene Palaeoecology and Palaeohydrology*. Wiley, Chichester.

65 Bernardi, R. de & Giussani, G. (1975) Population dynamics of three cladocerans of Lago Maggiore related to predation pressure by a planktophagous fish. *Verh. int. Verein theor. angew. Limnol.* **19**, 2906–12.

66 Bird, D.F. & Kalff, J. (1984) Empirical relationships between bacterial abundance and chlorophyll concentration in fresh and marine waters. *Can. J. Fish. Aquat. Sci.* **41**, 1015–23.

67 Bird, D.F. & Kalff, J. (1986) Bacterial grazing by planktonic lake algae. *Science* **231**, 493–5.

68 Bird, D.F. & Kalff, J. (1987) Algal phagotrophy: regulating factors and importance relative to photosynthesis in *Dinobryon* (Chrysophyceae). *Limnol. Oceanogr.* **32**, 277–84.

69 Birdsey, E.C. & Lynch, V.H. (1962) Utilization of nitrogen compounds by unicellular algae. *Science* **137**, 763–4.

70 Bjork, S. (1972) Swedish lake restoration program gets results. *Ambio* **1**(5), 153–65.

71 Bjork, S. (1985) Scandinavian lake restoration activities. In: *Lakes Pollution and Recovery*. European Water Pollution Control Association, Rome, pp. 293–301.

72 Blackburn, M.A. & Waldock, M.J. (1995) Concentrations of alkylphenols in rivers and estuaries in England and Wales. *Wat. Res.* **29**, 1623–9.

73 Blaustein, L. & Dumont, H.J. (1990) Typhloplanid flatworms (*Mesostoma* and related genera): mechanisms of predation and evidence that they structure aquatic invertebrate communities. *Hydrobiologia* **198**, 61–77.

74 Boar, R.R. & Crook, C.E. (1985) Investigations into the causes of reed-swamp regression in the Norfolk Broads. *Verh. int. Verein. theor. angew. Limnol.* **22**, 2916–19.

75 Boar, R.R., Crook, C.E. & Moss, B. (1989) Regression of *Phragmites australis* reedswamps and recent changes of water chemistry in the Norfolk Broadland. *Aquat. Bot.* **35**, 41–55.

76 Bogdan, K.G. & Gilbert, J.J. (1982) Seasonal patterns of feeding by natural populations of *Keratella, Polyarthra* and *Bosmina*: clearance rates, selectivities and contributions to community grazing. *Limnol. Oceanogr.* **27**, 918–34.

77 Booker, M.J. & Walsby, A.E. (1981) Bloom formation and stratification by a planktonic blue-green alga in an experimental water column. *Br. Phycol. J.* **16**, 411–21.

78 Boon, P.J., Holmes, N.T.H., Maitland, P.S., Rowell, T.A. & Davies, J. (1997) A system for evaluating rivers for conservation ('SERCON'): development, structure and function. In: Boon, P.J. & Howell, D.L. (eds) *Freshwater Quality: Defining the Indefinable*. HMSO, Edinburgh, pp. 299–326.

79 Boothby, J., Hull, A.P., Jeffreys, D.A. & Small, R.W. (1995) Wetland loss in North-West England: the conservation and management of ponds in Cheshire. In: Hughes, J.M.R. & Heathwaite, A.L. (eds) *Hydrology and Hydrochemistry of British Wetlands*. Wiley, Chichester, pp. 432–44.

80 Borgstrom, G. (1978) The contribution of freshwater fish to human food. In: Gerking, S.D. (ed.) *Ecology of Freshwater Fish Production*. Blackwell Scientific Publications, Oxford, pp. 469–91.

81 Boston, H.L. & Adams, M.S. (1983) Evidence of crassulacean acid metabolism in two North American isoetids. *Aquat. Bot.* **15**, 381–6.

82 Boston, H.L. & Adams, M.S. (1986) The contribution of crassulacean acid metabolism to the annual productivity of two aquatic vascular plants. *Oecologia* **68**, 615–22.

83 Bothwell, M.L. (1985) Phosphorus limitation of lotic periphyton growth rates: an intersite comparison using continuous-flow troughs (Thompson River System, British Columbia). *Limnol. Oceanogr.* **30**, 527–42.

84 Boulton, A.J. & Boon, P.I. (1991) A review of methodology used to measured leaf litter decomposition in lotic environments: time to turn over an old leaf? *Austr. J. mar. freshwat. Res.* **42**, 1–43.

85 Brabrand, A. & Faafeng, B. (1994) Habitat shift in roach (*Rutilus rutilus*) induced by the introduction of pikeperch (*Stizostedion lucioperca*). *Verh. int. Verein. theor. angew Limnol.* **25**, 2123.

86 Bradbury, I.K. & Grace, J. (1983) Primary production in wetlands. In: Gore, A.J.P. (ed.) *Ecosystems of the World 4A: Mires: Swamp, Bog, Fen and Moor*. Elsevier, Amsterdam, pp. 285–310.

87 Bradbury, J.P., Leyden, B., Salgado Labouriau, M. *et al.* (1981) Late Quaternary environmental history of Lake Valencia, Venezuela. *Science* **214**, 1299–305.

88 Brett, M.T. & Goldman, C.R. (1996) A meta-analysis of the freshwater trophic cascade. *Proc. Nat. Acad. Sci. USA* **93**, 7723–6.

89 Bright, C. (1997) Tracking the ecology of climate change. In: Brown, L.R. (ed.) *State of the World 1997*. Earthscan, London, pp. 78–94.

90 Brimblecome, P. & Stedman, D.H. (1982) Historical evidence for a dramatic increase in the nitrate component of acid rain. *Nature* **298**, 460–2.

91 Brinkhurst, R.O. (1974) *The Benthos of Lakes*. Macmillan, London.

92 Brix, H. & Schierup, H.-H. (1990) Soil oxygenation in constructed reed beds: the role of macrophyte and soil–atmosphere interface oxygen transport. In: Cooper, P.F. & Findlater, B.C. (eds) *Constructed Wetlands in Water Pollution Control*. Pergamon, Oxford, pp. 53–67.

93 Brock, M.A. (1982) Biology of the salinity tolerant genus *Ruppia* L. in saline lakes in South Australia. 1. Morphological variation within and between species and ecophysiology. *Aquat. Bot.* **13**, 219–48.

94 Brock, T.D. (1967) Life at high temperatures. *Science* **158**, 1012–19.

95 Brock, T.D. (1978) *Thermophilic Microorganisms and Life at High Temperatures*. Springer-Verlag, New York.

96 Bronmark, C. (1985) Interactions between macrophytes, epiphytes and herbivores: an experimental approach. *Oikos* **45**, 26–30.

97 Bronmark, C. (1989) Interactions between epiphytes, macrophytes and freshwater snails: a review. *J. moll. Stud.* **55**, 299–311.

98 Brooker, M.P. (1981) The impact of impoundments on the downstream fisheries and general ecology of rivers. *Adv. Appl. Biol.* **4**, 91–152.

99 Brookes, A. (1992) Recovery and restoration of some engineered British river channels. In: Boon, P.J., Calow, P. & Petts, G.E. (eds) *River Conservation and Management*. Wiley, Chichester, pp. 337–52.

100 Brookes, A. (1995a) Challenges and objectives for geomorphology in UK river management. *Earth Surf. Proc. Landf.* **20**, 593–610.

101 Brookes, A. (1995b) River channel restoration: theory and practice. In: Gurnell, A. & Petts, G. (eds) *Changing River Channels*. Wiley, Chichester, pp. 369–88.

102 Brookes, A., Gregory, K.J. & Dawson, F.H. (1983) An assessment of river channelization in England and Wales. *Sci. Total Environ.* **27**, 97–111.

103 Brooks, J.L. & Dodson, S.I. (1965) Predation, body size and composition of plankton. *Science* **150**, 28–35.

104 Brown, A.W.A. (1962) A survey of *Simulium* control in Africa. *Bull. World Health Org.* **27**, 511.

105 Brown, L.R. (1997) Facing the prospect of food scarcity. In: Brown, L.R. (ed.) *State of the World 1997* Earthscan, London, pp. 23–41.

106 Brown, L.R. *et al.* (eds) (1997) *State of the World 1997*. Earthscan Publications, London.

107 Brown, S. & Colman, B. (1963) Oscillaxanthin in lake sediments. *Limnol. Oceanogr.* **8**, 352–3.

108 Bryant, M.D. (1983) The role and management of woody debris in west coast salmonid nursery streams. *N. Am. J. Fish. Man.* **3**, 322–30.

109 Brylinsky, M. (1980) Estimating the productivity of lakes and reservoirs. In: Le Cren, E.D. & Lowe-McConnell, R.H. (eds) *The Functioning of Freshwater Ecosystems*. Cambridge University Press, Cambridge, pp. 411–54.

110 Brylinsky, M. & Mann, K.H. (1973) An analysis of factors governing productivity in lakes and reservoirs. *Limnol. Oceanogr.* **18**, 1–14.

111 Burgis, M.J. & Morris, P. (1987) *The Natural History of Lakes*. Cambridge University Press, Cambridge.

112 Burns, C. (1989) Parasitic regulation in a population of *Boeckella hamata* Brehm (Copepoda: Calanoida). *Freshwat. Biol.* **21**, 421–6.

113 Burns, C. (1994) Predation on ciliates by calanoid copepods. *Verh. int. Verein. theor. angew. Limnol.* **25**, 2445.

114 Burns, C.W. & Rigler, F.H. (1967) Comparison of filtering rates of *Daphnia rosea* in lake water and suspensions of yeast. *Limnol. Oceanogr.* **12**, 97–105, 492–502.

115 Burton, T.M., King, D.L. & Ervin, J.L. (1979) Aquatic plant harvesting as a lake restoration technique. In: *Lake Restoration*. USEPA, Washington.

116 Cadwallader, P.L. (1978) Some causes of the decline in range and abundance of native fish in the Murray–Darling river system. *Proc. Roy. Soc. Victoria* **90**, 211–24.

117 Calow, P. (1992) The three Rs of ecotoxicology. *Funct. Ecol.* **6**, 617–19.

118 Cannell, M.G.R. (1990) Carbon dioxide and the global carbon cycle. In: Cannell, M.G.R. & Hooper, M.D. (eds) *The Greenhouse Effect and Terrestrial Ecosystems of the UK*. Institute of Terrestrial Ecology Research Publication 4, HMSO, London, pp. 6–9.

119 Canter, H.M. (1979) Fungal and protozoan parasites and their importance in the ecology of phytoplankton. *Ann. Rep. Freshwat. Biol. Assoc.* **47**, 43–50.

120 Canter-Lund, H. & Lund, J.W.G. (1995) *Freshwater Algae*. Biopress, Bristol.

121 Cantrell, M.A. & McLachlan, A.J. (1982) Habitat duration and dipteran larvae in tropical rain pools. *Oikos* **38**, 343–8.

122 Carey, J. (1987) *The Faber Book of Reportage*. Faber & Faber, London.

123 Carignan, R. & Kalff, J. (1982) Phosphorus release by submerged macrophytes: significance

to epiphyton and phytoplankton. *Limnol. Oceanogr.* **27**, 419–27.

124 Carling, P. (1995) Implications of sediment transport for instream flow modelling of aquatic habitats. In: Harper, D. & Ferguson, A.J. D. (eds) *The Ecological Basis of River Management.* Wiley, Chichester, pp. 17–32.

125 Carney, H.J. (1992) Biodiversity, agriculture and water: change in the unique tropical Lake Titicaca landscape, Peru/Bolivia. *Verh. int. Verein, theor. angew Limnol.* **25**, 918–22.

126 Carpenter, S.R. (1983) Submersed macrophyte community structure and internal loading: relationship to lake ecosystem productivity and succession. In: Taggart, J. (ed.) *Lake Restoration, Protection and Management.* USEPA, Washington, pp. 105–11.

127 Carpenter, S.R. & Kitchell, J.F. (1992) *The Trophic Cascade in Lakes.* Cambridge University Press, Cambridge.

128 Carpenter, S.R. & Lodge, D.M. (1986) Effects of submersed macrophytes on ecosystem processes. *Aquat. Bot.* **26**, 341–70.

129 Carpenter, S.R. & McCreary, N.J. (1985) Effects of fish nests on pattern and zonation of submersed macrophytes in a softwater lake. *Aquat. Bot.* **22**, 21–32.

130 Carpenter, S.R., Kitchell, J.F. & Hodgson, J.R. (1985) Cascading trophic interactions and lake productivity. *BioScience* **10**, 634–9.

131 Carpenter, S.R., Chisholm, S.W., Krebs, C.J., Schindler, D.W. & Wright, R.F. (1995) Ecosystem experiments. *Science* **269**, 324–7.

132 Carter, G.S. (1955) *The Papyrus Swamps of Uganda.* Heffer, Cambridge University Press, Cambridge.

133 Carter, J.C.H. & Goudie, K.A. (1986) Diel vertical migrations and horizontal distributions of *Limnocalanus macrurus* and *Senecella calanoides* (Copepoda, Calanoida) in lakes of Southern Ontario in relation to planktivorous fish. *Can. J. Aquat. Sci.* **43**, 2508–14.

134 Carvalho, L. & Moss, B. (1995) The current status of a sample of English sites of special scientific interest subject to eutrophication. *Aquat. Cons. Mar. Freshwat. Ecos.* **5**, 191–204.

135 Carvalho, L., Beklioglu, M. & Moss, B. (1995) Changes in a deep lake following sewage diversion – a challenge to the orthodoxy of external phosphorus control as a restoration strategy? *Freshwat. Biol.* **34**, 399–410.

136 Casey, H.C. & Downing, A. (1976) Levels of inorganic nutrients in *Ranunculus penicillatus* var. *calcareus* in relation to water chemistry. *Aquat. Bot.* **2**, 75–80.

137 Caswell, H. (1972) On instantaneous and finite birth rates. *Limnol. Oceanogr.* **17**, 787–91.

138 Cattaneo, A. & Kalff, J. (1979) Primary production of algae growing on natural and artificial aquatic plants: a study of interactions between epiphytes and their substrate. *Limnol. Oceanogr.* **24**, 1031–7.

139 Cattaneo, A. & Mousseau, B. (1995) Empirical analysis of the removal rate of periphyton by grazers. *Oecologia* **103**, 249–54.

140 Caufield, C. (1982a) *Tropical Moist Forests.* International Institute for Environment and Development, London.

141 Caufield, C. (1982b) Brazil, energy and the Amazon. *N. Sci.* 28 Oct., 240.

142 Caufield, C. (1985a) The Yangtze beckons the Yankee dollar. *N. Sci.* 5 Dec., 26–7.

143 Caufield, C. (1985b) *In the Rainforest.* Heinemann, London.

144 Caulfield, P. (1971) *Everglades.* Sierra Club, Ballantyne, New York.

145 Cederholm, C.J. & Peterson, N.P. (1985) The retention of coho salmon (*Oncorhynchus kisutch*) carcasses by organic debris in small streams. *Can. J. Fish. Aquat. Sci.* **42**, 1222–5.

146 Chambers, P.A. (1987) Light and nutrients in the control of aquatic plant community structure. II. *In situ* observations. *J. Ecol.* **75**, 621–8.

147 Charlson, R.J. & Rodhe, H. (1982) Factors controlling the acidity of natural rainwaters. *Nature* **295**, 683–5.

148 Charnock, A. (1983) A new course for the Nile. *N. Sci.* 27 Oct., 285–8.

149 Christiansen, R., Friis, N.J.S. & Söndergaard, M. (1985) Leaf production and nitrogen and phosphorus tissue content of *Littorella uniflora* (L.) Aschers in relation to nitrogen and phosphorus enrichment of the sediment in oligotrophic Lake Hampen, Denmark. *Aquat. Bot.* **23**, 1–11.

150 Christie, W.J. (1974) Changes in the fish species composition of the Great Lakes. *J. Fish. Res. Board Can.* **31**, 827–54.

151 Chudbya, H. (1965) *Cladophora glomerata* and accompanying algae in the Skawa River. *Acta Hydrobiol.* **7**, 93–126.

152 Codd, G.A. & Bell, S.G. (1985) Eutrophication and toxic Cyanobacteria in freshwaters. *Wat. Poll. Contr.* **84**, 225–32.

153 Codd, G.A., Edwards, C., Beattie, K.A., Lawton, L.A., Campbell, D.L. & Bell, S.G. (1995) Toxins from cyanobacteria (blue-green algae). In: Wiessner, W., Schnepf, E. & Starr, R.C. (eds) *Algae, Environment and Human Affairs.* Biopress, Bristol, pp. 1–17.

154 Cogbill, C.V. & Likens, G.E. (1974) Acid precipitation in the northeastern United States. *Wat. Res.* **10**, 1133–7.

155 Cole, J.J., Findlay, S. & Pace, M.L. (1988) Bacterial production in fresh and saltwater ecosystems: a cross system overview. *Mar. Ecol. Prog. Ser.* **43**, 1–10.

156 Coleman, M.J. & Hynes, H.B.N. (1970) The vertical distribution of the invertebrate fauna in the bed of a stream. *Limnol. Oceanogr.* **15**, 31–40.

157 Collingwood, R.W. (1977) *A Survey of Eutrophication in Britain and Its Effects on Water Supplies.* Technical Report TR40, Water Research Centre, Medmenham.

158 Cook, C.D.K., Gut, B.J., Rix, T.M., Schneller, J. & Seitz, M. (1974) *The Waterplants of the World*. Junk, The Hague.

159 Cooke, A.S. (1987) Disturbance by anglers of birds at Graffham Water. In: Maitland, P.S. & Turner, A. (eds) *Angling and Wildlife in Freshwaters*. ITE Symposium 19, Institute of Terrestrial Ecology, Huntingdon, pp. 15–22.

160 Cooke, G.D., McComas, M.R., Waller, D.W. & Kennedy, R.H. (1977) The occurrence of internal phosphorus loading in two small eutrophic glacial lakes in northeastern Ohio. *Hydrobiologia* **56**, 129–35.

161 Cooper, P.F. & Findlater, B.C. (eds) (1990) *Constructed Wetlands in Water Pollution Control*. Pergamon Press, Oxford.

162 Coops, H. (1996) Helophyte zonation: impact of water depth and wave exposure. Doctoral thesis, University of Nijmegen, the Netherlands.

163 Coops, H. & Geilen, N. (1996) *Helophytes*. Institution for Inland Water Management and Waste Water Treatment (RIZA), Lelystad, the Netherlands.

164 Costanza, R., d'Arge, R., de Groot, R. *et al.* (1997) The value of the world's ecosystem services and natural capital. *Nature* **387**, 253–60.

165 Coulter, G.W., Allanson, B.R., Bruton, M.W. *et al.* (1986) Unique qualities and special problems of the African Great lakes. *Environ. Biol. Fish.* **17**, 161–84.

166 Coveney, M.F. & Wetzel, R.G. (1995) Biomass, production and specific growth rate of bacterioplankton and coupling to phytoplankton in an oligotrophic lake. *Limnol. Oceanogr.* **40**, 1187–200.

167 Cranwell, P.A. (1973) Branched chain and cyclopropanoid acids in a recent sediment. *Chem. Geol.* **11**, 307–13.

168 Cranwell, P.A. (1974) Monocarboxylic acids in lake sediments: indicators derived from terrestrial and aquatic biota of palaeoenvironmental trophic levels. *Chem. Geol.* **14**, 1–14.

169 Cranwell, P.A. (1976) Organic geochemistry of lake sediments. In: Nriagu, J.O. (ed.) *Environmental Biogeochemistry*, Vol. I. *Carbon, Nitrogen, Phosphorus, Sulphur and Selenium Cycles*. Ann Arbor Science, Ann Arbor, Michigan, pp. 75–88.

170 Crawford, R.M.M. (1966) The control of anaerobic respiration as a determining factor in the distribution of the genus *Senecio J. Ecol.* **54**, 403–13.

171 Crawford, R.M.M. (1982) Root survival in flooded soils, Part A. Analytical Studies. In: Gore, A.J.P. (ed.) *Mires: Swamp, Bog, Fen and Moor*. Elsevier, Amsterdam, pp. 257–83.

172 Crawford, R.M.M. (1989) *Studies in Plant Survival*. Blackwell Scientific Publications, Oxford.

173 Crawford, R.M.M. & McManmon, M. (1968) Inductive responses of alcohol and malic dehydrogenase in relation to flooding tolerance in roots. *J. Exp. Bot.* **19**, 435–41.

174 Crawford, R.M.M. & Tyler, P. (1969) Organic acid metabolism in relation to flooding tolerance in roots. *J. Ecol.* **57**, 235–44.

175 Cresswell, R.C. (1989) Conservation and management of brown trout, *Salmo trutta*, stocks in Wales by the Welsh Water Authority. *Freshwat. Biol.* **21**, 115–24.

176 Crisp, D.T. (1984) Effects of Cow Green reservoir upon downstream fish populations. *Ann. Rep. Freshwat. Biol. Assoc.* **52**, 47–62.

177 Crisp, D.T. (1993) The environmental requirements of salmon and trout in fresh water. *Freshwat. Forum* **3**(3), 176–202.

178 Croome, R.L. (1986) Biological studies on meromictic lakes. In: De Deckker, P. & Williams, W.D. (eds) *Limnology in Australia*. Junk, Dordrecht, pp. 113–30.

179 Crumley, J. (1993) When oafs kick ass in Eden. *Independent*. 23 Jan.

180 Cryer, M., Linley, N.W., Ward, R.M., Stratford, O. & Randerson, P.F. (1987) Disturbance of overwintering wildfowl by anglers at two reservoir sites in South Wales. *Bird Study* **34**, 191–9.

181 Cummins, K.W. (1974) Structure and function of stream ecosystems. *Bioscience* **24**, 631–41.

182 Cummins, K.W., Petersen, R.C., Howard, F.O., Wuycheck, J.C. & Holt, V.I. (1973) The utilization of leaf litter by stream detritivores. *Ecology* **54**, 336–45.

183 Cummings, B.J. (1990) *Dam the Rivers, Damn the People*. Earthscan, London.

184 Cushing, D.H. (1968) *Fisheries Biology*. University of Wisconsin Press, Madison.

185 Cushing, D.H. (1975) *Marine Ecology and Fisheries*. Cambridge University Press, Cambridge.

186 Cushing, D.H. (1977) *Science and the Fisheries*. Arnold, London.

187 Cyr, H. & Downing, J.A. (1988) The abundance of phytophilous invertebrates on different species of submerged macrophytes. *Freshwat. Biol.* **20**, 365–74.

188 Dacey, J.W.H. (1980) Internal wind in water lilies: an adaptation for life in anaerobic sediments. *Science* **210**, 1017–19.

189 Dacey, J.W.H. (1981) Pressurized ventilation in the yellow water lily. *Ecology* **62**, 1137–47.

190 Darch, J.P. (1983) *Drained Field Agriculture in Central and South America*. British Archaeological Reports International Series, 189. British Society for Archaeology, London.

191 Darch, J.P. (1988) Drained field agriculture in tropical latin America: parallels from past to present. *J. Biogeogr.* **15**, 87–95.

192 Davies, B.R., Thoms, M. & Meador, M. (1992) An assessment of the ecological impacts of inter-basin transfers, and their threats to river basin integrity and conservation. *Aquat. Cons. Mar. Freshwat. Ecos.* **2**, 325–50.

193 Davies, R.W. & Reynoldson, T.B. (1971) The

incidence and intensity of predation on lake-dwelling triclads in the field. *J. Anim. Ecol.* **40**, 191–214.

194 Davies, M.B. (1968) Pollen grains in lake sediments: redeposition caused by seasonal water circulation. *Science* **162**, 796.

195 Davis, R.B. (1974) Stratigraphic effects of tubificids in profoundal lake sediments. *Limnol. Oceanogr.* **19**, 466–88.

196 Dawidowicz, P. & Loose, C.J. (1992) Metabolic costs during predator-induced diel vertical migration of *Daphnia*. *Limnol. Oceanogr.* **37**, 1589–95.

197 Dawkins, R. (1976) *The Selfish Gene.* Oxford University Press, Oxford.

198 Dawson, F.H. (1976) The annual production of the aquatic macrophyte, *Ranunculus penicillatus* var. *calcarus* (R.W. Butcher) C.D.K. Cook. *Aquat. Bot.* **2**, 51–74.

199 Dawson, F.H. (1978) Aquatic plant management in seminatural streams: the role of marginal vegetation. *J. Environ. Man.* **6**, 213–21.

200 Dawson, F.H. (1979) *Ranunculus calcareus* and its role in lowland streams. *Ann. Rep. Freshwat. Biol. Assoc.* **47**, 60–9.

201 Dawson, F.H. & Haslam, S.M. (1983) The management of river vegetation with particular reference to shading effects of marginal vegetation. *Landscape Plan.* **10**(147) 69–158.

202 Dawson, F.H. & Kern-Hansen, V. (1979) The effect of natural and artificial shade on the macrophytes of lowland streams and the use of shade as a management technique. *Int. Rev. ges. Hydrobiol.* **64**, 437–55.

203 Dawson, F.H., Castellano, E. & Ladle, M. (1978) Concept of species succession in relation to river vegetation and management. *Verh. int. Verein. theor. angew. Limnol.* **20**, 1429–34.

204 Degens, E.T., Von Herzen, R.P. and Wong, H.-K. (1971) Lake Tanganyika: water chemistry, sediments, geological structure. *Naturwissenschaften* **58**, 229–41.

205 De Haan, H., De Boer, T., Voerman, J., Kramer, H.A. & Van Tongeren, O.F.R. (1990) Size class distribution of dissolved (< 200 nm) nutrients and essential metals in shallow, eutrophic and humic lakes. *Verh. int. Verein. theor. angew. Limnol.* **24**, 298–301.

206 DeMelo, R., France, R. & McQueen, D.J. (1992) Biomanipulation: hit or myth. *Limnol. Oceanogr.* **37**, 192–207.

207 DeMott, W.R. (1989) The role of competition in zooplankton succession. In: Sommer, U. (ed.) *Plankton Ecology – Succession in Planktonic Communities.* Springer Verlag, Berlin, pp. 195–252.

208 Denny, P. (1980) Solute movement in submerged angiosperms. *Biol. Rev.* **50**, 65–92.

209 Dierberg, F.E. & Brezonik, D.L. (1983) Nitrogen and phosphorus mass balances in natural and sewage-enriched cypress domes. *J. Appl. Ecol.* **20**, 323–37.

210 Dillon, P.J. (1975) The application of the phosphorus-loading concept to eutrophication research. *Environ. Can. Sci. Ser.* **46**.

211 Dillon, P.J. & Rigler, F.H. (1974) A test of a simple method for predicting the capacity of a lake for development based on lake trophic status. *J. Fish. Res. Bd. Can.* **32**, 1519–31.

212 Dimond, J.B. (1967) Evidence that drift of stream benthos is density related. *Ecology* **48**, 855–7.

213 Dobb, M. (1987) Marsh arabs who face extinction. *Guardian* 16 Jan., 7.

214 Dobson, M.J. (1980) 'Marsh fever' – the geography of malaria in England. *J. Hist. Geogr.* **6**, 357–89.

215 Dodson, S.I. (1974) Adaptive changes in plankton morphology in response to size-selective predation: a new hypothesis of cyclomorphosis. *Limnol. Oceanogr.* **19**, 721–9.

216 Dodson, S.I. & Egger, D.L. (1980) Selective feeding of red phalaropes on zooplankton of Arctic ponds. *Ecology* **61**, 755–63.

217 Dokulil, M. (1974) Der Neusiedler See (Österreich). *Ber. Naturhist. Ges.* **118**, 205–11.

218 Douglas, M.S. (1947) *The Everglades: River of Grass.* Ballantine, New York.

219 Downing, J.A. (1986) A regression technique for the estimation of epiphytic invertebrate populations. *Freshwat. Biol.* **6**, 161–73.

220 Downing, J.A., Plante, C. & Lalonde, S. (1990) Fish production correlated with primary productivity, not the morphoedaphic index. *Can. J. Fish. Aquat. Sci.* **47**, 1929–36.

221 Drake, J.A. (1989) Communities as assembled structures: do rules govern pattern? *TREE* **5**, 159–63.

222 Drake, J.C. & Heaney, S.I. (1987) Occurrence of phosphorus and its potential remobilization in the littoral sediments of a productive English lake. *Freshwat. Biol.* **17**, 513–24.

223 Dugan, P. (ed.) (1993) *Wetlands in Danger.* Mitchell Beazley, London.

224 Dugan, P. (ed.) (1994) *Wetland Conservation: a Review of Current Issues and Required Action.* IUCN, Gland, Switzerland.

225 Duplaix, N. (1990) South Florida water: paying the price. *Nat. Geogr.* **178**, 89–114.

226 Duthie, J.R. (1972) Detergents: nutrient considerations and total assessment. In: Likens, G.E. (ed.) *Nutrients and Eutrophication.* American Society for Limnology and Oceanography, Lawrence, Kansas, pp. 205–16.

227 Dvorak, J. & Best, E.P.H. (1982) Macro-invertebrate communities associated with the macrophytes of Lake Vechten: structural and functional relationships. *Hydrobiologia* **95**, 115–26.

228 Eccles, D.H. (1985) Lake flies and sardines – a cautionary note. *Biol. Cons.* **33**, 309–33.

229 Edmondson, W.T. (1960) Reproductive rates of rotifers in natural populations. *Mem. Ist. Ital. Idrobiol.* **12**, 21–77.

230 Edmondson, W.T. (1970) Phosphorus, nitrogen

and algae in Lake Washington after diversion of sewage. *Science* **169**, 690–1.

231 Edmondson, W.T. (1974) Secondary production. *Mitt. Verein. int. theor. angew. Limnol.* **20**, 229–72.

232 Edmondson, W.T. (1979) Lake Washington and the predictability of limnological events. *Erg. Limnol. Archiv. Hydrobiol. Beih.* **13**, 234–41.

233 Edmondson, W.T. (1992) Eulogy for G. Evelyn Hutchinson (1903–1991). *Verh. int. Verein. theor. angew. Limnol.* **25**, 49–55.

234 Edmondson, W.T. (1991) *The Uses of Ecology, Lake Washington and Beyond*. University of Washington Press, Seattle, London.

235 Edmondson, W.T., Anderson, G.C. & Peterson, D.R. (1956) Artificial eutrophication of Lake Washington. *Limnol. Oceanogr.* **1**, 47–53.

236 Edwards, A.M.C. (1971) Aspects of the chemistry of four East Anglian rivers. PhD thesis, University of East Anglia, Norwich.

237 Edwards, R.W & Cryer, M. (1987) Angler litter. In: Maitland, P.S. & Turner, A. (eds) *Angling and Wildlife in Freshwaters*. ITE Symposium 19, Institute of Terrestrial Ecology, Huntingdon, pp. 7–14.

238 Edwards, R.W., Densem, J.W. & Russell, P.A. (1979) An assessment of the importance of temperature as a factor controlling the growth rate of brown trout in streams. *J. Anim. Ecol.* **48**, 501–8.

239 Elliott, J.M. (1971) The distances travelled by drifting invertebrates in a Lake District stream. *Oecologia* **6**, 350–79.

240 Elliott, J.M. (1976) The energetics of feeding, metabolism and growth of brown trout (*Salmo trutta* L.) in relation to body weight, water temperature and ration size. *J. Anim. Ecol.* **45**, 923–48.

241 Elliott, J.M. (1977) Feeding, metabolism and growth of brown trout. *Ann. Rep. Freshwat. Biol. Assoc.* **45**, 70–7.

242 Elliott, J.M. (1989) Wild brown trout *Salmo trutta*: an important national and international resource. *Freshwat. Biol.* **21**, 1–6.

243 Elser, J.J. & Carpenter, S.R. (1988) Predation driven dynamics of zooplankton and phytoplankton communities in a whole lake experiment. *Oecologia* **76**, 148–54.

244 Emerson, J.W. (1971) Channelization, a case study. *Science* **173**, 325–6.

245 Eminson, D.F. & Moss, B. (1980) The composition and ecology of periphyton communities in freshwaters. I. The influence of host type and external environment on community composition. *Br. Phycol. J.* **15**, 429–46.

246 Entz, B.A.G., DeWitt, J.W., Massoud, A. & Khallaf, E.S. (1971) Lake Nasser, United Arab Republic. *Afr. J. Trop. Hydrobiol. Fish.* **1**, 69–83.

247 Environmental Data Services (1996) Phosphate pollution shock for arable farmers. *ENDS Rep.* **263**, 10.

248 Eriksson, F., Hornstrom, E., Mossberg, P. &

Nyberg, P. (1983) Ecological effects of lime treatment of acidified lakes and rivers in Sweden. *Hydrobiologia* **101**, 145–64.

249 Estep, K.W., Davis, P.G., Keller, M.D. & Sieburth, J. McN. (1986) How important are oceanic algal nannoflagellates in bacterivory? *Limnol. Oceanogr.* **31**, 646–50.

250 Etherington, J.R. (1982) *Environment and Plant Ecology*. Wiley, Chichester.

251 Etherington, J.R. (1983) *Wetland Ecology*. Arnold, London.

252 Ettlinger, M. & Ferch, H. (1978) Synthetic zeolites as new builders for detergents. *Manuf. Chem. Aerosol News* Oct., 51–66.

253 Evans, M.S. (1990) Large scale responses to declines in the abundance of a major fish planktivore – the Lake Michigan example. *Can. J. Fish. Aquat. Sci.* **47**, 1738–54.

254 Evans-Pritchard, E.E. (1940) *The Nuer*. Oxford University Press, Oxford.

255 Faafeng, B. & Brabrand, A. (1990) Biomanipulation of a small, urban lake – removal of fish exclude bluegreen algal blooms. *Verh. int. Verein. theor. angew. Limnol.* **24**, 597–602.

256 Fahnenstiel, G.L., Lang, G.A., Nalepa, T.F. & Johengen, T.H. (1995a) Effects of zebra mussel (*Dreissena polymorpha*) colonisation on water quality parameters in Saginaw Bay, Lake Huron. *J. Gt Lakes Res.* **21**, 435–48.

257 Fahnenstiel, G.L., Bridgeman, T.B., Lang, G.A., McCormick, M.J. & Nalepa, T.F. (1995b) Phytoplankton productivity in Saginaw Bay, Lake Huron: effects of zebra mussel (*Dreissena polymorpha*) colonization. *J. Gt Lakes Res.* **21**, 465–75.

258 Fanslow, D.L., Nalepa, T.F. & Lang, G.A. (1995) Filtration rates of the zebra mussel (*Dreissena polymorpha*) on natural seston from Saginaw Bay, Lake Huron. *J. Gt Lakes Res.* **21**, 489–500.

259 Fawell, J.K. & Wilkinson, M.J. (1994) Oestrogenic substances in water: a review. *J. Water SRT Aqua* **43**, 219–21.

260 Fay, P., Stewart, W.D.P., Walsby, A.E. & Fogg, G.E. (1968) Is the heterocyst the site of nitrogen fixation in blue-green algae? *Nature* **200**, 810–12.

261 Feldman, R.S. & Conor, E.F. (1992) The relationship between pH and community structure of invertebrates in streams of the Shenandoah National Park, Virginia, USA. *Freshwat. Biol.* **27**, 261–76.

262 Fenchel, T. (1969) The ecology of marine microbenthos IV. Structure and function of the benthic system, its chemical and physical factors and the microfauna communities with special reference to the ciliated Protozoa. *Opehelia* **6**, 1–182

263 Fenchel, T. & Finlay, B.J. (1995) *Ecology and Evolution in Anoxic Worlds*. Oxford University Press, Oxford.

264 Ferguson, A. (1989) Genetic differences among

brown trout, *Salmo trutta* stocks and their importance for conservation and management of the species. *Freshweat. Biol.* **21**, 35–46.

265 Fernando, C.H., Tudorancea, C. & Mengestou, S. (1990) Invertebrate zooplankter predator composition and diversity in tropical lentic waters. *Hydrobiologia* **198**, 13–31.

266 Finlay, B.J. (1980) Temporal and vertical distribution of ciliophoran communities in the benthos of a small eutrophic loch with particular reference to the redox profile. *Freshwat. Biol.* **10**, 15–34.

267 Finlay, B.J. (1982) Effects of seasonal anoxia on the community of benthic ciliated protozoa in a productive lake. *Arch. Protist.* **15**, 215–22.

268 Finlay, B.J., Fenchel, T. & Gardener, S. (1986) Oxygen perception and oxygen toxicity in the freshwater ciliated protozoon *Loxodes*. *J. Protozool.* **33**, 157–65.

269 Fisher, S.G. & Likens, G.E. (1973) Energy flow in Bear Brook, New Hampshire: an integrative approach to stream ecosystem metabolism. *Ecol. Monogr.* **43**, 421–39.

270 Fitkau, E.J. (1970) Role of caimans in the nutrient regime of mouth-lakes in Amazon effluents (a hypothesis). *Biotropica* **2**, 138–42.

271 Fitkau, E.J. (1973) Crocodiles and the nutrient metabolism of Amazonian water. *Amazoniana* **4**, 101–33.

272 Fitzgerald, G.P. (1969) Some factors in the competition or antagonism among bacteria, algae and aquatic weeds. *J. Phycol.* **5**, 351–9.

273 Flanders, M. & Swann, D. (undated) *The Hippopotamus Song*.

274 Flavin, C. (1997) The legacy of Rio. In: Brown, L.R. (ed.) *State of the World 1997*. Earthscan, London, pp. 3–22.

275 Flecker, A.S. (1992) Fish predation and the evolution of invertebrate drift periodicity: evidence from Neotropical streams. *Ecology* **73**, 438–48.

276 Flower, R.J. (1987) The relationship between surface sediment diatom assemblages and pH in 33 Galloway Lakes. *Hydrobiologia* **143**, 93–104.

277 Flower, R.J. & Battarbee, R.N. (1983) Diatom evidence for recent acidification of two Scottish Lochs. *Nature* **305**, 130–2.

278 Flower, R.J., Battarbee, R.W. & Appleby, P.G. (1987) The recent palaeolimnology of acid lakes in Gallaway, southwest Scotland: diatom analysis, pH trends, and the role of afforestation. *J. Ecol.* **75**, 797–824.

279 Fogg, G.E. (1971) Extracellular products of algae in freshwater. *Arch. Hydrobiol. Peih. Erg. Limnol.* **5**, 1–25.

280 Fogg, G.E. & Westlake, D.F. (1955) The importance of extracellular products of algae in freshwater. *Verh. int. Verein. theor. angew. Limnol.* **12**, 219–32.

281 Forbes, I.J. (1986) The quantity of lead shot, nylon line and other litter discarded at a coarse fishing lake. *Biol. Conserv.* **38**, 31–4.

282 Forbes, V.E. & Depledge, M.H. (1992) Predicting population response to pollutants: the significance of sex. *Funct. Ecol.* **6**, 376–81.

283 Forsberg, C. & Petersen, R.C., Jr (1990) A darkening of Swedish lakes due to increased humus inputs during the last 15 years. *Verh. int. Verein. theor. angew. Limnol.* **24**, 289–92.

284 Foster, J., Harrison, D. & MacLaren, I.S. (eds) (1992) *Buffalo*, University of Alberta Press, Edmonton.

285 Fowler, D. (1990) Methane, ozone, nitrous oxide and chlorofluorocarbons. In: Cannell, M.G.R. & Hooper, M.D. (eds) *The Greenhouse Effect and Terrestrial Ecosystems of the UK*. Institute of Terrestrial Ecology Research Publication **4**, HMSO, London, pp. 10–13.

286 Fox, G.A. (1993) What have biomarkers told us about the effects of contaminants on the health of fish-eating birds in the Great Lakes? The theory and a literature review. *J. Gt Lakes Res.* **9**, 722–36.

287 Fox, G.A., Kennedy, S.W., Norstrom, R.J. & Wigfield, D.C. (1988) Porphyria in herring gulls: a biochemical response to chemical contamination of Great lakes food chains. *Environ. Tox. Chem.* **7**, 831–9.

288 Foy, R.H. (1985) Phosphorus inactivation in a eutrophic lake by the direct addition of ferric aluminium sulphate: impact on iron and phosphorus. *Freshwat. Biol.* **15**, 613–30.

289 Frantz, T.C. & Cordone, A.J. (1976) Observations on deep water plants in Lake Tahoe, California and Nevada. *Ecology* **48**, 709–14.

290 French, H.F. (1997) Learning from the ozone experience. In Brown, L.R. (ed.) *State of the World 1997*. Earthscan, London, pp. 151–72.

291 From, P.O. (1980) A review of some physiological and toxicological responses of freshwater fish to acid stress. *Environ. Biol. Fish.* **5**, 79–93.

292 Frost, W.E. & Brown, M.E. (1967) *The Trout*. Collins, London.

293 Fry, G.L.A. & Cooke, A.S. (1984) *Acid Deposition and its Implications for Nature Conservation in Britain*. Nature Conservancy Council, Shrewsbury.

294 Fryer, G. (1959) The trophic interrelationships and ecology of some littoral communities of Lake Nyasa with especial reference to the fishes and a discussion of the evolution of a group of rock-frequenting Cichlidae. *Proc. Zool. Soc. London* **132**, 153–281.

295 Fryer, G. (1960) Concerning the proposed introduction of Nile perch into Lake Victoria. *E. Afr. Agric. J.* **25**, 267–70.

296 Fryer, G. (1968) Evolution and adaptive radiation in the Chydoridae (Crustacea: Cladocera): a study in comparative functional morphology and ecology. *Phil. Trans. Roy. Soc. (B)* **254**, 221–385.

297 Fryer, G. (1971) Functional morphology and niche specificity in chydorid and macrothricid cladocerans. *Trans. Am. Micros. Soc.* **90**, 103–4.

298 Fryer, G. (1973) The Lake Victoria fisheries: some facts and fallacies. *Biol. Cons.* **5**, 304–8.

299 Fryer, G. (1977a) The atyid prawns of Dominica. *Ann. Rep. Freshwat. Biol. Assoc.* **45**, 48–54.

300 Fryer, G. (1977b) Studies on the functional morphology and ecology of the atyid prawns of Dominica. *Phil. Trans. Roy. Soc. (B)* **277**, 57–129.

301 Fryer, G. (1987) Quantitative and qualitative: numbers and reality in the study of living organisms. *Freshwat. Biol.* **17**, 177–90.

302 Fryer, G. (1991) *A Natural History of the Lakes, Tarns and Streams of the English Lake District.* Freshwater Biological Association, Ambleside.

303 Fryer, G. & Iles, T.D. (1969) Alternative routes to evolutionary success as exhibited by African cichlid fishes of the genus *Tilapia* and the species flocks of the Great Lakes. *Evolution* **23**, 359–69.

304 Fryer, G. & Iles, T.D. (1972) *The Cichlid Fishes of the Great Lakes of Africa.* Oliver and Boyd, Edinburgh.

305 Fuhrman, J.A. & Noble, R.T. (1995) Viruses and protists cause similar bacterial mortality in coastal seawater. *Limnol. Oceanogr.* **40**, 1236–42.

306 Furch, K. (1984) Water chemistry of the Amazon Basin: the distribution of chemical elements among freshwaters. In: Sioli, H. (ed.) *The Amazon.* Junk, Dordrecht, pp. 167–200.

307 Gaarder, T. & Gran, H.H. (1927) Investigations of the production of plankton in the Oslo Fjord. *J. Conseil* **42**, 1–48.

308 Ganf, G.G. (1983) An ecological relationship between *Aphanizomenon* and *Daphnia pulex*. *Austr. J. Mar. Freshwat. Res.* **34**, 755–73.

309 Ganf, G.G. & Blazka, P. (1974) Oxygen uptake, ammonia and phosphate excretion by zooplankton of a shallow equatorial lake (Lake George, Uganda). *Limnol. Oceanogr.* **19**, 313–25.

310 Ganf, G.G. & Oliver, R.L. (1982) Vertical separation of light and available nutrients as a factor causing replacement of green algae by blue-green algae in the plankton of a stratified lake. *J. Ecol.* **70**, 529–844.

311 Ganf, G.F. & Shiel, R.J. (1985) Particle capture by *Daphnia carinata*. *Austr. J. Mar. Freshwat. Res.* **36**, 69–86.

312 Ganf, G.G. & Viner, A.B. (1973) Ecological stability in a shallow equatiorial lake (Lake George, Uganda). *Proc. Roy. Soc. (B)* **184**, 321–46.

313 Gardner, W.S. & Lee, G.F. (1975) The role of amino acids in the nitrogen cycle of Lake Mendota. *Limnol. Oceanogr.* **20**, 379–88.

314 Garrick, L.D. & Lang, J.W. (1977) The alligator revealed. *Nat. Hist.* **86**, 54–61.

315 Garrod, D.J. (1961a) The rational exploitation of the *Tilapia esculenta* stock of the North Buvuma island area, Lake Victoria. *E. Afr. Agric. Forest. J.* **27**, 69–76.

316 Garrod, D.J. (1961b) The history of the fishing industry of Lake Victoria, East Africa, in relation to the expansion of marketing facilities. *E. Afr. Agric. Forest. J.* **27**, 95–9.

317 Gash, J.H.C., Oliver, H.R., Stuttleworth, W.J. & Stewart, J.B. (1978) Evaporation from forests *J. Inst. Wat. Eng. Sci.* **32**, 104–10.

318 Geertz, C. (1959) Form and variation in Balinese village structure. *Am. Anthropol.* **61**, 991–1012.

319 Geller, W. & Müller, H. (1981) The filtration apparatus of Cladocera: filter mesh sizes and their implications on food selectivity. *Oecologia* **49**, 316–21.

320 Gerloff, G.C. & Krombholz, P.H. (1966) Tissue analysis as a measure of nutrient availability for the growth of aquatic plants. *Limnol. Oceanogr.* **11**, 529–37.

321 Gersberg, R.M., Elkins, B.W. & Goldman, C.R. (1983) Nitrogen removal in artifical wetlands. *Wat. Res.* **17**, 1009–14.

322 Gessner, F. (1952) Der Druck in seiner Bedeutung fur das Wachstum submerser Wasserpflanzen. *Planta* **40**, 391–7.

323 Ghedotti, M.J., Smihula, J.C. & Smith, G.R. (1995) Zebra mussel predation by round gobies in the laboratory. *J. Gt Lakes Res.* **21**, 665–9.

324 Gibbs, R.J. (1970) Mechanisms controlling world water chemisty. *Science* **170**, 1088–90.

325 Gierlowski-Kordesch, E. & Kelts, K. (1994) Introduction. In: Gierlowski-Kordesch, E. & Kelts, K. (eds) *Global Geological Record of Lake Basins* Vol. 1. Cambridge University Press, Cambridge, pp. xviii–xxiii.

326 Gilbert, J.J. (1966) Rotifer ecology and embryological induction *Science* **151**, 1234–7.

327 Gilbert, J.J. (1973) Induction and ecological significance of gigantism in the rotifer *Asplanchna sieboldi. Science* **181**, 63–6.

328 Gilbert, J.J. (1988) Suppression of rotifer populations by *Daphnia*: a review of the evidence, the mechanisms, and the effects on zooplankton community structure. *Limnol. Oceanogr.* **33**, 1286–303.

329 Gilbert, P.A. & De Jong, A.L. (1978) The use of phosphate in detergents and possible replacements for phosphate. In: *Phosphorus in the Environment: Its Chemistry and Biochemistry.* CIBA Foundation Symposium 57, Elsevier, Amsterdam, pp. 94–102.

330 Giles, N. (1987) Differences in the ecology of wet-dug and dry-dug gravel pit lakes. *Game Cons. Ann. Rev.* **18**, 130–3.

331 Giles, N. (1989) Assessing the status of British wild brown trout, *Salmo trutta*, stocks: a pilot study utilizing data from game fisheries. *Freshwat. Biol.* **21**, 125–34.

332 Giles, N. (1992) *Wildlife after Gravel: Twenty Years of Practical Research by the Game Conservancy and ARC.* Game Conservancy, Fordingbridge, Hants.

333 Gilwicz, Z.M. (1977) Food size selection and seasonal succession of filter feeding zooplankton in an eutrophic lake. *Ekol. Polsk.* **25**, 179–225.

334 Gliwicz, Z.M. (1980) Filtering rates, food size

selection and feeding rates of Cladocerans – another aspect of interspecific commpetition in filter feeding zooplankton. In: Kerfoot, W. (ed.) *Evolution and Ecology of Zooplankton Communities.* University of New England Press, Hanover, pp. 282–91.

335 Gliwicz, Z.M. (1986) Predation and the evolution of vertical migration behaviour in zooplankton. *Nature* **320**, 746–8.

336 Gliwicz, Z.M. & Jachner, A. (1993) Lake restoration by manipulating the behaviour of planktivorous fish with counterfeit information on risk to predation? *Verh. int. Verein. theor. angew. Limnol.* **25**, 666–70.

337 Gliwicz, M.S. & Sieniawska, A. (1986) Filtering activity of *Daphnia* in low concentrations of a pesticide. *Limnol. Oceanogr.* **31**, 1132–7.

338 Gliwicz, Z.M. & Wurtsbaugh, W.A. (in press) Predation-induced diversity in a simple ecosystem: the Great Salt Lake, Utah.

339 Goedkoop, W. & Johnson, R.K. (1992) Modelling the importance of sediment bacterial carbon for profundal macroinvertebrates along a lake nutrient gradient. *Neth. J. Aquuat. Ecol.* **26**, 477–83.

340 Goldman, C.R. & Home, A.J. (1983) *Limnology.* McGraw-Hill, New York.

341 Goldman, J.C., McCarthy, J.J. & Peavey, D.G. (1979) Growth rate influence on the chemical composition of phytoplankton in oceanic waters. *Nature* **179**, 210–15.

342 Goldschmidt, T. (1996) *Darwin's Dreampond.* MIT Press, Cambridge, Massachusetts.

343 Goldschmidt, T. & Witte, F. (1992) Explosive speciation and adaptive radiation of haplochromine cichlids from Lake Victoria: an illustration of the scientific value of a lost species flock. *Mitt. int. Verein. theor. angew. Limnol.* **23**, 101–8.

344 Goldschmidt, T., Witte, F. & Wanink, J. (1993) Cascading effects of the introduced Nile perch on the detritivorous/phytoplanktivorous species in the sublittoral areas of Lake Victoria. *Cons. Biol.* **7**, 686–700.

345 Goldsmith, E. & Hildyard, N. (1985) *The Social and Environmental Effects of Large Dams. 1. Overview.* Wadebridge Ecological Centre, Wadebridge.

346 Goldsmith, F. & Hildyard, N. (eds) (1986) *The Social and Environmental Effects of Large Dams. 2. Case Studies.* Wadebridge Ecological Centre, Wadebridge.

347 Good, R.E., Whigham, D.F., Simpson, R.L. & Jackson, C.G., Jr (1978) *Freshwater Wetlands.* Academic Press, New York.

348 Goodwin, P. (1976) Volta ten years on. *N. Sci.* **1976**, 596–7.

349 Gophen,M. & Geller, W. (1984) Filter mesh size and food particle uptake by *Daphnia. Oecologia* **64**, 408–12.

350 Gophen, M., Ochumba, P.B.O., Pollinger, U. & Kaufman, L.S. (1993) Nile perch (*Lates niloticus*) invasion in Lake Victoria (East Africa). *Verh. int.*

Verein. theor. angew. Limnol. **25**, 856–9.

351 Gorham, E. (1958) The influence and importance of daily weather conditions in the supply of chloride, sulphate and other ions to freshwater from atmospheric precipitation. *Phil. Trans. Roy. Soc. (B)* **247**, 147–78.

352 Gorham, E. (1961) Factors influencing supply of major ions to inland waters with special reference to the atmosphere. *Geol. Soc. Am. Bull.* **72**, 795–840.

353 Goulden, C.E. (1964) The history of the Cladoceran fauna of Esthwaite Water (England) and its limnological significance. *Arch. Hydrobiol.* **60**, 1–52.

354 Goulding, M. (1980) *The Fishes and the Forest: Explorations in Amazonian Natural History.* University of California Press, Los Angeles.

355 Goulding, M. (1981) *Man and Fisheries on an Amazon Frontier.* Junk, The Hague.

356 Graham, R. (1986) Ghana's Volta resettlement scheme. In: Goldsmith, F. & Hildyard, N. (eds) *The Social and Environmental Effects of Large Dams.* Wadebridge Ecological Centre, Wadebridge, pp. 131–9.

357 Green, J. (1960) Zooplankton of the River Sokoto: the Rotifers. *Proc. Zool. Soc. London* **135**, 491–523.

358 Green, J. (1967) The distribution and variation of *Daphnia lumholtzi* (Crustacea: Cladocera) in relation to fish predation in Lake Albert, East Africa. *J. Zool.* **151**, 181–97.

359 Green, J., Corbet, S.A. and Betney, E. (1973) Ecological studies on crater lakes in West Cameroon: the blood of endemic cichlids in Barombi Mbo in relation to stratification and their feeding habits. *J. Zool.* **170**, 299–308.

360 Green, J., Corbet, S.A., Watts, E. & Lan, O.B. (1976) Ecological studies on Indonesian lakes: overturn and restratification of Ranu Lamongan. *J. Zool.* **180**, 315–54.

361 Greenberg, L.A., Paskowski, C.A. & Tonn, W.M. (1995) Effects of prey species composition and habitat structure on foraging by two functionally distinct piscivores. *Oikos* **74**, 522–32.

362 Greenwood, O.H. (1981) Species-flock and explosive evolution. In: Forey, P.L. (ed.) *The Evolving Biosphere.* British Museum and Cambridge University Press, pp. 169–78.

363 Gregory, K.J. & Walling, D.E. (1973) *Drainage Basin Form and Process.* Arnold, London.

364 Griffin, J.J. & Goldberg, E.D. (1981) Sphericity as a characteristic of solids from fossil fuel burning in a Lake Michigan sediment. *Geochem. Cosmochem. Acta* **45**, 763–9.

365 Griffiths, M. & Edmondson, W.T. (1975) Burial of oscillaxanthin in the sediment of Lake Washington. *Limnol. Oceanogr.* **20**, 945–52.

366 Griffiths, M., Perrott, P.S. & Edmondson W.T. (1969) Oscillaxanthin in the sediment of Lake Washington. *Limnol. Oceanogr.* **14**, 317–26.

367 Gruber, D., Frago, C.H. & Rasnake, W.J. (1994) Automated biomonitors – first line of defence. *J. Aquat. Ecosyst. Health* **3**, 87–92.

368 Gulland, J.A. (1969) *Manual of Methods for Fish Stock Assessment. Part 1. Fish Population Analysis.* FAO, Rome.

369 Gulland, J.A. (1970) The fish resources of the oceans. *FAO Fish. Tech. Paper* **97**, 1–425.

370 Gurnell, A.M., Gregory, K.J. & Petts, G.E. (1995) The role of coarse woody debris in forest aquatic habitats: implications for management. *Aquat. Cons.* **5**, 143–66.

371 Haarg, W.R., Berg, D.J., Garton, D.W. & Farris, J.L. (1993) Reduced survival and fitness in native bivalves in response to fouling by the introduced zebra mussel (*Dreissena polymorpha*) in Western Lake Erie. *Can. J. Fish. Aquat. Sci.* **50**, 13–19.

372 Hall, C.A.S. & Moll, R. (1975) Methods of assessing aquatic primary productivity. In: Leith, H. & Whittaker, R.H. (eds) *Primary Productivity of the Biosphere.* Springer-Verlag, New York, pp. 19–54.

373 Hall, D.J. (1964) An experimental approach to the dynamics of a natural population of *Daphnia galeata mendotae. Ecology* **45**, 94–111.

374 Hall, D.J., Threlkeld, S.T., Burns, C.W. & Crowley, P.H. (1976) The size and efficiency hypothesis and the size structure of zooplankton communities. *Ann. Rev. Ecol. Syst.* **7**, 177–208.

375 Hall, R.J., Likens, G.E., Fiance, S.B. & Hendry, G.R. (1980) Experimental acidification of a stream in the Hubbard Brook Experimental Forest, New Hampshire. *Ecology* **61**, 976–89.

376 Halliday, W.M. (1935) *Potlack and Totem.* Dent, London.

377 Hanazato, T. & Ooi, T. (1992) Morphological responses of *Daphnia ambigua* to different concentrations of a chemical extract from *Chaoborus flavicans. Freshwat. Biol.* **27**, 379–85.

378 Harding, D. (1966) Lake Kariba, the hydrology and development of fisheries. In: Lowe-McConnell, R. (ed.) *Man-made Lakes.* Academic Press, London, pp. 7–20.

379 Hargeby, A. (1990) Macrophyte associated invertebrates and the effect of habitat permanence. *Oikos* **57**, 338–46.

380 Harriman, R. & Morrison, B.R.S. (1981) Forestry, fisheries and acid rain in Scotland. *Scot. Forest.* **36**, 89–95.

381 Harriman, R. & Morrison, B.R.S. (1982) Ecology of streams draining forested and non-forested catchments in an area of central Scotland subject to acid precipitation. *Hydrobiologia* **88**, 251–63.

382 Harrison, P. (1993) *The Third Revolution: Population, Environment and a Sustainable World.* Penguin, Harmondsworth.

383 Hart, D. (1968) *The Volta River Project.* Edinburgh University Press, Edinburgh.

384 Hart, R.C. (1986a) Zooplankton abundance, community structure and dynamics in relation to inorganic turbidity, and their implications for a potential fishery in subtropical Lake Le Roux, South Africa. *Freshwat. Biol.* **16**, 351–72.

385 Hart, R.C. (1986b) *Plankton, Fish and Man – a Triplet in Limnology.* Inaugural Speech, Rhodes University, Grahamstown.

386 Hartman, R.T. & Brown, D.L. (1967) Changes in internal atmosphere of submersed vascular hydrophytes in relation to photosynthesis. *Ecology* **48**, 252–8.

387 Hartmann, J. (1977) Fischereiliche Veranderungen in kulturbedingt eutrophierenden Seen. *Schweiz. Zh. Hydrol.* **39**, 243–54.

388 Harvey, G. (1997) *The Killing of the Countryside.* Cape, London.

389 Harvey, H.H., Dillon, P.J., Kramer, J.R., Pierce, R.C. & Whelpdale, D.M. (1981) *Acidification in the Canadian Aquatic Environment: Scientific Criteria for an Assessment of the Effects of Acidic Deposition on Aquatic Ecosystems.* National Research Council of Canada, Ottawa,

390 Haslam, S.M. (1978) *River Plants.* Cambridge University Press, Cambridge.

391 Havel, J.E. (1987) Predator-induced defences: a review. In: Kerfoot, W.C. & Sih, A. (eds) *Predation: Direct and Indirect Impacts on Aquatic Communities.* University Press of New England, Hanover, New Hampshire, pp. 263–78.

392 Havel, J.E. & Dodson, S.I. (1985) Environmental cues for cyclomorphosis in *Daphnia retrocurva* Forbes. *Freshwat. Biol.* **15**, 469–78.

393 Haworth, E.Y. (1969) The diatoms of a sediment core from Blea Tarn, Langdale. *J. Ecol.* **57**, 429–39.

394 Haworth, E.Y. (1972) Diatom succession in a core from Pickerel Lake, Northeastern South Dakota. *Geol. Soc. Am. Bull.* **83**, 157–72.

395 Haworth, E.Y. (1980) Comparison of continuous phytoplankton records with the diatom stratigraphy in the recent sediments of Blelham Tarn. *Limnol. Oceanogr.* **25**, 1093–103.

396 Hayes, C.R. & Greene, C.A. (1984) The evolution of eutrophication impact in public water supply reservoirs in East Anglia. *Wat. Poll. Cont.* **1984**, 42–51.

397 Healey, F.P. & Hendzel, L.L. (1980) Physiological indicators of nutrient deficiency in lake phytoplankton. *Can. J. Fish. Aquat. Sci.* **37**, 442–53.

398 Heaney, S.I. (1971) The toxicity of *Microcystis aeruginosa* Kratz from some English reservoirs. *Wat. Treat. Exam.* **20**, 235–44.

399 Heath, D.J. & Whitehead, A. (1992) A survey of pond loss in Essex, South-east England. *Aquat. Cons.* **2**, 267–73.

400 Heathwaite, A.L., Johnes, P.J. & Peters, N.E. (1996) Trends in nutrients. *Hydrol. Proc.* **10**, 263–93.

401 Hebert, P.D.N. (1978) The adaptive significance of cyclomorphosis in *Daphnia*: more possibilities. *Freshwat. Biol.* **8**, 313–20.

402 Hecky, R.E. (1992) The eutrophication of Lake Victoria. *Verh. int. Verein. theor. angew. Limnol.* **25**, 39–48.

403 Hecky, R.E., Campbell, P. & Hendzel, L.L. (1993) The stoichiometry of carbon, nitrogen

and phosphorus in particulate matter of lakes and oceans. *Limnol. Oceanogr.* **38**, 709–24.

404 Hellawell, J.M. (1977) Change in natural and managed ecosystems: detection, measurement and assessment. *Proc. Roy. Soc. London (B)* **197**, 31–57.

405 Hellawell, J.M. (1978) *Biological Surveillance of Rivers.* Water Research Centre, Publication 355, Stevenage.

406 Henriksen, A., Dickson, W. & Brakke, D.F. (1986) Estimates of critical loads to surface waters. In: Nilsson, J. (ed.) *Critical Loads for Sulphur and Nitrogen.* Nordic Council of Ministers, Copenhagen, pp. 87–120.

407 Hessen, D.O. (1985) Filtering structures and particle size selection in co-existing Cladocera. *Oecologia* **60**, 368–72.

408 Hessen, D.O. (1992) Dissolved organic carbon in a humic lake: effects on bacterial production and respiration. *Hydrobiologia* **299**, 115–23.

409 Hickley, P., Marsh, C. & North, R. (1995) Ecological management of angling. In: Harper, D.M. & Ferguson, A.J.D. (eds) *The Ecological Basis for River Management.* Wiley, Chichester, pp. 415–26.

410 Hickling, C.F. (1961) *Tropical Inland Fisheries.* Longman, London.

411 Hickling, C.F. (1966) On the feeding processes of the white amur, *Ctenopharyngodon idella*, Val. *J. Zool.* **148**, 408–19.

412 Hildebrand, S.G. (1974) The relation of drift to benthos density and food level in an artificial stream. *Limnol. Oceanogr.* **19**, 951–7.

413 Hildrew, A.G. & Ormerod, S.J. (1995) Acidification: causes, consequences and solutions. In: Harper, D.M. & Ferguson, A.J.D. (eds) *The Ecological Basis for River Management.* Wiley, Chichester, pp. 147–60.

414 Hildrew, A.G., Townsend, C.R. & Francis, J. (1984) Community structure in some southern England streams: the influence of species interactions. *Freshwat. Biol.* **14**, 297–310.

415 Hobbie, J.E., Crawford, C.C. & Webb, K.L. (1968) Amino acid flux in an estuary. *Science* **159**, 1463–4.

416 Hobbs, H.H., Jass, J.P. & Huner, J.V. (1989) A review of global crayfish introductions with particular emphasis on two North American species (Decapoda, Cambaridae). *Crustaceana* **56**, 299–316.

417 Holdgate, M. (1990) *The Environment of Tomorrow's World.* The David Davies Memorial Institute of International Studies, London, 18 pp.

418 Holdren, G.C., Jr & Armstrong, D.E. (1980) Factors affecting phosphorus release from intact lake sediment cores. *Environ. Sci. Technol.* **14**, 79–87.

419 Holdway, P.A., Watson, R.A. & Moss, B. (1978) Aspects of the ecology of *Prymnesium parvum* (Haptophyta) and water chemistry in the Norfolk Broads, England. *Freshwat. Biol.* **8**, 295–311.

420 Holland, D.G. & Harding, J.P.C. (1984) Mersey.

In: Whitton, B.A. (ed.) *Ecology of European Rivers.* Blackwell Scientific Publications, Oxford.

421 Holland, R.E. (1993) Changes in planktonic diatoms and water transparency in Hatchery Bay, Bass Island area, Western Lake Erie since the establishment of the zebra mussel. *J. Gt Lakes Res.* **19** 617–24.

422 Holm, L.G., Weldon, L.W. & Blackburn, R.D. (1969) Aquatic weeds. *Science* **66**, 699–708.

423 Holm, N.P. & Shapiro, J. (1984) An examination of lipid reserves and the nutritional status of *Daphnia pulex* fed *Aphanizomenon flos-aquae. Limnol. Oceanogr.* **29**, 1137–40.

424 Holm, N.P., Ganf, G.G. & Shapiro, J. (1983) Feeding and assimilation rates of *Daphnia pulex* fed *Aphanizomenon flosaquae. Limnol. Oceanogr.* **28**, 677–87.

425 Holomuzki, J.R. (1989) Salamander predation and vertical distributions of zooplankton. *Freshwat. Biol.* **21**, 461–72.

426 Hopson, A.J. (1972) *A Study of the Nile Perch in Lake Chad.* Overseas Research Publications **19**, HMSO, London.

427 Horne, A.J. & Fogg, G.E. (1970) Nitrogen fixation in some English lakes. *Proc. Roy. Soc. (B)* **175**, 351–66.

428 Horne, A.J. & Goldman, C.R. (1972) Nitrogen fixation in Clear Lake, California, I. Seasonal variation and the role of heterocysts. *Limnol. Oceanogr.* **17**, 678–92.

429 Hornung, M. (1984) The impact of upland pasture improvement on solute outputs in surface waters. In: Jenkins, D. (ed.) *Agriculture and the Environment.* Institute of Terrestrial Ecology, Cambridge, pp. 150–5.

430 Hossell, J.C. & Baker, J.H. (1979) Epiphytic bacteria of the freshwater plant *Ranunculus penicillatus*: enumeration, distribution and identification. *Arch. Hydrobiol.* **86**, 332–7.

431 Houghton, J.T. (ed.) *Climate Change 1995: the Science of Climate Change. Contribution of Working Group I to the Second Assessment Report of the Intergovernmental Panel on Climate Change.* Cambridge University Press, Cambridge.

432 Howard-Williams, C. (1977) Swamp ecosystems. *Malay. Nat. J.* **31**, 113–25.

433 Howard-Williams, C. (1983) Wetlands and watershed management: the role of aquatic vegetation. *J. Limnol. Soc. South. Afr.* **9(2)**, 54–62.

434 Howard-Williams, C. (1985a) Studies on the ability of a *Potamogeton pectinatus* community to remove dissolved nitrogen and phosphorus compounds from lake water. *J. Appl. Ecol.* **18**, 619–37.

435 Howard-Williams, C. (1985b) Cycling and retention of nitrogen and phosphorus in wetlands: a theoretical and applied perspective. *Freshwat. Biol.* **15**, 391–431.

436 Howard-Williams, C. & Downes, M.T. (1984) Nutrient removal by stream bank vegetation. In: Wilcock, R.J. (ed.) *Land Treatment of Wastes.* National Water and Soil Conservation

Authority, Wellington, New Zealand, pp. 409–22.

437 Howard-Williams, C., Davis, B.R. & Cross, R.H.M. (1978) The influence of periphyton on the surface structure of a *Potamogeton pectinatus* L. leaf (an hypothesis). *Aquat. Bot.* **5**, 87–91.

438 Howard-Williams, C., Davies, J. & Pickmere, S. (1982) The dynamics of growth, the effects of changing area and nitrate uptake by watercress *Nasturtium officinale* R. Br. in a New Zealand Stream. *J. Appl. Ecol.* **19**, 589–601.

439 Howell, A.D. (1932) *Florida Bird Life.* Florida Department of Game and Freshwater Fish, Talahassee, Florida.

440 Howell, P., Lock, M. & Cobb, S. (1988) *Jonglei Canal: Impact and Opportunity.* Cambridge University Press, Cambridge.

441 Howells, G. & Dalziel, T. (1995) A decade of studies at Loch Fleet, Galloway (Scotland): a catchment liming project and restoration of a brown trout fishery. *Freshwat. Forum* **5**(1), 4–37.

442 Hoyer, M.V. & Canfield, D.E. (1991) A phosphorus–fish standing crop relationship for streams? *Lake Reserv. Man.* **7**, 25–32.

443 Hrbacek, J., Bvorakova, K., Korinek, V. & Prochazkova, L. (1961) Demonstration of the effect of the fish stock on the species composition of the zooplankton and the intensity of metabolism of the whole plankton association. *Verh. int. Verein. theor. angew. Limnol.* **14**, 192–5.

444 Hubschman, J.H. (1971) Lake Erie: pollution abatement, then what? *Science* **171**, 536–640.

445 Hughes, J.C. & Lund, J.W.G. (1962) The rate of growth of *Asterionella formosa* Hass in relation to its ecology. *Arch. Mikrobiol.* **42**, 117–29.

446 Hultberg, H. (1985) Changes in fish populations and water chemistry in Lake Gardsjön and neighbouring lakes during the last century. *Ecol. Bull.* **37**, 64–72.

447 Hunter, R.D. & Bailey, J.F. (1992) *Dreissena polymorpha* (zebra mussel) – colonization of soft substrata and some effects on unionid bivalves. *Nautilus* **106**, 60–7.

448 Hurlbert, S.H., Zedler, J. & Fairbanks, D. (1971) Ecosystem alteration by mosquito fish (*Gambusia affinis*) predation. *Science* **175**, 639–41.

449 Hurlbert, S.H., Loayza, W. & Moreno, T. (1986) Fish–flamingo–plankton interactions in the Peruvian Andes. *Limnol. Oceanogr.* **31**, 457–68.

450 Hutchinson, G.E. (1937) A contribution to the limnology of arid regions. *Trans. Conn. Acad. Arts Sci.* **33**, 47–132.

451 Hutchinson, G.E. (1975) *A Treatise on Limnology,* Vol. 1. *Geography, Physics, Chemistry.* Wiley, New York.

452 Hutchinson, G.E. (1965) *The Ecological Theatre and the Evolutionary Play.* Yale University Press, New Haven, Connecticut.

453 Hutchinson, G.E. (1967) *A Treatise on Limnology,* Vol. 2. Wiley, New York.

454 Hutchinson, G.E. (1973) Eutrophication. *Am. Sci.* **61**, 269–79.

455 Hutchinson, G.E. (1975) *A Treatise on Limnology,* Vol. 3. *Limnological Botany.* Wiley, New York.

456 Hutchinson, G.E. & Loffler, H. (1956) The thermal classification of lakes. *Proc. Nat. Acad. Sci. USA* **42**, 84–6.

457 Hutchinson, G.E., Bonatti, E., Cowgill, U.M. *et al.* (1970) Ianula: an account of the history and development of the Lago di Monterosi, Latium, Italy. *Trans. Am. Phil. Soc.* **60**, 1–178.

458 Hynes, H.B.N. (1970) *The Biology of Polluted Waters.* Liverpool University Press, Liverpool.

459 Hynes, H.B.N. (1979) *The Ecology of Running Waters.* Liverpool University Press, Liverpool.

460 Hynes, H.B.N., Kaushik, N.K., Lock, M.A. *et al.* (1974) Benthos and allochthonous organic matter in streams. *J. Fish. Res. Bd Can.* **31**, 545–63.

461 Iannotta, B. (1996) Mystery of the Everglades. *N. Sci.* 9 Nov., 35–7.

462 Imevbore, A.M.A. & Adegoke, O.S. (eds) (1975) *The Ecology of Lake Kainji.* University of Ife Press, Nigeria.

463 Imhof, G. (1973) Aspects of energy flow by different food chains in a reed bed – a review. *Polsk. Arch. Hydrobiol.* **20**, 165–8.

464 Imhof, G. & Burian, K. (1972) *Energy Flow Studies in a Wetland Ecosystem.* Special Publication, Austrian Academy of Sciences, Springer-Verlag, Vienna.

465 Ineson, P. (1994) Aerial deposition of ammonia from agriculture to forest. *NERC News* April, 4–6.

466 Infante, A. & Abella, S.E.B. (1985) Inhibition of *Daphnia* by *Oscillatoria* in Lake Washington. *Limnol. Oceanogr.* **30**, 6–52.

467 Infante, A. & Litt, A.H. (1985) Differences between two species of *Daphnia* in the use of 10 species of algae in Lake Washington. *Limnol. Ocenogr.* **30**, 1053–9.

468 Ingold, C.T. (1966) The tetraradiate fungal spore. *Mycologia* **58**, 43–56.

469 International Union for the Conservation of Nature (IUCN), United Nations Educational Programme & Worldwide Fund for Nature (1991) *Caring for the Earth: a Strategy for Sustainable Living.* Gland, Switzerland.

470 Irons, J.G., III, Bryant, J.P. & Oswood, M.W. (1991) Effects of moose browsing on decomposition rates of birch leaf litter in a subarctic stream. *Can. J. Fish. Aquat. Sci.* **48**, 442–4.

471 Irvine, K., Moss, B., Bales, M. & Snook, D. (1993) The changing ecosystem of a shallow, brackish lake, Hickling Broad, Norfolk, UK, I. Trophic relationships with special reference to the role of *Neomysis integer. Freshwat. Biol.* **29**, 119–40.

472 Istvanovics, V., Pettersson, K., Rodrigo, M.A., Pierson, D., Padisak, J. & Colom, W. (1994) The colonial cyanobacterium, *Gleotrichia echinulata* has a unique phosphorus uptake and life strategy. *Verh. int. Verein. theor, angew. Limnol.* **25**, 2230.

473 Iversen, T.M., Kronvang, B., Madsen, B.L., Markmann, P. & Nielsen, M.B. (1993) Re-establishment of Danish streams: restoration and maintenance measures. *Aquat. Cons. Mar. Freshwat. Ecosys.* **3**, 73–92.

474 Ivlev, V.S. (1961) *Experimental Ecology of the Feeding of Fishes.* Yale University Press, Newhaven.

475 Jack, J.D. & Gilbert, J.J. (1993) Susceptibilities of different-sized ciliates to direct suppression by small large cladocerans. *Freshwat. Biol.* **29**, 19–29.

476 Jackson, P.B.N. (1971) The African Great Lakes fisheries: past, present and future. *Afr. J. Trop. Hydrobiol. Fish.* **1**, 35–49.

477 Jacobs, J. (1967) Untersuchungen zur Funktion und Evolution der Zyklomorphose bei *Daphnia* mit besonderer Berucksichtigung der Selektion durch Fische. *Arch. Hydrobiol.* **62**, 467–541.

478 Jacobsen, D. & Sand-Jensen, K. (1992) Herbivory of invertebrates on submerged macrophytes from Danish waters. *Freshwat. Biol.* **28**, 301–8.

479 Jawed, M. (1969) Body nitrogen and nitrogenous excretion in *Neomysis rayii* Murdoch and *Euphausia pacifica* Hansen. *Limnol. Oceanogr.* **14**, 748–54.

480 Jenkin, P.M. (1942) Seasonal changes in the temperature of Windermere (English Lake District). *J. Anim. Ecol.* **11**, 248–69.

481 Jensen, J.P., Jeppesen, E., Olrik, K. & Kristensen, P. (1994) Impact of nutrients and physical factors on the shift from cyanobacterial to chlorophyte dominance in shallow Danish lakes. *Can. J. Fish. Aquat. Sci.* **51**, 1692–9.

482 Jeppesen, E., Kristensen, P., Jensen, J.P., Søndergaard, M., Mortensen, E. & Lauridsen, T. (1991) Recovery resilience following a reduction in external phosphorus loading of shallow, eutrophic Danish lakes: duration, regulating factors and methods for overcoming resilience. *Mem. Isti. Ital. Idrobiol.* **48**, 127–48.

483 Jeppesen, E., Lauridsen, T.L., Kairesalo, T. & Perrow, M.R. (1997) Impact of submerged macrophytes on fish–zooplankton interactions in lakes. In: Jeppesen, E., Søndergaard, Ma., Søndergaard, Mo. & Christoffersen, K. (eds) *The Structuring Role of Submerged Macrophytes in Lakes.* Springer-Verlag, New York.

484 Jewson, D.H. (1976) The interaction components controlling net phytoplankton photosynthesis in a well-mixed lake (Lough Neagh) Northern Ireland. *Freshwat. Biol.* **6**, 551–76.

485 Jhingran, V.G. (1975) *Fish and Fisheries of India.* Hindustan Publishing Corporation, Delhi.

486 Jobling, S. & Sumpter, J.P. (1993) Detergent components in sewage effluent are weakly oestrogenic to fish: an *in vitro* study using rainbow trout (*Onchorhynchus mykiss*) hepatocytes. *Aquat. Toxicol.* **27**, 361–72.

487 Johannes, R.E. (1965) Influence of marine protozoa on nutrient regeneration. *Limnol. Oceanogr.* **10**, 434–42.

488 Johanson, K. & Nyberg, P. (1981) *Acidification of Surface Waters in Sweden – Effects and Extent 1980.* Publications of the Institute of Freshwater Research, Drottningholm, Swedish with English summary.

489 Jehengen, T.H., Nalepa, T.F., Fahnenstiel, G.L. & Goudy, G. (1995) Nutrient changes in Saginaw Bay, Lake Huron, after the establishment of the zebra mussel (*Dreissena polymorpha*). *J. Gt Lakes Res.* **21**, 449–64.

490 Johnes, P.J. (1996) Evaluation and management of the impact of land use change on the nitrogen and phosphorus load delivered to surface waters: the export coefficient modelling approach. *J. Hydrol.* **183**, 323–49.

491 Johnes, P., Moss, B. & Philips, G. (1996) The determination of total nitrogen and total phosphorus concentrations in freshwaters from land use, stock headage and population data: testing of a model for use in conservation and water quality management. *Freshwat. Biol.* **36**, 451–73.

492 Johnson, L.E. & Padilla, D.K. (1996) Geographic spread of exotic species – ecological lessons and opportunities from the invasion of the zebra mussel *Dreissena polymorpha.* *Biol. Cons.* **78**, 23–33.

493 Johnson, M.G. & Brinkhurst, R.O. (1971a) Associations and species diversity in benthic macroinvertebrates of Bay of Quinte and Lake Ontario. *J. Fish. Res. Bd Can.* **28**, 1683–97.

494 Johnson, M.G. & Brinkhurst, R.O. (1971b) Production of benthic macroinvertebrates of Bay of Quinte and Lake Ontario. *J. Fish. Res. Bd Can.* **28**, 1699–714.

495 Johnson, M.G. & Brinkhurst, R.O. (1971c) Benthic community metabolism in Bay of Quinte and Lake Ontario. *J. Fish. Res. Bd Can.* **28**, 1715–25.

496 Johnson, R.K. (1985) Feeding efficencies of *Chironomus plumosus* (L.) and *C. anthracinus* Zett. (Diptera: Chironomidae) in mesotrophic Lake Erken. *Freshwat. Biol.* **15**, 605–12.

497 Johnson, R.K., Boström, B. & van de Bund, W. (1989) Interactions between *Chironomus plumosus* (L.) and the microbial community in surficial sediments of a shallow, eutrophic lake. *Limnol. Oceanogr.* **34**, 993–1003.

498 Jonasson, P.M. (1972) Ecology and production of the profundal benthos in relation to phytoplankton in Lake Esrom. *Oikos* Suppl. **14**, 1–148.

499 Jonasson, P.M. (1977) Lake Esrom research, 1867–1977. *Fol. Limnol. Scand.* **17**, 67–90.

500 Jonasson, P.M. (1978) Zoobenthos of lakes. *Verh. int. Verein. theor. angew. Limnol.* **20**, 13–37.

501 Jonasson, P.M. (1996) Limits for life in the lake ecosystem. *Verh. int. Verein. theor. angew. Limnol.* **26**, 1–34.

502 Jones, J.G. (1979) Microbial activity in lake

sediments with particular reference to electrode potential gradients. *J. Gen. Microbiol.* **115**, 19–26.

503 Jones, J.G. (1985) Microbes and microbial processes in sediments. *Phil. Trans. Roy. Soc. London. (A)* **315**, 3–19.

504 Jones, J.I., Moss, B. & Young, J.O. (1997) Periphyton–nonmolluscan invertebrate–fish interactions in standing freshwaters. In: Jeppesen, E., Søndergaard, Ma., Søndergaard, Mo. and Christoffersen, K. (eds) *The Structuring Role of Submerged Macrophytes in Lakes.* Springer-Verlag, New York.

505 Jones, J.I., Young, J.O., Haynes, G.M., Moss, B., Eaton, J.W. & Hardwick, K.J. (in press) Mutualism in submerged plant communities – the roles of nutrients, grazing and plant type. I Effects on growth of plants and on periphyton biomass and nutritional quality. *J. Ecol.*

506 Jones, J.R.E. (1940) A study of the zinc polluted river Ystwyth in North Cardiganshire, Wales. *Ann. Appl. Biol.* **27**, 368–78.

507 Jones, J.R.E. (1964) *Fish and River Pollution.* Butterworth, London.

508 Jones, R.I. Salonen, K. & De Haan, H. (1988) Phosphorus transformations in the epilimnion of humic lakes: abiotic interactions between dissolved humic materials and phosphate. *Freshwat. Biol.* **19**, 357–69.

509 Jones, R.I., Laybourn-Parry, J., Walton, M.C. & Young, J.M. (1996a) The forms and distribution of carbon in a deep, oligotrophic lake (Loch Ness, Scotland). *Verth. int. Verein. theor. angew. Limnol.*

510 Jones, R.I. Young, J.M., Hartley, A.M. & Bailey-Watts, A.E. (1996b) Light limitation of phytoplankton development in an oligotrophic lake – Loch Ness, Scotland. *Freshwat. Biol.* **35**, 533–43.

511 Junk, W.J. & da Silva, C.J. (1995) Neotropical floodplains: a comparison between the Pantanal of Matto Grosso and the large Amazonian river floodplains. In: Tundisi, J.G., Bicudo, C.E.M. & Matsumara Tundisi, T. (eds) *Limnology in Brazil.* ABC/SBL. Rio de Janeiro, pp. 195–218.

512 Kadlec, R.H. & Tilton, D.L. (1979) The use of fresh-water wetlands as a tertiary wastewater treatment alternative. *Crit. Rev. Environ. Control* **9**, 185–212.

513 Kairesalo, T. & Koskimies, I. (1987) Grazing by oligochaetes and snails on epiphytes. *Freshwat. Biol.* **17**, 317–24.

514 Kalff, J. (1991) The utility of latitude and other environmental factors as predictors of nutrients, biomass and production in lakes worldwide: problems and alternatives. *Verh. int. Verein. theor. angew. Limnol.* **24**, 1235–9.

515 Kalk, M., McLachlan, A. & Howard Williams, C. (1979) *Lake Chilwa: Studies of Change in a Tropical Ecosystem.* Junk, The Hague.

516 Kamp-Nielsen, L. (1974) Mud–water exchange of phosphate and other ions in undisturbed sediment cores and factors affecting the

exchange rates. *Arch. Hydrobiol.* **73**, 218–37.

517 Kaushik. N.K. & Hynes, H.B.N. (1968) Experimental study on the role of autumn shed leaves in aquatic environments. *J. Ecol.* **56**, 229–43.

518 Kaushik, N.K. & Hynes, H.B.N. (1971) The fate of dead leaves that fall into streams. *Arch. Hydrobiol.* **68**, 465–515.

519 Keating, K.I. (1979) Blue-green algal inhibition of diatom growth: transition from mesotrophic to eutrophic community structure. *Science* **199**, 971–3.

520 Keeley, J.E. (1979) Population differentiation along a flood frequency gradient: physical adaptations to floods in *Nyssa sylvatica. Ecol. Monogr.* **49**, 89–108.

521 Keeley, J.E. (1982) Distribution of diurnal acid metabolism in the genus Isoetes. *Am. J. Bot.* **69**, 254–7.

522 Kemf, E. (ed.) (1994) *The Law of the Mother: Protecting Indigenous Peoples in Protected Areas.* Sierra Club Books, San Francisco.

523 Kerfoot, W.C. (1977) Competition in cladoceran communities: the cost of evolving defenses against copepod predation. *Ecology* **58**, 303–13.

524 Kershner, M.W. & Lodge, D.M. (1990) Effect of substrate architecture on aquatic gastropod-substrate associations. *J. North Am. Benth. Soc.* **9**, 319–26.

525 Kidron, M. & Segal, R. (1991) *The New State of the World* 4th edn. Simon & Schuster, New York.

526 Kilham, P. & Kiham, S.S. (1990) Endless summer: internal loading processes dominate nutrient cycling in tropical lakes. *Freshwat. Biol.* **23**, 379–89.

527 Kirk, J.T.O. (1983) *Light and Photosynthesis in Aquatic Ecosystems.* Cambridge University Press, Cambridge.

528 Klemer, A.R., Cullen, J.J., Mageau, M.T., Hanson, K.M. & Sundell, R.A. (1996) Cyano-bacterial buoyancy regulation: the paradoxical roles of carbon. *J. Phycol.* **32**, 47–53.

529 Kohler, S.L. (1985) Identification of stream drift mechanisms: an experimental and observational approach. *Ecology* **56**, 1749–61.

530 Kolata, A. (1991) The technology and organization of agricultural production in the Tiwanaku state. *Lat. Am. Antiq.* **2**, 99–125.

531 Korinkova, J. (1971) Quantitative relations between submerged macrophytes and populations of invertebrates in a carp pond. *Hidrobiologia (Bucharest)* **12**, 377–82.

532 Kornijow, R. (1989) Macrofauna of elodeids of two lakes of different trophy. I Relationships between plants and structure of fauna colonizing them. *Ekol. Pol.* **37**, 31–48.

533 Kornijow, R. (1992) Seasonal migration by larvae of an epiphytic chironomid. *Freshwat. Biol.* **27**, 85–9.

534 Kramer, J. & Tessier, A. (1982) Acidification of aquatic systems: a critique of chemical approaches. *Environ. Sci. Technol.* **16**, 606A–615A.

535 Krecker, F.H. (1939) A comparative study of the animal population of certain submerged aquatic plants, *Ecology* **20**, 552–62.

536 Kristensen, P. & Ole Hansen, H. (1994) *European Rivers and Lakes: Assessment of Their Environmental State.* Enivronment Monographs 1, European Environment Agency, Copenhagen.

537 Kuznetsov, S.I. (1977) Trends in the development of ecological microbiology. *Adv. Aquat. Microbiol.* **1**, 1–48.

538 Lack, T.J. (1981) Advances in the management of eutrophic reservoirs. *Notes Wat. Res.* **27**, 1–4

539 Ladle, M. & Welton, S. (1992) The rise and fall of the Blandford fly. *NERC News* Jan. 28–9.

540 Lake, P.S., Barmuta, L.A., Boulton, A.J., Campbell, I.C. & St Clair, R.M. (1985) Australian streams and Northern Hemisphere stream ecology comparisons and problems. *Proc. Ecol. Soc. Austr.* **14**, 61–82.

541 Lamarra, V. (1975) Digestive activities of carp as a major contributor to the nutrient loading of lakes. *Verh. Int. Verein. theor. angew. Limnol.* **19**, 2461–8.

542 Lamberti, G.A., Gregory, S.V., Hawkins, C.P., Wildman, R.C., Ashkenas, L.R. & Denicola, D.M. (1992) Plant–herbivore interactions in streams near Mount St Helens. *Freshwat. Biol.* **27**, 237–48.

543 Lampert, W. (1978) a field study on the dependence of the fecundity of *Daphnia* species on food concentration. *Oecologia* **36**, 363–9.

544 Lampert, W. (1981) Inhibiting and toxic effects of blue-green algae on *Daphnia. Int. Rev. Ges. Hydrobiol.* **66**, 285–98.

545 Lampert, W. & Taylor, B.E. (1984) *In situ* grazing rates and particle selection by zooplankton: effects of vertical migration. *Verh. int. Verein. theor. angew. Limnol.* **22**, 943–6.

546 Lampert, W. & Taylor, B.E. (1985) Zooplankton grazing in a eutrophic lake: implications of diel vertical migration. *Ecology* **66**, 68–82.

547 Lampert, W., Fleckner, W., Rai, H. & Taylor, B.E. (1986) Phytoplankton control by grazing zooplankton: a study on the spring clear-water phase. *Limnol. Oceanogr.* **31**, 478–90.

548 Lasenby, D.C., Northcote, T.G. & Farst, M. (1986) Theory, practice and effects of *Mysis relicta* introductions to North American and Scandinavian lakes. *Can. J. Fish. Aquat. Sci.* **43**, 1277–84.

549 Lauridsen, T.L. & Lodge, D. (1996) Avoidance of fish and macrophytes by *Daphnia magna*: chemical cues and predator-mediated use of macrophyte habitat. *Limnol. Oceanogr.* **22**, 805–10.

550 Lauridsen, T.L., Pedersen, L.J., Jeppesen, E. & Søndergaard, M. (1996) The importance of macrophyte bed size for cladoceran composition and horizontal migration in a shallow lake. *J. Plankton Res.* **18**, 2283–94.

551 Lavergne, M. (1986) The seven deadly sins of Egypt's Aswan High Dam. In: Goldsmith, F. & Hildyard, N. (eds) *The Social and Environmental Effects of Large Dams.* Wadebridge Ecological Centre, Wadebridge, pp. 181–3.

552 Lavrentyev, P.J., Gardner, W.S., Cavaletto, J.F. & Beaver, J.R. (1995) Effects of the zebra mussel (*Dreissena polymorpha* Pallas) on protozoa and phytoplankton from Saginaw bay, Lake Huron. *J. Gt Lakes Res.* **21**, 545–57.

553 Lawson, D.L., Klug, M.J. & Merritt, R.W. (1984) The influence of physical, chemical, and microbiological characteristics of decomposing leaves on the growth of the detritivore *Tipula abdominalis* (Diptera: Tipulidae). *Can. J. Zool.* **62**, 2339–43.

554 Lawton, L.A. & Codd, G.A. (1991) Cyanobacterial (blue green algal) toxins and their significance in UK and European waters. *J. Inst. Wat. Environ. Man.* **5**, 460–5.

555 Laybourn-Parry, J. (1992) *Protozoan Plankton Ecology.* Chapman & Hall, London.

556 Laybourn-Parry, J., Walton, M., Young, J., Jones, R.I. & Shine, A. (1994) Protozooplankton and bacterioplankton in a large oligotrophic lake – Loch Ness, Scotland. *J. Plankton Res.* **16**, 1655–70.

557 Leah, R.T., Moss, B. & Forrest, D.E. (1980) The role of predation in causing major changes in the limnology of a hypereutrophic lake. *Int. Rev. ges. Hydrobiol.* **65**, 223–47.

558 Lean, D.R.S. (1973) Phosphorus dynamics in lake waters. *Science* **179**, 678–80.

559 Le Cren, E.D. (1972) Fish production in freshwaters. *Symp. Zool. Soc. London* **29**, 115–33.

560 Le Cren, E.D. & Lowe-McConnell, R.H. (1980) *The Functioning of Freshwater Ecosystems.* Cambridge University Press, Cambridge.

561 Lee, G.F. (1977) Significance of oxic vs. anoxic conditions for Lake Mendota sediment phosphorus release. In: Golterman, H.L. (ed.) *Interactions between Sediments and Freshwater.* Junk, The Hague, pp. 294–306.

562 Leedale, G.F. (1967) *Euglenoid Flagellates.* Prentice-Hall, Englewood Cliffs, New Jersey.

563 Lehman, J.T. (1976) Ecological and nutritional studies on *Dinobryon* Ehrenb.: seasonal periodicity and the phosphate toxicity problem. *Limnol. Oceanogr.* **21**, 646–58.

564 Lehman, J.T. & Branstrator, D.K. (1993) Effects of nutrients and grazing on the phytoplankton of Lake Victoria. *Verh. int. Verein. theor. angew. Limnol.* **25**, 850–5.

565 Lehman, J.T. & Caceres, C. (1993) Food-web responses to species invasion by a predatory invertebrate: *Bythotrephes* in Lake Michigan. *Limnol. Oceanogr.* **38**, 879–91.

566 Lehman, J.T. & Scavia, D. (1982) Microscale patchiness of nutrients in plankton communities. *Science* **216**, 729–30.

567 Lennox, L.J. (1984) Lough Ennell: laboratory studies on sediment phosphorus release under varying mixing, aerobic and anaerobic conditions. *Freshwat. Biol.* **14**, 183–7.

568 Leopold, L.B. (1974) *Water, A Primer.* Freeman, San Francisco.

569 Levi-Strauss, C. (1994) *Saudades do Brasil.* Companha das Letras, São Paulo.

570 Libes, S.M. (1992) *An Introduction to Marine Biogeochemistry.* Wiley, New York.

571 Likens, G.E. & Bormann, F.H. (1974) Acid rain: a serious regional environmental problem. *Science* **184**, 1176–9.

572 Likens, G.E. & Bormann, F.H. (1975) An experimental approach in New England landscapes. In: Hasler, A.D. (ed.) *Coupling of Land and Water Systems.* Springer-Verlag, New York, pp. 7–29.

573 Likens, G.E., Bormann, F.H., Pierce, R.S. Eaton, J.S. & Johnson, N.M. (1977) *Biogeochemistry of a Forested Ecosystem.* Springer-Verlag, New York.

574 Linacre, E.T., Hicks, B.B., Sainty, G.R. & Grauze, G. (1970) The evaporation from a swamp. *Agric. Meteorol.* **7**, 375–86.

575 Lindegaard, C. (1992) Zoobenthos ecology of Thingvallavatn: vertical distribution, abundance, population dynamics and production. *Oikos* **64**, 257–304.

576 Lindegaard, C. & Jonasson, P.M. (1979) Abundance, population dynamics and production of zoobenthos in Lake Myvatn, Iceland. *Oikos* **32**, 202–27.

577 Linthurst, R.H. (1983) *The Acidic Deposition Phenomenon and its Effects.* Critical Assessment Review Papers II, US Environmental Protection Agency, Washington, DC.

578 Lodge, D.M. (1985) Macrophyte–gastropod associations: observations and experiments on macrophyte choice by gastropods. *Freshwat, Biol.* **15**, 695–708.

579 Lovelock, J.E. (1979) *Gaia: a New Look at Life on Earth.* Oxford University Press, Oxford.

580 Lowe-McConnell, R.H. (ed.) (1966) *Man-made Lakes.* Symposia of the Institute of Biology of London No. **15**, Academic Press, New York.

581 Lowe-McConnell, R.H. (1975) *Fish Communities in Tropical Freshwaters.* Longman, London.

582 Lowe-McConnell, R.H. (1987) *Ecological Studies in Tropical Fish Communities.* Cambridge University Press, Cambridge.

583 Lowe-McConnell, R.H. (1994) The changing ecosystem of lake Victoria, East Africa. *Freshwat. Forum* **4**(2), 75–88.

584 Lund, J.W.G. (1949) Studies on *Asterionella.* I. The origin and nature of the cells producing seasonal maxima. *J. Ecol.* **37**, 389–419.

585 Lund, J.W.G. (1950) Studies on *Asterionella formosa* Hass. II. Nutrient depletion and the spring maximum. *J. Ecol.* **38**, 1–4, 15–35.

586 Lund, J.W.G. (1954) The seasonal cycle of the plankton diatom *Melosira italica* (Ehr.) Kutz susp. *subarctica* O. Mull. *J. Ecol.* **42**, 151–79.

587 Lund, J.W.G. (1964) Primary productivity and periodicity of phytoplankton. *Verh. int. Verein. theor. angew. Limnol.* **15**, 37–56.

588 Lund, J.W.G. (1971) The seasonal periodicity of three planktonic desmids in Lake Windermere.

Mitt. int. Verein. theor. angew. Limnol. **19**, 3–25.

589 Lund, J.W.G. (1975) The uses of large experimental tubes in lakes. In: *The Effects of Storage on Water Quality.* Water Research Centre, Medmenham, pp. 291–311.

590 Lund, J.W.G. & Reynolds, C.S. (1982) The development and operation of large limnetic enclosures in Blelham Tarn, English Lake District, and their contribution to phytoplankton ecology. *Prog. Phycol. Res.* **1**, 2–65.

591 Lynch, M. (1979) Predation, competition, and zooplankton community structure: an experimental study. *Limnol. Oceanogr.* **24**, 253–72.

592 Lynch, M. (1980) *Aphanizomenon* blooms: alternate control and cultivation by *Daphnia pulex.* In: Kerfoot, W.C. (ed.) *Evolution and Ecology of Zooplankton Communities.* University Press of New England, Hanover, pp. 299–304.

593 Lynch, M. & Shapiro, J. (1980) Predation, enrichment and phytoplankton community structure. *Limnol. Oceanogr.* **26**, 86–102.

594 Maberley, S.C. & Spence, D.H.N. (1983) Photosynthetic inorganic carbon use by freshwater plants. *J. Ecol.* **71**, 705–24.

595 Macan, T.T. (1963) *Freshwater Ecology.* Longman, London.

596 Macan, T.T. (1970) *Biological Studies of the English Lakes.* American Elsevier.

597 Macan, T.T. (1973) *Ponds and Lakes.* Allen and Unwin, London.

598 Macan, T.T. (1976) A twenty-one-year study of the water-bugs in a moorland fishpond. *J. Anim. Ecol.* **45**, 913–22.

599 Macan, T.T. (1977) The influence of predation on the composition of freshwater animal communities. *Biol. Rev.* **52**, 45–70.

600 Macan, T.T. & Kitching, A. (1972) Some experiments with artificial substrata. *Verh. int. Verein. theor. angew. Limnol.* **18**, 213–20.

601 McCarthy, J.F. & Shugart, L.R. (eds) (1990) *Biological Markers for Environmental Contamination.* Lewis, Boca Raton.

602 McCauley, E. & Kalff, J. (1981) Empirical relationships between phytoplankton and zooplankton biomass in lakes. *Can. J. Fish. Aquat. Sci.* **38**, 458–63.

603 McComb, A.J. & Lake, P.S. (1990) *Australian Wetlands.* Collins, Angus & Robertson, North Ryde, New South Wales.

604 McGowan, S. (1997) Ancient cyanophyte blooms – studies on the paleolimnology of White Mere and Colemere. PhD thesis, University of Liverpool.

605 Mackenzie, D. (1986) Geology of Cameroon's gas catastrophe. *N. Sci.* 4 Sept., 26–7.

606 Mackereth, F.J.H. (1957) Chemical analysis in ecology illustrated from lake district tarns and lakes. 1. Chemical analysis. *Proc. Linn. Soc. London* **67**(1954–1955), 159–64.

607 Mackereth, F.J.H. (1958) A portable core

sampler for lake depositis. *Limnol. Oceanogr.* **3**, 181–91.

608 Mackereth, F.J.H. (1965) Chemical investigations of lake sediments and their interpretation. *Proc. Roy. Soc. (B)* **161**, 293–375.

609 Mackereth, F.J.H. (1966) Some chemical observations on post-glacial lake sediments. *Phil. Trans. Roy. Soc. (B)* **250**, 165–213.

610 Mackie, G.L. (1991) Biology of the exotic zebra mussel, *Dreissena polymorpha*, in relation to native bivalves and its potential impact in Lake St Clair. *Hydrobiologia* **219**, 251–68.

611 McLachlan, A.J. (1974a) Recovery of the mud substrate and its associated fauna following a dry phase in a tropical lake. *Limnol. Oceanogr.* **19**, 74–83.

612 McLachlan, A.J. (1974b) Development of some lake ecosystems in tropical Africa, with special reference to the invertebrates. *Biol. Rev.* **49**, 365–97.

613 McLachlan, A.J. (1981a) Food sources and foraging tactics in tropical rain pools. *Zool. J. Linn. Soc.* **71**, 265–77.

614 McLachlan, A.J. (1981b) Interaction between insect larvae and tadpoles in tropical rain pools. *Ecol. Enomol.* **6**, 175–82.

615 McLachlan, A.J. (1983) Life history tactics of rain-pool dwellers. *J. Anim. Ecol.* **51**, 545–61.

616 McLachlan, A.J. & Cantrell, M.A. (1980) Survival strategies in tropical rain pools. *Oecologia* **47**, 344–51.

617 McLachlan, A.J., Pearce, L.J. & Smith, J.A. (1979) Feeding interactions and cycling of peat in a bog lake. *J. Anim. Ecol.* **48**, 851–61.

618 McLachlan, J.A. (1985) *Estrogens in the Environment II.* Elsevier, North-Holland, New York.

619 McMahon, R.F. (1996) The physiological ecology of the zebra mussel. *Dreissena polymorpha* in North America and Europe. *Am. Zool.* **36**, 339–63.

620 McNeish, A.S., Leah, R.T., Connor, L. & Johnson, M.S. (1994) Methyl-hexachlorocyclohexane in mussels (*Mytilus edulis*) from the Mersey estuary. *Mar. Poll. Bull.* **28**, 254–8.

621 McQueen, D.G., Post, J.R. & Mills, E.L. (1986) Trophic relationships in freshwater pelagic ecosystems. *Can J. Fish. Aquat. Sci.* **43**, 1571–81.

622 Madsen, B.L. (1995) *Danish Watercourses – Ten Years with the New Watercourse Act.* Ministry of Environment and Energy, Danish Environment Protection Agency, Copenhagen, Denmark.

623 Madsen, B.L., Bengtson, J. & Butz, I. (1973) Observations on upstream migrations by imagines of some Plecoptera and Ephemeroptera. *Limnol. Oceanogr.* **18**, 678–81.

624 Madsen, T.V. Sand-Jensen, K. (1991) Photosynthetic carbon assimilation in aquatic macrophytes. *Aquat. Bot.* **41**, 5–40.

625 Mahan, D.C. & Cummings, K.W. (undated) *A Profile of Augusta Creek in Kalamazoo and Barry Counties, Michigan.* Technical Report No. 3, W.K. Kellogg Biological Station, Michigan State University, Michigan.

626 Maitland, P.S. (1995) Ecological impact of angling, In: Harper, D.M. & Ferguson, A.J.D. (eds) *The Ecological Basis for River Management.* Wiley, Chichester, pp. 443–52.

627 Makarewicz, J.C. & Likens, G.E. (1975) Niche analysis of a zooplankton community. *Science* **190**, 1000–2.

628 Malley, D.F. (1980) Decreased survival and Ca uptake by the crayfish, *Orconectes virilis* in low pH. *Can. J. Fish. Aquat. Sci.* **37**, 364–72.

629 Maltby, E. (1986) *Waterlogged Wealth.* Earthscan, London.

630 Maltby, L. & Calow, P. (1990) The application of bioassays in the resolution of environmental problems: past, present and future. *Hydrobiologia* **188/189**, 65–76.

631 Manny, B.A., Miller, M.C. & Wetzel, R.G. (1971) Ultraviolet combustion of dissolved organic compounds in lake waters. *Limnol. Oceanogr.* **16**, 71–85.

632 Marsden, M.W. (1989) Lake restoration by reducing external phosphorus loading: the influence of sediment release. *Freshwat. Biol.* **21**, 139–62.

633 Marshall, B.E. & Langerman, J.E. (1979) The Tanganyika sardine in Lake Kariba. *Rhodesia Sci. News* **13**, 104–5.

634 Marshall, T.K. (1977) Morphological, physiological and ethological differences between walleye (*Stizostedion vitreum*) and pikeperch (*S. lucioperca*). *J. Fish. Res. Bd Can.* **34**, 1515–23.

635 Martel, A. (1993) Dispersal and recruitment of zebra mussel (*Dreissena polymorpha*) in a near shore area in west-central Lake Erie – the significance of postmetamorphic drifting. *Can. J. Fish. Aquat. Sci.* **50**, 3–12.

636 Martens, E. von (1858) On the occurrence of marine animal forms in freshwater. *Ann. Nat. Hist. (Ser. 3)* **1**, 50–63.

637 Mason, C.F. (1966) *Biology of Freshwater Pollution.* Longman, London.

638 Mason, C.F. & Bryant, R.J. (1975) Periphyton production and grazing by chironomids in Alderfen Broad, Norfolk. *Freshwat. Biol.* **5**, 271–7.

639 Mathew, C.P. & Westlake, D.F. (1969) Estimation of production by populations of higher plants subject to high mortality. *Oikos* **20**, 156–60.

640 Matthews, G.V.T. (1993) *The Ramsar Convention on Wetlands: its History and Development.* Ramsar Convention Bureau, Gland.

641 Matveer, V. (1993) An investigation of allelopathic effects of *Daphnia*. *Freshwat. Biol.* **29**, 99–105.

642 Maxwell, G. (1957) *A Reed Shaken by the Wind.* Longmans, London. (1983) Penguin, Harmondsworth.

643 Maybury-Lewis, D. (1992) *Millennium, Tribal*

Wisdom and the Modern World. Viking Press, New York.

644 Meier, M.L., de Haan, M.W., Breukelaaw, A.W. & Buitenveld, H. (1990) Is reduction of the benthivorous fish an important cause of high transparency following biomanipulation in shallow lakes? *Hydrobiologia* **200/201**, 303–15.

645 Melack, J.M. (1979) Temporal variability of phytoplankton in tropical lakes. *Oecologia* **44**, 1–7.

646 Melack, J.M. & Kilham, P. (1974) Photosynthetic rates of phytoplankton in East African alkaline, saline lakes. *Limnol. Oceanogr.* **19**, 743–55.

647 Melack, J.M., Kilham, P. & Fisher, T.R. (1982) Responses of phytoplankton to experimental fertilization with ammonium and phosphate in an African soda lake. *Oecologia* **52**, 321–6.

648 Mellina, E. & Rasmussen, J.B. (1994a) Occurrence of zebra mussel (*Dreissena polymorpha*) in the intertidal region of the St Lawrence estuary. *J. Freshwat. Ecol.* **9**, 81–4.

649 Milbrink, G. (1977) On the limnology of two alkaline lakes (Nakura and Naivasha) in the east rift valley system in Kenya. *Int. Rev. ges. Hydrobiol.* **62**, 1–17.

650 Mills, D.H. (1967) A study of trout and young salmon populations in forest streams with a view to management. *Forestry* **40**, 85–90.

651 Mills, D.H. (1987) Predator control. In: Maitland, P.S. & Turner, A. (eds) *Angling and Wildlife in Freshwaters.* ITE Symposium 19, Institute of Terrestrial Ecology, Huntingdon, pp. 53–6.

652 Mills, D.H. (1989) Conservation and management of brown trout, *Salmo trutta*, in Scotland: an historical review and the future. *Freshwat. Biol.* **21**, 87–98.

653 Mills, S. (1982a) Salmon: demise of the landlord's fish. *N. Sci.* **1982**, 364–7.

654 Mills, S. (1982b) Britain's native trout is floundering. *N. Sci.* **1982**, 498–501.

655 Ministry of Agriculture, Fisheries and Food & Welsh Office Agriculture Department (1991) *Code of Good Agricultural Practice for the Protection of Water.* MAFF Publications, London.

656 Ministry of the Environment, Denmark (1991) *Environmental Impact of Nutrient Emissions in Denmark.* Redegorelse fra Miljøstyrelsen 1, National Agency of Environmental Protection, Copenhagen.

657 Mitsch, W.J. (1994) Preface. In: Mitsch, W.J. (ed.) *Global Wetlands: Old World and New.* Elsevier, Amsterdam, pp. v–ix.

658 Moeller, R.G., Burkholder, J.M. & Wetzel, R.G. (1988) Significance of sedimentary phosphorus to a rooted submersed macrophyte (*Naias flexilis* (Willd.) Rostk & Schmidt) and its algal epiphytes. *Aquat Bot.* **32**, 261–81.

659 Moncur, A. (1986) Volcanic gas kills 1500 villagers. *Guardian* 26 Aug., 1.

660 Monod, J. (1942) *Recherches sur la croissance des cultures bactériennes.* Hermann, Paris.

661 Moore, J.W. (1980) Zooplankton and related phytoplankton cycles in a eutrophic lake. *Hydrobiologia* **74**, 99–104.

662 Morgan, A. & Kalk, M. (1970) Seasonal changes in the waters of Lake Chilwa (Malawi) in a drying phase 1966–1968. *Hydrobiologia* **36**, 81–103.

663 Morgan, J.A.W. (1990) Genetic engineering of microorganisms: free release into the environment. *Ann. Rep. Freshwat. Biol. Assoc.* **58**, 27–32.

664 Moriarty, D.J.W. (1973) The physiology of digestion of blue-green algae in the cichlid fish *Tilapia nilotica. J. Zool.* **171**, 25–39.

665 Moriarty, D.J.W., Darling, J.P.E.C. Dunn, I.G., Moriarty, C.M. & Tevlin, M.P. (1973) Feeding and grazing in Lake George. Uganda. *Proc. Roy. Soc. London (B)* **184**, 227–346.

666 Morrison, B.R.S. (1987) Uses and effects of piscicides. In: Maitland, P.S. & Turner, A.K. (eds) *Angling and Wildlife in Fresh Waters.* ITE Symposium 19, Institute of Terrestrial Ecology, Huntingdon, pp. 47–52.

667 Mortimer, C.H. (1941–42) The exchange of dissolved substances between mud and water in lakes. *J. Ecol.* **29**, 280–329, **30**, 147–201.

668 Mortimer, C.H. (1954) Models of the flow pattern in lakes. *Weather* **9**, 177–84.

669 Mortimer, C.H. (1973) The Loch Ness monster – limnology or paralimnology? *Limnol. Oceanogr.* **18**, 343–4.

670 Morton, W. (1982) Comparative catches and food habits of dolly varden and charrs, *Salvelinus malma* and *S. alpinus* at Karluk, Alaska, in 1939–1941. *Environ. Biol. Fish.* **7**, 7–29.

671 Moss, B. (1969) Limitation of algal growth in some Central African waters. *Limnol. Oceanogr.* **14**, 591–601.

672 Moss, B. (1972a) Studies on Gull Lake, Michigan. I. Seasonal and depth distribution of phytoplankton. *Freshwat. Biol.* **2**, 289–307.

673 Moss, B. (1972b) The influence of environmental factors on the distribution of freshwater algae: an experimental study. I. Introduction and the influence of calcium concentration. *J. Ecol.* **60**, 917–32.

674 Moss, B. (1973a) The influence of environmental factors on the distribution of freshwater algae: an experimental study. II. The role of pH and the carbon dioxide–bicarbonate system. *J. Ecol.* **61**, 157–77.

675 Moss, B. (1973b) The influence of environmental factors on the distribution of freshwater algae: an experimental study. IV. Growth of test species in natural lake waters, and conclusion. *J. Ecol.* **61**, 193–211.

676 Moss, B. (1977) Adaptations of epipelic and epipsammic freshwater algae. *Oecologia* **27**, 103–8.

677 Moss, B. (1983) The Norfolk Broadland: experiments in the restoration of a complex wetland. *Biol. Rev.* **58**, 521–61.

678 Moss, B. (1986) Restoration of lakes and

lowland rivers. In: Bradshaw, A.D. Goode, D.A. & Thorp, E. (eds) *Ecology and Design in Landscape.* Blackwell Scientific Publications, Oxford, pp. 399–415.

679 Moss, B. (1989) Water pollution and the management of ecosystems: a case study of science and scientist. In: Grubb, P.J. & Whittaker, J.H. (eds) *Toward a More Exact Ecology.* Thirtieth Symposium of the British Ecological Society, Blackwell Scientific Publications, Oxford, pp. 401–22.

680 Moss, B. (1992) The scope for biomanipulation in improving water quality. In: Sutcliffe, D.W. & Jones, J.G. (eds) *Eutrophication: Research and Application to Water Supply.* Freshwater Biological Association, Ambleside, pp. 73–81.

681 Moss, B. (1995) The microwaterscape – a four-dimensional view of interactions among water chemistry, phytoplankton, periphyton, macrophytes, animals and ourselves. *Wat. Sci. Technol.* **32**, 105–16.

682 Moss, B. (1996) A land awash with nutrients – the problem of eutrophication. *Chem. Ind.* 3 June.

683 Moss, B. & Leah, R.T. (1982) Changes in the ecosystem of a guanotrophic and brackish shallow lake in Eastern England: potential problems in its restoration. *Int. Rev. ges. Hydrobiol.* **67**, 625–59.

684 Moss, B. Wetzel, R.G. & Lauff, G.H. (1980) Annual productivity and phytoplankton changes between 1969 and 1974 in Gull Lake, Michigan. *Freshwat. Biol.* **10**, 113–21.

685 Moss, B., Balls, H., Booker, I., Manson, K. & Timms, M. (1984) The River Bure, United Kingdom: patterns of change in chemistry and phytoplankton in a slow flowing fertile river. *Verh. int. Verein. theor. angew. Limnol.* **22**, 1959–64.

686 Moss, B., Balls, H.R. & Irvine, K. (1985) Management of the consequences of eutrophication in lowland lakes in England engineering and biological solutions. In: Lester, J.N. & Kirk, P.W.W. (eds) *Management Strategies for Phosphorus in the Environment.* Selper, London, pp. 180–5.

687 Moss, B., Balls, H., Irvine, K. & Stansfield, J. (1986) Restoration of two lowland lakes by isolation from nutrient rich water sources with and without removal of sediment. *J. Appl. Ecol.* **23**, 391–414.

688 Moss, B., Stansfield, J. & Irvine, K. (1991) Development of daphnid communities in diatom- and cyanophyte-dominated lakes and their relevance to lake restoration by biomanipulation. *J. Appl. Ecol.* **28**, 586–602.

689 Moss, B., Johnes, P. & Phillips, G. (1994a) August Thienemann and Loch Lomond – an approach to the design of a system for monitoring the state of north-temperate standing waters. *Hydrobiologia* **290**, 1–12.

690 Moss, B., McGowan, S. & Carvalho, L. (1994b) Determination of phytoplankton crops by top-down and bottom-up mechanisms in a group of English lakes, the West Midland Meres. *Limnol. Oceanogr.* **39**, 1020–30.

691 Moss, B., Beklioglu, M., Carvalho, L., Kilinc, S., Mcgowan, S. & Stephen, D. (1996a) Vertically challenged limnology: contrasts between deep and shallow lakes. *Hydrobiologia* **342/343**, 257–267.

692 Moss, B., Johnes, P. & Phillips, G. (1996b) The monitoring of ecological quality and the classification of standing waters in temperate regions: a review and proposal based on a worked scheme for British waters. *Biol. Rev.* **71**, 301–39.

693 Moss, B., Madgwick, J. & Phillips, G. (1996c) *A Guide to the Restoration of Nutrient-enriched Shallow Lakes.* Broads Authority, Environment Agency and EU Life Programme, Norwich.

694 Moss, B., Stansfield, J., Irvine, K., Perrow, M. & Phillips, G. (1996d) Progressive restoration of a shallow lake; a 12-year experiment in isolation, sediment removal and biomanipulation. *J. Appl. Ecol.* **33**, 71–86.

695 Moss, B., Johnes, P.J. & Phillips, G.L. (1997) New approaches to monitoring and classifying standing waters. In: Boon, P.J. & Howell, D.L. (eds) *Freshwater Quality: Defining the Indefinable 1.* Stationery Office, Edinburgh,, pp. 118–33.

696 Mudge, G.P. (1983) The incidence and significance of ingested lead pellet poisoning in British wildfowl. *Biol. Cons.* **27**, 333–72.

697 Mugidde, R. (1993) The increase in phytoplankton primary productivity and biomass in Lake Victoria (Uganda). *Verh. int. Verein. theor. angew. Limnol.* **25**, 846–9.

698 Müller, K. (1954) Investigations on the organic drift in North Swedish streams. *Rep. Inst. Freshwater Res. Drottingholm* **35**, 133–48.

699 Müller, K. (1965) Field experiments on periodicity of freshwater invertebrates. In: Aschoff, J. (ed.) *Circadian Clocks.* North-Holland, Amsterdam, pp. 314–17.

700 Müller, K. (1966) Zur Periodik von *Gammarus pulex.* *Oikos* **17**, 207–11.

701 Müller, K. (1974) Stream drift as a chronobiological phenomenon in running water ecosystems. *Ann. Rev. Ecol. Syst.* **5**, 309–23.

702 Munawar, M. & Munawar, I.F. (1996) *Phytoplankton Dynamics in the North American Great Lakes,* Vol. 1. *Lakes Ontario, Erie and St Clair.* SBS Publishing, Amsterdam.

703 Munk, W.H. & Riley, G.A. (1952) Absorption of nutrients by aquatic plants. *J. Mar. Res.* **11**, 215–40.

704 Munro, A.L.S. and Brock, R.S. (1968) Distinction between bacterial and algal utilization of soluble substances in the sea. *J. Gen. Microbiol.* **51**, 35–42.

705 Murphy, K.J., Hanbury, R.G. & Eaton, J.W. (1981) The ecological effects of 2-methylthiotriazine herbicides used for aquatic weed control in navigable canals. 1. Effects on aquatic flora and water chemistry *Arch. Hydrobiol.* **91**, 294–331.

706 Murphy, M.L., Heiftez, J., Johnson, S.W., Koski, K.V. & Thedinga, J.F. (1986) Effects of clear-cut logging with and without buffer strips on juvenile salmonids in Alaskan streams. *Can. J. Fish. Aquat. Sci.* **43**, 1521–33.

707 Murtaugh, P.A. (1981a) Size-selective predation on *Daphnia* by *Neomysis mercedis*. *Ecology* **62**, 894–900.

708 Murtaugh, P.A. (1981b) Selective predation by *Neomysis mercedis* in Lake Washington. *Limnol. Oceanogr.* **26**, 445–53.

709 Myers, N. (ed.) (1985) *The Gaia of Planet Management*. Pan, London.

710 Nalewajko, C. & Paul, P. (1985) Effects of manipulations of aluminium concentrations and pH on phosphate uptake and photosythesis of planktonic communities in two Precambrian shield lakes. *Can. J. Fish. Sci.* **42**, 1946–53.

711 National Rivers Authority (1990) *Toxic Blue Green Algae*. Water Quality Series 2, National Rivers Authority, HMSO, London.

712 National Rivers Authority (1995a) *The Mersey Estuary: a Report on Environmental Quality*. Water Quality Series 23, National Rivers Authority, HMSO, London.

713 National Rivers Authority (1995b) *Pesticides in the Aquatic Environment*. Water Quality Series 26, National Rivers Authority, HMSO, London.

714 National Rivers Authority (1996) *River Habitats in England and Wales: a National Overview*. River Habitat Survey Report 1, NRA, Bristol.

715 National Water Council (1981) *River Quality – the 1980 Survey and Future Outlook*. National Water Council, London.

716 Nature Conservancy Council (1984) *Nature Conservation in Great Britain*. Nature Conservancy Council, Shrewsbury.

717 Neary, B.P. & Leach, J.H. (1992) Mapping the potential spread of the zebra mussel (*Dreissena polymorpha*) in Ontario. *Can. J. Fish. Aquat. Sci.* **49**, 406–15.

718 Nelson, J.S. (1984) *Fishes of the World*, 2nd edn. Wiley, New York.

719 Newbold, C., Purseglove, J. & Holmes, N. (1983) *Nature Conservation and River Engineering*. Nature Conservancy Council, Shrewsbury.

720 Nicholls, K.H. & Hopkins, G.J. (1993) Recent changes in Lake Erie (North Shore) phytoplankton – cumulative impacts of phosphorus loading reductions and the zebra mussel introduction. *J. Gt. Lakes Res.* **19**, 637–47.

721 Noordwijk, M. van (1984) *Ecology Textbook for the Sudan*. Khartoum University Press, Khartoum.

722 North, A. (1994) Saddam drains life from Arab marshes. *Independent* 17 May, 13.

723 Novitski, R.P. (1978) Hydrologic characteristics of Wisconsin's wetlands and their influence on floods, stream flow and sediment. In: *Wetland Functions and Values: the State of our Understanding*. American Water Resources Association, Minneapolis, 323–337.

724 Nyholm, N.E.I. (1981) Evidence of involvement of aluminium in causation of defective formation of eggshells and of impaired breeding in wild passerine birds. *Environ. Res.* **26**, 363–71.

725 Nyholm, N.E.I. & Myhrberg, H.E. (1977) Severe eggshell defects and impaired reproductive capacity in small passerines in Swedish Lappland. *Oikos* **29**, 336–41.

726 Nystrom, P. & Strand, J.A. (1996) Grazing by a native and an exotic crayfish on aquatic macrophytes. *Freshwat. Biol.* **36**, 673–82.

727 Obeng, L.E. (1977) Should dams be built? The Volta Lake example. *Ambio* **6** 50.

728 O'Brien, W.J. & Vinyard, G.L. (1978) Polymorphine predation: the effect of invertebrate predation on the distribution of two varieties of *Daphnia carinata* in South India ponds. *Limnol. Oceanogr.* **23**, 452–60.

729 Ochumba, P.B.O. (1990) Massive fish kills within the Nyanza Gulf of Lake Victoria. *Hydrobiologia* **208**, 93–9.

730 Odum, H.T. (1956) Primary production in flowing waters. *Limnol. Oceanogr.* **1**, 102–17.

731 Office of Technology Assessment (1984) *Wetlands – their Use and Regulation*. Publication OTA-0-207, US Congress.

732 Oguta-Ohwayo, R. (1992) The purpose, costs and benefits of fish introductions: With specific reference to the Great Lakes of Africa. *Mitt. int. Verein. theor. angew. Limnol.* **23**, 37–44.

733 Ohwada, K. & Taga, N. (1973) Seasonal cycles of vitamin B12, thiamine and biotin in Lake Sagami: patterns in their distribution and ecological significance. *Int. Rev. ges. Hydrobiol.* **58**, 851–71.

734 O'Keefe, J.H. & Davies, B.R. (1991) Conservation and management of the rivers of the Kruger National Park: suggested methods for calculating instream flow needs. *Aquat. Cons.* **1**, 55–72.

735 Olsen, S. (1964) Phosphate equilibrium between reduced sediments and water: laboratory experiments with radioactive phosphorus. *Verh. int. Verein. theor. angew. Limnol.* **15**, 333–41.

736 Omernik, J.M. (1976) *The Influence of Land Use on Stream Nutrient Levels*. Report EPA-600/3-76-014, US Environmental Protection Agency, Corvallis, Oregon.

737 Ormerod, S.J. (1995) Modelling biological responses, present and future. In: Critical Loads Advisory Group (ed.) *Critical Loads of Acid Deposition for United Kingdom Freshwaters*. Institute of Terrestrial Ecology, Penicuik, pp. 47–55.

738 Ormerod, S.V., Tyler, S.J. & Lewis, J.M.S. (1985) Is the breeding distribution of dippers influenced by stream acidity? *Bird Study* **32**, 32–9.

739 Ormerod, S.V., Allinson, N., Hudson, D. & Tyler, S.J. (1986) The distribution of breeding dippers (*Cinclus cinclus* (L.)) Aves in relation to

stream acidity in upland Wales. *Freshwat. Biol.* **16**, 501–8.

740 Ormerod, S.V., Boole, P., McCahon, C.P., Weatherly, N.S., Pascoe, D. & Edwards, R.W. (1987) Short term experimental acidification of a Welsh stream: comparing the biological effects of hydrogen ions and aluminium. *Freshwat. Biol.* **17**, 341–56.

741 Orth, R.J. & van Montfrans, J. (1984) Epiphyte–sea grass relationships with an emphasis on the role of micrograzing: a review. *Aquat. Bot.* **18**, 43–69.

742 Ortloff, C.R. & Kolata, A.L. (1993) Climate and collapse: agroecological perspectives on the decline of the Tiwanaku state. *J. Archaeol. Sci.* **20**, 195–221.

743 Osborne, C. (1982) *The World Theatre of Wagner.* Phaidon, Oxford.

744 Osborne, P.L. (1980) Prediction of phosphorus and nitrogen concentrations in lakes from both internal and external loading rates. *Hydrobiologia* **69**, 229–33.

745 Osborne, P.L. & McLachlan, A.J. (1985) The effect of tadpoles on alga growth in temporary rain-filled rock pools. *Freshwat. Biol.* **15**, 77–88.

746 Osborne, P.L. & Moss, B. (1977) Palaeolimnology and trends in the phosphorus and iron budgets of an old man made lake, Barton Broad, Norfolk. *Freshwat. Biol.* **7**, 213–34.

747 Osborne, P.L. & Phillips, G.L. (1978) Evidence for nutrient release from the sediments of two shallow and productive lakes. *Verh. int. Verein. theor. angew. Limnol.* **20**, 654–8.

748 Ostrofsky, M.L. & Zettler, E.R. (1986) Chemical defences in aquatic plants. *J. Ecol.* **73**, 279–88.

749 Pace, M.L. & Cole, J.J. (1994) Primary and bacterial production in lakes: are they coupled over depth? *J. Plankton Res.* **16**, 661–72.

750 Paerl, H.W. & Ustach, J.F. (1982) Blue-green algae scums: an explanation for their occurence during freshwater blooms. *Limnol. Oceanogr.* **27**, 212–17.

751 Paffenhofer, G.-A., Strickler, J.R. & Alcaroz, M. (1982) Suspension-feeding by herbivorous calanoid copepods: a cinematographic study. *Mar. Biol.* **67**, 193–9.

752 Page, P., Ouellet, M., Hillaire-Marcel, C. & Dickman, M. (1984) Isotopic analysis (^{18}O ^{13}C ^{14}C) of two meromictic lakes in the Canadian Arctic archipelago. *Limnol. Oceanogr.* **29**, 564–73.

753 Pain, B.F. (1991) Ammonia volatilisation from livestock buildings, slurry stores and grassland. In: *International Conference on N, P and Organic Matter: Contributions by Invited International Experts.* Ministry of the Environment, Denmark, Copenhagen, pp. 17–27.

754 Pain, S. (1987) Australian invader threatens Britain's waterways. *N. Sci.* 23 July, 26.

755 Paloheimo, J.E. (1974) Calculation of instantaneous birth rate. *Limnol. Oceanogr.* **19**, 692–4.

756 Parejko, K. & Dodson, S. (1990) Progress towards characterization of a predator/prey kairomone: *Daphnia pulex* and *Chaoborus americanus. Hydrobiologia* **198**, 51–9.

757 Parker, J.I. & Edgington, D.N. (1976) Concentration of diatom frustules in Lake Michigan sediment cores. *Limnol. Oceanogr.* **21**, 887–93.

758 Paton, A. (1976) Dams and their interfaces. *Proc. Roy. Soc. London (A)* **351**, 1–17.

759 Pearce, F. (1995a) The biggest dam in the world. *N. Sci.* 28 Jan., 25–9.

760 Pearce, F. (1995b) Poisoned waters. *N. Sci.* 21 Oct. 1995, 29–33.

761 Penhale, P.A. & Thayer, G.W. (1980) Uptake and transfer of carbon and phosphorus by eelgrass (*Zostera marina* L.) and its epiphytes. *J. Exp. Mar. Biol. Ecol.* **42**, 113–23.

762 Pennington, W. (1984) Long-term natural acidification of upland sites in Cumbria: evidence from post-glacial lake sediments. *Ann. Rep. Freshwat. Biol. Assoc.* **52**, 28–46.

763 Pennington, W. & Lishman, J.P. (1971) Iodine in lake sediments in North England and Scotland. *Biol. Rev.* **46**, 279–313.

764 Peters, R.H. (1984) Methods for the study of feeding, grazing and assimilation by zooplankton. In: Downing, J.A. & Rigler, F.H. (eds) *A Manual on Methods for the Assessment of Secondary Productivity in Freshwaters.* Blackwell Scientific Publications, Oxford, pp. 336–412.

765 Peters, R.H. (1986) The role of prediction in limnology. *Limnol. Oceanogr.* **31**, 1143–59.

766 Peters, R.H. & Rigler, F.H. (1973) Phosphorus release by *Daphnia. Limnol. Oceanogr.* **18**, 821–39.

767 Petersen, R.C. (1992) The RCE: riparian, channel, and environmental inventory for small streams in the agricultural landscape. *Freshwat. Biol.* **27**, 295–306.

768 Peterson, R.H., Daye, P.G. & Metcalfe, J.L. (1980) Inhibition of Alantic salmon (*Salmo salar*) hatching at low pH. *Can. J. Fish. Aquat. Sci.* **37**, 770–4.

769 Petr, T. (1969) Fish population changes in Volta Lake over the period January 1965–September 1966. In: Obeng, L.E. (ed.) *Man-made Lakes, the Accra Symposium.* Ghana University Press, Accra.

770 Phillips, G.L., Eminson, D.F. & Moss, B. (1978) A mechanism to account for macrophyte decline in progressively eutrophicated freshwaters. *Aquat. Bot.* **4**, 103–26.

771 Piddocke, S. (1965). The potlack system of the southern Kwakiutl – a new perspective. *Southwestern J. Anthropol.* **21**, 244–64.

772 Pigott, C.D. & Pigott, M.E. (1963) Late-glacial and post-glacial deposits at Malham, Yorkshire. *N. Phytol.* **62**, 317–34.

773 Pigott, M.E. & Pigott, C.D. (1959) Stratigraphy and pollen analysis of Malham tarn and Tarn Moss. *Field Studies* **1**, 1–17.

774 Pijanowska, J. (1990) Cyclomorphosis in

Daphnia: an adaptation to avoid invertebrate predation. *Hydrobiologia* **198**, 41–50.

775 Pister, E.P. (1985) Desert pupfishes: reflections on reality, desirability and conscience. *Environ. Biol Fish.* **12**, 3–12.

776 Pitcher, T.J. & Hart, J.B. (1982) *Fisheries Ecology*. Croom Helm, London.

777 Pollingher, U. & Berman, T. (1991) Phytoplankton composition and activity in lakes of the warm belt. *Verh. int. Verein. theor. angew. Limnol.* **24**, 1230–4.

778 Por, F.D. (1995) *The Pantanal of Mato Grosso (Brazil)*. Kluwer, Dordrecht.

779 Porter, K.G. (1973) Selective grazing and differential digestion of algae by zooplankton. *Nature* **244**, 179–80.

780 Porter, K.G. (1976) Enhancement of algal growth and productivity by grazing zooplankton. *Science* **192**, 1332–4.

781 Porter, K.G. & McDonough, R. (1984) The energetic cost of response to bluegreen algal filaments by cladocerans. *Limnol. Oceanogr.* **29**, 365–9.

782 Porter, K.G. & Orcutt, J.D., Jr (1980) Nutritional adequacy, manageability and toxicity as factors that determine the food quality of green and blue-green algae for *Daphnia*. In: Kerfoot, W.C. (ed.) *Evolution and Ecology of Zooplankton Communities*. University of New England Press, Hanover, pp. 258–81.

783 Porter, K.G., Pace, M.L. & Battey, J.F. (1979) Ciliate protozoans as links in freshwater planktonic food chains. *Nature* **277**, 563–5.

784 Porter, K.G. Feig, T.S. & Vetter, E.F. (1983) Morphology, flow regimes, and filtering rates of *Daphnia*, *Ceriodaphnia* and *Bosmina* fed natural bacteria. *Oecologia* **58**, 156–63.

785 Potts, W.T.W. (1954) The energetics of osmotic regulation in brackish- and freshwater animals. *J. Exp. Biol.* **31**, 618–30.

786 Potts, W.T.W. & Parry, G. (1964) *Osmotic and Ionic Regulations in Animals*. Pergamon, Oxford.

787 Pounds, J.A. & Crump, M.L. (1994) Amphibian declines and climate disturbance: the case of the golden toad and the harlequin frog. *Cons. Biol.* **8**, 72–85.

788 Power, M.E. (1990) Effects of fish in river food webs. *Science* **250**, 811–14.

789 Prepas, E. & Rigler, F.H. (1978) The enigma of *Daphnia* death rates *Limnol. Oceanogr.* **23**, 970–88.

790 Proctor, H.C. & Pritchard, G. (1990) Prey detection by the water mite *Unionicola crassipes* (Acari: Unionicolidae). *Freshwat. Biol.* **23**, 271–80.

791 Prowse, G.A. (1959) Relationships between epiphytic algal species and their marcrophyte hosts. *Nature* **183** 1204–5.

792 Pullen, R. (1985) Tilapias: 'everyman's fish. *Biologist* **32**, 84–8.

793 Quinn, T.P. & Adams, D.J. (1996) Environmental changes affecting the migratory timing of American shad and sockeye salmon. *Ecology* **77**, 115–62.

794 Qureshi, A.A. & Patel, J. (1976) Adenosine triphosphate (ATP) levels in microbial cultures and a review of the ATP biomass estimation technique. *Environ. Can. Sci. Ser.* **63**, 1–33.

795 Raisewell, R., Brimblecombe, P., Dent, D.L. & Liss, P.S. (1980) *Environmental Chemistry*. Arnold, London.

796 Randall, R.G., Kelso, J.R.M. & Minns, C.K. (1995) Fish production in freshwaters: are rivers more productive than lakes? *Can. J. Fish. Aquat. Sci.* **52**, 631–43.

797 Rashid, M.A. (1985) *Geochemistry of Marine Humic Compounds*. Springer-Verlag, Heidelberg, Germany.

798 Raskin, I. & Kende, H. (1985) Mechanism of aeration in rice. *Science* **228**, 327–9.

799 Rasmussen, J.B. (1984) The life-history, distribution, and production of *Chironomus riparius* and *Glyptotendipes paripes* in a prairie pond. *Hydrobiologia* **119**, 65–72.

800 Raven, J.A. (1970) Exogenous inorganic carbon sources in plant photosynthesis. *Biol. Rev.* **45**, 167–202.

801 Raven, J.A. & Glidewell S.M. (1975) Photosynthesis, respiration and growth in the shade alga, *Hydrodictyon africanum*. *Photosynthetica* **9**, 361–71.

802 Reader, J. (1988) *Man on Earth*. Collins, London.

803 Redfield, A.C. (1934) On the proportion of organic derivatives in sea water and their relation to the composition of plankton. In: *James Johnstone Memorial Volume*. Liverpool University Press, Liverpool, pp. 176–92.

804 Reinertsen, H., Jensen, A., Langeland, A. & Olsen, Y. (1986) Algal competition for phosphorus: the influence of zooplankton and fish. *Can. J. Fish. Aquat. Sci.* **43**, 1135–41.

805 Renberg, I. (1981) Improved methods for sampling, photographing and varve-counting of varved lake sediments. *Boreas* **10**, 255–8.

806 Renberg, I. & Hellberg, T. (1982) The pH history of lakes in south-western Sweden, as calculated from the subfossi diatom flora of the sediments. *Ambio* **11**, 30–3.

807 Reynolds, C.S. (1995) River plankton: the paradigm regained. In: Harper, D.M. & Ferguson, A.J.D. (eds) *The Ecological Basis for River Management*. Wiley, Chichester, pp. 161–74.

808 Reynolds, C.S. (1996) The 1996 Founders' Lecture: Potamoplankters do it on the side. *Eur. J. Phycol.* **31**, 111–16.

809 Reynolds, C.S. & Butterwick, C. (1979) Algal bioassay of unfertilized and artificially fertilized lake water, maintained in Lund tubes. *Arch. Hydrobiol. Suppl.* **56**, 166–83.

810 Reynolds, C.S. & Walsby, A.E. (1975) Water-blooms. *Biol. Rev.* **50**, 437–81.

811 Reynolds, C.S., Wiseman, S.W. & Clarke, M.J.O. (1984) Growth- and loss rate responses of

phytoplankton to intermittent artificial mixing and their potential application to the control of planktonic algal biomass. *J. App. Ecol.* **21**, 11–39.

812 Reynolds, C.S., Harris, G.P. & Gouldney, D.N. (1985) Comparison of carbon-specific growth rates and rates of cellular increase of phytoplankton in large limnetic enclosures. *J. Plankton Res.* **7**, 791–820.

813 Reynoldson, T.B. (1966) The distribution and abundance of lake-dwelling triclads – towards a hypothesis. *Adv. Ecol. Res.* **3**, 1–71.

814 Reynoldson, T.B. (1983) The population biology of Turbellaria with special reference to the freshwater triclads of the British Isles. *Adv. Ecol. Res.* **13**, 235–326.

815 Ribbink, A.J. (1987) African lakes and their fishes: conservation scenarios and suggestions. *Environ. Biol. Fish.* **19**, 3–26.

816 Richardson, J.L. & Richardson, A.E. (1972) History of an African rift lake and its climatic implications. *Ecol. Monogr.* **42**, 499–534.

817 Richardson, J.S. (1992) Food, microhabitat, or both? Macroinvertebrate use of leaf accumulations in a montane stream. *Freshwat. Biol.* **27**, 169–76.

818 Richardson, K., Griffiths, H., Reed, M.L., Raven, J.A. & Griffths, N.M. (1984) Inorganic carbon assimilation in the isoetids, *Isoetes lacustris* L. and *Lobelia dortmanna* L. *Oecologia* **61**, 115–21.

819 Rigler, F.H. (1964) The phosphorus fractions and the turnover time of inorganic phosphorus in different types of lakes. *Limnol. Oceanogr.* **9**, 511–18.

820 Rigler, F.H. & Downing, J.A. (1984) The calculation of secondary productivity. In: *A Manual on Methods for the Assessment of Secondary Productivity in Freshwaters.* Blackwell Scientific Publications, Oxford, pp. 19–58.

821 Rimes, C.A., Farmer, A.M. & Howell, D. (1994) A survey of the threat of surface water acidification to the nature conservation interest of freshwaters on Sites of Special Scientific Interest in Britain. *Aquat. Cons. Mar. Freshwat. Ecosyst.* **4**, 31–44.

822 Ritter, J.A. & Porter, T.R. (1980) Issues and promises for Atlantic salmon management in Canada. In: Went, A.E.J. (ed.) *Atlantic Salmon: its Future.* Fishing News Books, Farnham, pp. 108–27.

823 Robarts, R. (1985) Dam troubles. *Scientiae* **26**, 17–23.

824 Roberts, J. & Sainty, G. (1996) *Listening to the Lachlan.* Sainty & Associates, Potts Point, New South Wales, Australia.

825 Robertson, W.B., Jr (1955) *A Survey of the Effects of Fire in Everglades National Park.* US National Park Service, Homestead, Florida, USA.

826 Robertson, W.B., Jr & Kushlan, J.A. (1974) The South Florida avifauna. In: Gleason, P.J. (ed.) *Environments of South Florida: Present and Past.* Memoir 2, Miami Geological Society, Miami, pp. 414–52.

827 Rogers, K.H. & Breen, C.M. (1981) Effects of epiphyton on *Potamogeton crispus* L. leaves. *Microbial Ecol.* **7**, 351–63.

828 Rogers, F.E.J., Rogers, K.H. & Buzer, J.S. (1985) *Wetlands for Wastewater Treatment with Special Reference to Municipal Wastewaters.* Witwatersrand University Press, Johannesburg.

829 Rosemond, A.D., Reice, S.R., Elwood, J.W. & Mulholland, P.J. (1992) The effects of stream acidity on benthic invertebrate communities in the south-eastern United States. *Freshwat. Biol.* **27**, 193–210.

830 Ross-Craig, S. (1948) *Drawings of British Plants*, Part I. *Ranunculaceae.* Bell & Sons, London.

831 Round, F.E. (1961) The diatoms of a core from Esthwaite Water. *N. Phytol.* **60**, 43–59.

832 Rounick, J.S. & Winterbourn, M.J. (1983) The formation, structure, and utilization of stone surface organic layers in two New Zealand streams. *Freshwat. Biol.* **13**, 57–72.

833 Rowland, S.J. (1989) Aspects of the history and fishery of the Murray cod, *Macullochella peelii* (Mitchell) (Percichthyidae). *Proc. Linn. Soc. N. South Wales* **111**, 201–13.

834 Royal Commission on Environmental Pollution (1994) *Eighteenth Report: Transport and the Environment.* Cmnd. 2674, HMSO, London.

835 Royal Society (1983) *The Nitrogen Cycle of the United Kingdom.* Royal Society, London.

836 Rubec, C.D.A. (1994) Canada's federal policy on wetland conservation: a global model. In: Mitsch, W.J. (ed.) *Global Wetlands: Old World and New.* Elsevier, Amsterdam, pp. 909–18.

837 Ruby, S.M., Aczel, J. & Craig, G.R. (1977) The effects of depressed pH on oogenesis in flagfish (*Jordanella floridae*). *Wat. Res.* **11**, 757–62.

838 Russell, E.S. (1931) Some theoretical considerations on the 'overfishing problem'. *J. Cons. Inf. Explor. Mer.* **6**, 3–20.

839 Ruttner, F. (1953) *Fundamentals of Limnology*, transl. D.G. Frey & F.E.J. Fry. University of Toronto Press, Toronto.

840 Ryder, R.A. (1978) Fish yield assessment of large lakes and reservoirs – a prelude to management. In: Gerking, S.D. (ed.) *Ecology of Freshwater Fish Production.* Blackwell Scientific Publication, Oxford, pp. 403–23.

841 Ryder, R.A. & Henderson, H.F. (1975) Estimates of potential fish yield for the Nasser Reservoir, Arab Republic of Egypt. *J. Fish. Res. Bd. Can.* **32**, 2137–51.

842 Ryder, R.A. & Pesendorfer, J. (1989) Large rivers are more than flowing lakes: a comparative review. In: Dodge, D.P. (ed.) *Proceedings of the International Large Rivers Symposium. Can Spec. Publ. Fish. Aquat. Sci.* **106**, 65–85.

843 Ryder, R.A., Kerr, S.R., Loftus, K.H. & Regier, H.A. (1974) The morphoedaphic index, a fish yield estimator – review and evaluation. *J. Fish. Res. Bd. Can.* **31**, 663–88.

844 Rzoska, J. (1976) A controversy reviewed. *Nature* **261**, 444–5.

845 Saether, O.A. (1979) Chironomid communities as water quality indicators. *Holarct. Ecol.* **2**, 65–74.

846 Salonen, K., Jones, R.I. & Arvola, L. (1984) Hypolimnetic phosphorus retrieval by diel vertical migrations of lake phytoplankton. *Freshwat. Biol.* **14**, 431–8.

847 Sand-Jensen, K. (1978) Metabolic adaptation and vertical zonation of *Littorella uniflora* (L.) Ashers and *Isoetes lacustris* L. *Aquat. Bot.* **4**, 1–10.

848 Sand-Jensen, K. & Borum, J. (1984) Epiphyte shading and its effect on photosynthesis and diel metabolism of *Lobelia dortmanna* L. during the spring bloom in a Danish lake. *Aquat. Bot.* **20**, 109–19.

849 Sand-Jensen, K. & Prahl. C. (1982) Oxygen exchange with the lacunae and access leaves and roots of the submerged vascular macrophyte *Lobelia dortmanna* L. *New. Phytol.* **91**, 103–20.

850 Sand-Jensen, K. & Söndergaard, M. (1981) Phytoplankton and epiphyte development and their shading effect on submerged macrophytes in lakes of different nutrient status. *Int. Rev. ges. Hydrobiol.* **66**, 529–52.

851 Sand-Jensen, K., Prahl, C. & Stockholm, H. (1982) Oxygen release from roots of submerged aquatic macrophytes. *Oikos* **38**, 349–54.

852 Sand-Jensen, K., Pedersen, M.F. & Nielsen, S.L. (1992) Photosynthetic use of inorganic carbon among primary and secondary water plants in streams. *Freshwat. Biol.* **27**, 283–93.

853 Sandlund, O.T., Gunnarsson, K., Jonasson, P.M. *et al.* (1992) The arctic charr *Salvelinus alpinus* in Thingvallavatn. *Oikos* **64**, 305–51.

854 Sanger, J.E. & Gorham, E. (1970) The diversity of pigments in lake sediments and its ecological significance. *Limnol. Oceanogr.* **15**, 59–69.

855 Sanni, S. & Waervagen, S.B. (1990) Oligotrophication as a result of planktivorous fish removal with rotenone in the small, eutrophic Lake Mosvatn, Norway. *Hydrobiologia* **200/201**, 263–74.

856 Saunders, J.R. & Saunders, V.A. (1997) The impact of molecular biology on assessment of water quality: advantages and limitations of current techniques. In: Sutcliffe, D.W. (ed.) *The Microbiological Quality of Water.* Freshwater Biological Association, Ambleside, pp. 11–18.

857 Scavia, D., Fahnenstiel, G.L., Evans, M.S., Jude, D.J. & Lehman, J.T. (1986a) Influence of salmonid predation and weather on long-term water quality trends in Lake Michigan. *Can. J. Fish. Aquat. Sci.* **43**, 435–43.

858 Scavia, D., Laird, G.A. & Fahnenstiel, G.L. (1986b) Production of planktonic bacteria in Lake Michigan. *Limnol. Oceanogr.* **31**, 612–26.

859 Scheffer, M., Hosper, S.H., Meijer, M.L., Moss, B. & Jeppesen, E. (1993) Alternative equilibria in shallow lakes. *Trends Ecol. Evol.* **8**, 275–9.

860 Scheider, W. & Wallis, P. (1973) An alternate method of calculating the population density of monsters in Loch Ness. *Limnol, Oceanogr.* **18**, 343.

861 Schelske, C.L. & Hodell, D.A. (1995) Using carbon isotopes of bulk sedimentary organic matter to reconstruct the history of nutrient loading and eutrophication in Lake Erie. *Limnol. Oceanogr.* **40**, 918–29.

862 Schierup, H.H. (1978) Biomass and primary production in *Phragmites communis* Trin. swamp in North Jutland, Denmark. *Verh. int. Verein. theor. angew. Limnol.* **20**, 93–9.

863 Schindler, D.W. (1974) Eutrophication and recovery in experimental lakes: implications for lake management. *Science* **184**, 897–8.

864 Schindler, D.W. (1977) The evolution of phosphorus limitation in lakes. *Science* **195**, 260–2.

865 Schindler, D.W. (1978) Factors regulating phytoplankton production and standing crop in the world's freshwaters. *Limnol. Oceanogr.* **23**, 478–86.

866 Schindler, D.W. & Fee, E.J. (1974) Experimental lakes area: whole lake experiments in eutrophication. *J. Fish. Res. Bd. Can.* **31**, 937–53.

867 Schindler, D.W., Welch, H.E., Kalff, J., Brunskill, G.J. & Kritsch, N. (1974) Physical and chemical limnology of Char Lake, Cornwallis Island (75°N lat.). *J. Fish. Res. Bd. Can.* **31**, 585–607.

868 Schindler, D.W., Curtis, P.J., Parker, B.R. & Stainton, M.P. (1996) Consequences of climate warming and lake acidification for UV-B penetration in North American boreal lakes. *Nature* **379**, 705–8.

869 Schindler, J.E. (1971) Food quality and zooplankton nutrition. *J. Anim. Ecol.* **40**, 589–95.

870 Schloesser, D.W., Nalepa, T.F. & Mackie, G.L. (1996) Zebra mussel infestation of unionid bivalves (Unionidae) in North America. *Am. Zool.* **36**, 300–10.

871 Schmidt, J.A. & Andren, A.W. (1984) Deposition of airborne metals into the Great Lakes: an evaluation of past and present estimates. In: Nriagu, J.O. & Simmons, M.S. (eds) *Toxic Contaminants in the Great Lakes.* Wiley, New York, pp. 81–104.

872 Schmidt-Nielsen, K. (1983) *Animal Physiology.* Cambridge University Press, Cambridge.

873 Schmitt, M.R. & Adams, M.S. (1981) Dependence of rates of apparent photosynthesis on tissue phosphorus concentrations in *Myriophyllum spicatum* L. *Aquat. Bot.* **11**, 379–87.

874 Schneider, S.H. & Boston, P. (eds) *Scientists on Gaia.* MIT Press, Cambridge, Massachusetts.

875 Schoenberg, S.A. & Carlson, R.E. (1984) Direct and indirect effects of zooplankton grazing on phytoplankton in a hypereutrophic lake. *Oikos* **42**, 291–302.

876 Schriver, P., Bøgestrand, J., Jeppesen, E. & Søndergaard, M. (1995) Impact of submerged

macrophytes on the interactions between fish, zooplankton and phytoplankton: large scale enclosure experiments in a shallow lake. *Freshwat. Biol.* **33**, 255–70.

877 Scullion, J., Parish, C.A., Morgan, N. & Edwards, R.W. (1982) Comparison of benthic macroinvertebrate fauna and substratum composition in riffles and pools in the impounded River Elan and the unregulated R. Wye, mid-Wales. *Freshwat. Biol.* **12**, 579–96.

878 Sculthorpe, C.D. (1967) *The Biology of Aquatic Vascular Plants*. Arnold, London.

879 Sear, D.A. (1994) River restoration and geomorphology. *Aquat. Cons. Mar. Freshwat. Ecosyst.* **4**, 169–78.

880 Setaro, F.V. & Melack, J.M. (1984) Responses of phytoplankton to experimental enrichment in an Amazon floodplain lake. *Limnol. Oceanogr.* **29**, 972–84.

881 Sevalrud, I.H., Muniz, I.P. & Kalvenes, S. (1980) Loss of fish populations in southern Norway: dynamics and magnitude of the problem. In: Drablos, D. & Tollan, A. (eds) *Proceedings of an International Conference on the Ecological Impact of Acid Precipitation*. Sandefiord, Norway, pp. 350–1.

882 Shapiro, J. (1958) The freeze corer – a new sampler for lake sediments. *Ecology* **39**, 758.

883 Shapiro, J. (1980a) The need for more biology in lake restoration. In: *Lake Restoration: Proceedings of a National Conference, August 1978*. 444/5-79-001, USEPA, Minneapolis, pp. 263–285.

884 Shapiro, J. (1980b) The importance of trophic-level interactions to the abundance and species composition of algae in lakes. In: Barica, J. & Mur, L.R. (eds) *Hypertrophic Ecosystems*. Junk, The Hague.

885 Shapiro, J. (1990) Current beliefs regarding dominance by blue-greens: the case for the importance of CO_2 and pH. *Verh. int. Verein. theor. angew. Limnol.* **24**, 38–54.

886 Shapiro, J. & Wright, D.I. (1984) Lake restoration by biomanipulation: Round Lake, Minnesota, the first two years. *Freshwat. Biol.* **14**, 371–83.

887 Shapiro, J., Lamarra, V. & Lynch, M. (1975) Biomanipulation: an ecosystem approach to lake restoration. In: Brezonik, P.L. & Fox, J.L. (eds) *Water Quality Management through Biological Control*. Report ENV-07-75-1, University of Florida, Gainsville.

888 Shearer, K.D. & Mulley, J.C. (1978) The introduction and distribution of carp, *Cyprinus carpio* Linnaeus, in Australia. *Austr. J. Mar. Freshwat. Res.* **29**, 551–63.

889 Sheldon, R.W. & Kerr, S.R. (1972) The population density of monsters in Loch Ness. *Limnol. Oceanogr.* **17**, 796–8.

890 Sheldon, R.W., Prakash, A. & Sutcliffe, W.H., Jr (1972) The size distribution of particles in the ocean. *Limnol. Oceanogr.* **17**, 327–40.

891 Shoard, M. (1980) *The Theft of the Countryside*. Temple Smith, London.

892 Short, F.T., Burdick, D.M. & Kaldy, J.E., III (1995) Mesocosm experiments quantify the effects of eutrophication on eelgrass, *Zostera marina*. *Limnol. Oceanogr.* **40**, 740–9.

893 Sieburth, J.McN., Smetacek, V. & Lenz, J. (1978) Pelagic ecosystem structure: heterotophic compartments of the plankton and their relationship to plankton size fractionation. *Limnol. Oceanogr.* **23**, 1256–63.

894 Sikora, L.J. & Keeney, D.R. (1983) Further aspects of soil chemistry under anaerobic conditions. In: Gore, A.J.P. (ed.) *Mires: Swamp, Bog, Fen and Moor: General Studies*. Elsevier, Amsterdam, pp. 247–56.

895 Simpson, P.S. & Eaton, J.W. (1986) Comparative studies of the photosynthesis of the submerged macrophyte, *Elodea canadensis* and the filamentous algae *Cladophora glomerata* and *Spirogyra* sp. *Aquat. Bot.* **14**, 1–12.

896 Sklar, L. & Williams, P. (1992) One dozen problems dam builders can't solve. *World Rivers Rev.* May/June, 8–9.

897 Small, J.W., Jr, Richard, D.I. & Osborne, J.A. (1985) The effects of vegetation removal by grass carp and herbicides on water chemistry of four Florida lakes. *Freshwat. Biol.* **15**, 587–96.

898 Smid, P. (1975) Evaporation from a reed swamp. *J. Ecol.* **63**, 299–309.

899 Smith, A.M. & apRees, T. (1979) Pathways of carbohydrate fermentation in the roots of marsh plants. *Planta* **146**, 327–34.

900 Smith, S.I. (1972a) Factors of ecologic succession in oligotrophic fish communities of the Laurentian Great Lakes. *J. Fish. Res. Bd Can.* **29**, 717–30.

901 Smith, S.I. (1972b) The future of salmonid communities in the Laurentian Great Lakes. *J. Fish. Res. Bd Can.* **29**, 951–7.

902 Smith, V.H. (1983) Low nitrogen to phosphorus ratios favour dominance by blue-green algae in lake phytoplankton. *Science* **221**, 669–71.

903 Smith, V.H. & Wallsten, M. (1986) Prediction of emergent and floating leaved macrophyte cover in central Swedish lakes. *Can. J. Fish. Aquat. Sci.* **43**, 2519–23.

904 Smyly, W.J.P. (1979) Population dynamics of *Daphnia hyalina* Leydig (Crustacea: Cladocera) in a productive and an unproductive lake in the English Lake District. *Hydrobiologia* **64**, 269–78.

905 Sollas, W.J. (1884) On the origin of freshwater faunas: a study in evolution. *Sci. Trans. Roy. Dublin Soc. (Ser. 2)* **3**, 87–118.

906 Sommaruga-Wögrath, S., Koinig, K.A., Schmidt, R., Sommaruga, R., Tessadri, R. & Psenner, R. (1997) Temperature effects on the acidity of remote alpine lakes. *Nature* **387**, 64–7.

907 Sommer, U. & Gliwicz, Z.M. (1986) Long range vertical migration of *Volvox* in tropical lake Cahora Bassa (Mozambique). *Limnol. Oceanogr.* **31**, 650–3.

908 Sommer, U. & Stabel, H.H. (1983) Silicon consumption and population density changes of

dominant planktonic diatoms in Lake Constance. *J. Ecol.* **73**, 119–30.

909 Sozska, G.P. (1975) Ecological relations between invertebrates and submerged macrophytes in the lake littoral. *Ekol. Polsk.* **23**, 593–615.

910 Spence, D.H.N. (1964) The macrophytic vegetation of lochs, swamps and associated fens. In: Burnett, J.H. (ed.) *The Vegetation of Scotland.* Oliver & Boyd, Edinburgh.

911 Spence, D.H.N. (1976) Light and plant response in freshwater. In: Evans, G.C., Bainbridge, R. & Rackham, O. (eds) *Light as an Ecological Factor II.* Blackwell Scientific Publications, Oxford, pp. 49–59.

912 Spence, D.H.N. (1982) The zonation of plants in freshwater lakes. *Adv. Ecol. Res.* **12**, 37–125.

913 Spence, D.H.N. & Chrystal, J. (1970a) Photosynthesis and zonation of freshwater macrophytes. I. Depth distribution and shade tolerance. *N. Phytol.* **69**, 205–15.

914 Spence, D.H.N. & Chrystal, J. (1970b) Photosynthesis, zonation of freshwater macrophytes. II. Adaptability of species of deep and shallow waters. *N. Phytol.* **69**, 217–27.

915 Spence, D.H.N., Milburn, T.R., Nalawula-Senyimba, M. & Roberts, E. (1971) Fruit biology and germination of two tropical *Potamogeton* species. *New. Phytol.* **70**, 197–212.

916 Stark, F. & Werner, H. (1976) Natural history and management of Everglades National Park. In: Nteta, D.N. (ed.) *Proceedings of the Symposium on the Okavango Delta and its Future Utilization.* Botswana Society, Gabarone, pp. 263–75.

917 Starkweather, P.L. & Bogdan, K.G. (1990) Detrital feeding in natural zooplankton communities: discrimination between live and dead algal foods. *Hydrobiologia* **73**, 83–5.

918 Steedman, R.J. & Anderson, N.H. (1985) Life history and ecological role of the xylophagous aquatic beetle, *Lara avara* LeConte (Dryopoidea: Elminidae). *Freshwat. Biol.* **15**, 535–46.

919 Steel, J.A. (1972) The application of fundamental limnological research in water supply system design and management. *Symp. Zool. Soc. London* **29**, 41–67.

920 Steeman Nielsen, E. (1952) The use of radioactive carbon (C14) for measuring organic production in the sea. *J. Cons. Int. Explor. Mer* **18**, 117–40.

921 Steinhorn, I. (1985) The disappearance of the long term meromictic stratification of the Dead Sea. *Limnol. Oceanogr.* **30**, 451–72.

922 Steinhorn, I., Assaf, G., Gat, J.R. *et al.* (1979) The Dead Sea: deepening of the mixolimnion signifies the overture to overturn of the water column. *Science* **206**, 55–7.

923 Stenson, J.A.E. (1985) Biotic structures and relations in the acidified Lake Gordsjön system – a synthesis. *Ecol. Bull.* **37**, 319–26.

924 Stevenson, A.C., Birks, H.J.B., Flower, R.J. & Battarbee, R.W. (1988) Diatom-based pH

reconstruction of lake acidification using Canonical Correspondence Analysis. *Ambio* **18**, 228–33.

925 Stewart, W.D.P. & Lex, M. (1970) Nitrogenase activity in blue-green alga, *Plectonema boryanum* strain 594. *Arch. Mikrobiol.* **73**, 250–60.

926 Stich, H.B. & Lampert, W. (1981) Predator evasion as an explanation of diurnal vertical migration of zooplankton. *Nature* **293**, 396–8.

927 Stockner, J.G. (1968) Algal growth and primary productivity in a thermal stream. *J. Fish. Res. Bd Can.* **25**, 2037–58.

928 Stockner, J.G. (1988) Phototrophic picoplankton: an overview from marine and freshwater systems. *Limnol. Oceanogr.* **33**, 765–75.

929 Stockner, J.G. & Antia, N.J. (1986) Algal picoplankton from marine and freshwater ecosystems: a multidisciplinary perspective. *Can. J. Fish. Aquat. Sci.* **43**, 2472–503.

930 Stockner, J.G. & Benson, W.W. (1967) The succession of diatom assemblages in the recent sediment of Lake Washington. *Limnol. Oceanogr.* **12**, 512–32.

931 Stoner, J.H., Gee, A.S. & Wade, K.R. (1983) *The Effects of Acid Precipitation and Land Use on Water Quality and Ecology in the Upper Tywi Catchment in West Wales.* Report to the Welsh Water Authority, Cardiff.

932 Strayer, D. (1985) The benthic micrometazoans of Mirror Lake, New Hampshire. *Arch. Hydrobiol. Suppl.* **72**, 287–426.

933 Strayer, D.L. (1991) Projected distribution of the zebra mussel, *Dreissena polymorpha*, in North America. *Can. J. Fish. Aquat. Sci.* **48**, 1389–95.

934 Strayer, D.L., Powell, J., Ambrose, P., Smith, L.C., Pace, M.L. & Fischer, D.T. (1996) Arrival, spread and early dynamics of a zebra mussel (*Dreissena polymorpha*) population in the Hudson River estuary. *Can. J. Fish. Aquat. Sci.* **53**, 1143–9.

935 Sumpter, J.P. & Wood, C.R.C. (1981) The trout. *Biologist* **28**, 219–24.

936 Suren, A.M. (1991) Bryophytes as invertebrate habitat in two New Zealand alpine streams. *Freshwat. Biol.* **26**, 399–418.

937 Suren, A.M. & Winterbourne, M.J. (1992) The influence of periphyton, detritus and shelter on invertebrate colonisation of aquatic bryophytes. *Freshwat. Biol.* **27**, 327–39.

938 Suttles, W. (1968) Coping with abundance: subsistence on the northwest coast. In: Lee, R.B. & DeVore, I. (eds) *Man the Hunter.* Aldine, New York, pp. 56–68.

939 Svensson, B.W. (1974) Population movements of adult Trichoptera at a south Swedish stream. *Oikos* **25**, 157–75.

940 Talling, J.F. (1965) The photosynthetic activity of phytoplankton in East African lakes. *Int. Rev. ges. Hydrobiol.* **50**, 1–32.

941 Talling, J.F. (1966) The annual cycle of stratification and phytoplankton growth in Lake Victoria (E. Africa). *Int. Rev. ges. Hydrobiol.* **51**, 545–621.

942 Talling, J.F. (1969) The incidence of vertical mixing, and some biological and chemical consequences, in tropical African lakes. *Verh. int. Verein. theor. angew. Limnol.* **17**, 998–1012.

943 Talling, J.F. (1971) The underwater light climate as a controlling factor in the production ecology of freshwater phytoplankton. *Mitt. int. Verein. theor. angew. Limnol.* **19**, 214–43.

944 Talling, J.F. (1976) The depletion of carbon dioxide from lake water by phytoplankton. *J. Ecol.* **64**, 79–121.

945 Talling, J.F. (1986) The seasonality of phytoplankton in African lakes. *Hydrobiologia* **138**, 139–60.

946 Talling, J.F. & Talling, I.B. (1965) The chemical composition of African lake water. *Int. Rev. ges. Hydrobiol.* **50**, 421–63.

947 Talling, J.F., Wood, R.B., Prosser, M.V. & Baxter, R.M. (1973) The upper limit of photosynthetic productivity by phytoplankton: evidence from Ethiopian soda lakes. *Freshwat. Biol.* **3**, 53–76.

948 Tarapchak, S.J. & Nalewajko, C. (1986a) Introduction: phosphorus–plankton dynamics symposium. *Can. J. Fish. Aquat. Sci.* **43**, 293–301.

949 Tarapchak, S.J. & Nalewajko, C. (1986b) Synopsis: phosphorus–plankton dynamics symposium. *Can. J. Fish. Aquat. Sci.* **43**, 416–19.

950 Taylor, B.E. (1988) Analyzing population dynamics of zooplankton. *Limnol. Oceanogr.* **33**, 1266–73.

951 Teal, J.M. (1980) Primary production of benthic and fringing plant communities. In: Barnes, R.S.K. & Mann, K.H. (eds) *Fundamentals of Aquatic Ecosystems*. Blackwell Scientific Publication, Oxford, pp. 67–83.

952 Templeton, R. (ed.) (1984) *Freshwater Fisheries Management*. Fishing News Books, Farnham.

953 ter Braak, C.J.F. & van Dam, H. (1989) Inferring pH from diatoms: a comparison of old and new calibration methods. *Hydrobiologia* **178**, 209–23.

954 Theis, T.L. & McCabe, P.J. (1978) Phosphorus dynamics in hypereutrophic lake sediments. *Wat. Res.* **12**, 677–85.

955 Thesiger, W. (1964) *The Marsh Arabs.* Longmans, London. (1967) Penguin, Harmondsworth.

956 Thesiger, W. (1979) *Desert, Marsh and Mountain: the World of a Nomad.* Collins, London.

957 Thompson, K., Shewry, P.R. & Woolhouse, H.W. (1979) Papyrus swamp development in the Upemba Basin, Zaire: studies of population structure in *Cyperus papyrus* stands. *Bot. J. Linn. Soc.* **78**, 299–316.

958 Thompson, R. (1973) Palaeolimnology and palaeomagnetism. *Nature* **242**, 182–4.

959 Threlkeld, S. (1979) Estimating cladoceran birth rates: the importance of egg mortality and the egg age distribution. *Limnol. Oceanogr.* **24**, 601–12.

960 Tilman, D. (1977) Resource competition between planktonic algae: an experimental and theoretical approach. *Ecology* **58**, 338–48.

961 Timms, R.M. & Moss, B. (1984) Prevention of growth of potentially dense phytoplankton populations by zooplankton grazing, in the presence of zooplanktivorous fish, in a shallow wetland ecosystem. *Limnol. Oceanogr.* **29**, 472–86.

962 Tippett, R. (1964) An investigation into the nature of the layering of deep water sediments in two eastern Ontario lakes. *Can. J. Bot.* **42**, 1693–709.

963 Tippett, R. (1970) Artificial surfaces as a method of studying populations of benthic micro-algae in freshwaters. *Br. Phycol. J.* **5**, 187–99.

964 Titman, D. (1976) Ecological competition between algae: experimental confirmation of resource-based competition theory. *Science* **192**, 463–5.

965 Torsvik, V., Salte, K., Sorheim, R. & Goksoyr, J. (1990a) Comparison of phenotypic diversity and DNA heterogeneity in a population of soil bacteria. *Appl. Environ. Microbiol.* **56**, 776–81.

966 Torsvik, V., Goksoyr, J. & Daae, F.L. (1990b) High diversity in DNA of soil bacteria. *Appl. Environ. Microbiol.* **56**, 782–7.

967 Toth, L. (1972) Reeds control eutrophication of Balaton Lake. *Wat. Res.* **6**, 1533–9.

968 Townsend, C.R. & Hildrew, A.G. (1976) Field experiments on the drifting, colonization, and continuous redistribution of stream benthos. *J. Anim. Ecol.* **45**, 759–72.

969 Townsend, C.R., Hildrew, A.G. & Francis, J. (1983) Community structure in some southern English streams: the influence of physicochemical factors. *Freshwat. Biol.* **13**, 521–44.

970 Tranvik, L. (1992) Allochthonous dissolved matter as an energy source for pelagic bacteria and the concept of the microbial loop. *Hydrobiologia* **229**, 107–14.

971 Trewavas, E., Green, J. & Corbet, S.A. (1972) Ecological studies on crater lakes in West Cameroon: fishes of Barombi Mbo. *J. Zool.* **167**, 41–95.

972 Tudge, C. (1993) *The Engineer in the Garden.* Pimlico, London.

973 Tuite, C.H. (1981) Standing crop densities and distribution of *Spirulina* and benthic diatoms in East African alkaline saline lakes. *Freshwat. Biol.* **11**, 345–60.

974 Turner, J.L. (1982) Lake flies, water fleas and sardines. In: *Biological Studies on the Pelagic Ecosystem of Lake Malawi.* Technical Report 1, Fishery Expansion Project, Malawi, FAO/UNDP FI.DP.MLW/75/019, Rome.

975 Turner, S.M. & Liss, P.S. (1985) Measurement of various sulfur gases in a coastal marine environment. *J. Atmos. Chem.* **2**, 223–32.

976 Tyler, C., Sumpter, J. & Jobling, S. (1995) Environmental oestrogens and sexual development in fish. *Freshwat. Forum* **5**, 154–7.

977 Tyler, P.A. (1986) Anthropological limnology in the Land of Moinee. In: De Deckker, P. & Williams, W.D. (eds) *Limnology in Australia.* Junk, Dordrecht, pp. 523–38.

978 Underwood, G.J.C. (1991) Growth enhancement of the macrophyte *Ceratophyllum demersum* in the presence of the snail *Planorbis planorbis*: the effect of grazing and chemical conditioning. *Freshwat. Biol.* **26**, 325–34.

979 Vallentyne, J.R. (1969) Sedimentary organic matter and palaeolimnology. *Mitt. int. Ver. theor. angew. Limnol.* **17**, 104–10.

980 Vanni, M.J. (1986) Competition in zooplankton communities: suppression of small species by *Daphnia pulex. Limnol. Oceanogr.* **31**, 1039–56.

981 Vannote, R.L., Minshall, G.W., Cummings, K.W., Sedell, J.R. & Cushing, C.E. (1980) The river continuum concept. *Can. J. Fish. Aquat. Sci.* **37**, 120–37.

982 Van Vierssen, W. & Prins, T.C. (1985) On the relationship between the growth of algae and aquatic macrophytes in brackish water. *Aquat. Bot.* **21**, 165–79.

983 Vickery, J. (1992) The reproductive success of the dipper *Cinclus cinclus* in relation to the acidity of streams in south-west Scotland. *Freshwat. Biol.* **28**, 195–205.

984 Vincent, W.F., Wurtsbaugh, W., Vincent, C.L. & Richerson, P.J. (1984) Seasonal dynamics of nutrient limitation in a tropical high-altitude lake (Lake Titicaca, Peru–Bolivia): application of physiological bioassays. *Limnol. Oceanogr.* **29**, 540–52.

985 Viner, A.B. (1973) Responses of a mixed phytoplankton population to nutrient enrichments of ammonia and phosphate and some associated ecological implications. *Proc. Roy. Soc. London (B)* **183**, 351–70.

986 Viner, A.B. & Smith, I.R. (1973) Geographical, historical and physical aspects of Lake George. *Proc. Roy. Soc. London (B)* **184**, 235–70.

987 Vollenweider, R.A. (1975) Input output models with special reference to the phosphorus loading concept in limnology. *Schweiz. Zh. Hydrol.* **37**, 53–84.

988 Vollenweider, R.A. & Kerekes, J.J. (1981) Appendix I. Background and Summary results of the OECD cooperative programme on eutrophication. In: Janus, L.L. & Vollenweider, R.A. (comp.) *The OECD Cooperative Programme on Eutrophication: Canadian Contribution* Scientific Series 131, Environment Canada, Ottawa, pp. 251–386.

989 Vrieling, E.G. & Anderson, D.M. (1996) Immunofluorescence in phytoplankton research: applications and potential. *J. Phycol.* **32**, 1–16.

990 Wager, R. & Jackson, P. (1993) *The Action Plan for Australian Freshwater Fishes.* Australian Nature Conservation Agency, Canberra.

991 Wakeford, T. & Walters, R. (eds) (1994) *Science for the Earth.* Wiley, Chichester.

992 Walker, D. & Flenley, J.R. (1979) Late Quaternary vegetational history of the Enga Province of upland Papua New Guinea. *Phil. Trans. Roy. Soc.* **286**, 265–344.

993 Walker, K.F. & Likens, G.E. (1975) Meromixis and a reconsidered typology of lake circulation patterns. *Verh. int. Verein. theoret. angew. Limnol.* **19**, 442–58.

994 Walsby, A.E. (1965) Biochemical studies on the extracellular polypeptides of *Anabaena cylindrica* Lemm. *Br. Phycol. Bull.* **1**, 514–15.

995 Ward, F. (1990) Florida's coral reefs are imperiled. *Nat. Geogr.* **178**, 115–32.

996 Ward, R.C. (1975) *Principles of Hydrology.* McGraw-Hill, London.

997 Waters, T.F. (1961) Standing crop and drift of stream bottom organisms. *Ecology* **42**, 352–7.

998 Waters, T.F. (1965) Interpretation of invertebrate drift in streams. *Ecology* **46**, 327–34.

999 Waters, T.F. (1972) The drift of stream insects. *Ann. Rev. Entomol.* **17**, 253–72.

1000 Watson, R.A. & Osborne, P.L. (1979) An algal pigment ratio as an indicator of the nitrogen supply to phytoplankton in three Norfolk Broads. *Freshwat. Biol.* **9**, 585–94.

1001 Webster, J.R. & Benfield, E.F. (1986) Vascular plant breakdown in freshwater ecosystems. *Ann. Rev. Ecol. Syst.* **17**, 567–94.

1002 Webster, K.E. & Peters, R.H. (1978) Some size-dependent inhibitions of larger cladoceran filterers in filamentous suspensions. *Limnol. Oceanogr.* **23**, 1238–45.

1003 Welcomme, R.L. (1979) *Fisheries Ecology of Floodplain Rivers.* Longman, London.

1004 Went, A.E.J. (ed.) (1980) *Atlantic Salmon: its Future.* Proceedings 2nd International Atlantic Salmon Association, Edinburgh, Fishing News Books, Farnham.

1005 Werner, E.E., Hall, D.J., Laughlin, D.R., Wagner, D.T., Wilsmann, L.A. & Funk, F.C. (1977) Habitat partitioning in a freshwater fish community. *J. Fish. Res. Bd. Can.* **34**, 360–70

1006 Westlake, D.F. (1967) Some effects of low velocity currents on the metabolism of aquatic macrophytes. *J. Exp. Bot.* **18**, 187–205.

1007 Westlake, D.F. (1978) Rapid exchange of oxygen between plant and water. *Verh. int. Verein. theor. angew. Limnol.* **20**, 2363–7.

1008 Westlake, D.F. (1982) The primary productivity of water plants. In: Symoens, J.J., Hooper, S.S. & Compere, P. (eds) *Studies on Aquatic Vascular Plants.* Royal Botanical Society of Belgium, Brussels, pp. 165–80.

1009 Westlake, D.F. *et al.* (1980) Primary production. In: Le Cren, E.D. & Lowe-McConnell, R.H. (eds) *The Functioning of Freshwater Ecosystems.* Cambridge University Press, Cambridge, pp. 141–246.

1010 Wetzel, R.G. (1979) The role of the littoral zone and detritus in lake metabolism. *Arch. Hydrobiol. Beih. Ergben. Limnol.* **13**, 145–61.

1011 Wetzel, R.G. (1987) *Limnology.* Saunders, Philadelphia.

1012 Wetzel, R.G. (1990) Land–water interfaces: metabolic and limnological regulators. *Verh. int. Verein. theor. angew. Limnol.* **24**, 6–24.

1013 Wetzel, R.G. & Manny, B.A. (1972) Secretion of dissolved organic carbon and nitrogen by aquatic macrophytes. *Verh. int. Verein. theor. angew. Limnol.* **18**, 162–70.

1014 Whitford, L.A. & Schumacher, G.J. (1961) Effect of current on mineral uptake and respiration by a freshwater alga. *Limnol. Oceanogr.* **6**, 423–5.

1015 Whitton, B.A. & Say, P.J. (1975) Heavy metals. In: Whitton, B.A. (ed.) *River Ecology.* Blackwell Scientific Publications, Oxford, pp. 286–311.

1016 Williams, W.D. (1986) Limnology, the study of inland waters: a comment on perceptions of studies of salt lakes, past and present. In: De Deckker, P. & Williams, W.D. (eds) *Limnology in Australia.* Junk, Dordrecht, pp. 471–86.

1017 Williams, W.D. (1992) The worldwide occurrence and limnological significance of falling water levels in large permanent saline lakes. *Verh. int. Verein. theor. angew. Limnol.* **25**, 980–3.

1018 Williams, W.D. & Aladin, N.V. (1991) The Aral Sea: recent limnological changes and their conservation significance. *Aquat. Cons. Mar. Freshwat. Ecosyst.* **1**, 3–24.

1019 Winfield, I.J. (1990) Predation pressure from above: observations on the activities of piscivorous birds at a shallow eutrophic lake. *Hydrobiologia* **191**, 223–31.

1020 Winterbourn, M.J., Rounick, J.S. & Cowie, B. (1981) Are New Zealand stream ecosystems really different? *N. Zealand J. Mar. Freshwat. Res.* **15**, 321–8.

1021 Winterbourn, M.J., Hildrew, A.G. & Box, A. (1985) Structure and grazing of stone surface organic layers in some acid streams of southern England. *Freshwat. Biol.* **15**, 363–74.

1022 Witte, F., Goldschmidt, T., Wanink, J. *et al.* (1992a) The destruction of an endemic species flock: quantitative data on the decline of the haplochromine cichlids of lake Victoria. *Environ. Biol. Fish.* **34**, 1–28.

1023 Witte, F., Goldschmidt, T., Goudswaard, P.C., Ligtvoet, W. Vanoijen, M.J.P. & Wanink, J.H. (1992b) *Neth. J. Zool.* **42**, 214–32.

1024 Wium-Andersen, S. (1971) Photosynthetic uptake of free CO_2 by roots of *Lobelia dortmanna. Physiol. Plant.* **25**, 245–8.

1025 Wium-Andersen, S. & Andersen, J.M. (1972) The influence of vegetation on the redox profile of the sediment of Grane Langsø, a Danish *Lobelia* lake. *Limnol. Oceanogr.* **17**, 948–52.

1026 Wium-Andersen, S., Anthoni, U., Christophersen, C. & Houen, G. (1982) Allelopathic effects on phytoplankton by substances isolated from aquatic macrophytes (Charales). *Oikos* **39**, 187–90.

1027 Wood, M. (1993) Saddam drains the life of the Marsh Arabs. *Independent* 28 Aug., 12.

1028 Woodwell, G.M. (1983) Aquatic systems as part of the biosphere. In: Barnes, R.S.K. & Mann, K.H. (eds) *Fundamentals of Aquatic Ecosystems.* Blackwell Scientific Publications, Oxford, pp. 201–15.

1029 Wootton, R.J. (1984) *A Functional Biology of Sticklebacks.* Croom Helm, Beckenham.

1030 Worthington, S. & Worthington, E.B. (1933) *Inland Waters of Africa.* Macmillan, London.

1031 Wright, H.E. (1967) A square rod piston sampler for lake sediments. *J. Sed. Petrol.* **37**, 975–6.

1032 Wright, J.F. (1995) Development and use of a system for predicting the macroinvertebrate fauna in flowing waters. *Austr. J. Ecol.* **20**, 181–97.

1033 Wright, J.F., Moss, D., Armitage, P.D. & Furse, M.T. (1984) A preliminary classification of running water sites in Great Britain based on macro-invertebrate species and the prediction of community type using environmental data. *Freshwat. Biol.* **14**, 221–56.

1034 Wright, J.F., Armitage, P.D., Furse, M.T. & Moss, D. (1985) The classification and prediction of macroinvertebrate communities in British rivers. *Ann. Rep. Freshwat. Biol. Assoc.* **53**, 80–93.

1035 Wright, S.J.L., Redhead, K. & Maudsley, H. (1981) *Acanthamoeba castellami*, a predator of Cyanobacteria. *J. Gen. Microbiol.* **125**, 293–300.

1036 Wurtsbaugh, W. (1992) Food-web modification by an invertebrate predator in the Great Salt Lake (USA). *Oecologia* **89**, 168–75.

1037 Wyatt, J.T. & Silvey, J.K.G. (1969) Nitrogen fixation by *Gloeocapsa. Science* **165**, 908–9.

1038 Yalden, D.W. (1992) The influence of recreational disturbance on common sandpipers (*Actitis hypoleucos*), breeding by an upland reservoir in England. *Biol. Cons.* **61**, 41–9.

1039 Young, G. & Wheeler, N. (1977) *Return to the Marshes.* Collins, London.

1040 Zaret, T.M. (1969) Predation-balanced polymorphism of *Ceriodaphnia cornuta* Sars. *Limnol. Oceanogr.* **14**, 301–3.

1041 Zaret, T.M. & Paine, R.T. (1973) Species introduction in a tropical lake. *Science* **218**, 444–5.

1042 Zaret, T.M. & Suffern, J.S. (1976) Vertical migration in zooplankton as a predator avoidance mechanism. *Limnol. Oceanogr.* **21**, 804–13.

Index